ADVANCES IN DEA
THEORY AND
APPLICATIONS

Wiley Series in

Operations Research and Management Science

Operations Research and Management Science (ORMS) is a broad, interdisciplinary branch of applied mathematics concerned with improving the quality of decisions and processes and is a major component of the global modern movement towards the use of advanced analytics in industry and scientific research. The *Wiley Series in Operations Research and Management Science* features a broad collection of books that meet the varied needs of researchers, practitioners, policy makers, and students who use or need to improve their use of analytics. Reflecting the wide range of current research within the ORMS community, the Series encompasses application, methodology, and theory and provides coverage of both classical and cutting edge ORMS concepts and developments. Written by recognized international experts in the field, this collection is appropriate for students as well as professionals from private and public sectors including industry, government, and nonprofit organizations who are interested in ORMS at a technical level. The Series is comprised of four sections: Analytics; Decision and Risk Analysis; Optimization Models; and Stochastic Models.

Advisory Editors • Optimization Models
Lawrence V. Snyder, *Lehigh University*
Ya-xiang Yuan, Chinese Academy of Sciences

Founding Series Editor
James J. Cochran, University of Alabama

Analytics

Yang and Lee • *Healthcare Analytics: From Data to Knowledge to Healthcare Improvement*

Forthcoming Titles

Attoh-Okine • *Big Data and Differential Privacy: Analysis Strategies for Railway Track Engineering*
Kong and Zhang • *Decision Analytics and Optimization in Disease Prevention and Treatment*

Decision and Risk Analysis

Barron • *Game Theory: An Introduction,* Second Edition
Brailsford, Churilov, and Dangerfield • *Discrete-Event Simulation and System Dynamics for Management Decision Making*
Johnson, Keisler, Solak, Turcotte, Bayram, and Drew • *Decision Science for Housing and Community Development: Localized and Evidence-Based Responses to Distressed Housing and Blighted Communities*
Mislick and Nussbaum • *Cost Estimation: Methods and Tools*

Forthcoming Titles

Aleman and Carter • *Healthcare Engineering*

Optimization Models

Ghiani, Laporte, and Musmanno • *Introduction to Logistics Systems Management,* Second Edition

Forthcoming Titles

Smith • *Learning Operations Research Through Puzzles and Games*

Stochastic Models

Ibe • *Random Walk and Diffusion Processes*

Forthcoming Titles

Donohue, Katok, and Leider • *The Handbook of Behavioral Operations*
Matis • *Applied Markov Based Modelling of Random Processes*

ADVANCES IN DEA THEORY AND APPLICATIONS

With Extensions to Forecasting Models

Edited by

KAORU TONE
National Graduate Institute for Policy Studies, Japan

Registered Offices
John Wiley & Sons, Inc., 111 River Street, Hoboken, NJ 07030, USA
John Wiley & Sons Ltd, The Atrium, Southern Gate, Chichester, West Sussex, PO19 8SQ, UK

Editorial Office
9600 Garsington Road, Oxford, OX4 2DQ, UK

For details of our global editorial offices, customer services, and more information about Wiley products visit us at www.wiley.com.

Wiley also publishes its books in a variety of electronic formats and by print-on-demand. Some content that appears in standard print versions of this book may not be available in other formats.

Library of Congress Cataloging-in-Publication Data

Names: Tone, Kaoru, 1931– editor.
Title: Advances in DEA theory and applications : with extensions to
    forecasting models / edited by Kaoru Tone, National Graduate Institute for
    Policy Studies, Japan.
Description: First edition. | Hoboken, NJ : John Wiley & Sons, Inc., 2017. |
    Series: Wiley series in operations research and management science |
    Includes bibliographical references and index. | Description based on print
    version record and CIP data provided by publisher; resource not viewed.
Identifiers: LCCN 2016053367 (print) | LCCN 2017012217 (ebook) | ISBN 9781118946701 (pdf) |
    ISBN 9781118946695 (epub) | ISBN 9781118945629 (cloth)
Subjects: LCSH: Data envelopment analysis.
Classification: LCC HA31.38 (ebook) | LCC HA31.38 .A38 2017 (print) |
    DDC 338.501/51972–dc23
LC record available at https://lccn.loc.gov/2016053367

Cover design by Wiley
Cover image: © HeliRy/Gettyimages

Set in 10/12pt Times by SPi Global, Pondicherry, India
Printed and bound in Malaysia by Vivar Printing Sdn Bhd

10 9 8 7 6 5 4 3 2 1

To
DEA lovers and my family
KT

CONTENTS

16 DEA in the Transport Sector — 192

Ming-Miin Yu and Li-Hsueh Chen

17 Dynamic Network Efficiency of Japanese Prefectures — 216

Hirofumi Fukuyama, Atsuo Hashimoto, Kaoru Tone and William L. Weber

18 A Quantitative Analysis of Market Utilization in Electric Power Companies — 231

Miki Tsutsui and Kaoru Tone

28 Time Series Benchmarking Analysis for New Product Scheduling: Who Are the Competitors and How Fast Are They Moving Forward? **443**

Dong-Joon Lim and Timothy R. Anderson

29 DEA Score Confidence Intervals with Past–Present and Past–Present–Future-Based Resampling **459**

Kaoru Tone and Jamal Ouenniche

LIST OF CONTRIBUTORS

Timothy R. Anderson Department of Engineering and Technology Management, Portland State University, Portland, OR, USA

Skarleth Carrales Business School, University of Edinburgh, Edinburgh, UK

Tsung-Sheng Chang Department of Transportation and Logistics Management, National Chiao Tung University, Hsinchu, Taiwan

Li-Hsueh Chen Department of Transportation Science, National Taiwan Ocean University, Keelung, Taiwan

Hirofumi Fukuyama Faculty of Commerce, Fukuoka University, Fukuoka, Japan

Atsuo Hashimoto Fukuoka Girls' Commercial High School, Chikushi-gun, Fukuoka, Japan

Masayoshi Hayashi Graduate School of Economics, University of Tokyo, Tokyo, Japan

Bo Hsiao Department of Information Management, Chang Jung Christian University, Taiwan

Andrew L. Johnson Department of Industrial and Systems Engineering, Texas A&M University, College Station, TX, USA

Hiroyuki Kawaguchi Economics Faculty, Seijo University, Setagaya-ku, Tokyo, Japan

Chia-Yen Lee Institute of Manufacturing Information and Systems, National Cheng Kung University, Tainan City, Taiwan

Dong-Joon Lim Portland State University, Portland, OR, USA

Mohammad M. Mousavi Business School, University of Edinburgh, Edinburgh, UK

Jamal Ouenniche Business School, University of Edinburgh, Edinburgh, UK

Joseph C. Paradi Centre for Management of Technology and Entrepreneurship, University of Toronto, Toronto, ON, Canada

Biresh K. Sahoo Xavier Institute of Management, Xavier University, Bhubaneswar, India

Kaoru Tone National Graduate Institute for Policy Studies, Tokyo, Japan

Miki Tsutsui Central Research Institute of Electric Power Industry, Tokyo, Japan

William L. Weber Southeast Missouri State University, Cape Girardeau, USA

Chen-Hui Wu Department of Accounting and Information Technology, National Chung Cheng University, Chia-yi County, Taiwan

Bing Xu School of Social Sciences, Heriot-Watt University, Edinburgh, UK

Xiaopeng Yang Centre for Management of Technology and Entrepreneurship, University of Toronto, Toronto, ON, Canada

Ming-Miin Yu Department of Transportation Science, National Taiwan Ocean University, Keelung, Taiwan

ABOUT THE AUTHORS

Timothy R. Anderson (tim.anderson@pdx.edu) is an Associate Professor of Engineering and Technology Management at Portland State University. He earned an Electrical Engineering degree from the University of Minnesota, as well as both an MS and a PhD degree in Industrial and Systems Engineering from the Georgia Institute of Technology. He has been the Program Chair or Co-Chair 13 times for PICMET, the Portland International Conference on the Management of Engineering and Technology, since 1997, with over 35 refereed publications, and current research interests in benchmarking, technology forecasting, data mining, and new product development.

Skarleth Carrales (s1471551@sms.ed.ac.uk) is a PhD candidate at the University of Edinburgh, UK. She holds a bachelor's degree with honours in Business Administration from the University of La Salle in Mexico, and an MSc degree in Finance from the Instituto Tecnológico de Estudios Superiores de Monterrey, Mexico. Her research focuses on the performance evaluation of banks as decision-making units with data envelopment analysis. She worked for six years as a subdirector in different areas for the Secretary of Agriculture, Animal Husbandry, Rural Development, Fisheries and Food in Mexico. In her last job she published a 'General manual of organization' in the *Official Gazette of Mexico* for that Secretary, and a 'Technical guide for the update and development of the organization and procedures manuals' in the Institutional Library of Regulations and 30 organizational manuals as well. She is currently a Teaching Assistant in the Management Science and Business Economics group at the University of Edinburgh. She participated as the Selection Process Officer in the 14th

Symposium of Mexican Students and Studies. Currently, she is Treasurer of the Doctoral Society of the Business School at the University of Edinburgh.

Tsung-Sheng Chang (tsc@nctu.edu.tw) received a PhD degree in Transportation Systems Engineering (in the School of Civil and Environmental Engineering) from Cornell University. He is a Professor in the Department of Transportation and Logistics Management at National Chiao Tung University, Taiwan. His research focuses on developing optimization and modelling procedures for large-scale, complex transportation and logistics problems. In addition, he also works on developing various new DEA models. His research has been published in various journals, such as *Transportation Science*, *Transportation Research Part B* and *European Journal of Operational Research*.

Li-Hsueh Chen (yoannachen@gmail.com) is a Postdoctoral Fellow in the Department of Transportation Science, National Taiwan Ocean University, Taiwan. She holds MA and PhD degrees in Economics from National Chengchi University, Taiwan. She has published papers on DEA in international journals, such as *Omega*, *Cornell Hospitality Quarterly*, *Journal of Business Research* and *Journal of Air Transport Management*. Her recent research interests include dynamic DEA with network structure, resource allocation and target setting in DEA, and scale elasticity in DEA.

Hirofumi Fukuyama (fukuyama@fukuoka-u.ac.jp) is a Professor in the Faculty of Commerce at Fukuoka University, Japan. He received a PhD in Economics from Southern Illinois University at Carbondale, USA, in 1988. His research interests include efficiency/productivity measurement in the financial services industry, as well as the theory and applications of data envelopment analysis. His work has appeared in refereed journals on operations research, finance, management and economics. These journals include *European Journal of Operational Research*; *Omega*; *Annals of Operations Research*; *International Journal of Information Technology and Decision Making*; *Journal of Optimization Theory and Applications*; *Applied Soft Computing*; *Journal of Intelligent & Fuzzy Systems*; *Journal of Productivity Analysis*; *Journal of the Operational Research Society* (UK); *Operational Research*; *Journal of the Operations Research Society of Japan*; *International Financial Markets, Institutions & Money*; *Pacific-Basin Finance Journal*; *Managerial and Decision Economics*; *Socio-Economic Planning Sciences*; and *International Journal of Applied Management*. He is an Associate Editor for *Data Envelopment Analysis Journal*, and is on the Editorial Board of *International Journal of Information Systems and Social Change*, *International Journal of Applied Management*, *International Journal of Society Systems Science* and *Open Management Journal*.

Atsuo Hashimoto (munekana76@ybb.ne.jp) is a teacher of information technology at Fukuoka Girls' Commercial High School and a part-time lecturer in operations research at Fukuoka University in Japan. He received his PhD (Doctor of Commercial

Science) from Fukuoka University in March 2015. His research focuses on the productivity evaluation of Japanese prefectures and their sustainability. He has published several journal articles. His articles have appeared in *Communications of the Operations Research Society of Japan* and *Japan Academic Society of Business Education*.

Masayoshi Hayashi (hayashim@e.u-tokyo.ac.jp) is a Professor of Economics in the Graduate School of Economics, University of Tokyo, Japan. He received a BA and MA in Political Science from the School of International Politics, Economics and Business (SIPEB), Aoyama Gakuin University, Tokyo, and an MA and PhD in Economics from Queen's University, Kingston, Ontario, Canada. His research interests centre on issues in public finance and public policy. He was a policy analyst at the Sanwa Research Institute Corporation (now Mitsubishi-UFJ Research & Consulting Co., Ltd), Tokyo, Japan, and a Principal Economist at the Policy Research Institute of the Japanese Ministry of Finance. He also served as an associate professor at Meiji Gakuin University and Hitotsubashi University. He has published a number of studies on various topics in public finance, including fiscal federalism, social policy, taxation and cost–benefit analysis. His publications in English have appeared in *Socio-Economic Planning Sciences*, *International Tax and Public Finance*, *International Journal of Health Care Finance and Economics*, *Canadian Journal of Economics*, *Canadian Public Policy* and *Journal of the Japanese and International Economies*. He is now the editor in chief of *Studies in Applied Economics*.

Bo Hsiao (bhsiao@mail.cjcu.edu.tw) is an Associate Professor in the Department of Information Management at Chang Jung Christian University, Taiwan. He received his PhD degree in Information Management from National Taiwan University, Taiwan. His research interests include manufacturing information systems, data envelopment analysis, project management, the knowledge economy and pattern recognition. Before joining academia, he was employed by the Taiwan Semiconductor Manufacturing Company (TSMC) and the Industrial Technology Research Institute (ITRI), and worked as an engineer, associate research fellow and project manager. He has published his research in *Applied Ergonomics*, *Maritime Policy & Management*, *Journal of Business Research*, *Emergence Markets Finance and Trade*, *Pattern Recognition*, *Journal of International Management Science*, *Decision Support Systems*, *Omega* and *Computers in Human Behavior*. He also serves on the Editorial Board for *Journal of Reviews on Global Economics*.

Andrew L. Johnson (ajohnson@tamu.edu) is an Associate Professor in the Department of Industrial and Systems Engineering at Texas A&M University and a Visiting Associate Professor at Osaka University. He obtained his BS from the Grado Department of Industrial and Systems Engineering at Virginia Tech and his MS and PhD from the H. Milton Stewart School of Industrial and Systems Engineering at Georgia Tech. His research interests include productivity and efficiency measurement, benchmarking, and production economics. He is an associate editor of *IIE Transactions* and a member of IIE, INFORMS, the National Eagle Scout Association and the German Club of Virginia Tech. He was a co-organizer of the 2016 NSF workshop to redefine broader impacts. For more information, see his website andyjohnson.guru.

Hiroyuki Kawaguchi (kawaguchi@seijo.ac.jp) is a Professor in the Economics Faculty at Seijo University, Japan. He holds a Master of Science degree in Health Economics from the University of York, UK and a PhD in Economics from Hitotsubashi University, Japan. He started his career at the Industrial Bank of Japan and worked there for 11 years in several divisions, including the industrial research division, where he worked as an economist of the healthcare industry. He served as a professor at the International University of Health and Welfare for 12 years and also worked at Seijo University for six years. His main work has been in health economics and health policy. He authored an introductory textbook on health economics, *Economics of Health Care: Evaluation of Health Policy from an Economic Point of View*. He has also published several papers on health economics and health policy in international journals, such as *Health Care Management Science*, *BMC Health Services Research*, *International Journal of Health Geographies* and *Japanese Journal of Health Economics and Policy*. He continues with research activity on the economic aspects of Japanese health policy. Current interests include the application of data envelopment analysis in the healthcare field, the economic effects of disease management methods, and risk adjustment methods for healthcare finance systems.

Chia-Yen Lee (cylee@mail.ncku.edu.tw) is an Associate Professor at the Institute of Manufacturing Information and Systems, National Cheng Kung University (NCKU), Taiwan. He received a BS in Mathematical Sciences and a BBA in Management Information Systems from National Chengchi University in 2002, an MS in Industrial Engineering and Engineering Management from National Tsing Hua University in 2006, and a PhD degree in Industrial and Systems Engineering from Texas A&M University, USA, in 2012. He is also a Co-Principal Investigator at the Semiconductor Technologies Empowerment Partners Consortium for Big Data Analytics and Optimization Technologies, Taiwan. He has received several grants from an industry–academia cooperation related to fault detection and classification, robust capacity planning, multi-objective job-shop stochastic scheduling optimization, and other things. His research interests include productivity and efficiency analysis, manufacturing-data science, stochastic optimization, multi-objective decision analysis, and data mining. He has published several papers in *European Journal of Operational Research*, *IEEE Transactions on Power Systems*, *Energy Economics*, *Annals of Operations Research*, *Journal of Optimization Theory and Applications* and other journals. His recent innovations include work on effectiveness, meta-DEA, Nash–profit efficiency and the directional shadow price of air pollutant emissions. He has received a Best Paper Award from the Chinese Institute of Industrial Engineers (CIIE), Rising Star Research Grants from NCKU, and Outstanding Young Scholar Grants from the Ministry of Science and Technology, Taiwan. He serves as an Editorial Board member for *Flexible Services and Manufacturing Journal*.

Dong-Joon Lim (dongjoon@pdx.edu) is currently a data scientist at Nike, Inc., USA. He has participated in a variety of new product development projects, including compression suits, cooling materials, golf shoes and running suits. He earned a PhD in Engineering and Technology Management from Portland State University, USA, as well as both a BS and an MS degree in Industrial Engineering from Sungkyunkwan University, Korea. He has published many papers on technometrics in international journals, such as *European Journal of Operational Research*, *Omega*, *International Transactions in Operational Research*, *Mathematical and Computer Modelling*, *R&D Management*, *Technological Forecasting and Social Change* and *Advances in Business and Management Forecasting*. His current research interests include multivariate data analysis, data visualization, perceptual mapping, experiment design, predictive modelling and reliability testing. He is also a developer of the open source R package DJL, which makes it possible to reproduce most of his research findings freely. He has served as a peer reviewer for international journals, including *Technological Forecasting and Social Change*, *International Journal of Energy Technology and Policy*, *International Transactions in Operational Research*, *Journal of the Knowledge Economy* and *International Journal of Transitions and Innovation Systems*, among others.

Mohammad M. Mousavi (smousavi@kean.edu) is a Lecturer in Finance at Kean University, Wenzhou, China and a PhD student in Business Economics at the University of Edinburgh Business School, UK. He holds a BS and MA in Financial Management from Imam Sadiq University, Tehran, Iran, and an MSc in Finance from Essex Business School, UK. His research interests cover a variety of topics, including the design and performance evaluation of bankruptcy prediction models, credit scoring, corporate finance, and international business. He has published a number of papers in international journals, such as *International Review of Financial Analysis*, *Journal of Developing Areas* and *Journal of Economics, Business and Management*. Furthermore, he has over five years' work experience as a capital market analyst at the Stock Exchange.

Jamal Ouenniche (jamal.ouenniche@ed.ac.uk) is a Reader in Management Science at the Business School at the University of Edinburgh, United Kingdom, and Head of the Management Science and Business Economics group. He holds a BSc in Mathematics and an MSc in Operational Research from the University of Montreal, Canada, and a PhD in Operations Management from Laval University, Quebec, Canada. His research portfolio encompasses a broad range of applications and a variety of research methodologies in descriptive, predictive and prescriptive analytics, and tackles important managerial issues in energy, manufacturing, transport, banking and public sector policy. Some of his research is concerned with methodological contributions to the fields of optimization, artificial intelligence, data envelopment analysis and forecasting. With respect to forecasting, he pioneered research on the performance evaluation of competing forecasting models under multiple criteria using both DEA and MCDA methodologies. He acts as a referee for over 20 academic journals, several international conferences, and three funding bodies, namely, the Social Sciences and Humanities Research Council of Canada (SSHRC), the Portuguese Foundation for Science and Technology (FCT) and the Czech Science Foundation (GACR). He also serves as an Editorial Board member for eight journals and is an associate editor for *Journal of Optimization Theory and Applications*. His research has been published in *Operations Research, European Journal of Operational Research, Computers and Operations Research, Journal of Optimization Theory and Applications, International Journal of Production Economics, International Journal of Production Research, Expert Systems with Applications, International Review of Financial Analysis, Applied Financial Economics, Applied Economics Letters, Energy Economics, Applied Energy* and *The Journal of Developing Areas*, amongst others.

Joseph C. Paradi (paradi@mie.utoronto.ca) is a Professor Emeritus at the Centre for Management of Technology and Entrepreneurship (CMTE) at the University of Toronto, Canada. He immigrated to Canada from Budapest, Hungary, as a youngster. Dr Paradi has spent over five decades 'building' something – first, an education at the University of Toronto, where he obtained a degree in Chemical Engineering, followed by MASc and PhD degrees. He is a member of the Professional Engineers of Ontario and a Fellow of the Canadian Academy of Engineers. After graduation, he entered the business world and founded Dataline Inc. in 1968. The company was very successful, engaged in the business of time-sharing of computer services and grew to a $25 million (~$45 million today) large Canadian company, which he sold in late 1987 and left in January 1989 after 20 years at the helm. He started teaching on a part-time basis during the early 1980s when the first lectures in 'Innovation and Entrepreneurship' were delivered. Today, he is the Executive Director of the CMTE. He is the holder of the Chair in Information Engineering in the Faculty of Engineering. He teaches/ organizes eight courses in entrepreneurship and business, and small business management. During the past 25 years he has advised 17 PhD candidates, supervised over 60 MASc research students and over 180 undergraduate theses involving 250 students, and advised seven postdoctoral fellows. His research is focused on the financial services industry, particularly banking, a field in which he and his students have developed many innovative approaches to using DEA. He is a co-author of 58 peer-reviewed papers, five books and five chapters in books. He has participated in over 150 conference sessions and acted as chairman, moderator and keynote speaker in 50 events.

Biresh K. Sahoo (biresh@ximb.ac.in) is presently a Professor of Economics at Xavier Institute of Management, Xavier University, Bhubaneswar, India. He is also a Japan Society for the Promotion of Science (JSPS) Fellow at GRIPS, Tokyo, Japan, a Lise Meitner Fellow at Vienna University of Economics and Business, Vienna (WU-Wien), Austria, and a Visiting Professor at Jawaharlal Nehru University, New Delhi, India. He holds a PhD in Economics from the Indian Institute of Technology Kharagpur, India, and an MPhil and MA in Economics from the University of Hyderabad,

India. His research work has appeared in *European Journal of Operational Research*, *International Journal of Production Economics*, *Omega*, *Ecological Economics*, *Annals of Operations Research*, *International Journal of Systems Science*, *International Transactions in Operational Research*, *Socio-Economic Planning Sciences*, *Journal of the Operations Research Society of Japan*, *Opsearch* and others. He has also co-authored with Professor Jati Sengupta (University of California, Santa Barbara, USA) a book entitled *Efficiency Models in Data Envelopment Analysis: Techniques of Evaluation of Productivity of Firms in a Growing Economy*, published by Palgrave Macmillan, UK. He specializes in applied production frontier analysis, and his research interests are in the areas of efficiency and productivity performance of firms and benchmarking.

Kaoru Tone (tone@grips.ac.jp) is a Professor Emeritus in the National Graduate Institute for Policy Studies (GRIPS), Japan. He holds a BS in Mathematics from the University of Tokyo and a PhD in Operations Research from Keio University, Japan. He has served as a professor at GRIPS, Saitama University and Keio University for over 40 years. He was President of the Operations Research Society of Japan from 1996 to 1998. His contribution to DEA is manifested in a variety of achievements. He authored a classic book *Data Envelopment Analysis: A Comprehensive Text with Models, Applications, References and DEA-Solver Software*, with co-authors Professor Cooper (University of Texas) and Professor Seiford (University of Michigan). He has also published many papers on DEA in international journals, such as *European Journal of Operational Research*, *Omega*, *Journal of the Operational Research Society*, *Journal of Productivity Analysis*, *Socio-Economic Planning Sciences*, *Annals of Operations Research*, *International Transactions of Operational Research* and *Journal of Optimization Theory and Applications*. He opened up a new avenue for performance evaluation, called the slacks-based measure (SBM). This model is widely utilized all over the world. His recent innovations include network SBM, dynamic SBM, dynamic DEA with network structure, congestion in DEA, returns-to-growth in DEA, ownership-specified network DEA, non-convex frontier DEA, past–present–future inter-temporal DEA, resampling DEA and SBM-Max. He has served as an Editorial Board member for *Omega*, *The Journal of Data Envelopment Analysis*, *Socio-Economic Planning Sciences* and *Journal of Optimization Theory and Applications*, among others. Currently, he has no teaching, no meetings and hence no power, but is just enjoying research and the violin. It has been said 'He is in noisy Paradise.'

Miki Tsutsui (miki@criepi.denken.or.jp) is a researcher in the Socio-Economic Research Center at the Central Research Institute of Electric Power Industry (CRIEPI) in Japan and has been engaged in productivity and efficiency analysis for the electric power industry. She is also involved in research on network and dynamic DEA models. She holds a PhD in Operations Research from the National Graduate Institute for Policy Studies (GRIPS), Tokyo, Japan. Her research has been published in *Omega*, *European Journal of Operational Research*, *Energy Policy*,

Energy Economics and *Socio-Economic Planning Sciences*. She is a member of the Operation Research Society of Japan.

William L. Weber (wlweber@semo.edu) is a Professor in the Department of Economics and Finance at Southeast Missouri State University. He earned his PhD in Economics at Southern Illinois University Carbondale. His research interests lie in using production theory to measure performance for various kinds of financial institutions, public schools and universities, manufacturing firms, and firms that produce polluting by-products. He is fortunate to have great colleagues. He has published research papers in journals such as *Management Science*, *Review of Economics and Statistics*, *Journal of Econometrics*, *European Journal of Operations Research*, *Journal of the Operational Research Society*, *Journal of Productivity Analysis* and *Journal of Urban Economics*. His recent textbook *Production, Growth and the Environment: An Economic Approach* introduces students to production theory methods when undesirable by-products are jointly produced with desirable outputs.

Chen-Hui Wu (chenhui@ccu.edu.tw) is an Associate Professor in the Department of Accounting and Information Technology, National Chung Cheng University. Dr Wu earned her PhD from the National Sun Yat-sen University. Earlier, she passed both the certified public accountant (CPA) and certified internal auditor (CIA) examinations in Taiwan. She has also worked in one of the Big Four CPA firms for two years. Dr Wu taught at the National Dong Hwa University, and received Faculty Research Awards in 2009–2011 and an Excellence in Teaching Award from the College of Management in 2013. Her area of research includes financial accounting, corporate finance, behavioural finance and corporate governance. Dr Wu has published academic articles in *Academia Economic Papers*, *European Journal of Operational Research*, *Journal of Financial Studies*, *Journal of Multinational Financial Management*, *Journal of the Operational Research Society* and *Pacific-Basin Finance Journal*.

Bing Xu (b.xu@hw.ac.uk) is an Associate Professor in the School of Social Sciences at Heriot-Watt University. She holds an MA (Hons) in Business Studies and Accounting and a PhD in Management, both from the University of Edinburgh. Her research concerns banks' lending behaviour, energy economics, applied energy, data envelopment analysis and multi-criteria decision-making analysis (MCDA). Bing has also worked on a number of externally funded research projects and collaborated with a wide range of industrial and government partners. For example, she was a co-investigator in 'E-Harbours', funded by EU Interreg IVB North Sea Region, and 'Efficient Sustainable Energy Management with the Abattoir and Dairy Industries in Scotland', funded by the Scottish Funding Council. Currently, Bing is working on an EPSRC-funded project on low-carbon aviation biofuel through integration of novel technologies for co-valorization of CO_2 and biomass (EP/N009924/1), and she is the work package 5 leader on policy, public engagement and regulation. Dr Xu serves as a referee for over 10 academic journals. She has published in journals such as *Journal of Financial Stability*, *International Review of Financial Analysis*,

Applied Energy, Energy Economics, Energy Journal, Economics Letters, Transportation Research Part A: Policy and Practice and *Expert Systems with Applications*, among others.

Xiaopeng Yang (xiaopeng.yang@utoronto.ca) is currently a Financial Analyst at Softek, Canada. He received his PhD in Operations Research from Osaka University, Japan in 2012. After that, he did postdoctoral research in decision analysis and system optimization at the Centre for Management of Technology and Entrepreneurship (CMTE) at the University of Toronto. His primary research interests involve the integration of game theory methodologies for decision making in the banking industry, data envelopment analysis, data mining and related research areas in operations research. He has tackled a number of projects, including efficiency evaluation of Japanese regional banks, staff allocation in a large Canadian bank's branches, corporate failure prediction, company valuation, neurorehabilitation benchmarking and efficiency measurement of Canadian hospitals. One of the projects he has worked on at the University of Toronto focused on building a benchmarking system for evaluating inpatients' recovery status for Bridgepoint Hospital in Toronto. This dynamic benchmarking system can update the database in a timely way and recalculate a new discharge benchmarking criterion to judge whether an inpatient is qualified for being discharged from hospital, and at the same time it can automatically provide a continued rehabilitation plan for the inpatient. He has published his research in *European Journal of Operational Research, Omega, Health Services Management Research* and other international journals.

Ming-Miin Yu (yumm@mail.ntou.edu.tw) is a Distinguished Professor in the Department of Transportation Science, National Taiwan Ocean University, Taiwan. He received his PhD from National Taiwan University, Taiwan. His research interests include transportation economics and management, and logistics, particularly productivity and efficiency analysis of transportation and logistics. He has published many papers on DEA in international journals, such as *Omega, Transportation Research Part A: Policy and Practice, Transportation Research Part E: Logistics and Transportation Review, Tourism Management, Cornell Hospitality Quarterly, Transport Reviews, Journal of Air Transport Management, Transportation Planning and Technology, Journal of Advanced Transportation, International Journal of Sustainable Transportation, Central European Journal of Operations Research, Decision Support Systems, Current Issues in Tourism, Maritime Policy & Management, Annals of Operations Research, China Economic Review, Applied Economics Letters, Agricultural Economics, Applied Mathematics and Computation, International Journal of Uncertainty, Fuzziness and Knowledge-Based Systems, International Journal of Information Technology & Decision Making, Journal of Managerial and Financial Accounting, Emerging Markets Finance and Trade, International Journal of Transport Economics, Journal of Advanced Transportation, Journal of Environmental Management, China Agricultural Economic Review, Journal of Environmental Management, Journal of Civil Engineering and Management, Expert Systems with*

Applications and *The Service Industries Journal*. He is a member of the Eastern Asia Society for Transportation, the Chinese Institute of Transportation, and the Taiwan Efficiency and Productivity Society. He is on the Editorial Board of *Modern Traffic and Transportation Engineering Research*, *Periodica Polytechnica Series Transportation Engineering*, *Journal of Management Studies* and *Journal of Hotel & Business Management*.

PREFACE

A TRIBUTE TO THE LATE PROFESSORS ABRAHAM CHARNES AND WILLIAM W. COOPER

I dedicate this volume to the late Professors Abraham Charnes (1917–1992) and William W. Cooper (1910–2012), who opened the door to this wonderful land of research in efficiency and productivity.

Memoir of Abe Charnes

It was in August 1984 when I visited Abe for the first time in Austin. I was invited to his home and we talked until midnight. At the end of my visit, Karmarkar's LP algorithm appeared in the magazine *Science*. Abe was strongly against the projective transformation that Karmarkar was reported to employ and against the way the article was disclosed. In January 1987, I was invited to Austin, for the second time, in order to collaborate on research with Abe. The sudden visit was opened by an international telegram from Austin to my home in Tokyo, beginning with the phrase 'No Karmarkar, no, no, no.' I saw him again in 1988 at the 13th International Symposium on Mathematical Programming in Tokyo and in 1990 at IFORS in Greece. Each time, it was impressive to touch his strong and warm personality even when he showed his likes and dislikes directly.

In this volume, I have added a memorial unpublished paper by Abe and me, 'DEA models with infinitely many DMUs', which was written in January 1987 when I visited Abe in Austin.

Memoir of Bill Cooper

I cannot help but say how I miss Bill. I met Bill for the first time in 1987 at Dr Charnes' office in Austin. In 1993, Bill visited Aoyama-Gakuin in Tokyo, where we agreed to write a textbook on DEA. I began to write the first draft in 1996 and the book was published in late 1999 by Kluwer (now Springer) under the names of Cooper, Seiford and Tone. I will talk about something that happened during work on this publication. We exchanged a memorandum on writing this book. First, we agreed it should be a textbook but not a monograph. At that time, we had no Windows or e-mail. So, I wrote the first draft in TeX and sent the dvi file as printed matter to Bill by airmail. It took about one week to reach Austin. Bill carefully read my draft and responded to me by revising it with his handwritten material. It was a wonderful experience for me that, even if I wrote only a few lines on some subject, he expanded it to several pages! His sentences were long with no periods but with much ornamentation. When I was an undergraduate student, I read Immanuel Kant's *Prolegomena zu einer jeden künftigen Metaphysik, die als Wissenschaft wird auftreten können*, in the Reclam edition. I wondered how the great philosopher was able to express his thoughts in continuous long sentences in a multi-stratified manner. I felt the same surprise at Bill's writing. I first learnt to write such long sentences just like composing a symphony. Bill's brain was full of polyphonic structure. Moreover, his handwritten letters were difficult to decipher, as many acquaintances know. He said that when he was a schoolboy he won an award in penmanship. However, after the invention of the ballpoint pen, he came to write speedily to express his flowing ideas one after another. So, his cacography was caused by the ballpoint pen!

No words can express the deep sorrow I felt when I heard of his demise.

About This Book

This book is a product of the DEA Workshop 2015 held on 1 and 2 December 2015 at the National Graduate Institute for Policy Studies (GRIPS) in Tokyo, Japan. The workshop was supported by the Japan Society for Promotion of Science (JSPS), Grant-in-Aid for Scientific Research (B), #25282090, titled 'Studies in Theory and Applications of DEA for Forecasting Purposes'. I hope DEA will be utilized not only for evaluation of the efficiency of past and present achievements but also for future prospects.

I thank all authors for contributing their valuable work.

This book consists of three parts: Part I, DEA Theory; Part II, DEA Applications (Past–Present Scenario); and Part III, DEA for Forecasting and Decision Making (Past–Present–Future Scenario).

I acknowledge great support from the GRIPS staff, particularly Ms Kyoko Hirose, Ms Akiko Sawaji, Mr Tohru Takahashi and Dr Xing Zhang, for their efforts in holding the Workshop. Special thanks are due to Mr Takahashi. In great measure, this book could not have been completed without his extraordinary efforts to edit the many manuscripts by many authors into the present volume.

I wish to thank the people at Wiley for their support for this project, especially Shivana Raj, Jeba Paul Sharon, Rajitha Selvarajan and, most importantly, Douglas Meekison, who as a copyeditor did an excellent job of polishing the content and style of this book. I believe that this book would never have appeared without their kind and patient collaboration.

Last but not least, I thank Miki Tsutsui. She has been my continual colleague for a long time.

Kaoru Tone
June 2016

PART I

DEA THEORY

1

RADIAL DEA MODELS

KAORU TONE

National Graduate Institute for Policy Studies, Tokyo, Japan

1.1 INTRODUCTION

Data envelopment analysis (DEA) models started from the seminal paper by Charnes, Cooper and Rhodes [1] (hereafter referred to as CCR). This opened up fertile territory for efficiency evaluation. This paper has been cited by more than 20 000 papers as of the publication date of this book. CCR extended Farrell's work [2] to models with multiple inputs and multiple outputs by utilizing linear programming technology and succeeded in establishing DEA as a powerful basis for efficiency analysis.

1.2 BASIC DATA

DEA compares the relative efficiency of a set of enterprises, called DMUs (decision-making units), which have common input and output factors. Let the numbers of DMUs, inputs and outputs be n, m and s, respectively. We denote input i and output r of DMU$_j$ by x_{ij} $(i=1,\ldots,m; j=1,\ldots,n)$ and y_{rj} $(r=1,\ldots,s; j=1,\ldots,n)$, respectively. The input and output vectors for DMU$_h$ $(h=1,\ldots,n)$ are defined as $\mathbf{x}_h=(x_{1h},\ldots,x_{mh})^T$ and $\mathbf{y}_h=(y_{1h},\ldots,y_{sh})^T$. The input and output matrices are defined

Advances in DEA Theory and Applications: With Extensions to Forecasting Models,
First Edition. Edited by Kaoru Tone.
© 2017 John Wiley & Sons Ltd. Published 2017 by John Wiley & Sons Ltd.

as $\mathbf{X} = (x_{ij})(i = 1, \ldots, m; j = 1, \ldots, n)$ and $\mathbf{Y} = (y_{rj})(r = 1, \ldots, s; j = 1, \ldots, n)$. We assume $\mathbf{X} > \mathbf{0}$ and $\mathbf{Y} > \mathbf{0}$.[1] For the input, smaller is better, while for the output, larger is better. We evaluate DMUs by the ratio scale of output/input.

1.3   INPUT-ORIENTED CCR MODEL

Let the weights of the inputs and outputs be $\mathbf{v} = (v_1, \ldots, v_m) \geq \mathbf{0}$ and $\mathbf{u} = (u_1, \ldots, u_m) \geq \mathbf{0}$. The input-oriented CCR model evaluates the efficiency of a DMU $(\mathbf{x}_h, \mathbf{y}_h)(h = 1, \ldots, n)$ by solving the following fractional programming problem:

[Ratio form]

$$\max_{\mathbf{v}, \mathbf{u}} \theta = \frac{u_1 y_{1h} + \cdots + u_s y_{sh}}{v_1 x_{1h} + \cdots + v_m x_{mh}} \tag{1.1}$$

$$\text{s.t.} \quad \frac{u_1 y_{1j} + \cdots + u_s y_{sj}}{v_1 x_{1j} + \cdots + v_m x_{mj}} \leq 1 \quad (j = 1, \ldots, n) \tag{1.2}$$

$$\mathbf{v} \geq \mathbf{0}, \mathbf{u} \geq \mathbf{0}$$

This fractional program can be transformed into the following equivalent linear program:

[Multiplier form]

$$\max_{\mathbf{v}, \mathbf{u}} \theta = u_1 y_{1h} + \cdots + u_s y_{sh} \tag{1.3}$$

$$\text{s.t.} \quad v_1 x_{1h} + \cdots + v_m x_{mh} = 1$$

$$v_1 x_{1j} + \cdots + v_m x_{mj} - u_1 y_{1j} - \cdots - u_s y_{sj} \geq 0 \quad (j = 1, \ldots, n) \tag{1.4}$$

$$\mathbf{v} \geq \mathbf{0}, \mathbf{u} \geq \mathbf{0}$$

The dual to the above LP can be described as follows:

[Envelopment form]

$$\min_{\lambda, s^-, s^+} \theta \tag{1.5}$$

$$\text{s.t.} \quad x_{i1}\lambda_1 + \cdots + x_{in}\lambda_n - \theta x_{ih} \leq 0 \text{ or } x_{i1}\lambda_1 + \cdots + x_{in}\lambda_n - \theta x_{ih} + s_i^- = 0 \quad (i = 1, \ldots, m)$$

$$y_{r1}\lambda_1 + \cdots + y_{rn}\lambda_n \geq y_{rh} \text{ or } y_{r1}\lambda_1 + \cdots + y_{rn}\lambda_n - s_r^+ = y_{rh} \quad (r = 1, \ldots, s)$$

$$\lambda_j \geq 0 \ (\forall j), \ s_i^- \geq 0 \ (\forall i), \ s_r^+ \geq 0 \ (\forall r) \tag{1.6}$$

---

[1] In some models, we can relax these assumptions.

$\boldsymbol{\lambda}$, $\mathbf{s}^-$ and $\mathbf{s}^+$ are the intensity, input-slack and output-slack vectors, respectively. This model aims at minimizing inputs while producing at least the given output level.

Let an optimal solution to [Envelopment form] be $\theta^*, \boldsymbol{\lambda}^*, \mathbf{s}^{-*}$ and $\mathbf{s}^{+*}$.

Definition 1.1 (CCR score)
The CCR score of DMU$_h$ is defined by θ^*.

Definition 1.2 (Strongly efficient)
DMU$_h$ is *strongly* CCR efficient if $\theta^* = 1$ and $(\mathbf{s}^{-*} = \mathbf{0}$ and $\mathbf{s}^{+*} = \mathbf{0})$ for all optimal solutions to [Envelopment form].

Definition 1.3 (Weakly efficient)
DMU$_h$ is *weakly* CCR efficient if $\theta^* = 1$ and $(\mathbf{s}^{-*} \neq \mathbf{0}$ or $\mathbf{s}^{+*} \neq \mathbf{0})$ for some optimal solutions to [Envelopment form].

Definition 1.4 (Inefficient)
DMU$_h$ is CCR inefficient if $\theta^* < 1$.

Definition 1.5 (Production possibility set)
From the data matrices $\mathbf{X}$ and $\mathbf{Y}$, we define the production possibility set P by

$$P = \{(\mathbf{x}, \mathbf{y}) \mid \mathbf{x} \geq \mathbf{X}\boldsymbol{\lambda}, \mathbf{y} \leq \mathbf{Y}\boldsymbol{\lambda}, \boldsymbol{\lambda} \geq \mathbf{0}\} \qquad (1.7)$$

Figure 1.1 shows a typical production possibility set in two dimensions for the single-input and single-output case. In this example, the possibility set is determined by B and the ray from the origin through B is the efficient frontier. DMU A is inefficient and its input-oriented score is $PQ/PA = 0.5$.

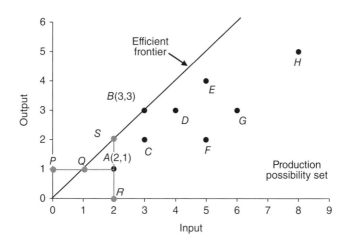

Figure 1.1 Production possibility set for the CCR model.

1.3.1 The CRS Model

This model is called the *constant-returns-to-scale* (CRS) model.

Definition 1.6 (Reference set)
For an optimal solution $(\theta^*, \lambda^*, \mathbf{s}^{-*}, \mathbf{s}^{+*})$ to [Envelopment form], we define the reference set of DMU$_h$ by

$$E(h) = \left\{ i \mid \lambda_j^* > 0;\, j = 1, \ldots, n \right\} \tag{1.8}$$

The reference set is not always uniquely determined.

Definition 1.7 (CCR projection)
The CCR projection is defined as

$$\hat{\mathbf{x}}_h = \theta^* \mathbf{x}_h - \mathbf{s}^{-*}, \quad \hat{\mathbf{y}}_h = \mathbf{y}_h + \mathbf{s}^{+*} \tag{1.9}$$

Theorem 1.1   The projected $(\hat{\mathbf{x}}_h, \hat{\mathbf{y}}_h)$ is strongly CCR efficient.

1.4 THE INPUT-ORIENTED BCC MODEL

The envelopment form of the BCC (Banker–Charnes–Cooper) model [3] is defined as follows:

[Envelopment form of the BCC model]

$$\min_{\lambda, \mathbf{s}^-, \mathbf{s}^+} \theta \tag{1.10}$$

s.t. $x_{i1}\lambda_1 + \cdots + x_{in}\lambda_n - \theta x_{ih} \leq 0$ or $x_{i1}\lambda_1 + \cdots + x_{in}\lambda_n - \theta x_{ih} + s_i^- = 0$ $(i = 1, \ldots, m)$

$y_{r1}\lambda_1 + \cdots + y_{rn}\lambda_n \geq y_{rh}$ or $y_{r1}\lambda_1 + \cdots + y_{rn}\lambda_n - s_r^+ = y_{rh}$ $(r = 1, \ldots, s)$

$$\lambda_1 + \cdots \lambda_n = 1 \tag{1.11}$$

$\lambda_j \geq 0$ $(\forall j)$, $s_i^- \geq 0$ $(\forall i)$, $s_r^+ \geq 0$ $(\forall r)$

The multiplier form is as follows:

[Multiplier form]

$$\max_{\mathbf{v}, \mathbf{u}, u_0} \theta = u_1 y_{1h} + \ldots + u_s y_{sh} - u_0 \tag{1.12}$$

s.t. $v_1 x_{1h} + \cdots + v_m x_{mh} = 1$

$$v_1 x_{1j} + \cdots + v_m x_{mj} - u_1 y_{1j} - \cdots - u_s y_{sj} + u_0 \geq 0 \quad (j = 1, \ldots, n) \tag{1.13}$$

$\mathbf{v} \geq \mathbf{0}, \mathbf{u} \geq \mathbf{0}, u_0$ free in sign

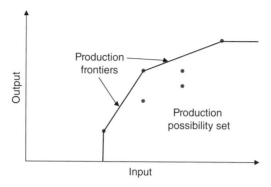

Figure 1.2 Production possibility set for the BCC model.

The equivalent BCC fractional program is obtained from the multiplier form as follows:

[Ratio form of the BCC model]

$$\max_{\mathbf{v},\mathbf{u},u_0} \theta = \frac{u_1 y_{1h} + \cdots + u_s y_{sh} - u_0}{v_1 x_{1h} + \cdots + v_m x_{mh}} \tag{1.14}$$

$$\text{s.t.} \quad \frac{u_1 y_{1j} + \cdots + u_s y_{sj} - u_0}{v_1 x_{1j} + \cdots + v_m x_{mj}} \leq 1 \quad (j = 1, \ldots, n) \tag{1.15}$$

$$\mathbf{v} \geq \mathbf{0}, \mathbf{u} \geq \mathbf{0}, u_0 \text{ free in sign}$$

Figure 1.2 shows a typical production possibility set for the BCC model.

1.4.1 The VRS Model

This model is called the *variable-returns-to-scale* (VRS) model.

1.5 THE OUTPUT-ORIENTED MODEL

This model attempts to maximize the outputs while using no more than the observed amount of any input:

$$\eta^* = \max \eta \tag{1.16}$$

$$\text{s.t.} \quad x_{i1}\lambda_1 + \cdots + x_{in}\lambda_n \leq x_{ih}0 \text{ or } x_{i1}\lambda_1 + \cdots + x_{in}\lambda_n + s_i^- = x_{ih} \quad (i = 1, \ldots, m)$$

$$y_{r1}\lambda_1 + \cdots + y_{rn}\lambda_n \geq \eta y_{rh} \text{ or } y_{r1}\lambda_1 + \cdots + y_{rn}\lambda_n - s_r^+ = \eta y_{rh} \quad (r = 1, \ldots, s) \tag{1.17}$$

$$\lambda_j \geq 0 \; (\forall j), \; s_i^- \geq 0 \; (\forall i), \; s_r^+ \geq 0 \; (\forall r)$$

We define the output-oriented efficiency θ^* as the inverse of η^*:

$$\theta^* = 1/\eta^* \tag{1.18}$$

In Figure 1.1, DMU A has $\eta^* = RS/RA = 2$ and hence its output-oriented score is 0.5. In the CCR model, the input- and output-oriented scores are identical, whereas in the BCC model they are usually different.

1.6 ASSURANCE REGION METHOD

In the optimal weight (v_i^*, u_j^*) of a DEA model, we may see many zeros – showing that the DMU has a weakness in the corresponding items compared with other (efficient) DMUs. Large differences in weights from item to item may also be a concern. This leads to the assurance region method, which imposes constraints on the relative magnitudes of the weights for special items. For example, we may add a constraint on the ratio of weights for Input 1 and Input 2 as follows:

$$L_{12} \leq v_2/v_1 \leq U_{12} \tag{1.19}$$

where L_{12} and U_{12} are lower and upper bounds that the ratio v_2/v_1 may assume. See [4] for details.

1.7 THE ASSUMPTIONS BEHIND RADIAL MODELS

These models assume a proportional reduction of the inputs (such as $\theta^* \mathbf{x}_h$) and a proportional expansion of the outputs (such as $\eta^* \mathbf{y}_h$). In some instances, these assumptions are too restrictive. This has led to the development of non-radial models.

1.8 A SAMPLE RADIAL MODEL

We show an example of a radial model here. Table 1.1 represents 12 hospitals with two inputs, Doctor and Nurse, and two outputs, Outpatient and Inpatient, where (I) and (O) indicate input and output, respectively.

Table 1.2 reports scores for the hospital example, both input-oriented (CCR-I, BCC-I) and output-oriented (CCR-O, BCC-O), while Figure 1.3 shows a graphical comparison. The scores for CCR-I and CCR-O are identical.[2]

---

[2] Software for the CCR, BCC and other models is included in DEA-Solver Pro V13 (http://www.saitech-inc.com). See also Appendix A.

TABLE 1.1   A hospital example.

Hospital	(I) Doctor	(I) Nurse	(O) Outpatient	(O) Inpatient
A	20	151	100	90
B	19	131	150	50
C	25	160	100	55
D	27	168	180	72
E	25	158	80	66
F	55	255	150	60
G	33	235	170	70
H	31	206	130	60
I	30	244	110	60
J	50	290	250	100
K	53	306	230	110
L	38	284	150	90

TABLE 1.2   Efficiency scores obtained by radial models.

Hospital	CCR-I	CCR-O	BCC-I	BCC-O
A	1	1	1	1
B	1	1	1	1
C	0.6915	0.6915	0.8344	0.6916
D	1	1	1	1
E	0.7208	0.7208	0.8797	0.7332
F	0.5490	0.5490	0.5555	0.6524
G	0.7048	0.7048	0.7676	0.8693
H	0.6366	0.6366	0.6602	0.7253
I	0.5651	0.5651	0.6417	0.6809
J	0.8046	0.8046	1	1
K	0.7694	0.7694	1	1
L	0.6362	0.6362	0.7556	0.8919

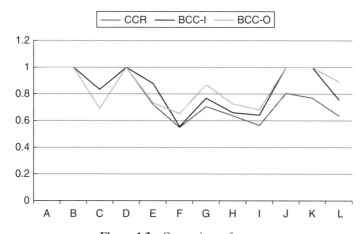

Figure 1.3   Comparison of scores.

REFERENCES

[1] Charnes, A., Cooper, W.W. and Rhodes, E. (1978) Measuring the efficiency of decision making units. *European Journal of Operational Research*, **2**, 429–444.

[2] Farrell, M.J. (1957) The measurement of production efficiency. *Journal of the Royal Statistical Society A*, **120**, 253–281.

[3] Banker, R., Charnes, A. and Cooper, W.W. (1984) Some models for estimating technical and scale inefficiencies in data envelopment analysis. *Management Science*, **30**, 1078–1092.

[4] Cooper, W.W., Seiford, L.M. and Tone, K. (2007) *Data Envelopment Analysis: A Comprehensive Text with Models, Applications, References and DEA-Solver Software*, 2nd edn, Springer, New York.

2

NON-RADIAL DEA MODELS

KAORU TONE

National Graduate Institute for Policy Studies, Tokyo, Japan

2.1 INTRODUCTION

There are two types of model in data envelopment analysis (DEA): radial and non-radial. Radial models are represented by the CCR model. Basically, they deal with proportional changes of inputs or outputs. As such, the CCR score reflects the proportional maximum input (or output) reduction (or expansion) rate which is common to all inputs (or outputs). However, in real-world businesses, not all inputs (or outputs) behave in a proportional way. For example, if we employ labour, materials and capital as inputs, some of them are substitutional and do not change proportionally. Another shortcoming of radial models is the neglect of slacks in reporting the efficiency score. In many cases, we find a lot of remaining non-radial slacks. So, if these slacks have an important role in evaluating managerial efficiency, the radial approaches may mislead the decision process if we utilize the efficiency score as the only index for evaluating the performance of decision-making units (DMUs).

In contrast, non-radial SBM (slacks-based measure) models put aside the assumption of proportional changes in inputs and outputs, and deal with slacks directly. This

---

Advances in DEA Theory and Applications: With Extensions to Forecasting Models,
First Edition. Edited by Kaoru Tone.
© 2017 John Wiley & Sons Ltd. Published 2017 by John Wiley & Sons Ltd.

may discard varying proportions of the original inputs and outputs. SBM models are designed to meet the following two conditions:

1. *Units-invariant:* the measure should be invariant with respect to the units of the data.
2. *Monotone:* the measure should be monotonically decreasing in each slack in the input and output.

The rest of this chapter organized as follows. Section 2.2 introduces SBM models in the input-, output- and non-oriented cases under the constant-returns-to-scale assumption. We present an illustrative example in Section 2.3. We observe the dual side of these models in Section 2.4. We extend them to the variable-returns-to-scale environment and to weighted-SBM models in Section 2.5. Section 2.6 concludes the chapter.

2.2   THE SBM MODEL

The SBM model was introduced by Tone [1] (see also Pastor *et al.* [2]). It has three variations, namely input-, output- and non-oriented. The non-oriented model is both input- and output-oriented.

Let the set of DMUs be $J = \{1, 2, \ldots, n\}$, each DMU having m inputs and s outputs. We denote the vectors of inputs and outputs for DMU$_j$ by $\mathbf{x}_j = (x_{1j}, x_{2j}, \ldots, x_{mj})^T$ and $\mathbf{y}_j = (y_{1j}, y_{2j}, \ldots, y_{sj})^T$, respectively. We define input and output matrices $\mathbf{X}$ and $\mathbf{Y}$ by

$$\mathbf{X} = (\mathbf{x}_1, \mathbf{x}_2, \cdots, \mathbf{x}_n) \in R^{m \times n} \text{ and } \mathbf{Y} = (\mathbf{y}_1, \mathbf{y}_2, \cdots, \mathbf{y}_n) \in R^{s \times n} \tag{2.1}$$

We assume that all data are positive, that is, $\mathbf{X} > \mathbf{0}$ and $\mathbf{Y} > \mathbf{0}$.

The production possibility set is defined using a non-negative combination of the DMUs in the set J as

$$P = \left\{ (\mathbf{x}, \mathbf{y}) \middle| \mathbf{x} \geq \sum_{j=1}^{n} \lambda_j \mathbf{x}_j, \ \mathbf{0} \leq \mathbf{y} \leq \sum_{j=1}^{n} \lambda_j \mathbf{y}_j, \ \lambda \geq \mathbf{0} \right\} \tag{2.2}$$

$\lambda = (\lambda_1, \lambda_2, \ldots, \lambda_n)^T$ is called the intensity vector.

The inequalities in (2.2) can be transformed into equalities by introducing slacks as follows:

$$\mathbf{x} = \sum_{j=1}^{n} \lambda_j \mathbf{x}_j + \mathbf{s}^-$$
$$\mathbf{y} = \sum_{j=1}^{n} \lambda_j \mathbf{y}_j - \mathbf{s}^+ \tag{2.3}$$
$$\mathbf{s}^- \geq \mathbf{0}, \ \mathbf{s}^+ \geq \mathbf{0}$$

where $\mathbf{s}^- = (s_1^-, s_2^-, \ldots, s_m^-)^T \in R^m$ and $^+ = (s_1^+, s_2^+, \ldots, s_s^+)^T \in R^s$ are called the input and output slacks, respectively.

2.2.1   Input-Oriented SBM

In order to evaluate the relative efficiency of $DMU_h = (\mathbf{x}_h, \mathbf{y}_h)$, we solve the following linear program. This process is repeated n times for $h = 1, \ldots, n$:

[SBM-I-C] (Input-oriented SBM under constant-returns-to-scale assumption)

$$\rho_I^* = \min_{\lambda, \mathbf{s}^-, \mathbf{s}^+} 1 - \frac{1}{m} \sum_{i=1}^m \frac{s_i^-}{x_{ih}}$$

subject to

$$x_{ih} = \sum_{j=1}^n x_{ij} \lambda_j + s_i^- \quad (i = 1, \ldots, m) \tag{2.4}$$

$$y_{rh} = \sum_{j=1}^n y_{rj} \lambda_j - s_r^+ \quad (r = 1, \ldots, s)$$

$$\lambda_j \geq 0 \ (\forall j), \ s_i^- \geq 0(\forall i), \ s_r^+ \geq 0(\forall r)$$

ρ_I^* is called the SBM-input efficiency.

Proposition 2.1 ρ_I^* is units-invariant, that is, it is independent of the units in which the inputs and outputs are measured.

Let an optimal solution of [SBM-I-C] be $(\lambda^*, \mathbf{s}^{-*}, \mathbf{s}^{+*})$.

Definition 2.1 (SBM-input-efficient)
A $DMU_h = (\mathbf{x}_h, \mathbf{y}_h)$ is called SBM-input-efficient if $\rho_I^* = 1$ holds.

This means $\mathbf{s}^{-*} = \mathbf{0}$, that is, all input slacks are zero. However, output slacks may be non-zero.

Definition 2.2 (Projection)
Using an optimal solution $(\lambda^*, \mathbf{s}^{-*}, \mathbf{s}^{+*})$, we define a projection of $DMU_h = (\mathbf{x}_h, \mathbf{y}_h)$ by

$$(\bar{\mathbf{x}}_h, \bar{\mathbf{y}}_h) = (\mathbf{x}_h - \mathbf{s}^{-*}, \mathbf{y}_h + \mathbf{s}^{+*}) \tag{2.5}$$

Proposition 2.2 The projected DMU is SBM-input-efficient.

Definition 2.3 (Reference set)
We define a reference set R of $DMU_h = (\mathbf{x}_h, \mathbf{y}_h)$ by

$$R = \left\{ j \mid \lambda_j^* > 0, \ j \in J \right\} \tag{2.6}$$

Thus, $(\mathbf{x}_h, \mathbf{y}_h)$ can be expressed as follows:

$$x_{ih} = \sum_{j \in R} x_{ij} \lambda_j^* + s_i^{-*} \quad (i = 1, \ldots, m)$$
$$y_{rh} = \sum_{j \in R} y_{rj} \lambda_j^* - s_r^{+*} \quad (r = 1, \ldots, s) \tag{2.7}$$

Proposition 2.3 DMUs in the reference set R of $(\mathbf{x}_h, \mathbf{y}_h)$ are SBM-input-efficient.

Proposition 2.4 The SBM-input-efficiency score is not greater than the CCR efficiency score. (See Tone [1]) for a proof.)

2.2.2 Output-Oriented SBM

The output-oriented SBM efficiency ρ_O^* of $\mathrm{DMU}_h = (\mathbf{x}_h, \mathbf{y}_h)$ is defined by
[SBM-O-C]

$$1/\rho_O^* = \max_{\lambda, \mathbf{s}^-, \mathbf{s}^+} 1 + \frac{1}{s} \sum_{r=1}^{s} \frac{s_r^+}{y_{rh}}$$

subject to

$$x_{ih} = \sum_{j=1}^{n} x_{ij}\lambda_j + s_i^- \quad (i = 1, \ldots, m) \tag{2.8}$$

$$y_{rh} = \sum_{j=1}^{n} y_{rj}\lambda_j - s_r^+ \quad (r = 1, \ldots, s)$$

$$\lambda_j \geq 0 \ (\forall j), \ s_i^- \geq 0 \ (\forall i), \ s_r^+ \geq 0 \ (\forall r)$$

Let an optimal solution of [SBM-O-C] be $(\boldsymbol{\lambda}^*, \mathbf{s}^{-*}, \mathbf{s}^{+*})$.

Definition 2.4 (SBM-output-efficient)

A $\mathrm{DMU}_h = (\mathbf{x}_h, \mathbf{y}_h)$ is called SBM-output-efficient if $\rho_O^* = 1$ holds.

This means $\mathbf{s}^{+*} = \mathbf{0}$, that is, all output slacks are zero. However, the input slacks may be non-zero.

Definition 2.5 (Projection)

Using an optimal solution $(\boldsymbol{\lambda}^*, \mathbf{s}^{-*}, \mathbf{s}^{+*})$, we define a projection of $\mathrm{DMU}_h = (\mathbf{x}_h, \mathbf{y}_h)$ by

$$(\bar{\mathbf{x}}_h, \bar{\mathbf{y}}_h) = (\mathbf{x}_h - \mathbf{s}^{-*}, \mathbf{y}_h + \mathbf{s}^{+*}) \tag{2.9}$$

Proposition 2.5 The projected DMU is SBM-output-efficient.

2.2.3 Non-Oriented SBM

The non-oriented or both-oriented SBM efficiency ρ_{IO}^* is defined by
[SBM-C]

$$\rho_{IO}^* = \min_{\lambda, \mathbf{s}^-, \mathbf{s}^+} \frac{1 - \dfrac{1}{m} \sum_{i=1}^{m} \dfrac{s_i^-}{x_{ih}}}{1 + \dfrac{1}{s} \sum_{r=1}^{s} \dfrac{s_r^+}{y_{rh}}}$$

subject to

$$x_{ih} = \sum_{j=1}^{n} x_{ij}\lambda_j + s_i^- \quad (i = 1, \ldots, m) \tag{2.10}$$

$$y_{rh} = \sum_{j=1}^{n} y_{rj}\lambda_j - s_r^+ \quad (r = 1, \ldots, s)$$

$$\lambda_j \geq 0 \ (\forall j), \ s_i^- \geq 0 \ (\forall i), \ s_r^+ \geq 0 \ (\forall r)$$

Definition 2.6 (SBM-efficient)

A $DMU_h = (\mathbf{x}_h, \mathbf{y}_h)$ is called SBM-efficient if $\rho_{IO}^* = 1$ holds.

This means $\mathbf{s}^- = \mathbf{0}$ and $\mathbf{s}^{+*} = \mathbf{0}$, that is, all input and output slacks are zero.

[SBM-C] can be transformed into a linear program using the Charnes–Cooper transformation as follows:

[SBM-C-LP]

$$\tau^* = \min_{t,\Lambda,\mathbf{S}^-,\mathbf{S}^+} t - \frac{1}{m}\sum\nolimits_{i=1}^{m} \frac{S_i^-}{x_{ih}}$$

subject to

$$1 = t + \frac{1}{s}\sum\nolimits_{r=1}^{s} \frac{S_r^+}{y_{rh}}$$

$$tx_{ih} = \sum\nolimits_{j=1}^{n} x_{ij}\Lambda_j + S_i^- \quad (i=1,\ldots,m) \qquad (2.11)$$

$$ty_{rh} = \sum\nolimits_{j=1}^{n} y_{rj}\Lambda_j - S_r^+ \quad (r=1,\ldots,s)$$

$$\Lambda_j \ge 0 \; (\forall j), \; S_i^- \ge 0 \; (\forall i), \; S_r^+ \ge 0 \; (\forall r), \; t > 0$$

Let an optimal solution be $(\tau^*, t^*, \Lambda^*, \mathbf{S}^{-*}, \mathbf{S}^{+*})$. Then, we have an optimal solution of [SBM-C] defined by

$$\rho^* = \tau^*, \; \lambda^* = \Lambda^*/t^*, \; \mathbf{s}^{-*} = \mathbf{S}^{-*}/t^*, \; \mathbf{s}^{+*} = \mathbf{S}^{+*}/t^* \qquad (2.12)$$

2.3 AN EXAMPLE OF AN SBM MODEL

Table 2.1 shows data for six DMUs using two inputs (x_1, x_2) to produce two outputs (y_1, y_2). We report the results obtained from the SBM models along with that from the CCR model in Table 2.2.[1]

TABLE 2.1   Data.

DMU	x_1	x_2	y_1	y_2
A	4	3	1	2
B	14	6	2	6
C	24	3	3	12
D	20	2	2	6
E	48	4	4	16
F	50	7.5	5	30

[1] Software for SBM models is included in DEA-Solver Pro V13 (http://www.saitech-inc.com). See also Appendix A.

TABLE 2.2 Scores and ranks of efficiency.

	CCR-I		SBM-I-C		SBM-O-C		SBM-C	
DMU	Score	Rank	Score	Rank	Score	Rank	Score	Rank
A	1	1	1	1	1	1	1	1
B	0.8085	6	0.75	6	0.8067	6	0.6923	6
C	1	1	1	1	1	1	1	1
D	1	1	0.9	4	0.8571	5	0.7714	5
E	1	1	0.8333	5	1	1	0.8333	4
F	1	1	1	1	1	1	1	1

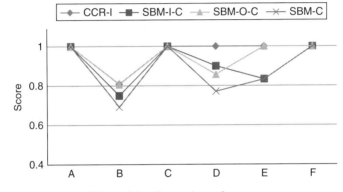

Figure 2.1 Comparison of scores.

TABLE 2.3 Optimal slacks for CCR-I and SBM-I-C.

	CCR-I					SBM-I-C			
DMU	θ^*	s_1^{-*}	s_2^{-*}	s_1^{+*}	s_2^{+*}	ρ_I^*	s_1^{-*}	s_2^{-*}	s_1^{+*}	s_2^{+*}
A	1	0	0	0	0	1	0	0	0	0
B	0.8085	0	0	0	0	0.75	0	3	0	1
C	1	0	0	0	0	1	0	0	0	0
D	1	4	0	0	2	0.9	4	0	0	2
E	1	16	0	0	0	0.8333	16	0	0	0
F	1	0	0	0	0	1	0	0	0	0

The CCR-I model found five DMUs out of six to be efficient. This caused by the radial nature of the model, although slacks remain in some of them. However, the SBM models deal with slacks directly and found DMUs D and E inefficient. In the SBM-O-C model, DMU E was judged to be efficient, since this DMU has no output slacks. Figure 2.1 compares the scores graphically.

Table 2.3 shows the optimal slacks for the CCR-I and SBM-I-C models. DMUs D and E have positive slacks in some input or output. The CCR model does not account for them in the efficiency measure. However, the SBM-I-C model accounts for the input slacks in the efficiency measurement, and DMUs D and E are judged inefficient.

2.4   THE DUAL PROGRAM OF THE SBM MODEL

The dual program of [SBM-C-LP] can be expressed as follows, with the dual variables $\mathbf{v} \in R^m$ and $\mathbf{u} \in R^s$:

[SBM-C-LP-Dual]

$$\max_{\xi, \mathbf{v}, \mathbf{u}} \xi$$

$$\text{subject to} \quad \xi + \mathbf{v}\mathbf{x}_h - \mathbf{u}\mathbf{y}_h = 1 \tag{2.13}$$

$$-\mathbf{v}X + \mathbf{u}Y \le 0, \quad \mathbf{v} \ge \frac{1}{m}[1/\mathbf{x}_h], \quad \mathbf{u} \ge \frac{\xi}{s}[1/\mathbf{y}_h]$$

where the notation $[1/\mathbf{x}_h]$ designates the row vector $(1/x_{1h}, 1/x_{2h}, \ldots, 1/x_{mh})$. By eliminating ξ from the above program, we have the following equivalent program:

$$\max_{\mathbf{v}, \mathbf{u}} \mathbf{u}\mathbf{y}_h - \mathbf{v}\mathbf{x}_h$$

$$\text{subject to} \quad -\mathbf{v}X + \mathbf{u}Y \le 0 \tag{2.14}$$

$$\mathbf{v} \ge \frac{1}{m}[1/\mathbf{x}_h], \quad \mathbf{u} \ge \frac{1 - \mathbf{v}\mathbf{x}_h + \mathbf{u}\mathbf{y}_h}{s}[1/\mathbf{y}_h]$$

The dual variables $\mathbf{v} \in R^m$ and $\mathbf{u} \in R^s$ can be interpreted as the virtual costs and prices of the input and output items, respectively. The dual program aims to find the optimal virtual costs and prices for DMU $(\mathbf{x}_h, \mathbf{y}_h)$ so that the profit $\mathbf{u}\mathbf{y}_j - \mathbf{v}\mathbf{x}_j$ does not exceed zero for any DMU (including $(\mathbf{x}_h, \mathbf{y}_h)$), and to maximize the profit $\mathbf{u}\mathbf{y}_h - \mathbf{v}\mathbf{x}_h$ for the target DMU $(\mathbf{x}_h, \mathbf{y}_h)$. Apparently, the optimal profit is at best zero and hence $\xi^* = 1$ for the SBM-C efficient DMUs.

2.5   EXTENSIONS OF THE SBM MODEL

In this section, we extend the SBM model to the variable-returns-to-scale (VRS) environment, and introduce the weighted-SBM model. See [3] for details.

2.5.1   Variable-Returns-to-Scale (VRS) Model

All models can be adjusted to the variable-returns-to-scale environment by adding the constraint $\mathbf{e}\lambda = 1$, where $\mathbf{e}$ denotes a row vector in which all elements are equal to one. Thus, the production possibility set is modified to

$$P_{\text{VRS}} = \left\{ (\mathbf{x}, \mathbf{y}) \,\middle|\, x_i \ge \sum_{j=1}^{n} x_{ij} (\forall i), \ 0 \le y_r \le \sum_{j=1}^{n} y_{rj} (\forall r), \ \mathbf{e}\lambda = 1, \ \lambda \ge \mathbf{0} \right\} \tag{2.15}$$

For example, input-oriented SBM under VRS can be defined as follows:
[SBM-I-V] (Input-oriented SBM under variable-returns-to-scale assumption)

$$\rho_I^* = \min_{\lambda,s^-,s^+} 1 - \frac{1}{m}\sum_{i=1}^m \frac{s_i^-}{x_{ih}}$$

subject to

$$x_{ih} = \sum_{j=1}^n x_{ij}\lambda_j + s_i^- \quad (i=1,\ldots,m)$$

$$y_{rh} = \sum_{j=1}^n y_{rj}\lambda_j - s_r^+ \quad (r=1,\ldots,s)$$

$$\sum_{j=1}^n \lambda_j = 1,\ \lambda_j \ge 0\ (\forall j),\ s_i^- \ge 0\ (\forall i),\ s_r^+ \ge 0\ (\forall r)$$

(2.16)

We can define [SBM-O-V] and [SBM-V] models similarly.

2.5.2 Weighted-SBM Model

We can assign weights to the input and output slacks in the objective function of (2.10) corresponding to the relative importance of items as follows:
[Weighted-SBM-C]

$$\rho_{IO}^* = \min_{\lambda,s^-,s^+} \frac{1 - \frac{1}{m}\sum_{i=1}^m \frac{w_i^- s_i^-}{x_{ih}}}{1 + \frac{1}{s}\sum_{r=1}^s \frac{w_r^+ s_r^+}{y_{rh}}}$$

subject to

$$x_{ih} = \sum_{j=1}^n x_{ij}\lambda_j + s_i^- \quad (i=1,\ldots,m)$$

$$y_{rh} = \sum_{j=1}^n y_{rj}\lambda_j - s_r^+ \quad (r=1,\ldots,s)$$

$$\lambda_j \ge 0\ (\forall j),\ s_i^- \ge 0\ (\forall i),\ s_r^+ \ge 0\ (\forall r)$$

(2.17)

with $\sum_{i=1}^m w_i^- = m$ and $\sum_{r=1}^m w_r^+ = s$. The weights should reflect the intentions of the decision-makers. We can define input- and output-oriented weighted-SBM models by neglecting the denominator and numerator, respectively, of the objective function in (2.17).

2.6 CONCLUDING REMARKS

In this chapter, we have introduced non-radial slacks-based measure of efficiency (SBM) models and their extensions. SBM models utilize the amount of slacks to the maximum extent in measuring efficiency. This can be a merit as well as a demerit. Weighted-SBM models serve to make models more reliable. This corresponds to the assurance region approach in radial models. Readers can learn more from the references cited in this chapter. In Chapter 22, we extend the ordinary SBM (-Min) model to an SBM-Max model which searches for nearly the closest point on the efficient frontiers. Thus, the projected point can be obtained with less input reduction (or output expansion).

REFERENCES

[1] Tone, K. (2001) A slacks-based measure of efficiency in data envelopment analysis. *European Journal of Operational Research*, **130**, 498–509.

[2] Pastor, J.T., Ruiz, J.L. and Sirvent, I. (1999) An enhanced DEA Russell graph efficiency measure. *European Journal of Operational Research*, **115**, 596–607.

[3] Cooper, W.W., Seiford, L.M. and Tone, K. (2007) *Data Envelopment Analysis: A Comprehensive Text with Models, Applications, References and DEA-Solver Software*, 2nd edn, Springer, New York.

3

DIRECTIONAL DISTANCE DEA MODELS

HIROFUMI FUKUYAMA

Faculty of Commerce, Fukuoka University, Fukuoka, Japan

WILLIAM L. WEBER

Southeast Missouri State University, Cape Girardeau, USA

3.1 INTRODUCTION

Luenberger [1, 2] formulated the benefit and shortage functions, and these functions were popularized as directional distance functions in production economics by Chambers, Chung, and Färe [3, 4] and by Färe and Grosskopf [5]. Shephard's [6, 7] distance functions are special cases of directional distance functions.

In this chapter, Section 3.2 presents the basics of the directional distance DEA (DD) model under constant returns to scale (CRS), while Section 3.3 extends the model to variable returns to scale (VRS). Section 3.4 introduces a slacks-based inefficiency model, and Section 3.5 discusses the choice of directional vectors.

3.2 DIRECTIONAL DISTANCE MODEL

This section formalizes a directional distance function methodology within a multi-output, multi-input setting. Let $\mathbf{y} \in \mathfrak{R}^s_+$ and $\mathbf{x} \in \mathfrak{R}^m_+$ denote the vectors

Advances in DEA Theory and Applications: With Extensions to Forecasting Models,
First Edition. Edited by Kaoru Tone.
© 2017 John Wiley & Sons Ltd. Published 2017 by John Wiley & Sons Ltd.

of outputs and inputs, respectively. The conceptual production technology is defined as

$$T = \left\{ (\mathbf{x}, \mathbf{y}) \in \mathfrak{R}_+^m \times \mathfrak{R}_+^s \mid \text{inputs } \mathbf{x} \text{ yield outputs } \mathbf{y} \right\} \tag{3.1}$$

which is the set of feasible inputs and outputs. The production technology (3.1) is assumed to be a nonempty, closed set, exhibiting free input and output disposability. In addition, the producible output set is assumed to be bounded for finite inputs. This boundedness property is sometimes called scarcity, and indicates that finite inputs cannot produce infinite outputs. Chambers *et al.* [3] introduced a directional (technology) distance function, which is a complete characterization of the production technology (3.1). This directional distance function is defined by

$$\vec{D}(\mathbf{x}, \mathbf{y}; \mathbf{g}) = \sup\{\beta | (\mathbf{x} - \beta \mathbf{g}^-, \mathbf{y} + \beta \mathbf{g}^+) \in T\} \tag{3.2}$$

where $\mathbf{g} = (\mathbf{g}^-, \mathbf{g}^+) = \left(g_1^-, \ldots, g_m^-, g_1^+, \ldots, g_s^+\right) \in \mathfrak{R}_+^m \times \mathfrak{R}_+^s$ is the directional vector that scale outputs and inputs to the frontier of the technology set. Since $\vec{D}(\mathbf{x}, \mathbf{y}; \mathbf{g}) \geq 0$ if and only if $(\mathbf{x}, \mathbf{y}) \in T$, the directional technology distance function (3.2) is a complete characterization of the production technology (3.1). Under regularity conditions, the following translation property always holds:

$$\vec{D}(\mathbf{x} - \sigma \mathbf{g}^-, \mathbf{y} + \sigma \mathbf{g}^+; \mathbf{g}) = \vec{D}(\mathbf{x}, \mathbf{y}; \mathbf{g}) - \sigma, \quad \text{for } \sigma \in \mathfrak{R} \tag{3.3}$$

We assume there are $j = 1, \ldots, n$ observations or decision-making units (DMUs). Relative to the unknown production technology T defined in (3.1), the DD model for DMU h is given by the following linear program:
[Envelopment form of DD model]

$$\max_{\beta, \lambda, s^-, s^+} \beta$$

$$\text{s.t. } x_{i1}\lambda_1 + \cdots + x_{in}\lambda_n + \beta g_i^- + s_i^- = x_{ih} \ (i = 1, \ldots, m)$$

$$y_{r1}\lambda_1 + \cdots + y_{rn}\lambda_n - \beta g_r^+ - s_r^+ = y_{rh} \ (r = 1, \ldots, s) \tag{3.4}$$

$$\lambda_j \geq 0 \ (\forall j), \ s_i^- \geq 0 \ (\forall i), \ s_r^+ \geq 0 \ (\forall r)$$

The optimal objective function value in (3.4) equals the directional distance (DEA) function. The dual to the envelopment form consisting of (3.4) is
[Multiplier form of DD model]

$$\min_{\mathbf{v}, \mathbf{u}} v_1 x_{1h} + \cdots + v_m x_{mh} - u_1 y_{1h} - \cdots - u_s y_{sh}$$

$$\text{s.t. } v_1 g_1^- + \cdots + v_m g_m^- + u_1 g_1^+ + \cdots + u_s g_s^+ = 1$$

$$v_1 x_{1j} + \cdots + v_m x_{mj} - u_1 y_{1j} - \cdots - u_s y_{sj} \geq 0 \ (j = 1, \ldots, n) \tag{3.5}$$

$$\mathbf{v} \geq \mathbf{0}, \mathbf{u} \geq \mathbf{0}$$

The variables $\mathbf{v}$ and $\mathbf{u}$ are virtual prices, with the objective equal to the virtual costs minus the virtual revenues. Under CRS, the objective in (3.5) equals the negative of the virtual profits. Relative shadow or support prices for inputs i and i' are obtained as $v_i/v_{i'}$. Shadow prices for outputs r and r' are obtained as $u_r/u_{r'}$. These shadow prices can be compared with actual prices to determine whether inputs/outputs are efficiently allocated. For the envelopment and multiplier forms, see for example Fukuyama [8]. Let β^*, λ, $\mathbf{s}^-$, $\mathbf{s}^+$ be an optimal solution to [Envelopment form of DD model]. We make the following definitions.

Definition 3.1 (DD score)
The DD score is represented by β^*, which takes a value greater than or equal to zero.

Definition 3.2 (Strong DD-efficiency)
DMU$_h$ is strongly DD-efficient if $\beta^* = 0$, $\mathbf{s}^{-*} = \mathbf{0}$, and $\mathbf{s}^{+*} = \mathbf{0}$ for all optimal solutions to [Envelopment form of DD model].

Definition 3.3 (Weak DD-efficiency)
DMU$_h$ is weakly DD-efficient if $\beta^* = 0$, $\mathbf{s}^{-*} \geq \mathbf{0}$, and $\mathbf{s}^{+*} \geq \mathbf{0}$ for some optimal solution to [Envelopment form of DD model].

Definition 3.4 (DD-inefficiency)
DMU$_h$ is DD-inefficient if $\beta^* > 0$.

Definition 3.5 (DD-efficient projection)
The DD projection expressed by $\hat{\mathbf{x}} = \mathbf{x}_h - \beta \mathbf{g}^- - \mathbf{s}^-$ and $\hat{\mathbf{y}} = \mathbf{y}_h - \beta^* \mathbf{g}^+ - \mathbf{s}^{+*}$ is strongly DD-efficient.

Figure 3.1 depicts the relationship between the DD measure and a directional vector for a single-input, single-output case. The observed DMUs are A, B, and C, where C is strongly DD-efficient. The points D and E are projection points of DMU A and

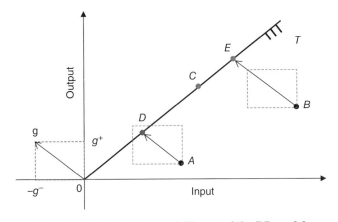

Figure 3.1   Production possibility set of the DD model.

DMU B, respectively. Given (g^-, g^+), the DD score for DMU A equals the ratio of the line segments $AD/0\mathbf{g} < 1$ and that for DMU B equals $BE/0\mathbf{g} > 1$.

3.3 VARIABLE-RETURNS-TO-SCALE DD MODELS

In this subsection we develop a variable-returns-to-scale DD model by adding the convexity constraint $\sum_{j=1}^{n} \lambda_j = 1$ to (3.4). The envelopment form of the DD model is defined as follows:
[Envelopment form of DD model]

$$
\begin{aligned}
\max_{\beta, \, \boldsymbol{\lambda}, \, \mathbf{s}^-, \, \mathbf{s}^+} \quad & \beta \\
\text{s.t.} \quad & x_{i1}\lambda_1 + \cdots + x_{in}\lambda_n + \beta g_i^- + s_i^- = x_{ih} \quad (i=1,\ldots,m) \\
& y_{r1}\lambda_1 + \cdots + y_{rn}\lambda_n - \beta g_r^+ - s_r^+ = y_{rh} \quad (r=1,\ldots,s) \\
& \lambda_1 + \cdots + \lambda_n = 1 \\
& \lambda_j \geq 0 \ (\forall j), \ s_i^- \geq 0 \ (\forall i), \ s_r^+ \geq 0 \ (\forall r)
\end{aligned}
\tag{3.6}
$$

The multiplier form of (3.6) is written as follows:
[Multiplier form]

$$
\begin{aligned}
\max_{\mathbf{v}, \, \mathbf{u}, \, u_0} \quad & v_1 x_{1h} + \cdots + v_m x_{mh} - u_1 y_{1h} - \cdots - u_s y_{sh} - u_0 \\
\text{s.t.} \quad & v_1 g_1^- + \cdots + v_m g_m^- + u_1 g_1^+ + \cdots + u_s g_s^+ = 1 \\
& v_1 x_{1j} + \cdots + v_m x_{mj} - u_1 y_{1j} - \cdots - u_s y_{sj} - u_0 \geq 0 \quad (j=1,\ldots,n) \\
& \mathbf{v} \geq \mathbf{0}, \mathbf{u} \geq \mathbf{0}, \ u_0 \text{ free in sign}
\end{aligned}
\tag{3.7}
$$

This model, consisting of (3.6) and (3.7), is called the variable-returns-to-scale DD model.

3.4 SLACKS-BASED DD MODEL

Fukuyama and Weber [9] introduced a slacks-based directional distance model as an extension and generalization of Tone's [10] slacks-based efficiency model. Under the assumption of variable returns to scale, the slacks-based directional distance model[1] $\mathrm{SDD}(\mathbf{x}_h, \mathbf{y}_h; \mathbf{g}^-, \mathbf{g}^+)$ takes the form

---

[1] Fukuyama and Weber [9] called (3.8) the slacks-based inefficiency.

[Envelopment form of SDD model]

$$\text{SDD}(\mathbf{x}_h, \mathbf{y}_h; \mathbf{g}^-, \mathbf{g}^-) = \max_{\lambda, \, \mathbf{s}^-, \, \mathbf{s}^+} \frac{1}{2} \left(\frac{1}{m} \sum_{i=1}^{m} \frac{s_i^-}{g_i^-} + \frac{1}{s} \sum_{r=1}^{s} \frac{s_r^+}{g_r^+} \right)$$

s.t. $x_{i1}\lambda_1 + \cdots + x_{in}\lambda_n + s_i^- = x_{ih} \ (i = 1, \ldots, m)$

$\quad\ y_{r1}\lambda_1 + \cdots + y_{rn}\lambda_n - s_r^+ = y_{rh} \ (r = 1, \ldots, s)$ (3.8)

$\quad\ \lambda_1 + \cdots + \lambda_n = 1$

$\quad\ \lambda_j \geq 0 \ (\forall j), \ s_i^- \geq 0 \ (\forall i), \ s_r^+ \geq 0 \ (\forall r)$

where $\mathbf{g}^- \in \Re_{++}^m$ and $\mathbf{g}^+ \in \Re_{++}^s$ are directional vectors that contract inputs and expand outputs. The directional vectors have the same units of measurement as the vectors of input slacks and output slacks, which allows the ratios of normalized slacks to be added. The objective of (3.8) maximizes the mean of two components that comprise the average input inefficiencies and the average output inefficiencies. When $\text{SDD}(\mathbf{x}_h, \mathbf{y}_h; \mathbf{g}^-, \mathbf{g}^-) = 0$, DMU h is strongly efficient.

The dual to (3.8) is

[Multiplier form of SDD model]

$$\min_{v, \, u} \ v_1 x_{1h} + \cdots + v_m x_{mh} - u_1 y_{1h} - \cdots - u_s y_{sh} - u_0$$

s.t. $v_1 g_1^- + \cdots + v_m g_m^- + u_1 g_1^+ + \cdots + u_s g_s^+ = 1$

$\quad\ v_1 x_{1j} + \cdots + v_m x_{mj} - u_1 y_{1j} - \cdots - u_s y_{sj} - u_0 \geq 0 \ (j = 1, \ldots, n)$

$\quad\ v_i \geq \dfrac{1}{2m \, g_i^-} \ (i = 1, \ldots, m)$ (3.9)

$\quad\ u_r \geq \dfrac{1}{2s \, g_r^+} \ (r = 1, \ldots, s)$

$\quad\ \mathbf{v} \geq \mathbf{0}, \mathbf{u} \geq \mathbf{0}, u_0 \text{ free in sign}$

The SDD model also generalizes the additive model of Bardhan *et al.* [11,12]. The objective function of the additive model equals the sum of the input slacks as a proportion of the actual inputs plus the sum of the output slacks as a proportion of the outputs $\left(\sum_{i=1}^{m} s_i^- / x_{ih} + \sum_{r=1}^{s} s_r^+ / y_{rh} \right)$ with exactly the same constraints as in (3.8).

The Farrell measures of input and output efficiency scale the inputs and outputs by the same multiplicative factor to either the input isoquant or the production possibility frontier. Färe and Lovell [13] introduced Russell measures of input and output efficiency that scaled inputs and outputs by varying multiplicative factors. Fukuyama and Weber [9] generalized the Russell measures by scaling outputs and inputs additively to the technology set for given directional vectors. Their Russell measure of inefficiency, called the directional Russell inefficiency, takes the form

$$\mathrm{RD}(\mathbf{x}_h,\mathbf{y}_h;\mathbf{g}^-,\mathbf{g}^+) = \max \; \frac{1}{2}\left(\frac{1}{m}\sum_{i=1}^{m}\sigma_i + \frac{1}{s}\sum_{r=1}^{s}\gamma_r\right)$$

$$\text{s.t. } x_{i1}\lambda_1 + \cdots + x_{in}\lambda_n + \sigma_i g_i^- = x_{ih} \; (i=1,\ldots,m)$$

$$y_{r1}\lambda_1 + \cdots + y_{rn}\lambda_n - \gamma_r g_r^+ = y_{rh} \; (r=1,\ldots,s)$$ \hfill (3.10)

$$\lambda_1 + \cdots + \lambda_n = 1$$

$$\lambda_j \geq 0 \; (\forall j), \; \sigma_i^- \geq 0 \; (\forall i), \; \gamma_r^+ \geq 0 \; (\forall r)$$

Setting $\sigma_i = s_i^-/g_i^-$ and $\gamma_r = s_r^+/g_r^+$, it is easy to see that $\mathrm{SDD}(\mathbf{x}_h,\mathbf{y}_h;\mathbf{g}^-,\mathbf{g}^-) = \mathrm{RD}(\mathbf{x}_h,\mathbf{y}_h;\mathbf{g}^-,\mathbf{g}^+)$. Thus, the multiplicative Russell efficiency measures of Färe and Lovell [13] can be extended to additive measures of inefficiency for any choice of directional vector.

3.5 CHOICE OF DIRECTIONAL VECTORS

Some reasonable candidates for the directional vectors $\mathbf{g} = (\mathbf{g}^-,\mathbf{g}^+)$ include (i) $\mathbf{g} = (\bar{\mathbf{x}},\bar{\mathbf{y}})$, where $\bar{\mathbf{x}}$ and $\bar{\mathbf{y}}$ are the averages of the observed inputs and outputs and the DD model objective function yields the expansion of outputs and contraction of inputs as proportions of the mean; (ii) $\mathbf{g} = (\mathbf{1}^m,\mathbf{1}^s)$, where $(\mathbf{1}^m,\mathbf{1}^s)$ are vectors of ones, so that the DD model objective yields a unit expansion of outputs and a unit contraction of inputs; (iii) $\mathbf{g} = (\mathbf{1}^m/2m,\mathbf{1}^s/2s)$, which was used by Färe and Grosskopf [5] in a slacks-based inefficiency model; (iv) $\mathbf{g} = (\mathbf{x}_h,\mathbf{y}_h)$, proposed by Briec [14,15] and employed by Fukuyama and Weber [9] so that the DD model objective yields the expansion of outputs and contraction of inputs as a proportion of the outputs and inputs of DMU h; (v) $\mathbf{g} = (\mathbf{x}^*,\mathbf{y}^*)$, where the inputs and outputs are chosen endogenously as in the work of Färe, Grosskopf, and Margaritis [16]; and (vi) $\mathbf{g} = (\mathbf{x}^{\max-\min},\mathbf{y}^{\max-\min})$, where $\mathbf{x}^{\max-\min}$ equals the range of the inputs and $\mathbf{y}^{\max-\min}$ equals the range of the outputs among the $j=1,\ldots,n$ DMUs.

For the directional vector (vi), Cooper, Park, and Pastor [17] introduced the RAM (range-adjusted measure) of inefficiency,[2] defined by

$$\mathrm{RAM}(\mathbf{x}_h,\mathbf{y}_h) = \max_{\lambda,s^-,s^+} \; \frac{1}{m+s}\left(\sum_{i=1}^{m}\frac{s_i^-}{x_i^{\max-\min}} + \sum_{r=1}^{s}\frac{s_r^+}{y_r^{\max-\min}}\right)$$

$$\text{s.t. } x_{i1}\lambda_1 + \cdots + x_{in}\lambda_n + s_i^- = x_{ih} \; (i=1,\ldots,m)$$

$$y_{r1}\lambda_1 + \cdots + y_{rn}\lambda_n - s_r^+ = y_{rh} \; (r=1,\ldots,s)$$ \hfill (3.11)

$$\lambda_1 + \cdots + \lambda_n = 1$$

$$\lambda_j \geq 0 \; (\forall j), \; s_i^- \geq 0 \; (\forall i), \; s_r^+ \geq 0 \; (\forall r)$$

---

[2] Cooper *et al.* [17] also defined the RAM efficiency as one minus the optimum objective value in (3.11).

If the numbers of outputs and inputs are equal (i.e., $m = s$) and

$$g_i^- = x_i^{\max-\min} = x_{ij}^{\max} - x_{ij}^{\min} \quad (\forall i) \quad \text{and} \quad g_r^+ = y_r^{\max-\max} = y_{rj}^{\max} - y_{rj}^{\min} \quad (\forall r) \qquad (3.12)$$

then the SDD measure (or, equivalently, the Russell directional measure) is equal to one half of the RAM of inefficiency. While the slacks-based directional distance measure can also be thought of as a weighted additive model [18], the directional vectors expressed in DD models give a direct indication of what the directions mean.

REFERENCES

[1] Luenberger, D.G. (1992) Benefit functions and duality. *Journal of Mathematical Economics*, **21**, 461–481.

[2] Luenberger, D.G. (1995) *Microeconomic Theory*. McGraw-Hill, New York.

[3] Chambers, R.G., Chung, Y., and Färe, R. (1996) Benefit and distance functions. *Journal of Economic Theory*, **70**(2), 407–419.

[4] Chambers, R.G., Chung, Y., and Färe, R. (1998) Profit, directional distance functions and Nerlovian efficiency. *Journal of Optimization Theory and Applications*, **98**(2), 351–364.

[5] Färe, R. and Grosskopf, S. (2010) Directional distance functions and slacks-based measures of efficiency. *European Journal of Operational Research*, **206**, 320–322.

[6] Shephard, R.W. (1953) *Cost and Production Functions. Princeton University Press, Princeton, NJ*.

[7] Shephard, R.W. (1970) *Theory of Cost and Production Functions*. Princeton University Press, Princeton, NJ.

[8] Fukuyama, H. (2003) Scale characterizations in a DEA directional technology distance function framework. *European Journal of Operational Research*, **144**(1), 108–127.

[9] Fukuyama, H. and Weber, W.L. (2009) A directional slacks-based measure of technical inefficiency. *Socio-Economic Planning Sciences*, **43**(4), 274–287.

[10] Tone, K. (2001) A slacks-based measure of efficiency in Data Envelopment Analysis. *European Journal of Operational Research*, **130**, 498–509.

[11] Bardhan, I., Bowlin, W.J., Cooper, W.W., and Sueyoshi, T. (1996) Models and measures for efficiency dominance in DEA, Part I. *Journal of the Operations Research Society of Japan*, **39**, 322–332.

[12] Bardhan, I., Bowlin, W.J., Cooper, W.W., and Sueyoshi, T. (1996) Models and measure for efficiency dominance in DEA: Part II. Free disposal hull and Russell measure approaches. *Journal of the Operations Research Society of Japan*, **39**, 333–344.

[13] Färe, R. and Lovell, C.A.K. (1978) Measuring the technical efficiency of production. *Journal of Economic Theory*, **19**, 150–162.

[14] Briec, W. (1997) A graph-type extension of Farrell technical efficiency measure. *Journal of Productivity Analysis*, **8**, 95–110.

[15] Briec, W. (2000) An extended Färe–Lovell technical efficiency measure. *International Journal of Production Economics*, **65**, 191–199.

[16] Färe, R., Grosskopf, S., and Margaritis, D. (2015) *Advances in Data Envelopment Analysis*. World Scientific Now.

[17] Cooper, W.W., Park, K.S., and Pastor, J.T. (1999) A range adjusted measure of inefficiency for use with additive models and relations to other models and measures in DEA. *Journal of Productivity Analysis*, **11**, 5–42.

[18] Charnes, A., Cooper, W.W., Golany, B., and Seiford, L. (1985) Foundations of data envelopment analysis for Pareto–Koopmans efficient empirical production functions. *Journal of Econometrics*, **30**, 91–107.

4

SUPER-EFFICIENCY DEA MODELS

Kaoru Tone

National Graduate Institute for Policy Studies, Tokyo, Japan

4.1 INTRODUCTION

In this chapter, we introduce super-efficiency models. Efficiency scores are obtained from these models by eliminating the data for the decision-making unit (DMU) DMU_h to be evaluated from the solution set. This can result in values which are regarded as according DMU_h the status of being 'super-efficient.' These values can then be used to rank the DMUs and thereby eliminate some (but not all) of the ties that occur for efficient DMUs.

4.2 RADIAL SUPER-EFFICIENCY MODELS

In this section, we introduce input-oriented and output-oriented super-efficiency models. See [1] for details.

4.2.1 Input-Oriented Radial Super-Efficiency Model

Using the notation in Chapter 2, the input-oriented radial super-efficiency model can be described as follows:

Advances in DEA Theory and Applications: With Extensions to Forecasting Models,
First Edition. Edited by Kaoru Tone.
© 2017 John Wiley & Sons Ltd. Published 2017 by John Wiley & Sons Ltd.

$$[\text{Radial Super-I-C}] \quad \theta^* = \min_{\theta,\lambda,s^-,s^+} \theta$$

$$\text{s.t.} \quad \theta\mathbf{x}_h = \sum_{j=1,j\neq h}^{n} \lambda_j \mathbf{x}_j + \mathbf{s}^-$$

$$\mathbf{y}_h = \sum_{j=1,j\neq h}^{n} \lambda_j \mathbf{y}_j - \mathbf{s}^+ \tag{4.1}$$

$$\lambda \geq 0, \mathbf{s}^- \geq 0, \mathbf{s}^+ \geq 0$$

This model is under the constant returns-to-scale assumption. If we add the following condition, we can get the variable-returns-to-scale (VRS) model: [Radial Super-I-V]

$$\sum_{j=1,j\neq h}^{n} \lambda_j = 1 \tag{4.2}$$

4.2.2 Output-Oriented Radial Super-Efficiency Model

The output-oriented radial super-efficiency model can be described as follows:

$$[\text{Radial Super-O-C}] \quad 1/\theta^* = \eta^* = \max_{\theta,\lambda,s^-,s^+} \eta$$

$$\text{s.t.} \quad \mathbf{x}_h = \sum_{j=1,j\neq h}^{n} \lambda_j \mathbf{x}_j + \mathbf{s}^-$$

$$\eta\mathbf{y}_h = \sum_{j=1,j\neq h}^{n} \lambda_j \mathbf{y}_j - \mathbf{s}^+ \tag{4.3}$$

$$\lambda \geq 0, \mathbf{s}^- \geq 0, \mathbf{s}^+ \geq 0$$

If we add the constraint (4.2), we can get the variable-returns-to-scale (VRS) model [Radial Super-O-V].

4.2.3 Infeasibility Issues in the VRS Model

By dint of the constraint $\sum_{j=1,j\neq h}^{n} \lambda_j = 1$, variable-returns-to-scale models may encounter infeasibility.

Proposition 4.1 [Radial Super-I-V] has no feasible solution if there exists r such that $y_{rh} > \max_{j\neq h}\{y_{rj}\}$, and [Super-Radial-O-V] has no feasible solution if there exists i such that $x_{ih} < \min_{j\neq h}\{x_{ij}\}$.

4.3   NON-RADIAL SUPER-EFFICIENCY MODELS

Non-radial slacks-based super-efficiency models have three variations: input-, output- and non-oriented. See [2] for details.

4.3.1 Input-Oriented Non-Radial Super-Efficiency Model

We solve the following program for an efficient DMU $(\mathbf{x}_h, \mathbf{y}_h)$ to measure the minimum ratio-scale distance from the efficient frontier excluding the DMU $(\mathbf{x}_h, \mathbf{y}_h)$. The input-oriented non-oriented model under the constant-returns-to-scale assumption is described by the following scheme:

$$[\text{Super-SBM-I-C}] \quad \delta^* = \min{}^0 1 + \frac{1}{m}\sum\nolimits_{i=1}^{m} \frac{s_i^-}{x_{ih}}$$

$$\text{subject to}$$

$$\mathbf{x}_h + \mathbf{s}^- = \sum\nolimits_{j=1, j\neq h}^{n} \mathbf{x}_j \lambda_j$$

$$\mathbf{y}_h - \mathbf{s}^+ = \sum\nolimits_{j=1, j\neq h}^{n} \mathbf{y}_j \lambda_j$$

$$\lambda \geq 0, \mathbf{s}^- \geq 0, \mathbf{s}^+ \geq 0$$

(4.4)

4.3.2 Output-Oriented Non-Radial Super-Efficiency Model

The output-oriented super-efficiency is measured by the following program:

$$[\text{Super-SBM-O-C}] \quad 1/\delta^* = \max 1 - \frac{1}{s}\sum\nolimits_{r=1}^{s} \frac{s_r^+}{y_{rh}}$$

$$\text{subject to}$$

$$\mathbf{x}_h + \mathbf{s}^- = \sum\nolimits_{j=1, j\neq h}^{n} \mathbf{x}_j \lambda_j$$

$$\mathbf{y}_h - \mathbf{s}^+ = \sum\nolimits_{j=1, j\neq h}^{n} \mathbf{y}_j \lambda_j$$

$$\lambda \geq 0, \mathbf{s}^- \geq 0, \mathbf{s}^+ \geq 0$$

(4.5)

4.3.3 Non-Oriented Non-Radial Super-Efficiency Model

The non-oriented model is described by the following program:

$$[\text{Super-SBM-C}] \quad \delta^* = \min \frac{1 + \frac{1}{m}\sum\nolimits_{i=1}^{m} \frac{s_i^-}{x_{ih}}}{1 - \frac{1}{s}\sum\nolimits_{r=1}^{s} \frac{s_r^+}{y_{rh}}}$$

$$\text{subject to}$$

$$\mathbf{x}_h + \mathbf{s}^- = \sum\nolimits_{j=1, j\neq h}^{n} \mathbf{x}_j \lambda_j$$

$$\mathbf{y}_h - \mathbf{s}^+ = \sum\nolimits_{j=1, j\neq h}^{n} \mathbf{y}_j \lambda_j$$

$$\lambda \geq 0, \mathbf{s}^- \geq 0, \mathbf{s}^+ \geq 0$$

(4.6)

4.3.4 Variable-Returns-to-Scale Models

By adding the constraint (4.2), we can define the models [Super-SBM-I-V], [Super-SBM-O-V] and [Super-SBM-V].

Proposition 4.2 [Super-SBM-I-V] and [Super-SBM-O-V] encounter the same infeasibility problem as [Proposition 4.1] does. However, [Super-SBM-V] is always feasible and has a finite optimum. (See Cooper *et al.* [3] and Tone [2].)

4.4 AN EXAMPLE OF A SUPER-EFFICIENCY MODEL

Here, we compare super-efficiency scores for the data presented in Table 4.1 and Figure 4.1. We compared the models [Super-Radial-I-C] and [Super-SBM-I-C], and the results are shown in Table 4.2.

DMUs A and E are judged efficient by the radial model, but inefficient by the SBM model. Figure 4.2 illustrates the case of DMU D. The radial super-efficiency of DMU D is

TABLE 4.1   Sample data.

	Input		Output
	x_1	x_2	y
A	2	6	1
B	2	4	1
C	4	2	1
D	8	1	1
E	10	1	1
F	4	4	1

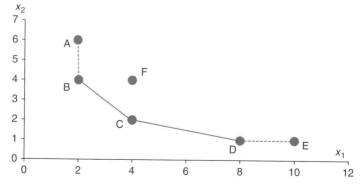

Figure 4.1   The unit isoquant spanned by the test data in Table 4.1.

TABLE 4.2 Super-efficiency scores.

DMU	Super-Radial-I-C	Super-SBM-I-C
A	1	0.8333
B	1.25	1.25
C	1.25	1.25
D	1.1429	1.125
E	1	0.9
F	0.75	0.75

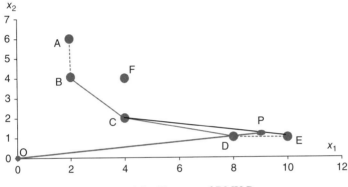

Figure 4.2 The case of DMU D.

measured as OP/OD = 1.1429, while its non-radial super-efficiency is given by $1 + DE/(2^*8) = 1.125$. E gives the minimum objective value of (4.4).[1]

REFERENCES

[1] Andersen, P. and Petersen, N.C. (1993) A procedure for ranking efficient units in data envelopment analysis. *Management Science*, **39**, 1261–1264.

[2] Tone, K. (2002) A slacks-based measure of super-efficiency in data envelopment analysis. *European Journal of Operational Research*, **143**, 32–41.

[3] Cooper, W.W., Seiford, L.M. and Tone, K. (2007) *Data Envelopment Analysis: A Comprehensive Text with Models, Applications, References and DEA-Solver Software*, 2nd edn, Springer, New York.

[1] Software for super-efficiency models is included in DEA-Solver Pro V13 (http://www.saitech-inc.com). See also Appendix A.

5

DETERMINING RETURNS TO SCALE IN THE VRS DEA MODEL

BIRESH K. SAHOO

Xavier Institute of Management, Xavier University, Bhubaneswar, India

KAORU TONE

National Graduate Institute for Policy Studies, Tokyo, Japan

5.1 INTRODUCTION

One of the most important aspects of the applied production analysis of organizational units (called decision-making units, or DMUs) is returns to scale (RTS), which helps in determining pricing policies and market structure, and consequently government policies toward both of these [1, 2]. It is therefore imperative that this concept be measured accurately. To assess the efficiency of DMUs, it is necessary to identify the nature of the RTS that characterize efficient production. In production economics, RTS are defined as the maximum proportional increase in all outputs (α) resulting from a given proportional increase in all inputs (ζ). Constant returns to scale (CRS) prevail if $\alpha = \zeta$, increasing returns to scale (IRS) prevail if $\alpha > \zeta$, and decreasing returns to scale (DRS) prevail if $\alpha < \zeta$.

Ever since the nonparametric methodology of data envelopment analysis (DEA) was introduced by Charnes *et al.* [3], the economic concept of RTS has been widely studied within two broader frameworks of DEA. The first framework, by Färe *et al.* [4], is aimed at characterizing the RTS of a DMU by considering the ratios of two

Advances in DEA Theory and Applications: With Extensions to Forecasting Models,
First Edition. Edited by Kaoru Tone.

radial efficiency measures under different RTS assumptions, that is, the ratio of the efficiency measure under CRS to either that under variable returns to scale (VRS) or that under nonincreasing returns to scale (NIRS). The second framework, which stems from the work of Banker *et al.* [5] and Banker and Thrall [6], proceeds by examining tangential planes to the VRS-based DEA production frontier at a given point. This is done either by looking at the constant term that represents the intercept of such a plane with the plane in which all inputs are set to zero, or by observing the weights of the corner points of the facet of the frontier associated with that plane.

This second framework can also be extended to both additive and multiplicative DEA models. Unlike the radial CCR and BCC models, the additive model of Cooper *et al.* [7] avoids the problem of choosing between input and output orientations. In the case of multiplicative models [8], where the piecewise linear frontiers usually employed in CCR and BCC models are replaced by the piecewise log-linear frontiers, RTS are obtained from the exponents of these piecewise log-linear functions for the different segments that form the underlying production frontier. Note that in both frameworks, the characterization of the RTS of a DMU depends on whether an input- or output-oriented model is used, since different orientations identify different points on the frontier from which evaluations are performed.

Since the DEA production technologies are not differentiable at extreme points, researchers have suggested determining both right- and left-hand RTS at these extreme points (see, e.g., [6, 9–31], among others).

As recently pointed out by Podinovski *et al.* [32], the existing methods of computing RTS apply only to the standard, VRS (BCC), and CRS (CCR) DEA production technologies, which are examples of a large class of polyhedral technologies. This large class also includes technologies with production trade-offs [33,34] and weight restrictions [35,36], technologies with negative inputs and outputs [37], technologies with weakly disposable undesirable outputs [38], and network DEA technologies [30,31]. Podinovski *et al.* suggested a unified linear programming approach to determining left- and right-hand characterizations of the RTS of technically efficient firms in any polyhedral technology.

In this chapter, however, we discuss the evaluation of RTS characterizations of firms in a VRS-based DEA production technology.

5.2 TECHNOLOGY SPECIFICATION AND SCALE ELASTICITY

5.2.1 Technology

We assume throughout that we are dealing with n observed firms; each uses m inputs to produce s outputs. Let $x_j = \left(x_{1j},\ldots,x_{mj}\right)^T \in \mathbb{R}_{\geq 0}^m$ and $y_j = \left(y_{1j},\ldots,y_{sj}\right)^T \in \mathbb{R}_{\geq 0}^s$ be the vectors of inputs and outputs, respectively, of firm j, and let J be the index set of all the observed firms, that is, $J = \{1,\ldots,n\}$.

The production technology that transforms an input vector $x \in \mathbb{R}^m_{\geq 0}$ to an output vector $y \in \mathbb{R}^s_{\geq 0}$ can be characterized by the technology set $T \subset \mathbb{R}^m_{\geq 0} \times \mathbb{R}^s_{\geq 0}$, defined as

$$T = \left\{ (x,y) \in \mathbb{R}^{m+s}_{\geq 0} \,\middle|\, x \in \mathbb{R}^m_{\geq 0} \text{ can produce } y \in \mathbb{R}^s_{\geq 0} \right\} \tag{5.1}$$

The neoclassical characterization of the production function is the transformation function $\psi(x, y)$, which decreases with y and increases with x such that

$$\psi(x,y) \leq 0 \text{ if and only if } (x,y) \in T \tag{5.2}$$

$\psi(x,y) = 0$ represents those input–output vectors that operate on the boundary of T and, hence, are technically efficient.

5.2.2 Measure of Scale Elasticity

The RTS, or scale elasticity (SE), is based on a relationship such that, for a given proportional expansion of all inputs (α), one can find the maximum proportional expansion of all outputs (β) such that

$$\psi(\alpha x, \beta y) = 0 \tag{5.3}$$

Assuming $\psi(\cdot)$ to be smooth, differentiation of (5.3) with respect to the input scaling factor α yields the following measure of SE $\varepsilon(x, y)$ [39]:

$$\frac{d\beta}{d\alpha} = \varepsilon(x,y) = -\frac{\displaystyle\sum_{i=1}^{m} \frac{\partial \psi(\cdot)}{\partial x_i} x_i}{\displaystyle\sum_{r=1}^{s} \frac{\partial \psi(\cdot)}{\partial y_r} y_r} \tag{5.4}$$

Proposition 5.1 The RTS defined at a point (x, y) are increasing (IRS), constant (CRS), and decreasing (DRS) if $\varepsilon(x, y) > 1$, $\varepsilon(x, y) = 1$, and $\varepsilon(x, y) < 1$, respectively.

5.2.3 Scale Elasticity in DEA Models

The DEA technology under the VRS specification [5] can be expressed as

$$T^{\text{DEA}}_{\text{VRS}} = \left\{ (x,y) : \sum_{j \in J} x_{ij}\lambda_j \leq x_i \ (\forall i), \ \sum_{j \in J} y_{rj}\lambda_j \geq y_r \ (\forall r), \ \sum_{j \in J} \lambda_j = 1, \lambda_j \geq 0 \ (\forall j) \right\} \tag{5.5}$$

Consider the evaluation of the input-oriented SE for any firm o ($o \in J$). The input-oriented technical efficiency of firm o can be obtained from the following linear programming (LP) problem:

$$\alpha(\beta) = \min\{\alpha : (\alpha x, \beta y) \in T_{\text{VRS}}^{\text{DEA}}; \beta = 1\} \tag{5.6}$$

Alternatively, the primal *envelopment*-form-based LP program (5.6) can be expressed in its dual *multiplier* form as

$$\alpha(1) = \max \sum_{r=1}^{s} u_r y_{ro} - u_o \tag{5.7}$$

$$\text{s.t.} \sum_{r=1}^{s} u_r y_{rj} - \sum_{i=1}^{m} v_i x_{ij} - u_o \leq 0, \quad \sum_{i=1}^{m} v_i x_{io} = 1, \quad u_r, v_i \geq 0 \ (\forall i, r); u_o : \text{free}$$

For any firm o ($o \in J$), the transformation function is the following:

$$\psi(\alpha(1)x_o, y_o) \sum_{r=1}^{s} u_r y_{ro} - \sum_{i=1}^{m} v_i(\alpha(1)x_{io}) - u_o = 0 \tag{5.8}$$

Using (5.4), the input-oriented SE of firm o can be obtained as

$$\varepsilon_i(x_o, y_o) = \frac{\alpha(1)}{\alpha(1) + u_o} = \frac{1}{1 + u_o/\alpha(1)} \tag{5.9}$$

It is well known that production technologies in DEA are not differentiable at extreme efficient points, owing to the existence of multiple optimal solutions for $u_o(v_o)$. Following Banker and Thrall [6], we therefore set up the following LP problems to find the maximum and minimum values of u_o for firm o as follows:

$$u_o^+ \ (u_o^-) = \max \ (\min) u_o \tag{5.10}$$

$$\text{s.t.} \sum_{r=1}^{s} u_r y_{ro} - u_o = \alpha(1), \quad \sum_{i=1}^{m} v_i x_{io} = 1$$

$$\sum_{r=1}^{s} u_r y_{rj} - \sum_{i=1}^{m} v_i x_{ij} - u_o \leq 0 \ (\forall j \neq o), v_i, u_r \geq 0 \ (\forall i, r), u_o : \text{free}$$

Based on the results of solving (5.10), one can determine the input-oriented right-hand SE $(\varepsilon_i^+ (\cdot))$ and left-hand SE $(\varepsilon_i^- (\cdot))$ for firm o as

$$\varepsilon_i^+ (x_o, y_o) = \frac{1}{1 + u_o^+/\alpha(1)} \quad \text{and} \quad \varepsilon_i^- (x_o, y_o) = \frac{1}{1 + u_o^-/\alpha(1)} \tag{5.11}$$

We have now our second proposition.

Proposition 5.2 Assuming alternate optima in u_o, the firm o in T_{VRS}^{DEA} exhibits (input-oriented) IRS $\left(\varepsilon_i^+(\cdot)>1\right)$ if $u_o^+<0$, (input-oriented) CRS $\left(\varepsilon_i^+(\cdot)\leq 1\leq\varepsilon_i^-(\cdot)\right)$ if $u_o^+\geq 0\geq u_o^-$, and (input-oriented) DRS $\left(\varepsilon_i^-(\cdot)<1\right)$ if $u_o^->0$.

5.3 SUMMARY

We have briefly provided a discussion of left- and right-hand RTS characterizations of efficient firms in a VRS DEA production technology. However, as has recently been demonstrated by Podinovski *et al.* [32], it is now possible to perform RTS characterizations of firms in any polyhedral technology, which is a larger class of technologies that includes, besides CRS and VRS DEA production technologies, technologies with production trade-offs and weight restrictions, technologies with negative inputs and outputs, technologies with weakly disposable undesirable outputs, and network DEA technologies.

REFERENCES

[1] Sahoo, B.K., Mohapatra, P.K.J., and Trivedi, M.L. (1999) A comparative application of data envelopment analysis and frontier translog production function for estimating returns to scale and efficiencies. *International Journal of Systems Science*, **30**, 379–394.

[2] Tone, K. and Sahoo, B.K. (2003) Scale, indivisibilities and production function in data envelopment analysis. *International Journal of Production Economics*, **84**, 165–192.

[3] Charnes, A., Cooper, W.W., and Rhodes, E. (1978) Measuring the efficiency of DMUs. *European Journal of Operational Research*, **2**, 429–444.

[4] Färe, R., Grosskopf, S., and Lovell, C.A.K. (1985) *The Measurement of Efficiency of Production,* Kluwer-Nijhoff, Boston, MA.

[5] Banker, R.D., Charnes, A., and Cooper, W.W. (1984) Some models for estimating technical and scale inefficiencies in data envelopment analysis. *Management Science*, **30**, 1078–1092.

[6] Banker, R.D. and Thrall, R.M. (1992) Estimation of returns to scale using data envelopment analysis. *European Journal of Operational Research*, **62**, 74–84.

[7] Cooper, W.W., Seiford, L.M., and Tone, K. (2007) *Data Envelopment Analysis: A Comprehensive Text with Models, Applications, References and DEA-Solver Software,* Springer, New York.

[8] Banker, R.D. and Maindiratta, A. (1986) Piecewise loglinear estimation of efficient production surfaces. *Management Science*, **32**, 126–135.

[9] Golany, B. and Yu, G. (1997) Estimating returns to scale in DEA. *European Journal of Operational Research*, **103**, 28–37.

[10] Fukuyama, H. (2000) Returns to scale and scale elasticity in data envelopment analysis. *European Journal of Operational Research*, **125**, 93–112.

[11] Fukuyama, H. (2001) Returns to scale and scale elasticity in Farrell, Russell and additive models. *Journal of Productivity Analysis*, **16**, 225–239.

[12] Fukuyama, H. (2003) Scale characterizations in a DEA directional technology distance function framework. *European Journal of Operational Research*, **144**, 108–127.

[13] Tone, K. and Sahoo, B.K. (2004) Degree of scale economies and congestion: A unified DEA approach. *European Journal of Operational Research*, **158**, 755–772.

[14] Tone, K. and Sahoo, B.K. (2005) Evaluating cost efficiency and returns to scale in the Life Insurance Corporation of India using data envelopment analysis. *Socio-Economic Planning Sciences*, **39**, 261–285.

[15] Tone, K. and Sahoo, B.K. (2006) Re-examining scale elasticity in DEA. *Annals of Operations Research*, **145**, 69–87.

[16] Førsund, F.R. and Hjalmarsson, L. (2004) Calculating scale elasticity in DEA models. *Journal of the Operational Research Society*, **55**, 1023–1038.

[17] Sengupta, J.K. and Sahoo, B.K. (2006) *Efficiency Models in Data Envelopment Analysis: Techniques of Evaluation of Productivity of Firms in a Growing Economy,* Palgrave Macmillan, London.

[18] Hadjicostas, P. and Soteriou, A.C. (2006) One-sided elasticities and technical efficiency in multi-output production: A theoretical framework. *European Journal of Operational Research*, **168**, 425–449.

[19] Førsund, F.R., Hjalmarsson, L., Krivonozhko, V., and Utkin, O.B. (2007) Calculation of scale elasticities in DEA models: Direct and indirect approaches. *Journal of Productivity Analysis*, **28**, 45–56.

[20] Sahoo, B.K., Sengupta, J.K., and Mandal, A. (2007) Productive performance evaluation of the banking sector in India using data envelopment analysis. *International Journal of Operations Research*, **4**, 1–17.

[21] Podinovski, V.V., Førsund, F.R., and Krivonozhko, V.E. (2009) A simple derivation of scale elasticity in data envelopment analysis. *European Journal of Operational Research*, **197**, 149–153.

[22] Podinovski, V.V. and Førsund, F.R. (2010) Differential characteristics of efficient frontiers in data envelopment analysis. *Operations Research*, **58**, 1743–1754.

[23] Sahoo, B.K. and Gstach, D. (2011) Scale economies in Indian commercial banking sector: Evidence from DEA and translog estimates. *International Journal of Information Systems and Social Change*, **2**, 13–30.

[24] Atici, K.B. and Podinovski, V.V. (2012) Mixed partial elasticities in constant-returns-to-scale production technologies. *European Journal of Operational Research*, **220**, 262–269.

[25] Sahoo, B.K., Kerstens, K., and Tone, K. (2012) Returns to growth in a non-parametric DEA approach. *International Transactions in Operational Research*, **19**, 463–486.

[26] Zelenyuk, V. (2013) A scale elasticity measure for directional distance function and its dual: Theory and DEA estimation. *European Journal of Operational Research*, **228**, 592–600.

[27] Sahoo, B.K. and Tone, K. (2013) Non-parametric measurement of economies of scale and scope in non-competitive environment with price uncertainty. *Omega*, **41**, 97–111.

[28] Sahoo, B.K. and Tone, K. (2015) Scale elasticity in non-parametric DEA approach, in *Data Envelopment Analysis: A Handbook of Models and Methods* (ed. J. Zhu), Springer, New York, pp. 269–290.

[29] Sahoo, B.K. and Sengupta, J.K. (2014) Neoclassical characterization of returns to scale in nonparametric production analysis. *Journal of Quantitative Economics*, **12**, 78–86.

[30] Sahoo, B.K., Zhu, J., Tone, K., and Klemen, B.M. (2014) Decomposing technical efficiency and scale elasticity in two-stage network DEA. *European Journal of Operational Research*, **233**, 584–594.

[31] Sahoo, B.K., Zhu, J., and Tone, K. (2014) Decomposing efficiency and returns to scale in two-stage network systems, in *Data Envelopment Analysis: A Handbook of Modeling Internal Structure and Network* (eds W.D. Cook and J. Zhu), Springer, New York, pp. 137–164.

[32] Podinovski, V.V., Chambers, R.G., Atici, K.B., and Deineko, I.D. (2016) Marginal values and returns to scale for nonparametric production frontiers. *Operations Research*, **64**, 236–250.

[33] Podinovski, V.V. (2004) Production trade-offs and weight restrictions in data envelopment analysis. *Journal of the Operational Research Society*, **55**, 1311–1322.

[34] Podinovski, V.V. and Bouzdine-Chameeva, T. (2013) Weight restrictions and free production in data envelopment analysis. *Operations Research*, **61**, 426–437.

[35] Tone, K. (2001) On returns to scale under weight restrictions in data envelopment analysis. *Journal of Productivity Analysis*, **16**, 31–47.

[36] Korhonen, P.J., Soleimani-damaneh, M., and Walleneus, J. (2011) Ratio-based RTS determination in weight-restricted DEA models. *European Journal of Operational Research*, **215**, 431–438.

[37] Sahoo, B.K., Khoveyni, M., Eslami, R., and Chaudhury, P. (2016) Returns to scale and most productive scale size in DEA with negative data. *European Journal of Operational Research*, **255**, 245–258.

[38] Kousmanen, T. (2005) Weak disposability in nonparametric productivity analysis with undesirable outputs. *American Journal of Agricultural Economics*, **87**, 1077–1082.

[39] Hanoch, G. (1970) Homotheticity in joint production. *Journal of Economic Theory*, **2**, 423–426.

6

MALMQUIST PRODUCTIVITY INDEX MODELS

KAORU TONE

National Graduate Institute for Policy Studies, Tokyo, Japan

MIKI TSUTSUI

Central Research Institute of Electric Power Industry, Tokyo, Japan

6.1 INTRODUCTION

The Malmquist index (MI) [1] evaluates the change in efficiency of a decision-making unit (DMU) between two time periods. It is defined as the product of catch-up (CU) and frontier shift (FS) terms. The CU term is related to the degree of effort that the DMU has made to improve its efficiency, while the FS term reflects the change in the efficient frontiers surrounding the DMU between the two time periods 1 and 2. We denote DMU_o in the time periods 1 and 2 by (x_o^1, y_o^1) and (x_o^2, y_o^2), respectively. The CU effect γ is measured by the following formula:

$$\gamma = \frac{\text{Efficiency of } \left(x_o^2, y_o^2\right) \text{ with respect to the period 2 frontier}}{\text{Efficiency of } \left(x_o^1, y_o^1\right) \text{ with respect to the period 1 frontier}}. \quad (6.1)$$

We evaluate each element (efficiency) of the above formula by non-parametric DEA models as described later. A simple single-input, single-output case is illustrated in Figure 6.1.

Advances in DEA Theory and Applications: With Extensions to Forecasting Models,
First Edition. Edited by Kaoru Tone.
© 2017 John Wiley & Sons Ltd. Published 2017 by John Wiley & Sons Ltd.

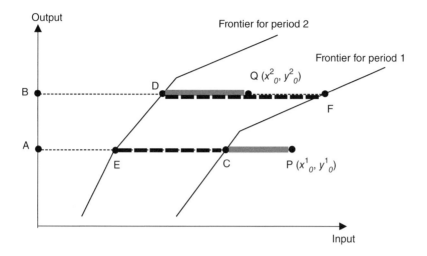

Figure 6.1 Catch-up and frontier shift.

The CU effect (in the input orientation) can be computed as

$$\gamma = \frac{{}^{BD}/_{BQ}}{{}^{AC}/_{AP}} \qquad (6.2)$$

Here, $\gamma > 1$ indicates progress in relative efficiency from period 1 to 2, while $\gamma = 1$ and $\gamma < 1$ indicate the status quo and regress in efficiency, respectively.

In addition to the CU term, we must take account of the FS effect in order to evaluate totally the efficiency change of the DMUs, since the CU term is determined by the efficiencies measured by the distances from the respective frontiers. In the simple case of Figure 6.1, this can be implemented as follows. The reference point C for (x_o^1, y_o^1) is moved to E on the frontier for period 2. Thus, the FS effect at (x_o^1, y_o^1) is evaluated from

$$\phi_1 = \frac{AC}{AE} \qquad (6.3)$$

This is equivalent to

$$\phi_1 = \frac{{}^{AC}/_{AP}}{{}^{AE}/_{AP}} = \frac{\text{Efficiency of } \left(x_o^1, \ y_o^1\right) \text{ with respect to the period 1 frontier}}{\text{Efficiency of } \left(x_o^1, \ y_o^1\right) \text{ with respect to the period 2 frontier}} \qquad (6.4)$$

The numerator of the right-hand side of (6.4) has already been obtained in (6.1). The denominator is measured by the distance from the period 2 production possibility set to (x_o^1, y_o^1). Likewise, the FS effect at (x_o^2, y_o^2) is expressed by

$$\phi_2 = \frac{{}^{BF}/_{BQ}}{{}^{BD}/_{BQ}} = \frac{\text{Efficiency of } \left(x_o^2, \; y_o^2 \right) \text{ with respect to the period 1 frontier}}{\text{Efficiency of } \left(x_o^2, \; y_o^2 \right) \text{ with respect to the period 2 frontier}} \qquad (6.5)$$

We can evaluate the numerator of the above by means of DEA models. Using ϕ_1 and ϕ_2, we define the FS effect ϕ by their geometric mean as

$$\phi = \sqrt{\phi_1 \phi_2} \qquad (6.6)$$

Now, the MI (μ) is obtained as the product of the CU (γ) and FS (ϕ) as

$$\mu = \gamma \times \phi \qquad (6.7)$$

This is an index representing the total factor productivity (TFP) of the DMU on moving from P (x_o^1, y_o^1) to Q (x_o^2, y_o^2) in Figure 6.1, in that it reflects progress or regress in the relative efficiency of the DMU along with progress or regress of the frontier technology.

We now employ the following notation for the efficiency score of DMU $(x_o, y_o)^{t_1}$ measured by use of the frontier technology t_2:

$$\delta^{t_2} \left((x_o, y_o)^{t_1} \right) \quad (t_1 = 1, 2 \text{ and } t_2 = 1, 2) \qquad (6.8)$$

Using this notation, the CU effect γ in (6.1) can be expressed as

$$\text{CU}: \quad \gamma = \frac{\delta^2 \left((x_o, y_o)^2 \right)}{\delta^1 \left((x_o, y_o)^1 \right)} \qquad (6.9)$$

The FS effect is described as

$$\text{FS}: \quad \phi = \left[\frac{\delta^1 \left((x_o, y_o)^1 \right)}{\delta^2 \left((x_o, y_o)^1 \right)} \times \frac{\delta^1 \left((x_o, y_o)^2 \right)}{\delta^2 \left((x_o, y_o)^2 \right)} \right]^{1/2} \qquad (6.10)$$

From the product of γ and ϕ, we obtain the following formula for the computation of the MI:

$$\text{MI}: \quad \mu = \left[\frac{\delta^1 \left((x_o, y_o)^2 \right)}{\delta^1 \left((x_o, y_o)^1 \right)} \times \frac{\delta^2 \left((x_o, y_o)^2 \right)}{\delta^2 \left((x_o, y_o)^1 \right)} \right]^{1/2} \qquad (6.11)$$

This last expression gives an another interpretation of the MI, that is, as the geometric mean of the two relative efficiency ratios, the first being the efficiency change measured by use of the period 1 technology and the other the efficiency change measured by use of the period 2 technology.

As can be seen from these formulas, the MI consists of four terms: $\delta^1((x_o, y_o)^1)$, $\delta^2((x_o, y_o)^2)$, $\delta^1((x_o, y_o)^2)$ and $\delta^2((x_o, y_o)^1)$. The first two are related to measurements *within* the same time period, while the last two are related to *intertemporal* comparison. If $\mu > 1$, this indicates progress in the total factor productivity of DMU_o from period 1 to 2, while $\mu = 1$ and $\mu < 1$ indicate the status quo and decay in the total factor productivity, respectively.

In the non-parametric framework, the MI is constructed by means of DEA techniques. There are a number of ways to compute the MI. First, Färe *et al.* [2] utilized an input/output-oriented radial DEA model to compute the MI. However, the radial models suffer from one shortcoming, that is, neglect of slacks. Second, the MI can be computed using slacks-based non-radial DEA models, which include both oriented and non-oriented cases.

6.2   RADIAL MALMQUIST MODEL

The input-oriented radial MI measures the *within* and *intertemporal* scores by means of the linear programs given below:

[Within score in input orientation]

$$\delta^s\left((x_o, y_o)^s\right) = \min_{\theta, \lambda} \theta \qquad (6.12)$$

subject to $\theta x_o^s \geq X^s \lambda, \quad y_o^s \leq Y^s \lambda, \quad L \leq e\lambda \leq U, \quad \lambda \geq 0$

where $X^s = \left(x_1^s, \ldots, x_n^s\right)$ and $Y^s = \left(y_1^s, \ldots, y_n^s\right)$ are the input and output matrices (observed data), respectively, for the period s. We solve this program for $s = 1$ and 2. It holds that $\delta^s((x_o, y_o)^s) \leq 1$, and $\delta^s((x_o, y_o)^s) = 1$ indicates that $(x_o, y_o)^s$ is on the technically efficient frontier of $(X, Y)^s$.

[Intertemporal score in input orientation]

$$\delta^s\left((x_o, y_o)^t\right) = \min_{\theta, \lambda} \theta \qquad (6.13)$$

subject to $\theta x_o^t \geq X^s \lambda, \quad y_o^t \leq Y^s \lambda, \quad L \leq e\lambda \leq U, \quad \lambda \geq 0$

We solve this program for the pairs $(s, t) = (1, 2)$ and $(2, 1)$. If $(x_o, y_o)^t$ is not enveloped by the technology in the period s, the score $\delta^s((x_o, y_o)^t)$, if exists, has a value greater than 1. This corresponds to the concept of super-efficiency proposed by Andersen and Petersen [3].

Although the above schemes are input-oriented, we can develop an output-oriented MI as well by means of output-oriented radial DEA models. This is explained below:

[Within score in output orientation]

$$\delta^s\left((x_o, y_o)^s\right) = \min_{\eta, \lambda} \frac{1}{\eta} \tag{6.14}$$

$$\text{subject to } x_o^s \geq X^s\lambda, \quad \eta y_o^s \leq Y^s\lambda, \quad L \leq e\lambda \leq U, \quad \lambda \geq 0$$

[Intertemporal score in output orientation]

$$\delta^s\left((x_o, y_o)^t\right) = \min_{\eta, \lambda} \frac{1}{\eta} \tag{6.15}$$

$$\text{subject to } x_o^t \geq X^s\lambda, \quad \eta y_o^t \leq Y^s\lambda, \quad L \leq e\lambda \leq U, \quad \lambda \geq 0$$

Remark 6.1 Inclusive or Exclusive Scheme
For evaluating the *within* score $\delta^s((x_o, y_o)^s)$, there are two schemes: 'inclusive' and 'exclusive'. The 'inclusive' scheme means that, when we evaluate $(x_o, y_o)^s$ with respect to the technology $(X, Y)^s$, the DMU $(x_o, y_o)^s$ is always included in the evaluator $(X, Y)^s$, thus resulting in a score not greater than 1. The 'exclusive' scheme employs a method in which the DMU $(x_o, y_o)^s$ is removed from the evaluator group $(X, Y)^s$. This method of evaluation is equivalent to that for super-efficiency evaluation, and the score, if exists, may be greater than 1. The *intertemporal* comparisons naturally utilize this 'exclusive' scheme. So, the adoption of this scheme even in the *within* evaluations is not unnatural and promotes discrimination power.

Remark 6.2 Infeasible-LP issues
In the BCC (VRS) model [$(L, U) = (1, 1)$: variable returns to scale], it may occur that the intertemporal LP (6.13) has no solution in its input or output orientation. In the case of the input-oriented model, (6.13) has no feasible solution if there exists i such that $y_{io}^t > \max_j\left\{y_{ij}^s\right\}$, whereas in the output-oriented case, (6.15) has no feasible solution if there exists i such that $x_{io}^t < \min_j\left\{x_{ij}^s\right\}$. In the IRS model [$(L, U) = (1, \infty)$: increasing returns to scale], it may occur that the output-oriented intertemporal LP has no solution, while the input-oriented case is always feasible. In the case of the DRS model [$(L, U) = (0, 1)$: decreasing returns to scale], it might be possible that the input-oriented problem (6.13) has no solution, while the output-oriented model is always feasible. However, the CRS (CCR) model does not suffer from any such trouble in its intertemporal measurements. One solution to avoid this difficulty is

to assign 1 to the score, since we have no means to evaluate the DMU within the evaluator group.

6.3   NON-RADIAL AND ORIENTED MALMQUIST MODEL

The radial approaches suffer from one general problem, that is, the neglect of slacks. In an effort to overcome this problem, Tone [4,5] has developed non-radial measures of efficiency and super-efficiency. Using these measures, we develop here a non-radial, slacks-based MI.

First, we introduce the input-oriented SBM (slacks-based measure) and super-SBM [4,5]. The SBM evaluates the efficiency of the examinee $(x_o, y_o)^s$ $(s = 1, 2)$ with respect to the evaluator set $(X, Y)^t$ $(t = 1, 2)$ with the help of the following LP:

[SBM-I]

$$\delta^t((x_o, y_o)^s) = \min_{\lambda, s^-} 1 - \frac{1}{m} \sum_{i=1}^{m} \frac{s_i^-}{x_{io}^s} \tag{6.16}$$

subject to $x_o^s = X^t \lambda + s^-$, $y_o^s \leq Y^t \lambda$, $L \leq e\lambda \leq U$, $\lambda \geq 0$, $s^- \geq 0$

Or, equivalently,

[SBM-I]

$$\delta^t((x_o, y_o)^s) = \min_{\theta, \lambda} \frac{1}{m} \sum_{i=1}^{m} \theta_i \tag{6.17}$$

subject to $\theta_i x_{io}^s \geq \sum_{j=1}^{n} x_{ij}^t \lambda_j (i = 1, \ldots, m)$, $y_o^s \leq Y^t \lambda$, $\theta_i \leq 1 (i = 1, \ldots, m)$,

$L \leq e\lambda \leq U$, $\lambda \geq 0$

where the vector $s^- \in R^m$ denotes the input slacks. The equivalence between (6.16) and (6.17) can be shown as follows. Define $\theta_i = \left(1 - s_i^- / x_{io}^s\right)$. Then it holds that $\theta_i \leq 1$ $(\forall i)$, and the equivalence follows straightforwardly. This model takes input slacks (surpluses) into account but not output slacks (shortfalls). Notice that, under the 'inclusive' scheme (see Remark 6.2 above), [SBM-I] is always feasible in the case where $s = t$. However, under the 'exclusive' scheme, we remove $(x_o, y_o)^s$ from the evaluator group $(X, Y)^s$ and hence [SBM-I] may have no feasible solution even in the case $s = t$. In this case, we solve [Super-SBM-I] below:

[Super-SBM-I]

$$\delta^t((x_o, y_o)^s) = \min_{\lambda, s^-} 1 + \frac{1}{m} \sum_{i=1}^{m} \frac{s_i^-}{x_{io}^s} \tag{6.18}$$

subject to $x_o^s \geq X^t \lambda - s^-$, $y_o^s \leq Y^t \lambda$, $L \leq e\lambda \leq U$, $\lambda \geq 0$, $s^- \geq 0$

Or, equivalently,

[Super-SBM-I]

$$\delta^t((x_o, y_o)^s) = \min_{\theta, \lambda} \frac{1}{m} \sum_{i=1}^m \theta_i \qquad (6.19)$$

subject to $\theta_i x_{io}^s \geq \sum_{j=1}^n x_{ij}^t \lambda_j (i=1,\ldots,m), \quad y_o^s \leq Y^t \lambda, \quad \theta_i \geq 1 (i=1,\ldots,m),$
$L \leq e\lambda \leq U, \quad \lambda \geq 0$

In this model, the score, if exists, satisfies $\delta^t((x_o, y_o)^s) \geq 1$.

In the output-oriented case, we solve the following LPs:

[SBM-O]

$$\delta^t((x_o, y_o)^s) = \min_{\lambda, s^+} \frac{1}{1 + \dfrac{1}{r} \displaystyle\sum_{i=1}^r \dfrac{s_i^+}{y_{io}^s}} \qquad (6.20)$$

subject to $x_o^s \geq X^t \lambda, \quad y_o^s = Y^t \lambda - s^+, \quad L \leq e\lambda \leq U, \quad \lambda \geq 0, \quad s^+ \geq 0$

where the vector $s^+ \in R^r$ denotes the output slacks. Or, equivalently,

[SBM-O]

$$\delta^t((x_o, y_o)^s) = \min_{\lambda, \eta} \frac{1}{\dfrac{1}{r} \displaystyle\sum_{i=1}^r \eta_i} \qquad (6.21)$$

subject to $x_o^s \geq X^t \lambda, \quad \eta_i y_{io}^s \leq \sum_{j=1}^n y_{ij}^t \lambda_j (i=1,\ldots,r), \quad \eta_i \geq 1 (i=1,\ldots,r), \quad L \leq e\lambda \leq U, \quad \lambda \geq 0$

[Super-SBM-O]

$$\delta^t((x_o, y_o)^s) = \min_{\lambda, s^+} \frac{1}{1 - \dfrac{1}{r} \displaystyle\sum_{i=1}^r \dfrac{s_i^+}{y_{io}^s}} \qquad (6.22)$$

subject to $x_o^s \geq X^t \lambda, \quad y_o^s \leq Y^t \lambda + s^+, \quad L \leq e\lambda \leq U, \quad \lambda \geq 0, \quad s^+ \geq 0$

Or, equivalently,

[Super-SBM-O]

$$\delta^t((x_o, y_o)^s) = \min_{\lambda, \eta} \frac{1}{\dfrac{1}{r} \displaystyle\sum_{i=1}^r \eta_i} \qquad (6.23)$$

subject to $x_o^s \geq X^t\lambda$, $\eta_i y_{io}^s \leq \sum_{j=1}^{n} y_{ij}^t \lambda_j (i=1,\ldots,r)$, $0 \leq \eta_i \leq 1 (i=1,\ldots,r)$,

$L \leq e\lambda \leq U$, $\lambda \geq 0$

The output-oriented models take all output slacks (shortfalls) into account, but not input slacks (surpluses).

The non-radial and slacks-based MI evaluates the four elements of the MI, $\delta^1((x_o, y_o)^1)$, $\delta^2((x_o, y_o)^2)$, $\delta^1((x_o, y_o)^2)$ and $\delta^2((x_o, y_o)^1)$, by means of the LPs [SBM-I] and [Super-SBM-I].

Remark 6.3 Infeasible-LP issues
These models may suffer from the same infeasibility troubles as the radial ones may encounter.

6.4   NON-RADIAL AND NON-ORIENTED MALMQUIST MODEL

The models in this category deal with input and output slacks. The models [SBM] and [Super-SBM] used for computing $\delta^t((x_o, y_o)^s)$ are represented by the following fractional programs:

[SBM]

$$\delta^t\big((x_o, y_o)^s\big) = \min_{\lambda, s^-, s^+} \frac{1 - \frac{1}{m}\sum_{i=1}^{m} \frac{s_i^-}{x_{io}^s}}{1 + \frac{1}{r}\sum_{i=1}^{r} \frac{s_i^+}{y_{io}^s}} \tag{6.24}$$

subject to $x_o^s = X^t\lambda + s^-$, $y_o^s = Y^t\lambda - s^+$, $L \leq e\lambda \leq U$, $\lambda \geq 0$, $s^- \geq 0$, $s^+ \geq 0$

Or, equivalently,

[SBM]

$$\delta^t\big((x_o, y_o)^s\big) = \min_{\theta, \eta, \lambda} \frac{\frac{1}{m}\sum_{i=1}^{m} \theta_i}{\frac{1}{r}\sum_{i=1}^{r} \eta_i} \tag{6.25}$$

subject to $\theta_i x_{io}^s \geq \sum_{j=1}^{n} x_{ij}^t \lambda_j (i=1,\ldots,m)$, $\eta_i y_{io}^s \leq \sum_{j=1}^{n} y_{ij}^t \lambda_j (i=1,\ldots,r)$,

$\theta_i \leq 1 \ (i=1,\ldots,m), \eta_i \geq 1 (i=1,\ldots,r)$, $L \leq e\lambda \leq U$, $\lambda \geq 0$

[Super-SBM]

$$\delta^t\left((x_o,y_o)^s\right) = \min_{\lambda,s^-,s^+} \frac{\frac{1}{m}\sum_{i=1}^m \frac{\bar{x}_i}{x_{io}^s}}{\frac{1}{r}\sum_{i=1}^r \frac{\bar{y}_i}{y_{io}^s}} \tag{6.26}$$

subject to $\bar{x} \geq X^t\lambda,\ \ \bar{y} \leq Y^t\lambda,\ \ \bar{x} \geq x_o^s,\ \ \bar{y} \leq y_o^s,\ \ L \leq e\lambda \leq U,\ \ \bar{y} \geq 0,\ \ \lambda \geq 0$

Or, equivalently,

[Super-SBM]

$$\delta^t\left((x_o,y_o)^s\right) = \min_{\theta,\eta,\lambda} \frac{\frac{1}{m}\sum_{i=1}^m \theta_i}{\frac{1}{r}\sum_{i=1}^r \eta_i} \tag{6.27}$$

subject to $\theta_i x_{io}^s \geq \sum_{j=1}^n x_{ij}^t\lambda_j (i=1,\ldots,m),\ \ \eta_i y_{io}^s \leq \sum_{j=1}^n y_{ij}^t\lambda_j (i=1,\ldots,r),$

$\theta_i \geq 1 (i=1,\ldots,m),\ \ 0 \leq \eta_i \leq 1\ (i=1,\ldots,r),\ \ L \leq e\lambda \leq U,\ \ \lambda \geq 0$

These fractional programs can be transformed into LPs [4]. This model, under the exclusive scheme (see Remark 6.1) evaluates the four components of the MI, $\delta^1((x_o, y_o)^1)$, $\delta^2((x_o, y_o)^2)$, $\delta^1((x_o, y_o)^2)$ and $\delta^2((x_o, y_o)^1)$, using [SBM], and, if the corresponding LP is found to be infeasible, we then apply [Super-SBM].

Remark 6.4 Infeasible-LP issues
For this non-oriented model, [Super-SBM] is always feasible, and has a finite minimum in any RTS environment under some mild conditions, that is, for each output $i (= 1, \ldots, q)$, at least two DMUs have positive values. This can be seen from the constraints in (6.27). See Tone [5] for details.

6.5 CUMULATIVE MALMQUIST INDEX (CMI)

Although the above MI is defined on a two-period base $(s \rightarrow t)$, we can find a cumulative Malmquist index (CMI) based on the first period and period t $(1 \rightarrow t)$ as follows:

$$\tilde{\mu}^{1 \rightarrow t} = \Pi_{\tau=1}^t \mu^{\tau \rightarrow \tau+1} (t=1,\ldots,T-1) \tag{6.28}$$

The value of the CMI in period 1 $(t = 1)$ is equal to one, since both the CU and the FS are in the status quo $(\gamma^{1 \rightarrow 1} = 1$ and $\phi^{1 \rightarrow 1} = 1)$. Therefore, we can easily capture the productivity change of DMU_o from the first period through multiple periods $(1 \rightarrow t)$, and compare the results among different DMUs.

In addition, the CMI turns out to be given by

$$\widetilde{\mu}^{1 \rightarrow t} = \mu^{1 \rightarrow 2} \times \widetilde{\mu}^{2 \rightarrow t} \tag{6.29}$$

Thus, the intertemporal productivity change between period 1 and period t is modified by the position at 2, ..., t.

Furthermore, the CMI can be decomposed into a cumulative FS (CFS) and the ratio of the efficiency scores between period 1 and period t as follows:

$$\begin{aligned}
\widetilde{\mu}^{1 \rightarrow t} &= \Pi_{\tau=1}^{t} \left(\phi^{\tau \rightarrow \tau+1} \cdot \gamma^{\tau \rightarrow \tau+1} \right) \\
&= \widetilde{\phi}^{1 \rightarrow t} \cdot \frac{\theta^t}{\theta^1}
\end{aligned} \tag{6.30}$$

where $\widetilde{\phi}^{1 \rightarrow t}$ indicates the CFS to period t from the base period.

6.6   ADJUSTED MALMQUIST INDEX (AMI)

The CMI captures the productivity change from the base period $(t = 1)$. However, the differences in the efficiency levels of the DMUs in the base period are ignored, since the initial scores for the CMI for all DMUs are equal to one. In order to take the efficiency levels of DMUs in the base period into account in the CMI, we calculate an adjusted Malmquist index (AMI) as the product of the CMI and the efficiency score in the first period as follows:

$$\widetilde{\xi}^{1 \rightarrow t} = \widetilde{\mu}^{1 \rightarrow t} \cdot \theta^1 \tag{6.31}$$

This can be transformed into the product of the CFS and the efficiency score in period t as follows:

$$\widetilde{\xi}^{1 \rightarrow t} = \widetilde{\phi}^{1 \rightarrow t} \cdot \theta^1 \tag{6.32}$$

which means that the AMI is an efficiency score (θ^t) incorporating the frontier shift effect.

The AMI is the same as the 'actual performance index' of Thore *et al.* [6]. This is a practical measure to capture both the relative efficiency of the DMUs in the base period and the productivity change from the base period to period t. This can help to evaluate unfortunate DMUs, which are scored relatively low in terms of efficiency even they achieve a large productivity change.

6.7   NUMERICAL EXAMPLE

Table 6.1 shows input and output data for eight DMUs for four periods as a numerical example. The outputs of all DMUs in each period have been set to one for the sake of convenience. In this sample, DMUs P, Q, R and S are efficient and form the frontier in each period. We focus on the intertemporal behaviour of the inefficient DMUs A, B, C and D, in order to clarify the differences in the related indices.

Figure 6.2 depicts the results for the SBM under CRS (SBM-C). DMU A has relatively good scores during all periods, while the scores of DMUs C and D improve period by period and finally reach the level of DMU A. In contrast, the scores of DMU B decrease. However, these are relative scores evaluated using the frontiers for each period. In order to correctly compare the efficiency trends, we must take frontier shift effects into account.

The rate of change of the SBM scores between two periods is the CU (Figure 6.3), which does not include the FS (Figure 6.4). After incorporating the FS into the CU, we can obtain the MI (Figure 6.5), which is a non-relative productivity index.

Here, we must be careful in interpreting the trends in the CU, FS and MI, which measure the change in the indices from the previous period. For instance, the FSs for DMU B show a decreasing trend. However, they are substantially larger than one. This means that the frontier is in an advancing status period by period, although its growth rate is gradually decreasing. Another instance where we must be careful is the MI for DMU C in the period t2 → t3, which has the largest value of all in this period. However, this is caused by the fact that the MI for the previous period (t1 → t2) is very poor (negative). Even though the growth rate is very large, it is problematic that the productivity level of DMU C exceeds that of the other DMUs in the period t2 → t3.

In order to observe the trends in the growth rate comparatively, cumulative indices such as the CFS, CCU and CMI are helpful. These indices indicate the growth from the first period, for which the values are standardized to one for all DMUs (Figures 6.6, 6.7 and 6.8). We can easily see that the frontier for DMU B is progressing by large amounts period by period, and the MI of DMU C cannot reach the productivity level of the other DMUs owing to the negative growth in the second period.

Furthermore, the AMI includes the relative efficiency level for the first period instead of having a value of one (Figure 6.9). As shown in (6.32), the AMI implies an SBM efficiency score incorporating the FS, and the results for it are different from those for the SBM (Figure 6.5), except for DMU D. In particular, the AMI of DMU B increases continuously, while the SBM decreases. The AMI of DMU D develops more than DMU A during this period, but the efficiency score in the first period is too low for it finally to reach the efficiency level of DMU A, although the SBM nearly reaches the level of DMU A in the last period. The trend in the AMI for DMU C is close to that for DMU D, although the SBM scores exceed those for DMU D.

TABLE 6.1 Dataset for numerical example.

DMU	Period 1			Period 2			Period 3			Period 4		
	Input 1	Input 2	Output	Input 1	Input 2	Output	Input 1	Input 2	Output	Input 1	Input 2	Output
A	4	5	1	3.4	6	1	2.8	5.7	1	2.2	6	1
B	3	12	1	2	10	1	1.8	8.8	1	1.5	8	1
C	9	3	1	10	3.5	1	8	2.8	1	8	2.3	1
D	10	8	1	8	7	1	7	5	1	5	3.5	1
P	7	1	1	10	2	1	10	2	1	8	2	1
Q	4	4	1	4	4	1	3	5	1	3	5	1
R	3	14	1	1	12	1	1	9	1	1	7	1
S	5	3	1	5	3	1	5	3	1	5	3	1

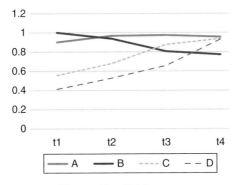

Figure 6.2   SBM scores.

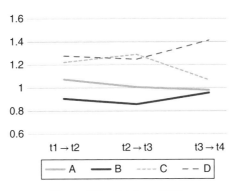

Figure 6.3   Catch-up (CU).

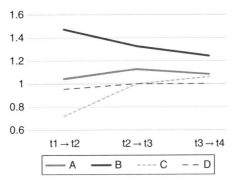

Figure 6.4   Frontier shift (FS).

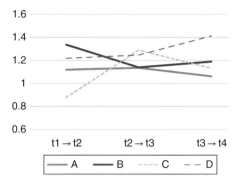

Figure 6.5 Malmquist index (MI).

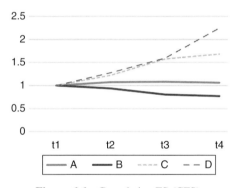

Figure 6.6 Cumulative FS (CFS).

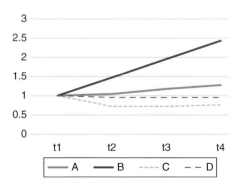

Figure 6.7 Cumulative CU (CCU).

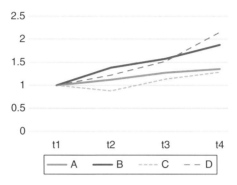

Figure 6.8   Cumulative MI (CMI).

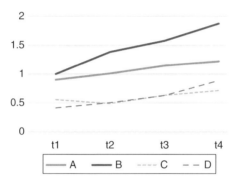

Figure 6.9   Adjusted Malmquist index (AMI).

6.7.1   DMU A

The frontier for DMU A increases slightly (the CFSs are more than one) and the DMU correspondingly follows it (the CCUs are nearly equal to one). As the result, the CMI increases only slightly. After the relative efficiency score in the first period is incorporated (AMI), the index values are less than those for DMU B, and hence DMU A takes second place, even though DMU A gets the best scores in in terms of SBM in the last three periods.

6.7.2   DMU B

The SBM score of DMU B decreases period by period. However, the frontier for DMU B is constantly progressing, and therefore the AMI of this DMU outperforms the others, even though DMU B cannot perfectly catch up with the frontier (the index is less than one).

6.7.3   DMU C

The SBM score of DMU C increases period by period. On the other hand, the frontier for DMU C regresses in the first period. DMU C catches up and gets closer to the frontier, and, finally, the CMI increases after a decrease in the second period. Nevertheless, the AMI of DMU C cannot reach the score of DMU A in the last period, since the efficiency score in the first period is relatively low.

6.7.4   DMU D

DMU D takes the worst place for the SBM score in the first period, even though that score increases period by period. The frontier for this DMU does not move during the period considered (the CFS is scored as nearly one), and therefore DMU D catches up and gets closer to the frontier very well. As a result, the CMI increases more than that of DMU B. However, because of the worse efficiency score in the first period, the AMI is as low as that of DMU C.

6.8   CONCLUDING REMARKS

In this chapter, we have briefly surveyed Malmquist index models. We have explained the cumulative Malmquist index (CMI) and the adjusted Malmquist index (AMI) in detail using graphical presentations. The former (CMI) assumes that all DMUs have equal status in the starting period, and hence they take part in a scratch (no handicap) race thereafter. The latter (AMI) accounts for different starting conditions in the first period and evaluates productivity changes thereafter. Both indices differ from the traditional MI in that the MI deals with productivity change between two consecutive periods, whereas the CMI and AMI evaluate the productivity change from the starting period. Hence, it should be noted that the selection of the starting period affects the whole of the results.[1]

REFERENCES

[1] Malmquist, S. (1953) Index numbers and indifference surfaces. *Trabajos de Estadistica*, **4**, 209–242.

[2] Färe, R., Grosskopf, S., Lindgren, B. and Roos, P. (1994) Productivity change in Swedish hospitals: A Malmquist output index approach, in *Data Envelopment Analysis: Theory, Methodology and Applications* (eds A. Charnes, W.W. Cooper, A.Y. Lewin and L.M. Seiford), Kluwer Academic, Boston, MA, pp. 253–272.

---

[1] Software for Malmquist index models is included in DEA-Solver Pro V13 (http://www.saitech-inc.com). See also Appendix A.

[3] Andersen, P. and Petersen, N.C. (1993) A procedure for ranking efficient units in data envelopment analysis. *Management Science*, **39**, 1261–1264.

[4] Tone, K. (2001) A slacks-based measure of efficiency in data envelopment analysis. *European Journal of Operational Research*, **130**, 498–509.

[5] Tone, K. (2002) A slacks-based measure of super-efficiency in data envelopment analysis. *European Journal of Operational Research*, **143**, 32–41.

[6] Thore, S., Kozmetsky, G. and Phillips F. (1994) DEA of financial statements data: The U.S. computer industry. *Journal of Productivity Analysis*, **5**(3), 229–248.

7

THE NETWORK DEA MODEL[1]

KAORU TONE

National Graduate Institute for Policy Studies, Tokyo, Japan

MIKI TSUTSUI

Central Research Institute of Electric Power Industry, Tokyo, Japan

7.1   INTRODUCTION

Traditional DEA models deal with measurements of the relative efficiency of DMUs with respect to multiple inputs or multiple outputs. One of the drawbacks of these models is the neglect of internal or linking activities. For example, many companies comprise several divisions that are linked as illustrated in Figure 7.1. In this example, the company has three divisions. Each division utilizes its own input resources to produce its own outputs. However, there are linking activities (or intermediate products) as shown by Link $1 \rightarrow 2$, Link $1 \rightarrow 3$, Link $2 \rightarrow 1$ and Link $2 \rightarrow 3$. Link $1 \rightarrow 2$ indicates that part of the output from division 1 is utilized as input to division 2.

In traditional DEA models, every activity must belong to either the input or the output but not to both. So, these models usually employ multiple steps in the evaluation, using intermediate products as outputs in one step and as inputs in another step. Thus, these models cannot deal with intermediate products directly in a single step. Although there may be many variants of this process flow, the existence of linking activities is an indispensable part of network DEA models.

---

[1] Part of the material in this chapter is adapted from *European Journal of Operational Research*, Vol. 197, Tone K. and Tsutsui M., Network DEA: A slacks-based measure approach, 243–252, 2009 [1], with permission from Elsevier Science.

Advances in DEA Theory and Applications: With Extensions to Forecasting Models,
First Edition. Edited by Kaoru Tone.
© 2017 John Wiley & Sons Ltd. Published 2017 by John Wiley & Sons Ltd.

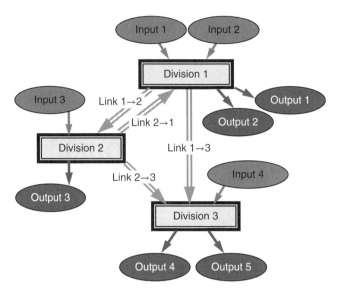

Figure 7.1   Example of network structure.

Network DEA evaluates the efficiencies of multidivisional organizations. This model solves for the comparative overall efficiency of the organization along with the divisional efficiencies in a unified framework.[2]

7.2   NOTATION AND PRODUCTION POSSIBILITY SET

We use the following notation to describe network DEA.

n: number of DMUs ($j = 1, \ldots, n$);
K: number of divisions ($k = 1, \ldots, K$);
m_k: number of inputs to division k ($i = 1, \ldots, m_k$);
r_k: number of outputs from division k ($i = 1, \ldots, r_k$);
p_{kh}: number of items in link from division k to division h ($l = 1, \ldots, p_{kh}$);
(k, h): link from division k to division h;
S: set of divisions which have no incoming links, i.e. starting divisions;
T: set of divisions which have no outgoing links, i.e. terminal divisions.

[2] Software for network DEA models is included in DEA-Solver Pro V13 (http://www.saitech-inc.com). See also Appendix A.

The observed data are as follows:

$\{x_{ijk} \in R_+\}$: input resource i to division k of DMU$_j$ ($i = 1, ..., m_k$, $\forall k$, $\forall j$);

$\{y_{ijk} \in R_+\}$: output product i from division k of DMU$_j$ ($i = 1, ..., r_k$, $\forall k$, $\forall j$);

$\{z_{j(k,h)l}^{\alpha} \in R_+\}$: linking internal output product l at division k to division h of DMU$_j$ ($\forall l$, $\forall (k,h)$, $\forall j$, $\alpha = 4$ types of links as explained in Section 7.3.2) = linking internal input resource l at division h from division k of DMU$_j$ ($\forall l$, $\forall (k,h)$, $\forall j$, $\forall \alpha$).

We assume

$$z_{j(k,h)l}^{\alpha} = 0 \ (\forall l, \forall j, \forall \alpha, h \in S) : \text{no linking input to starting divisions}$$
$$z_{j(k,h)l}^{\alpha} = 0 \ (\forall l, \forall j, \forall \alpha, k \in T) : \text{no linking output from terminal divisions}$$

(7.1)

The production possibility set $P = \{(\mathbf{x}_k, \mathbf{y}_k, \mathbf{z}_{(k,h)})\}$ is defined in vector notation by

$$\mathbf{x}_k \geq \sum_{j=1}^{n} \mathbf{x}_{jk} \lambda_j^k \ (\forall k)$$

$$\mathbf{y}_k \leq \sum_{j=1}^{n} \mathbf{y}_{jk} \lambda_j^k \ (\forall k)$$

$$\mathbf{z}_{(k,h)} \leq \sum_{j=1}^{n} \mathbf{z}_{j(k,h)} \lambda_j^k \ (\forall (k,h)) \ \text{(as outputs from } k)$$

$$\mathbf{z}_{(k,h)} \geq \sum_{j=1}^{n} \mathbf{z}_{j(k,h)} \lambda_j^h \ (\forall (k,h)) \ \text{(as inputs to } h)$$

$$\sum_{j=1}^{n} \lambda_j^k = 1 \ (\forall k)$$

$$\lambda_j^k \geq 0 \ (\forall k, \forall j)$$

(7.2)

where $\lambda_j^k \in R_+$ is the intensity variable for DMU$_j$ ($\forall j$) corresponding to division k ($\forall k$). The constraint $\sum_{j=1}^{n} \lambda_j^k = 1$ corresponds to the variable-returns-to-scale (VRS) assumption. If we delete this constraint, we have the constant-returns-to-scale (CRS) model.

7.3 DESCRIPTION OF NETWORK STRUCTURE

7.3.1 Inputs and Outputs

The inputs and outputs of DMU$_o$ ($o = 1, ..., n$) $\in P$ can be represented by

$$x_{iok} = \sum_{j=1}^{n} x_{ijk}\lambda_j^k + s_{ik}^- \quad (i = 1, \ldots, m_k, \forall k)$$

$$y_{iok} = \sum_{j=1}^{n} y_{ijk}\lambda_j^k - s_{ik}^+ \quad (i = 1, \ldots, r_k, \forall k)$$

$$\sum_{j=1}^{n} \lambda_j^k = 1 (\forall k) \tag{7.3}$$

$$\lambda_j^k \geq 0(\forall k, \forall j), \quad s_{ik}^- \geq 0(i = 1, \ldots, m_k, \forall k), \quad s_{ik}^+ \geq 0(i = 1, \ldots, r_k, \forall k)$$

where s_{ik}^- and s_{ik}^+ indicate input and output slacks (non-negative), respectively.

7.3.2 Links

As regards the linking constraints, we have several options, of which we present four possible cases. We can choose any of the cases below according to the nature of the links.

In all cases, we assume the following continuity condition for the links (k, h), which is critical for a network model connecting activities in two divisions:

$$\sum_{j=1}^{n} z_{j(k,h)l}^{\alpha}\lambda_j^k = \sum_{j=1}^{n} z_{j(k,h)l}^{\alpha}\lambda_j^h \left(l = 1, \ldots, p_{(k,h)}^{\alpha}, \forall (k,h) \right) \tag{7.4}$$

where α stands for 'free', 'fix', 'out' or 'in' in the equations below. $z_{j(k,h)l}^{\alpha}$ is the observed link l from division k to division h of DMU$_j$ ($\forall l$, $\forall (k,h)$, $\forall j$) in the case α.

7.3.2.1   The 'Free' Link Value Case   The linking activities are freely determined (discretionary), while keeping the continuity of the free links between division k and division h as formulated in (7.4). This case can be used to see whether or not the current link flow has an appropriate volume in the light of other DMUs, that is, the link flow may increase or decrease in the optimal solution of the linear programs introduced in Section 7.4. Between the current link value and the free link value, we have the following relationship:

$$z_{o(k,h)l}^{\text{free}} = \sum_{j=1}^{n} z_{j(k,h)l}^{\text{free}}\lambda_j^k + s_{(k,h)l}^{\text{free}} \left(l = 1, \ldots, p_{(k,h)}^{\text{free}}, \forall (k,h) \right) \tag{7.5}$$

where $s_{(k,h)l}^{\text{free}}$ is a slack of the free link l from division k to division h ($\forall l$, $\forall (k,h)$) and is free in sign.

7.3.2.2   The 'Fixed' Link Value Case   The linking activities are kept unchanged (non-discretionary):

$$z_{o(k,h)l}^{\text{fix}} = \sum_{j=1}^{n} z_{j(k,h)l}^{\text{fix}}\lambda_j^k \left(l = 1, \ldots, p_{(k,h)}^{\text{fix}}, \forall (k,h) \right)$$

$$z_{o(k,h)l}^{\text{fix}} = \sum_{j=1}^{n} z_{j(k,h)l}^{\text{fix}}\lambda_j^h \left(l = 1, \ldots, p_{(k,h)}^{\text{fix}}, \forall (k,h) \right) \tag{7.6}$$

This case corresponds to the situation where the intermediate products are beyond the control of DMUs or the discretion of the management.

7.3.2.3 The 'As Output' Link Value Case The linking activities for which a larger amount is regarded as favourable are treated 'as output' from the preceding division, and shortages are accounted for in the output inefficiency:

$$z_{o(k,h)l}^{\text{out}} = \sum_{j=1}^{n} z_{j(k,h)l}^{\text{out}} \lambda_j^k - s_{(k,h)l}^{\text{out}+} \left(l = 1, \ldots, p_{(k,h)}^{\text{out}}, \forall (k,h) \right) \tag{7.7}$$

where $s_{(k,h)l}^{\text{out}+}$ is an 'as output' link slack (non-negative).

7.3.2.4 The 'As Input' Link Value Case The linking activities for which a smaller amount is regarded as favourable are treated 'as input' to the succeeding division, and excesses are accounted for in the input inefficiency:

$$z_{o(k,h)l}^{\text{in}} = \sum_{j=1}^{n} z_{j(k,h)l}^{\text{in}} \lambda_j^k + s_{(k,h)l}^{\text{in}-} \left(l = 1, \ldots, p_{(k,h)}^{\text{in}}, \forall (k,h) \right) \tag{7.8}$$

where $s_{(k,h)l}^{\text{in}-}$ is an 'as input' link slack (non-negative).

7.4 OBJECTIVE FUNCTIONS AND EFFICIENCIES

We employ the non-radial (SBM) model and the following objective functions for each case, with the constraints presented in (7.3)–(7.8).

7.4.1 Input-Oriented Case

In the input-oriented case, the excess of inputs and the 'as input' links are evaluated as inefficiency:

$$\theta_o^* = \min \sum_{k=1}^{K} w_k \left[1 - \frac{1}{m_k + p_{(k,h)}^{\text{in}}} \left(\sum_{i=1}^{m_k} \frac{s_{ik}^-}{x_{iok}} + \sum_{l=1}^{p_{(k,h)}^{\text{in}}} \frac{s_{(k,h)l}^{\text{in}-}}{z_{o(k,h)l}^{\text{in}}} \right) \right] \tag{7.9}$$

where w_k is the weight of division k, which is supplied exogenously according to the importance of the division, and these weights satisfy the following condition:

$$\sum_{t=1}^{k} w_k = 1 \tag{7.10}$$

We define the divisional efficiency by

$$\theta_{ok}^* = 1 - \frac{1}{m_k + p_{(k,h)}^{\text{in}}} \left(\sum_{i=1}^{m_k} \frac{s_{iok}^{-*}}{x_{iok}} + \sum_{l=1}^{p_{(k,h)}^{\text{in}}} \frac{s_{o(k,h)l}^{\text{in}-*}}{z_{o(k,h)l}^{\text{in}}} \right) \tag{7.11}$$

where s_{iok}^{-*} and $s_{o(k,h)l}^{\text{in}-*}$ are the optimal slacks. The overall efficiency is the weighted arithmetic mean of the divisional efficiencies:

$$\theta_o^* = \sum_{k=1}^{K} w_k \theta_{ok}^* \tag{7.12}$$

The free link slacks are not directly incorporated into (7.9), because the signs of these slacks are free, and therefore we do not know whether there exists an excess or a shortfall beforehand. The free link values are related to the efficiency scores only through the link constraints (7.4). The same applies to the output-oriented and non-oriented models mentioned below.

7.4.2 Output-Oriented Case

In the output-oriented case, the shortfall of the outputs and the 'as output' links are evaluated as inefficiency:

$$\frac{1}{\eta_o^*} = \max \sum_{k=1}^{K} w_k \left[1 + \frac{1}{r_k + p_{(k,h)}^{\text{out}}} \left(\sum_{i=1}^{r_k} \frac{s_{ik}^+}{y_{iok}} + \sum_{l=1}^{p_{(k,h)}^{\text{out}}} \frac{s_{(k,h)l}^{\text{out}+}}{z_{o(k,h)l}^{\text{out}}} \right) \right] \tag{7.13}$$

In order to confine all scores within the range [0, 1], we define the efficiency score of division k by

$$\eta_{ok}^* = \frac{1}{1 + \dfrac{1}{r_k + p_{(k,h)}^{\text{out}}} \left(\displaystyle\sum_{i=1}^{r_k} \dfrac{s_{iok}^{+*}}{y_{iok}} + \displaystyle\sum_{l=1}^{p_{(k,h)}^{\text{out}}} \dfrac{s_{o(k,h)l}^{\text{out}+*}}{z_{o(k,h)l}^{\text{out}}} \right)} \tag{7.14}$$

where s_{iko}^{+*} and $s_{o(k,h)l}^{\text{out}+*}$ are the optimal slacks.

Hence, the overall efficiency η_o^* is not the weighted arithmetic mean of the divisional efficiencies but the weighted harmonic mean. Thus, we usually have

$$\eta_o^* \leq \sum_{k=1}^{K} w_k \eta_{ok}^* \tag{7.15}$$

7.4.3 Non-Oriented Case

In the non-oriented case, both the excess of inputs and 'as input' links and the shortfall of outputs and 'as output' links are evaluated as inefficiency:

$$\rho_o^* = \min \frac{\displaystyle\sum_{k=1}^{K} w_k \left[1 - \dfrac{1}{m_k + p_{(k,h)}^{\text{in}}} \left(\displaystyle\sum_{i=1}^{m_k} \dfrac{s_{ik}^-}{x_{iok}} + \displaystyle\sum_{l=1}^{p_{(k,h)}^{\text{in}}} \dfrac{s_{(k,h)l}^{\text{in}-}}{z_{o(k,h)l}^{\text{in}}} \right) \right]}{\displaystyle\sum_{k=1}^{K} w_k \left[1 + \dfrac{1}{r_k + p_{(k,h)}^{\text{out}}} \left(\displaystyle\sum_{i=1}^{r_k} \dfrac{s_{ik}^+}{y_{iok}} + \displaystyle\sum_{l=1}^{p_{(k,h)}^{\text{out}}} \dfrac{s_{(k,h)l}^{\text{out}+}}{z_{o(k,h)l}^{\text{out}}} \right) \right]} \tag{7.16}$$

In this case we define the efficiency score of division k by

$$\rho_{ok}^* = \frac{1 - \frac{1}{m_k + p_{(k,h)}^{in}}\left(\sum_{i=1}^{m_k}\frac{s_{iok}^{-*}}{x_{iok}} + \sum_{l=1}^{p_{(k,h)}^{in}}\frac{s_{o(k,h)l}^{in-*}}{z_{o(k,h)l}^{in}}\right)}{1 + \frac{1}{r_k + p_{(k,h)}^{out}}\left(\sum_{i=1}^{r_k}\frac{s_{iok}^{+*}}{y_{iok}} + \sum_{l=1}^{p_{(k,h)}^{out}}\frac{s_{o(k,h)l}^{out+*}}{z_{o(k,h)l}^{out}}\right)} \qquad (7.17)$$

Thus, the overall efficiency is neither the arithmetic nor the harmonic mean of the divisional efficiencies.

REFERENCE

[1] Tone, K. and Tsutsui, M. (2009) Network DEA: A slacks-based measure approach. *European Journal of Operational Research*, **197**(1), 243–252.

8

THE DYNAMIC DEA MODEL[1]

KAORU TONE

National Graduate Institute for Policy Studies, Tokyo, Japan

MIKI TSUTSUI

Central Research Institute of Electric Power Industry, Tokyo, Japan

8.1 INTRODUCTION

The measurement of intertemporal efficiency change has long been a subject of concern in data envelopment analysis (DEA). Window analysis and Malmquist index methods are representative methods. However, the models used in these methods do not account for the effect of carry-over activities between two consecutive periods. For each period, these models have inputs and outputs, but the connecting activities between periods are not accounted for explicitly.

The dynamic DEA model proposed by Färe and Grosskopf [2] was the first innovative scheme that formally dealt with these interconnecting activities. Traditional DEA models usually deal with the efficiency of input resources versus the output products of associated decision-making units (DMUs) within cross-sectional data. In contrast, the dynamic DEA model extends these models to dynamic situations as shown in Figure 8.1. In each period t, each DMU has inputs and outputs along with a carry-over

---

[1] Part of the material in this chapter is adapted from *Omega*, Vol. 38, Tone K. and Tsutsui M., Dynamic DEA: A slacks-based measure approach, 145–156, 2010 [1], with permission from Elsevier Science.

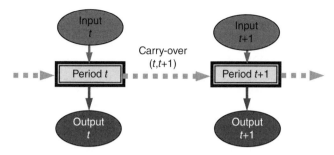

Figure 8.1 Dynamic structure.

to the next period, $t + 1$. What distinguishes dynamic DEA from ordinary DEA is the existence of carry-overs that connect two consecutive periods.

In this chapter, we extend the dynamic DEA model within the slacks-based measure framework. The dynamic DEA model can also be positioned as an extension of the network SBM described in Chapter 7 to dynamic structures.

Our model has the following features. (i) Since it is a dynamic model, we can compare the long-range performance of companies. (ii) The adoption of non-radial SBM models enables us to deal with inputs and outputs individually, and hence non-proportional changes in inputs and outputs are allowed. (iii) Carry-over activities are categorized into four types: discretionary (free), non-discretionary (fixed), desirable (good) and undesirable (bad), and hence we are able to correctly and properly cope with the demands of researchers and practitioners. (iv) We have developed three orientations for every model: input-, output- and non-oriented models. Thus, in accordance with the purpose of the research being carried out, we can choose appropriate models for evaluation. If input- or output-side efficiency is the main target, we can choose input- or output-oriented models, respectively. If both input and output efficiencies are to be evaluated concurrently, we can apply non-oriented models.[2]

8.2 NOTATION AND PRODUCTION POSSIBILITY SET

We use the following notation to describe dynamic DEA:

n: number of DMUs ($j = 1, \ldots, n$);
T: number of periods ($t = 1, \ldots, T$);
m: number of inputs ($i = 1, \ldots, m$);
r: number of outputs ($i = 1, \ldots, r$);

---

[2] Software for dynamic DEA models is included in DEA-Solver Pro V13 (hppt://www.saitech-inc.com). See also Appendix A.

q: number of items in carry-over from period t to period $t + 1$ $(c = 1, \ldots, q)$;

$(t, t + 1)$: carry-overs from period t to period $t + 1$.

The observed data are as follows:

$\{x_{ij}^t \in R_+\}$: input resource i to DMU$_j$ in period t ($\forall i$, $\forall j$, $\forall t$);

$\{y_{ij}^t \in R_+\}$: output product i from DMU$_j$ in period t ($\forall i$, $\forall j$, $\forall t$);

$\{\omega_{jc}^{\alpha,t} \in R_+\}$: carry-over c of DMU$_j$ from period t to period $t + 1$ ($\forall c$, $\forall j$, $\forall t$, $\alpha = 4$

types of carry-overs as explained in Section 8.3.2).

We postulate that we have homogeneous panel data throughout periods 1 to T. So, we look at the enterprises concerned as a continuum between period 1 and period T. In addition, we take the initial inputs for carry-overs in period 1 into account as follows:

$\omega_{jc}^{\alpha,0} (\forall j, \forall c, \forall \alpha)$: initial carry-over to period 1.

The production possibility set $P = \{(\mathbf{x}^t, \mathbf{y}^t, \boldsymbol{\omega}^t, \boldsymbol{\omega}^0)\}$ is defined in vector notation by

$$\mathbf{x}^t \geq \sum_{j=1}^{n} \mathbf{x}_j^t \lambda_j^t \ (\forall t)$$

$$\mathbf{y}^t \leq \sum_{j=1}^{n} \mathbf{y}_j^t \lambda_j^t \ (\forall t)$$

$$\boldsymbol{\omega}^t \leq \sum_{j=1}^{n} \boldsymbol{\omega}_j^t \lambda_j^t \ (t = 1, \ldots, T) \text{(carry-overs as outputs from period } t)$$

$$\boldsymbol{\omega}^t \geq \sum_{j=1}^{n} \boldsymbol{\omega}_j^t \lambda_j^{t+1} \ (t = 1, \ldots, T-1) \text{(carry-overs as inputs to period } t + 1) \qquad (8.1)$$

$$\boldsymbol{\omega}^0 \leq = \geq \sum_{j=1}^{n} \boldsymbol{\omega}_j^0 \lambda_j^1 \ \text{(initial carry-overs to period 1)}$$

$$\sum_{j=1}^{n} \lambda_j^t = 1 \ (\forall t)$$

$$\lambda_j^t \geq 0 (\forall j, \forall t)$$

where $\lambda_j^t \in R_+$ is the intensity variable for DMU$_j$ ($\forall j$) corresponding to period t ($\forall t$). The constraint $\sum_{j=1}^{n} \lambda_j^k = 1$ corresponds to the variable-returns-to-scale (VRS) assumption. If we delete this constraint, we have the constant-returns-to-scale (CRS) model.

8.3 DESCRIPTION OF DYNAMIC STRUCTURE

8.3.1 Inputs and Outputs

$\text{DMU}_o \ (o = 1, \ldots, n) \in P$ can be represented by

$$
\begin{aligned}
& x_{io}^t = \sum\nolimits_{j=1}^{n} x_{ij}^t \lambda_j^t + s_i^{t-} \ (i = 1, \ldots, m, \forall t) \\
& y_{io}^t = \sum\nolimits_{j=1}^{n} y_{ij}^t \lambda_j^t - s_i^{t+} \ (i = 1, \ldots, r, \forall t) \\
& \sum\nolimits_{j=1}^{n} \lambda_j^t = 1 (\forall t) \\
& \lambda_j^t \geq 0 (\forall j, \forall t), \ s_i^{t-} \geq 0 (i = 1, \ldots, m, \forall t), \ s_i^{t+} \geq 0 (i = 1, \ldots, r, \forall t)
\end{aligned}
\tag{8.2}
$$

where s_i^{t-} and s_i^{t+} indicate input and output slacks (non-negative), respectively.

8.3.2 Carry-Overs

We classify carry-over activities into four categories of 'free', 'fixed', 'good' and 'bad'. In all cases, the continuity of carry-overs between periods t and $t + 1$ can be guaranteed by the following condition:

$$
\sum\nolimits_{j=1}^{n} \omega_{jc}^{\alpha,t} \lambda_j^t = \sum\nolimits_{j=1}^{n} \omega_{jc}^{\alpha,t} \lambda_j^{t+1} \ (c = 1, \ldots, q^{\alpha}, t = 1, \ldots, T-1)
\tag{8.3}
$$

where the symbol α stands for 'free', 'fix', 'good' or 'bad' as explained below. $\omega_{jc}^{\alpha,t}$ is the observed carry-over c in the case α for DMU_j from period t ($\forall c, \forall j, \forall t$). This constraint is critical for the dynamic model, since it connects activities in period t and period $t + 1$.

8.3.2.1 The 'Free' Carry-over Case This corresponds to discretionary carry-overs that the DMU can handle freely while keeping the continuity between periods t and $t + 1$ described by (8.3). The values of the carry-overs can be increased or decreased from the observed values. The deviation from the current value is measured as a slack in (8.4):

$$
\omega_{oc}^{\text{free},t} = \sum\nolimits_{j=1}^{n} \omega_{jc}^{\text{free},t} \lambda_j^t + s_c^{\text{free},t} \ (c = 1, \ldots, q^{\text{free}}, \ \forall t)
\tag{8.4}
$$

where $s_c^{\text{free},t}$ is a slack of the free carry-over c in period t ($\forall c, \forall t$, free in sign), which is not directly reflected in the efficiency evaluation presented in Section 8.4, but the continuity condition between two periods in (8.3) exerts an indirect effect on the efficiency score. This slack can be directly incorporated into efficiency scores using mixed-integer programming (MIP) [1].

The initial condition in the free carry-over case is

$$\omega_{oc}^{\text{free},0} = \sum_{j=1}^{n} \omega_{jc}^{\text{free},0} \lambda_j^1 + s_c^{\text{free},0} \left(c = 1, \dots, q^{\text{free}} \right) \tag{8.5}$$

where $s_c^{\text{free},0}$ is free in sign.

8.3.2.2 The 'Fixed' Carry-over Case This indicates non-discretionary carry-overs that are beyond the control of the DMU. Their value is fixed at the observed level. Similarly to free carry-overs, fixed carry-overs affect the efficiency score indirectly through the continuity condition between two periods:

$$\omega_{oc}^{\text{fix},t} = \sum_{j=1}^{n} \omega_{jc}^{\text{fix},t} \lambda_j^t \left(c = 1, \dots, q^{\text{fix}}, t = 1, \dots, T \right)$$

$$\omega_{oc}^{\text{fix},t} = \sum_{j=1}^{n} \omega_{jc}^{\text{fix},t} \lambda_j^{t+1} \left(c = 1, \dots, q^{\text{fix}}, t = 1, \dots, T-1 \right) \tag{8.6}$$

The initial condition in the fixed carry-over case is

$$\omega_{oc}^{\text{fix},0} = \sum_{j=1}^{n} \omega_{jc}^{\text{fix},0} \lambda_j^1 \left(c = 1, \dots, q^{\text{fix}} \right) \tag{8.7}$$

8.3.2.3 The 'Good' Carry-over Case This indicates desirable carry-overs, for example profit carried forward and net earned surplus carried to the next period. In our model, desirable carry-overs are treated as outputs and their value is restricted to be not less than the observed value. A comparative shortage of carry-overs in this category is accounted for as inefficiency:

$$\omega_{oc}^{\text{good},t} = \sum_{j=1}^{n} \omega_{jc}^{\text{good},t} \lambda_j^t - s_c^{\text{good},t+} \left(c = 1, \dots, q^{\text{good}}, \ \forall t \right) \tag{8.8}$$

where $s_c^{\text{good},t+}$ is a slack (non-negative), which indicates a shortfall in a good carry-over c in period t ($\forall c$, $\forall t$).

The initial condition in the good carry-over case is

$$\omega_{oc}^{\text{good},0} \leq \sum_{j=1}^{n} \omega_{jc}^{\text{good},0} \lambda_j^1 \left(c = 1, \dots, q^{\text{good}} \right) \tag{8.9}$$

8.3.2.4 The 'Bad' Carry-over Case This indicates undesirable carry-overs, for example losses carried forward, bad debts and dead stock. In our model, undesirable carry-overs are treated as inputs and their value is restricted to be not greater than the observed value. A comparative excess in carry-overs in this category is accounted for as inefficiency:

$$\omega_{oc}^{\text{bad},t} = \sum_{j=1}^{n} \omega_{jc}^{\text{bad},t} \lambda_j^t + s_c^{\text{bad},t-} \left(c = 1, \ldots, q^{\text{bad}}, \forall t \right) \qquad (8.10)$$

where $s_c^{\text{bad},t-}$ is a slack (non-negative), which indicates an excess in a bad carry-over c in period t ($\forall c$, $\forall t$).

The initial condition in the bad carry-over case is

$$\omega_{oc}^{\text{bad},0} \geq \sum_{j=1}^{n} \omega_{jc}^{\text{bad},0} \lambda_j^1 \left(c = 1, \ldots, q^{\text{bad}} \right) \qquad (8.11)$$

8.4 OBJECTIVE FUNCTIONS AND EFFICIENCIES

In this study, the non-radial SBM model is employed in order to evaluate the overall efficiency of DMU$_o$ ($o = 1, \ldots, n$). We present the objective function for the following three orientations for each case subject to (8.2)–(8.11). Let an optimal solution be $\left(\{\lambda_o^{t*}\}, \{s_{io}^{t-*}\}, \{s_{io}^{t+*}\}, \{s_{oc}^{\text{free},t*}\}, \{s_{oc}^{\text{good},t+*}\}, \{s_{oc}^{\text{bad},t-*}\} \right)$.

8.4.1 Input-Oriented Case

The input-oriented overall efficiency θ_o^* is defined by

$$\theta_o^* = \min \sum_{t=1}^{T} W^t \left[1 - \frac{1}{m + q^{\text{bad}}} \left(\sum_{i=1}^{m} \frac{s_i^{t-}}{x_{io}^t} + \sum_{c=1}^{q^{\text{bad}}} \frac{s_c^{\text{bad},t-}}{\omega_{oc}^{\text{bad},t}} \right) \right] \qquad (8.12)$$

where W^t is the weight of period t, which is supplied exogenously according to the importance of the period, and the weights satisfy the following condition:

$$\sum_{t=1}^{T} W^t = 1 \qquad (8.13)$$

This objective function is based on the input-oriented SBM model and deals not only with excesses in input resources but also with undesirable (bad) carry-overs as the main targets of evaluation. Excesses in undesirable carry-overs are accounted for in the objective function in the same way as input excesses, because they have a similar nature to inputs, that is, a smaller amount is favourable. However, undesirable carry-overs are not inputs. They play the role of connections between two consecutive periods, as demonstrated by the constraint (8.3).

Each period in the expression in square brackets in (8.12) expresses the efficiency for the period t as measured by the relative slacks of the inputs and carry-overs, and this efficiency is equal to unity if all slacks are zero. The efficiency is units-invariant and its value is between 0 and 1. Hence, (8.12) is the weighted average of the period efficiencies over the whole set of periods, which we call the overall efficiency and which is also between 0 and 1.

We define the period efficiency θ_{ot}^* by

$$\theta_o^{t*} = 1 - \frac{1}{m + q^{\mathrm{bad}}} \left(\sum_{i=1}^{m} \frac{s_{io}^{t-*}}{x_{io}^t} + \sum_{c=1}^{q^{\mathrm{bad}}} \frac{s_{oc}^{\mathrm{bad},t-*}}{\omega_{oc}^{\mathrm{bad},t}} \right) \quad (\forall t) \tag{8.14}$$

This period efficiency expresses the input-oriented efficiency score for the period t. The overall efficiency for all the periods (θ_o^*) is the weighted average of the period efficiencies θ_o^{t*}, as demonstrated below:

$$\theta_o^* = \sum_{t=1}^{T} W^t \theta_o^{t*} \tag{8.15}$$

8.4.2   Output-Oriented Case

The output-oriented overall efficiency η_o^* is defined by

$$\frac{1}{\eta_o^*} = \max \sum_{t=1}^{T} W^t \left[1 + \frac{1}{r + q^{\mathrm{good}}} \left(\sum_{i=1}^{r} \frac{s_i^{t+}}{y_{io}^t} + \sum_{c=1}^{q^{\mathrm{good}}} \frac{s_c^{\mathrm{good},t+}}{\omega_{oc}^{\mathrm{good},t}} \right) \right]. \tag{8.16}$$

This objective function is an extension of the output-oriented SBM model and deals with shortfalls in output products and desirable (good) carry-overs as the main targets of evaluation. Shortfalls in desirable carry-overs are accounted for in the objective function in the same way as output shortfalls, because they have a similar nature to outputs, that is, a larger amount is favourable. However, desirable links are not outputs. They play the role of connections between two consecutive periods, as demonstrated by (8.3).

Each period in the expression in square brackets in (8.16) relates to the efficiency for the period t as measured by the relative slacks of the outputs and carry-overs, and this efficiency is equal to unity if all slacks are zero. The efficiency is units-invariant and its value is greater than or equal to 1. Hence, the right-hand side of (8.16) is the weighted average over the whole period, which is greater than or equal to 1. Since we define the overall efficiency by the reciprocal of this quantity, the output overall efficiency is between 0 and 1.

Using an optimal solution to (8.16), we define the output-oriented period efficiency η_o^{t*} by

$$\eta_o^{t*} = \frac{1}{1 + \dfrac{1}{r + q^{\mathrm{good}}} \left(\displaystyle\sum_{i=1}^{r} \frac{s_{io}^{t+*}}{y_{io}^t} + \sum_{c=1}^{q^{\mathrm{good}}} \frac{s_{oc}^{\mathrm{good},t+*}}{\omega_{oc}^{\mathrm{good},t}} \right)} \quad (\forall t) \tag{8.17}$$

The output-oriented overall efficiency for all the periods (η_o^*) is the weighted harmonic mean of the period efficiencies η_o^{t*}, as demonstrated below:

$$\frac{1}{\eta_o^*} = \sum_{t=1}^{T} \frac{W^t}{\eta_o^{t*}} \tag{8.18}$$

8.4.3 Non-Oriented Case

We define the non-oriented efficiency measure as a combination of the input- and output-oriented cases, by solving the program below:

$$\rho_o^* = \min \frac{\sum_{t=1}^{T} W^t \left[1 - \frac{1}{m + q^{\text{bad}}} \left(\sum_{i=1}^{m} \frac{s_i^{t-}}{x_{io}^t} + \sum_{c=1}^{q^{\text{bad}}} \frac{s_c^{\text{bad},t-}}{\omega_{oc}^{\text{bad},t}} \right) \right]}{\sum_{t=1}^{T} W^t \left[1 + \frac{1}{r + q^{\text{good}}} \left(\sum_{i=1}^{r} \frac{s_i^{t+}}{y_{io}^t} + \sum_{c=1}^{q^{\text{good}}} \frac{s_c^{\text{good},t+}}{\omega_{oc}^{\text{good},t}} \right) \right]} \tag{8.19}$$

subject to (8.2)–(8.11).

This objective function is an extension of the non-oriented SBM model and deals with excesses in both input resources and undesirable (bad) carry-overs, and with shortfalls in both output products and desirable (good) carry-overs in a single unified scheme. The numerator is the average input efficiency and the denominator is the inverse of the average output efficiency. We define the non-oriented overall efficiency as their ratio, which ranges between 0 and 1, and attains a value of 1 when all slacks are zero. This objective function is also units-invariant.

Using an optimal solution $\left(\{\lambda_o^{t*}\}, \{s_{io}^{t-*}\}, \{s_{io}^{t+*}\}, \{s_{oc}^{\text{free},t*}\}, \{s_{oc}^{\text{good},t+*}\}, \{s_{oc}^{\text{bad},t-*}\} \right)$ to (8.19), we define the non-oriented period efficiency as follows:

$$\rho_o^{t*} = \frac{1 - \frac{1}{m + q^{\text{bad}}} \left(\sum_{i=1}^{m} \frac{s_{io}^{t-*}}{x_{io}^t} + \sum_{c=1}^{q^{\text{bad}}} \frac{s_{oc}^{\text{bad},t-*}}{\omega_{oc}^{\text{bad},t}} \right)}{1 + \frac{1}{r + q^{\text{good}}} \left(\sum_{i=1}^{r} \frac{s_{io}^{t+*}}{y_{io}^t} + \sum_{c=1}^{q^{\text{good}}} \frac{s_{oc}^{\text{good},t+*}}{\omega_{oc}^{\text{good},t}} \right)} \quad (\forall t) \tag{8.20}$$

8.5 DYNAMIC MALMQUIST INDEX

The period efficiencies in the dynamic DEA model are measured relatively, based on the frontier in each period, and do not take the frontier shift during the study periods into account. Therefore, even if the period efficiencies of a certain DMU increase period by period, the absolute productivity of the DMU may not increase because of regress of the frontier. In order to capture the absolute productivity change in the dynamic DEA model, we can use the Malmquist productivity index.

The concept of the Malmquist productivity index is an index representing the total factor productivity (TFP) growth of a DMU, in that it reflects (i) progress or regress in efficiency along with (ii) progress or regress of the frontier technology. More details can be found in Chapter 7. In this section, we define a dynamic Malmquist index based on the period efficiency scores in the dynamic DEA model. We utilize period efficiency scores θ_o^{t*} measured in the input-oriented model, but the same procedure can be applied to output-oriented and non-oriented efficiency scores.

8.5.1   Dynamic Catch-up Index

We define the dynamic catch-up index (DCU) of DMU$_o$ as the ratio of the period efficiencies between t and $t + 1$, as follows:

$$\text{DCU}: \gamma_o^{t \to t+1} = \frac{\theta_o^{t+1*}}{\theta_o^{t*}} \quad (t = 1, \dots, T-1) \tag{8.21}$$

where $\gamma_o^{t \to t+1} > 1$, $\gamma_o^{t \to t+1} = 1$ and $\gamma_o^{t \to t+1} < 1$ indicate progress, the status quo and regress, respectively, in the dynamic catch-up effect.

8.5.2   Dynamic Frontier Shift Effect

We define the dynamic frontier shift effect (DFS) from t to $t + 1$ in the dynamic DEA model following the non-radial Malmquist model:

$$\text{DFS}: \phi_o^{t \to t+1} \quad (t = 1, \dots, T-1) \tag{8.22}$$

where $\phi_o^{t \to t+1} > 1$, $\phi_o^{t \to t+1} = 1$, and $\phi_o^{t \to t+1} < 1$ indicate progress, the status quo and regress, respectively, in the dynamic frontier shift effect.

8.5.3   Dynamic Malmquist Index

Using the above DCU and DFS, we define the dynamic Malmquist index (DMI) by their product as

$$\text{DMI}: \mu_o^{t \to t+1} = \gamma_o^{t \to t+1} \phi_o^{t \to t+1} \quad (t = 1, \dots, T-1) \tag{8.23}$$

which indicates the absolute productivity change of DMU$_o$ between two consecutive periods. We can compare DMIs among DMUs.

8.5.4   Dynamic Cumulative Malmquist Index

Although the above dynamic Malmquist index is defined on a two-period ($t \to t + 1$) base, we can obtain a dynamic cumulative Malmquist index based on the first period as

$$\text{DCMI}: \tilde{\mu}_o^{1 \to t} = \Pi_{\tau=1}^{t} \mu_o^{\tau \to \tau+1} (t = 1, \dots, T-1) \tag{8.24}$$

The DCMI also turns out to be given by

$$\widetilde{\mu}_o^{1\rightarrow t} = \mu_o^{1\rightarrow 2} \times \widetilde{\mu}_o^{2\rightarrow t} \tag{8.25}$$

Thus, the intertemporal productivity change between period 1 and period t is modified by the positions at $2,\ldots,T-1$.

Furthermore, the DCMI can be decomposed into a dynamic cumulative frontier shift (DCFS) and the ratio of the period efficiencies between period 1 and period t as

$$\widetilde{\mu}_o^{1\rightarrow t} = \Pi_{\tau=1}^t \left(\phi_o^{\tau\rightarrow\tau+1} \gamma_o^{\tau\rightarrow\tau+1} \right) = \widetilde{\phi}_o^{1\rightarrow t} \cdot \frac{\theta_o^{t*}}{\theta_o^{1*}} \tag{8.26}$$

where $\widetilde{\phi}_o^{1\rightarrow t}$ indicates the DCFS to period t from the base period.

8.5.5 Dynamic Adjusted Malmquist Index

The dynamic adjusted Malmquist index (DAMI), which can capture both the relative efficiency among DMUs in the base period and the productivity change from the base period to period t, can be obtained as

$$\text{DAMI}: \widetilde{\xi}_o^{1\rightarrow t} = \widetilde{\mu}_o^{1\rightarrow t} \cdot \theta_o^{1*} = \widetilde{\phi}_o^{1\rightarrow t} \cdot \theta_o^{t*} \tag{8.27}$$

REFERENCES

[1] Tone, K. and Tsutsui, M. (2010) Dynamic DEA: A slacks-based measure approach. *Omega*, **38**, 145–156.

[2] Färe, R. and Grosskopf, S. (1996) *Intertemporal Production Frontiers: With Dynamic DEA, Kluwer Academic, Boston.*

9

THE DYNAMIC NETWORK DEA MODEL[1]

KAORU TONE

National Graduate Institute for Policy Studies, Tokyo, Japan

MIKI TSUTSUI

Central Research Institute of Electric Power Industry, Tokyo, Japan

9.1 INTRODUCTION

The dynamic model with network structure (DNSBM) is a composite of the network SBM (NSBM) and the dynamic SBM (DSBM). Vertically, we deal with multiple divisions connected by links of the network structure within each period and, horizontally, we combine the network structures for different periods by means of carry-over activities between two succeeding periods. See Figure 9.1 for an example.

This model can evaluate (i) the overall efficiency over the entire observed period, (ii) dynamic changes in the period efficiency and (iii) dynamic changes in divisional efficiency. The model can be implemented in input-oriented, output-oriented and non-oriented (both input-oriented and output-oriented) forms under a constant-returns-to-

[1] Part of the material in this chapter is adapted from *Omega*, Vol. 42, Tone K. and Tsutsui M. [1], Dynamic DEA with network structure: A slacks-based measure approach, 124–131, 2014, with permission from Elsevier Science.

Advances in DEA Theory and Applications: With Extensions to Forecasting Models,
First Edition. Edited by Kaoru Tone.

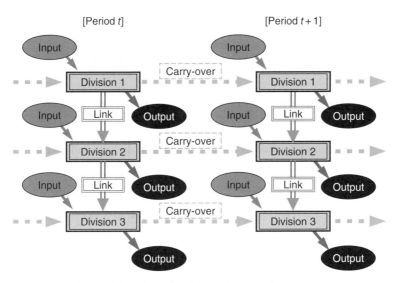

Figure 9.1 Example of dynamic network structure.

scale (CRS) or variable-returns-to-scale (VRS) assumption about the production possibility set. A Malmquist index can be developed.[2]

9.2 NOTATION AND PRODUCTION POSSIBILITY SET

9.2.1 Notation

We use the following notation to describe dynamic network DEA:

n: number of DMUs ($j = 1, \ldots, n$);

K: number of divisions ($k = 1, \ldots, K$);

T: number of periods ($t = 1, \ldots, T$);

m_k: number of inputs to division k ($i = 1, \ldots, m_k$);

r_k: number of outputs from division k ($i = 1, \ldots, r_k$) ;

p_{kh}: number of items in link from division k to division h ($l = 1, \ldots, p_{kh}$);

(k, h): from division k to division h;

q_k: number of items in carry-over for division k ($c = 1, \ldots, q_k$);

$(t, t + 1)$: carry-overs from period t to period $t + 1$;

S: set of divisions which have no incoming links, that is, starting divisions;

T: set of divisions which have no outgoing links, that is, terminal divisions.

---

[2] Software for dynamic network DEA models is included in DEA-Solver Pro V13 (http://www.saitech-inc. com). See also Appendix A.

The observed data are as follows:

$\left\{ x_{ijk}^t \in R_+ \right\}$: input resource i to division k of DMU$_j$ in period t $(i = 1, \ldots, m_k, \forall k,$
$\forall j, \forall t)$;

$\left\{ y_{ijk}^t \in R_+ \right\}$: output product i from division k of DMU$_j$ in period t $(i = 1, \ldots, r_k, \forall k,$
$\forall j, \forall t)$;

$\left\{ z_{j(k,h)l}^{\alpha,t} \in R_+ \right\}$: linking products l from division k to division h of DMU$_j$ in period t
$(\forall l, \forall (kh), \forall j, \forall t, \alpha = 4$ types of links as explained in Section 9.3.2);

$\left\{ \omega_{jkc}^{\alpha,t} \in R_+ \right\}$: carry-over c from period t to period $t + 1$ for division k of DMU$_j$ $(\forall c,$
$\forall k, \forall j, \forall t, \alpha = 4$ types of carry-overs as explained in Section 9.3.3).

We assume that

$$
\begin{aligned}
z_{j(k,h)l}^{\alpha,t} &= 0 (\forall j, \forall t, \forall l, \forall \alpha, h \in S) : \text{no linking input } \textit{to starting divisions} \\
z_{j(k,h)l}^{\alpha,t} &= 0 (\forall j, \forall t, \forall l, \forall \alpha, k \in T) : \text{no linking output from terminal divisions}
\end{aligned}
\tag{9.1}
$$

We postulate that we have homogeneous panel data throughout periods 1 to T. So, we look at the enterprises concerned as a continuum between period 1 and period T. In addition, we take the initial inputs for carry-overs in period 1 into account as follows:

$\omega_{jkc}^{\alpha,0} (\forall c, \forall k, \forall j, \forall \alpha)$: initial external input in period 1 for carry-over for division k.

The production possibility set $P = \left\{ \left(\mathbf{x}_k^t, \mathbf{y}_k^t, \mathbf{z}_{(kh)}^t, \boldsymbol{\omega}_k^t, \boldsymbol{\omega}_k^0 \right) \right\}$ is defined by

$$
\begin{aligned}
&\mathbf{x}_k^t \geq \sum_{j=1}^n \mathbf{x}_{jk}^t \lambda_{jk}^t (\forall k, \forall t) \\
&\mathbf{y}_k^t \leq \sum_{j=1}^n \mathbf{y}_{jk}^t \lambda_{jk}^t (\forall k, \forall t) \\
&\mathbf{z}_{(kh)}^t \leq \sum_{j=1}^n \mathbf{z}_{j(kh)}^t \lambda_{jk}^t (\forall (k,h), \forall t) (\text{links as outputs from division } k \text{ in term } t) \\
&\mathbf{z}_{(kh)}^t \geq \sum_{j=1}^n \mathbf{z}_{j(kh)}^t \lambda_{jh}^t (\forall (k,h), \forall t) (\text{links as inputs to division } h \text{ in term } t) \\
&\boldsymbol{\omega}_k^t \leq \sum_{j=1}^n \boldsymbol{\omega}_{jk}^t \lambda_{jk}^t (\forall k, \forall t) (\text{carry} - \text{overs as outputs for division } k \text{ from term } t) \\
&\boldsymbol{\omega}_k^t \geq \sum_{j=1}^n \boldsymbol{\omega}_{jk}^t \lambda_{jk}^{t+1} (\forall k, t = 1, \ldots, T-1) (\text{carry} - \text{overs as input for division } k \text{ to term } t+1) \\
&\boldsymbol{\omega}_k^0 \geq \sum_{j=1}^n \boldsymbol{\omega}_{jk}^0 \lambda_{jk}^1 (\text{initial external inputs for carry-overs for division } k \text{ into term 1}) \\
&\sum_{j=1}^n \lambda_{jk}^t = 1 (\forall k, \forall t) \\
&\lambda_{jk}^t \geq 0 (\forall k, \forall j, \forall t)
\end{aligned}
\tag{9.2}
$$

where $\lambda_{jk}^t \in R_+$ is the intensity variable corresponding to division k of DMU$_j$ for period t ($\forall k$, $\forall j$, $\forall t$).

We notice that the above model makes the VRS assumption about production. That is, the production frontiers are spanned by the convex hull of the existing DMUs. However, if we neglect the constraint $\sum_{j=1}^{n} \lambda_{jk}^t = 1$, we can deal with the CRS case as well.

9.3 DESCRIPTION OF DYNAMIC NETWORK STRUCTURE

9.3.1 Inputs and Outputs

DMU$_o$ $(o = 1, \ldots, n) \in P$ can be expressed as follows:

$$x_{iok}^t = \sum_{j=1}^{n} x_{ijk}^t \lambda_{jk}^t + s_{ik}^{t-} \ (\forall i, \forall k, \forall t)$$

$$y_{iok}^t = \sum_{j=1}^{n} y_{ijk}^t \lambda_{jk}^t - s_{ik}^{t+} \ (\forall i, \forall k, \forall t)$$

$$\sum_{j=1}^{n} \lambda_{jk}^t = 1 \ (\forall k, \forall t) \tag{9.3}$$

$$\lambda_{jk}^t \ge 0(\forall j, \forall k, \forall t), \ s_{ik}^{t-} \ge 0(\forall i, \forall k, \forall t), \ s_{ik}^{t+} \ge 0(\forall i, \forall k, \forall t)$$

where s_{ik}^{t-} and s_{ik}^{t+} indicate input and output slacks (non-negative), respectively.

9.3.2 Links

As regards the linking constraints, we have several options, of which we present four possible cases.

In all cases, we assume the following continuity condition for the links (k, h), which is critical for a network model connecting activities in two divisions:

$$\sum_{j=1}^{n} z_{j(k,h)l}^{\alpha,t} \lambda_{jk}^t = \sum_{j=1}^{n} z_{j(k,h)l}^{\alpha,t} \lambda_{jh}^t \left(l = 1, \ldots, p_{(k,h)}^\alpha, \ \forall(k,h), \ \forall t \right) \tag{9.4}$$

where α stands for 'free', 'fix', 'out' or 'in' as explained below. $z_{j(k,h)l}^{\alpha,t}$ is the observed link l from division k to division h of DMU$_j$ in period t ($\forall l$, $\forall(k, h)$, $\forall j$, $\forall t$) in the case α.

9.3.2.1 The Discretionary 'Free' Link Value Case The linking activities are freely determined (discretionary) while keeping continuity between division k and division h as formulated in (9.4). This case can be used to see whether or not the current link flow has an appropriate volume in the light of other DMUs; that is, the link flow may increase or decrease in the optimal solutions of the linear programs which

we will introduce in Section 9.4. We have the following relationship between the current link value and the free link value:

$$z_{o(k,h)l}^{\text{free},t} = \sum_{j=1}^{n} z_{j(k,h)l}^{\text{free},t} \lambda_{jk}^{t} + s_{(k,h)l}^{\text{free},t} \left(l = 1, \ldots, p_{(k,h)}^{\text{free}}, \quad \forall (k,h), \quad \forall t \right) \tag{9.5}$$

where $s_{(k,h)l}^{\text{free},t}$ is a free link slack and is free in sign.

9.3.2.2 The Non-discretionary 'Fixed' Link Value Case
The linking activities are kept unchanged (non-discretionary):

$$\begin{aligned} z_{o(k,h)l}^{\text{fix},t} &= \sum_{j=1}^{n} z_{j(k,h)l}^{\text{fix},t} \lambda_{jk}^{t} \left(l = 1, \ldots, p_{(k,h)}^{\text{fix}}, \quad \forall (k,h), \quad \forall t \right) \\ z_{o(k,h)l}^{\text{fix},t} &= \sum_{j=1}^{n} z_{j(k,h)l}^{\text{fix},t} \lambda_{jh}^{t} \left(l = 1, \ldots, p_{(k,h)}^{\text{fix}}, \quad \forall (k,h), \quad \forall t \right) \end{aligned} \tag{9.6}$$

This case corresponds to the situation where the intermediate products are beyond the control of the DMUs or the discretion of the management.

9.3.2.3 The 'As Output' Link Value Case
The linking activities for which a larger amount is regarded as favourable are treated 'as output' from the preceding division, and shortages are accounted for in the output inefficiency:

$$z_{o(k,h)l}^{\text{out},t} = \sum_{j=1}^{n} z_{j(k,h)l}^{\text{out},t} \lambda_{jk}^{t} - s_{(k,h)l}^{\text{out},t+} \left(l = 1, \ldots, p_{(k,h)}^{\text{out}}, \quad \forall (k,h), \quad \forall t \right) \tag{9.7}$$

where $s_{(k,h)l}^{\text{out},t+}$ is an 'as output' link slack (non-negative).

9.3.2.4 The 'As Input' Link Value Case
The linking activities for which a smaller amount is regarded as favourable are treated 'as input' to the succeeding division, and excesses are accounted for in the input inefficiency:

$$z_{o(k,h)l}^{\text{in},t} = \sum_{j=1}^{n} z_{j(k,h)l}^{\text{in},t} \lambda_{jk}^{t} + s_{(k,h)l}^{\text{in},t-} \left(l = 1, \ldots, p_{(k,h)}^{\text{in}}, \quad \forall (k,h), \quad \forall t \right) \tag{9.8}$$

where $s_{(k,h)l}^{\text{in},t-}$ is an 'as input' link slack (non-negative).

9.3.3 Carry-Overs

We classify carry-over activities into four categories of 'free', 'fixed', 'good' and 'bad'. In all cases, the continuity of carry-overs between periods t and $t+1$ can be guaranteed by the following condition:

$$\sum_{j=1}^{n} \omega_{jkc}^{\alpha,t} \lambda_{jk}^{t} = \sum_{j=1}^{n} \omega_{jkc}^{\alpha,t} \lambda_{jk}^{t+1} \left(c = 1, \ldots, q_{k}^{\alpha}, \quad \forall k, \quad t = 1, \ldots, T-1 \right) \tag{9.9}$$

where the symbol α stands for 'free', 'fix', 'good' or 'bad' as explained below. $\omega_{jkc}^{\alpha,t}$ is the observed carry-over c in the case α for division k of DMU_j from period t ($\forall c$, $\forall k$, $\forall j$, $\forall t$,). This constraint is critical for the dynamic model, since it connects activities in period t and period $t+1$.

9.3.3.1 The Discretionary 'Free' Carry-over Case

This corresponds to carry-overs that the DMU can handle freely while keeping the continuity between period t and $t+1$ described by (9.9). The values of the carry-overs can be increased or decreased from the observed values. The deviation from the current value is not directly reflected in the efficiency evaluation, but the continuity condition between two periods presented below exerts an indirect effect on the efficiency score:

$$\omega_{okc}^{\text{free},t} = \sum_{j=1}^{n} \omega_{jkc}^{\text{free},t} \lambda_{jk}^{t} + s_{kc}^{\text{free},t} \left(c = 1,\ldots,q_{k}^{\text{free}}, \ \forall k, \ \forall t \right) \tag{9.10}$$

where $s_{kc}^{\text{free},t}$ is a free link slack (free in sign), which is not directly reflected in the efficiency evaluation presented in Section 9.4, but the continuity condition between two periods in (9.9) exerts an indirect effect on the efficiency score. This slack can be directly incorporated into efficiency scores using mixed-integer programming (MIP) [2].

The initial condition in the free carry-over case is

$$\omega_{okc}^{\text{free},0} = \sum_{j=1}^{n} \omega_{jkc}^{\text{free},0} \lambda_{jk}^{1} + s_{kc}^{\text{free},0} \left(c = 1,\ldots,q_{k}^{\text{free}}, \ \forall k \right) \tag{9.11}$$

where $s_{kc}^{\text{free},0}$ is free in sign.

9.3.3.2 The Non-discretionary 'Fixed' Carry-over Case

This indicates carry-overs that are beyond the control of the DMU. Their value is fixed at the observed level. Similarly to the free carry-overs, fixed carry-overs affect the efficiency score indirectly through the continuity condition between two periods.

$$\omega_{okc}^{\text{fix},t} = \sum_{j=1}^{n} \omega_{jkc}^{\text{fix},t} \lambda_{jk}^{t} \left(c = 1,\ldots,q_{k}^{\text{fix}}, \ \forall k, \ t = 1,\ldots,T \right)$$

$$\omega_{okc}^{\text{fix},t} = \sum_{j=1}^{n} \omega_{jkc}^{\text{fix},t} \lambda_{jk}^{t+1} \left(c = 1,\ldots,q_{k}^{\text{fix}}, \ \forall k, \ t = 1,\ldots,T-1 \right) \tag{9.12}$$

The initial condition in the fixed carry-over case is

$$\omega_{okc}^{\text{fix},0} = \sum_{j=1}^{n} \omega_{jkc}^{\text{fix},0} \lambda_{jk}^{1} \left(c = 1,\ldots,q_{k}^{\text{fix}}, \ \forall k \right) \tag{9.13}$$

9.3.3.3 The Desirable 'Good' Carry-over Case

This indicates desirable carry-overs, for example profit carried forward and net earned surplus carried to the next period. In our model, desirable carry-overs are treated as outputs and their value is restricted to be not less than the observed value. A comparative shortage of carry-overs in this category is accounted for as inefficiency:

$$\omega_{okc}^{\text{good},t} = \sum_{j=1}^{n} \omega_{jkc}^{\text{good},t} \lambda_{jk}^{t} - s_{kc}^{\text{good},t+} \left(c = 1, \ldots, q_{k}^{\text{good}}, \ \forall k, \ \forall t \right) \qquad (9.14)$$

where $s_{kc}^{\text{good},t+}$ is a slack (non-negative), which indicates a shortfall in a good carry-over.

The initial condition in the good carry-over case is

$$\omega_{okc}^{\text{good},0} \leq \sum_{j=1}^{n} \omega_{jkc}^{\text{good},0} \lambda_{jk}^{1} \left(c = 1, \ldots, q_{k}^{\text{good}}, \ \forall k \right) \qquad (9.15)$$

9.3.3.4   The Undesirable 'Bad' Carry-over Case   This indicates undesirable carry-overs, for example losses carried forward, bad debts and dead stock. In our model, undesirable carry-overs are treated as inputs and their values are restricted to be not greater than the observed value. A comparative excess in carry-overs in this category is accounted for as inefficiency:

$$\omega_{okc}^{\text{bad},t} = \sum_{j=1}^{n} \omega_{jkc}^{\text{bad},t} \lambda_{jk}^{t} + s_{kc}^{\text{bad},t-} \left(c = 1, \ldots, q_{k}^{\text{bad}}, \ \forall k, \ \forall t \right) \qquad (9.16)$$

where $s_{kc}^{\text{bad},t-}$ is a slack (non-negative), which indicates an excess in a bad carry-over.

The initial condition in the bad carry-over case is

$$\omega_{okc}^{\text{bad},0} \geq \sum_{j=1}^{n} \omega_{jkc}^{\text{bad},0} \lambda_{jk}^{1} \left(c = 1, \ldots, q_{k}^{\text{bad}}, \forall k \right) \qquad (9.17)$$

9.4   OBJECTIVE FUNCTION AND EFFICIENCIES

This section deals with the overall, period and divisional efficiencies in the case of the non-oriented (i.e. both input- and output-oriented) model.

9.4.1   Overall Efficiency

The overall efficiency is evaluated by the following program:

$$\rho_{o}^{*} = \min \frac{\sum_{t=1}^{T} W^{t} \left[\sum_{k=1}^{K} w_{k} \left[1 - \frac{1}{m_{k} + p_{(k,h)}^{\text{in}} + q_{k}^{\text{bad}}} \left(\sum_{i=1}^{m_{k}} \frac{s_{ik}^{t-}}{x_{iok}^{t}} + \sum_{l=1}^{p_{(k,h)}^{\text{in}}} \frac{s_{(k,h)l}^{\text{in},t-}}{z_{o(k,h)l}^{\text{in},t}} + \sum_{c=1}^{q_{k}^{\text{bad}}} \frac{s_{kc}^{\text{bad},t-}}{\omega_{okc}^{\text{bad},t}} \right) \right] \right]}{\sum_{t=1}^{T} W^{t} \left[\sum_{k=1}^{K} w_{k} \left[1 + \frac{1}{r_{k} + p_{(k,h)}^{\text{out}} + q_{k}^{\text{good}}} \left(\sum_{i=1}^{r_{k}} \frac{s_{ik}^{t+}}{y_{iok}^{t}} + \sum_{l=1}^{p_{(k,h)}^{\text{out}}} \frac{s_{(k,h)l}^{\text{out},t+}}{z_{o(k,h)l}^{\text{out},t}} + \sum_{c=1}^{q_{k}^{\text{good}}} \frac{s_{kc}^{\text{good},t+}}{\omega_{okc}^{\text{good},t}} \right) \right] \right]}$$

$$(9.18)$$

with $\sum_{t=1}^{T} W^t = 1$, $\sum_{k=1}^{K} w_k = 1$, $W^t \geq 0$ $(\forall t)$ and $w_k \geq 0$ $(\forall k)$, and subject to (9.3)–(9.17), where W^t $(\forall t)$ is the weight of period t and w_k $(\forall k)$ is the weight of division k. These weights are supplied exogenously. The input- and output-oriented models can be defined by considering only the numerator and the denominator, respectively, of the above objective function.

Let an optimal solution be

$$\left(\{\lambda_{ok}^{t*}\}, \{s_{iok}^{t-*}\}, \{s_{iok}^{t+*}\}, \left\{s_{o(k,h)l}^{\text{free},t*}\right\}, \left\{s_{o(k,h)l}^{\text{out},t+*}\right\}, \left\{s_{o(k,h)l}^{\text{in},t-*}\right\}, \left\{s_{okc}^{\text{free},t*}\right\}, \left\{s_{okc}^{\text{good},t+*}\right\}, \left\{s_{okc}^{\text{bad},t-*}\right\} \right).$$

9.4.2 Period and Divisional Efficiencies

The period efficiency is defined by

$$\pi_o^{t*} = \frac{\sum_{k=1}^{K} w_k \left[1 - \dfrac{1}{m_k + p_{(k,h)}^{\text{in}} + q_k^{\text{bad}}} \left(\sum_{i=1}^{m_k} \dfrac{s_{iok}^{t-*}}{x_{iok}^t} + \sum_{l=1}^{p_{(k,h)}^{\text{in}}} \dfrac{s_{o(k,h)l}^{\text{in},t-*}}{z_{o(k,h)l}^{\text{in},t}} + \sum_{c=1}^{q_k^{\text{bad}}} \dfrac{s_{okc}^{\text{bad},t-*}}{\omega_{okc}^{\text{bad},t}} \right) \right]}{\sum_{k=1}^{K} w_k \left[1 + \dfrac{1}{r_k + p_{(k,h)}^{\text{out}} + q_k^{\text{good}}} \left(\sum_{i=1}^{r_k} \dfrac{s_{iok}^{t+*}}{y_{iok}^t} + \sum_{l=1}^{p_{(k,h)}^{\text{out}}} \dfrac{s_{o(k,h)l}^{\text{out},t+*}}{z_{o(k,h)l}^{\text{out},t}} + \sum_{c=1}^{q_k^{\text{good}}} \dfrac{s_{okc}^{\text{good},t+*}}{\omega_{okc}^{\text{good},t}} \right) \right]} \quad (\forall t) \tag{9.19}$$

where the variables on the right-hand side indicate optimal values for the overall efficiency ρ_o^*.

The divisional efficiency is defined by

$$\delta_{ok}^* = \frac{\sum_{t=1}^{T} W^t \left[1 - \dfrac{1}{m_k + p_{(k,h)}^{\text{in}} + q_k^{\text{bad}}} \left(\sum_{i=1}^{m_k} \dfrac{s_{iok}^{t-*}}{x_{iok}^t} + \sum_{l=1}^{p_{(k,h)}^{\text{in}}} \dfrac{s_{o(k,h)l}^{\text{in},t-*}}{z_{o(k,h)l}^{\text{in},t}} + \sum_{c=1}^{q_k^{\text{bad}}} \dfrac{s_{okc}^{\text{bad},t-*}}{\omega_{okc}^{\text{bad},t}} \right) \right]}{\sum_{t=1}^{T} W^t \left[1 + \dfrac{1}{r_k + p_{(k,h)}^{\text{out}} + q_k^{\text{good}}} \left(\sum_{i=1}^{r_k} \dfrac{s_{iok}^{t+*}}{y_{iok}^t} + \sum_{l=1}^{p_{(k,h)}^{\text{out}}} \dfrac{s_{o(k,h)l}^{\text{out},t+*}}{z_{o(k,h)l}^{\text{out},t}} + \sum_{c=1}^{q_k^{\text{good}}} \dfrac{s_{okc}^{\text{good},t+*}}{\omega_{okc}^{\text{good},t}} \right) \right]} \quad (\forall k) \tag{9.20}$$

Finally, the period-divisional efficiency is defined by

$$\rho_{ok}^{t*} = \frac{1 - \dfrac{1}{m_k + p_{(k,h)}^{\text{in}} + q_k^{\text{bad}}} \left(\sum_{i=1}^{m_k} \dfrac{s_{iok}^{t-*}}{x_{iok}^t} + \sum_{l=1}^{p_{(k,h)}^{\text{in}}} \dfrac{s_{o(k,h)l}^{\text{in},t-*}}{z_{o(k,h)l}^{\text{in},t}} + \sum_{c=1}^{q_k^{\text{bad}}} \dfrac{s_{okc}^{\text{bad},t-*}}{\omega_{okc}^{\text{bad},t}} \right)}{1 + \dfrac{1}{r_k + p_{(k,h)}^{\text{out}} + q_k^{\text{good}}} \left(\sum_{i=1}^{r_k} \dfrac{s_{iok}^{t+*}}{y_{iok}^t} + \sum_{l=1}^{p_{(k,h)}^{\text{out}}} \dfrac{s_{o(k,h)l}^{\text{out},t+*}}{z_{o(k,h)l}^{\text{out},t}} + \sum_{c=1}^{q_k^{\text{good}}} \dfrac{s_{okc}^{\text{good},t+*}}{\omega_{okc}^{\text{good},t}} \right)} \quad (\forall k, \forall t) \tag{9.21}$$

In the input- and output-oriented models, the numerator and the denominator, respectively, of the above formulas are applied. We notice that, although the overall efficiency is uniquely determined, the period, divisional and period-divisional efficiencies are not necessarily unique. Furthermore, in the input-oriented model, the overall efficiency is the weighted arithmetic mean of the period efficiencies and, in the output-oriented model, the overall efficiency is the weighted harmonic mean of the period efficiencies, whereas in the non-oriented model the overall efficiency is neither the arithmetic nor the harmonic mean of the period efficiencies.

9.5 DYNAMIC DIVISIONAL MALMQUIST INDEX

The period-divisional efficiencies in the dynamic network DEA model are measured relatively based on the frontier in each period for each division, and do not take the frontier shift during the study periods into account. Therefore, even if the period-divisional efficiency of division k of DMU_o increases period by period, the absolute productivity of the DMU may not increase because of regress of the frontier for division k. In order to capture the absolute productivity change of DMUs in the dynamic network DEA model, we can use the Malmquist index. In this section, we define a Malmquist index based on the period-divisional efficiency score as follows.

9.5.1 Dynamic Divisional Catch-up Index

We define the dynamic divisional catch-up index (DDCU) of DMU_o as the ratio of the period-divisional efficiencies between t and $t + 1$ for division k, as follows:

$$\text{DDCU}: \gamma_{ok}^{t \to t+1} = \frac{\rho_{ok}^{t+1^*}}{\rho_{ok}^{t^*}} \quad (t = 1, \ldots, T-1, \forall k) \tag{9.22}$$

where $\gamma_{ok}^{t \to t+1} > 1$, $\gamma_{ok}^{t \to t+1} = 1$ and $\gamma_{ok}^{t \to t+1} < 1$ indicate progress, the status quo and regress, respectively, in the catch-up effect.

9.5.2 Dynamic Divisional Frontier Shift Effect

We define the dynamic divisional frontier shift effect (DDFS) from t to $t + 1$ for division k following the non-radial Malmquist model described in Chapter 7:

$$\text{DDFS}: \phi_{ok}^{t \to t+1} \quad (t = 1, \ldots, T-1, \forall k) \tag{9.23}$$

If a division has no inputs or no outputs, its DDFS is unity.

9.5.3 Dynamic Divisional Malmquist Index

Using the above DDCU and DDFS, we define the dynamic divisional Malmquist index (DDMI) for division k by their product as

$$\text{DDMI}: \mu_{ok}^{t \to t+1} = \gamma_{ok}^{t \to t+1} \phi_{ok}^{t \to t+1} \quad (t = 1, \ldots, T-1, \forall k) \tag{9.24}$$

9.5.4 Dynamic Divisional Cumulative Malmquist Index

Although the above DDMI is defined on a two-period ($t \rightarrow t+1$) base, we can obtain a dynamic divisional cumulative Malmquist index (DDCMI) for division k based on the first period as

$$\text{DDCMI}: \widetilde{\mu}_{ok}^{1 \rightarrow t} = \Pi_{\tau=1}^{t} \mu_{ok}^{\tau \rightarrow \tau+1} \ (t=1,...,T-1, \forall k) \tag{9.25}$$

The DDCMI also turns out to be given by

$$\widetilde{\mu}_{ok}^{1 \rightarrow t} = \mu_{ok}^{1 \rightarrow 2} \times \widetilde{\mu}_{ok}^{2 \rightarrow t} \tag{9.26}$$

Furthermore, the DDCMI can be decomposed into a dynamic divisional cumulative frontier shift (DDCFS) and the ratio of the period-divisional efficiencies between period 1 and period t for division k as

$$\widetilde{\mu}_{ok}^{1 \rightarrow t} = \Pi_{\tau=1}^{t} \left(\phi_{ok}^{\tau \rightarrow \tau+1} \gamma_{ok}^{\tau \rightarrow \tau+1} \right)$$

$$= \widetilde{\phi}_{ok}^{1 \rightarrow t} \cdot \frac{\rho_{ok}^{t*}}{\rho_{ok}^{1*}} \tag{9.27}$$

where $\widetilde{\phi}_{ok}^{1 \rightarrow t}$ indicates the DDCFS to period t from the base period for division k.

9.5.5 Dynamic Divisional Adjusted Malmquist Index

The dynamic divisional adjusted Malmquist Index (DDAMI), which can capture both the relative efficiency among DMUs in the base period and the productivity change from the base period to period t for division k, can be obtained as

$$\text{DDAMI}: \widetilde{\xi}_{ok}^{1 \rightarrow t} = \widetilde{\mu}_{ok}^{1 \rightarrow t} \cdot \rho_{ok}^{1*}$$

$$= \widetilde{\phi}_{ok}^{1 \rightarrow t} \cdot \rho_{ok}^{t*} \tag{9.28}$$

9.5.6 Overall Dynamic Malmquist Index

We can calculate the overall dynamic Malmquist index (ODMI) as the weighted geometric mean of the DDMIs:

$$\text{ODMI}: \mu_{o}^{t \rightarrow t+1} = \Pi_{k=1}^{K} \left(\mu_{ok}^{t \rightarrow t+1} \right)^{w_k} \tag{9.29}$$

where $w_k \geq 0$ is the weight of division k, with $\sum_{k=1}^{K} w_k = 1$.

The ODMI can be decomposed into a weighted geometric mean of the DDCU and DDFS. These quantities can be calculated mathematically; however, it should be noted that the weighted geometric mean of the DDFS does not indicate an 'overall frontier shift', because we do not assume an overall frontier throughout all divisions. The weighted geometric mean of the DDCU is also not an 'overall catch-up', since it is calculated with reference to each divisional frontier, not to the 'overall frontier'. Therefore, we should understand that the ODMI is a supplemental index that is only calculated mathematically.

REFERENCES

[1] Tone, K. and Tsutsui, M. (2014) Dynamic DEA with network structure: A slacks-based measure approach. *Omega: The International Journal of Management Science*, **42**, 124–131.
[2] Tone, K. and Tsutsui, M. (2010) Dynamic DEA: A slacks-based measure approach. *Omega: The International Journal of Management Science*, **38**, 145–156.

10

STOCHASTIC DEA: THE REGRESSION-BASED APPROACH

ANDREW L. JOHNSON

Department of Industrial and Systems Engineering, Texas A&M University, College Station, TX, USA

10.1 INTRODUCTION

The papers of Charnes *et al.* [1] and Banker *et al.* [2] are considered the two seminal papers that established data envelopment analysis (DEA). Since the development of the DEA method there have been multiple attempts to generalize DEA to the stochastic setting. This chapter will briefly review the key developments in the field, focusing on the assumptions or postulates and data requirements. I will emphasize the regression-based approaches to stochastic DEA.

The original DEA models are deterministic and require strong assumptions in order for the efficiency measures to consistently estimate efficiency. These assumptions include the requirement that the model has been exhaustively specified and the data have been measured correctly. DEA has some notable features, however, such as axiomatic structure, ease of implementation via linear programming, and straightforward extensions to the multiple-input/multiple-output production case. Since the stochastic extension of DEA typically attempts to relax some aspects of the deterministic assumptions, we begin by listing the four properties (postulates) [2] that form the basis of DEA.

Postulate 10.1 (Convexity)
If $(x_1, y_1) \in T$ and $(x_2, y_2) \in T$, then for any scalar $\theta \in [0,1]$, $(\theta x_1 + (1-\theta)x_2, \theta y_1 + (1-\theta)y_2) \in T$.

Advances in DEA Theory and Applications: With Extensions to Forecasting Models,
First Edition. Edited by Kaoru Tone.

Convexity implies that if two production units, A and B, are observed, a third production unit, C, which is unobserved, can be constructed by mixing the operations of the two units observed. Specifically, we can create production unit C by using 80% of the inputs and producing 80% of the outputs of production unit A and by using 20% of the inputs and producing 20% of the outputs of production unit B, or any arbitrary percentages that sum to 100%. In general, C can be constructed from not just two observed production units, but an arbitrarily large number of production units. Varian [3] has argued that convexity can be divided into the two assumptions of divisibility and additivity. Divisibility means that an observed production unit can be proportionally reduced and operated independently, while additivity implies that feasible production processes can be combined without a loss or gain in productivity. If both divisibility and additivity hold, then the production possibility set satisfies not only convexity, but also constant returns to scale. If only divisibility and convexity hold, and additivity does not hold, then nonincreasing returns to scale is implied [2].

Postulate 10.2 (Monotonicity)
(a) If $(x,y) \in T$ and $x_1 \geq x$, then $(x_1,y) \in T$. (b) If $(x,y) \in T$ and $y_1 \leq y$, then $(x,y_1) \in T$.

Monotonicity implies that if a particular output level can be achieved with a given input vector, the same output level or greater should be achievable if additional input is given. Alternatively, Färe *et al.* [4] have proposed the concept of congestion. If there is too much input, the use of additional inputs could lead to less output, thus violating the monotonicity postulate. Production units typically operate in the monotonic region of the production function. Firms maximizing output per unit input operate at the most productive scale size, and the congestion region lies beyond the most productive scale size at even larger output levels, which implies significant optimization errors.

Postulate 10.3 (Inclusion)
The observed $(x_j,y_j) \in T$ for all DMUs $j=1,\ldots,n$.

Inclusion implies that all observed production units must be part of the production possibility set [4]. Specifically, all observations are below the estimated production function. Inclusion, however, makes the production function estimator a boundary estimator and sensitive to outliers, mismeasurement, model specification, and so on. The use of a stochastic model begins to address the issues of outliers, or sensitivity of the results to only a few observations. Allowing some observations to lie above the estimated production function makes the production function a more robust characterization of the production process.

Postulate 10.4 (Minimum Extrapolation)
If a production possibility set T_1 satisfies Postulates 10.1, 10.2, and 10.3, then $T_1 \subseteq T$.

Minimum extrapolation implies that a boundary estimator that is as close as possible to the data is selected. In regression-based techniques, a loss function that minimizes the sum of the deviations, adding both positive and negative deviations, measured from the observed data to the estimated function, assures that an estimated function is as close to the data as possible. While there are many functions that are monotonic and concave and define a production possibility set that includes all the observed production units, the set of functions that also minimizes the distance from the data, in terms of an L^1 norm, is smaller but still infinite. Regression-based stochastic DEA methods, such as the convex nonparametric least squares (CNLS) method described in Section 10.2.5, uses the minimum-extrapolation principle to uniquely identify a single functional estimate that minimizes the size of the production possibility set from among the infinite set of functions that minimizes the least squares criteria.

10.2 REVIEW OF LITERATURE ON STOCHASTIC DEA

Several discussions of stochastic DEA methods already exist in the literature; see, for example, Olesen and Petersen [5] and the references that lie within. However, we distinguish this review by focusing on the underlying assumptions and the data requirements, in contrast to Olesen and Petersen [5], who emphasize the importance of the management science perspective. Olesen and Petersen use Sherman and Zhu [6] as an example application, which had 5 inputs, 5 outputs, and 33 bank branch observations. After measuring efficiency using DEA, Olesen and Petersen discuss the use of questionnaires, field visits, and branch reviews, concluding that improvements are possible by standardizing management practices, reducing task mismatch, cross-training, and so on. While these are good consultancy recommendations and techniques, it is not clear how the DEA analysis really informed the later activities. Furthermore, applying a nonparametric estimator in such a high-dimensional space is unlikely to result in a meaningful functional estimation. When relatively complicated and flexible (nonparametric) statistical estimators such as DEA are used on small datasets for which the production process is difficult to define, there is a higher risk of obtaining meaningless efficiency estimates. Cases with limited data require more analysis of individual processes via consultancy practices. Flexible nonparametric statistical models are useful when rich, accurate data are available, but when this is not the case, the first steps become gathering measurable information about the production process, typically through observation and adding structure to the model.

The most basic property of an estimator is consistency. Consistency assures us that as more data are gathered, the estimator will converge to better solutions, so that when an infinite amount of data is gathered, the truth is recovered. Consistency of an estimator can only be shown for specific data generation processes that have assumptions associated with them. Therefore, when comparing estimators, it is important to make the assumptions clear so that it is possible to establish whether or not an estimator has the basic property of consistency. Typically, consistent estimators are preferred to inconsistent estimators, and estimators that are consistent under more general assumptions help avoid unnecessary modeling assumptions.

This chapter structures the literature on stochastic DEA and related topics into five categories: random sampling, imprecise measurement of data, uncertainty in the membership of observations with respect to the production technology, random production possibility sets, and random noise. The following five sections review the key characteristics of the methods in each category.

10.2.1 Random Sampling

The methods in this subsection consider the original DEA estimator. The research results account for the fact that a random sample of n production units is observed, and we would like to infer characteristics of the production technology that generated this set of observed production units. Specifically, consider estimating a multivariate concave and monotonic function from observations $\{(x_i, y_i)\}_{i=1}^n$, where $x_i \in \mathbb{R}^d, d \geq 1$ is a vector of random variables quantifying the inputs to the production process, or resources, of length d, and $y_i \in \mathbb{R}^q, q \geq 1$ is a vector of random variables quantifying the outputs of the production process. The production process is defined by $(x, y) \in T$, where T is often referred to as the technology, such that the resource vector x can produce y. Thus, the four postulates in the introduction still hold.

Banker [7] was the first to describe DEA as a maximum likelihood estimator in a single-output–multiple-input setting for a deterministic output-oriented model. The primary implication was that under deterministic modeling assumptions as the number of observations, n, approached infinity, DEA recovered the true production frontier. Korostelev *et al.* [8,9] found similar results and developed more rigorous definitions, proofs, and rates of convergence. Kneip *et al.* [10] proved consistency in the multiple-input–multiple-output setting and showed the rate of convergence.

These early papers provided the basis to apply other standard statistical methods, such as bootstrapping [11]. For example, Simar and Wilson [12,13] developed bootstrapping methods to estimate the finite sample bias and confidence intervals for DEA estimators. Although they were not the first to explore using bootstrapping methods with DEA [14–17], their methods are the most widely used today.

The papers mentioned in this section are often included in the discussion of stochastic DEA. However, the only uncertainty in these models is the random sample observed. Because the analysis is based on resampling methods, the data requirements are simply the input and output data.

10.2.2 Imprecise Measurement of Data

Cooper *et al.* [18] proposed the imprecise DEA model. Imprecise DEA models frequently specify a uniform or triangular distribution to characterize the probability density of the data over a specified bounded interval in which the data have been imprecisely measured. Cooper *et al.* [19] applied this method and demonstrated that imprecise DEA allows larger distinctions in efficiency to be measured. Imprecise DEA relaxes Postulate 10.3 to allow some of the data to lie above the frontier because those observations have been imprecisely measured.

Cooper *et al.* [18] formulated the imprecise DEA estimator as

$$\max \sum_{k=1}^{q} \mu_k y_{k0}$$

$$\text{s.t.} \sum_{k=1}^{q} \mu_k y_{kj} - \sum_{i=1}^{d} \omega_i x_{ij} \leq 0 \quad j = 1, \ldots, n$$

$$\sum_{i=1}^{d} \omega_i x_{i0} = 1 \tag{10.1}$$

$$y_k = (y_{kj}) \in D_k^+ \quad k = 1, \ldots, q$$

$$x_i = (x_{ij}) \in D_i^- \quad i = 1, \ldots, d$$

$$\mu = (\mu_k) \in A^+$$

$$\omega = (\omega_i) \in A^-$$

$$\mu, \omega \geq 0$$

There are q outputs $y_{1j}, \ldots, y_{qj}$ and d inputs $x_{1j}, \ldots, x_{dj}$ for production units $j = 1, \ldots, n$, where both the inputs and the outputs are random variables. The variables y_0 and x_0 represent the output and input vectors for production unit 0, which is under evaluation in this linear program. The variables μ and ω are often referred to as the weights or multipliers. These variables are determined within the optimization problem. The first constraints are for the standard form of a DEA multiplier model. The data y_{kj} and x_{ij} are assumed to be known imprecisely, meaning that the exact value is unknown, but that the value is known to lie within upper and lower bounds so that $\underline{y}_{kj} \leq y_{kj} \leq \bar{y}_{kj}$ and $\underline{x}_{ij} \leq x_{ij} \leq \bar{x}_{ij}$. The set D_k^+ defines an upper limit on the output variables, and D_i^- defines a lower limit on the input variables. Cooper *et al.* [18] included the concept of multiplier bounds or assurance regions in the imprecise DEA model to show the analogy between multipliers that were imprecise, but could be bounded, and production data that may also be imprecise but boundable. Similarly to D_k^+ and D_i^-, the sets A^+ and A^- define an upper and a lower limit, respectively, on the output weights μ and the input weights ω. Cooper *et al.* also assume that the input/output data are semipositive, which means that $y_k \geq 0$ and $x_i \geq 0$, $k = 1, \ldots, q$, $i = 1, \ldots, d$, and for at least one k and one i $y_k \neq 0$ and $x_i \neq 0$, which allows the last constraints to restrict the output and input multiplier weights to simply be nonnegative.

Imprecise DEA requires the probability density function for each of the input and output variables, which is not directly observable. Because imprecise DEA is closer to a sensitivity analysis approach, it is not clear how to interpret the results in terms of a true unobserved technology. Thus, I view imprecise DEA as distinct from other stochastic DEA methods, with a different purpose and different results. Furthermore, the data needed for imprecise DEA are not directly observable in cross-sectional data and therefore panel data are often used. However, this approach requires assuming that minimal change has occurred between time periods. When the time periods are short, this assumption is more tenable; however, typical production data are measured annually, creating a new set of challenges regarding how to correct the data for changes over time.

10.2.3 Uncertainty in the Membership of Observations

Uncertainty in the membership of a particular observation with respect to the technology leads to fuzzy DEA models. The use of fuzzy methods is the most common way to model this uncertainty in membership. However, fuzzy methods cover a wide range of modeling issues, some of which are very close to those mentioned in the previous subsections; see Hatami-Marbini et al. [20] for a detailed review of fuzzy methods. Fuzzy DEA models relax Postulate 10.3. Sengupta [21,22] was the first to introduce fuzzy DEA methods, and in the past 10 years these models have seen a rapid increase in attention. However, there have been extensive debates in the statistics literature related to the value and purpose of fuzzy methods [23].

Fuzzy methods use a continuous variable in the range of 0 to 1 to quantify the membership of a particular observation with respect to the production technology. Often the membership is not directly observable, but needs to be modeled as an unobserved latent variable or can be adjusted to perform a sensitivity analysis [24]. The Triantis and Girod [24] model can be described as a three-step model, as shown in Figure 10.1.

Here, the random output and input vectors for each production unit $j = 1, \ldots, n$ are $y_{1,j}, \ldots, y_{q,j}$ and $x_{1,j}, \ldots, x_{d,j}$, respectively. Similarly, $y_{1,j}^0, \ldots, y_{q,j}^0$ and $x_{1,j}^0, \ldots, x_{d,j}^0$,

Step 1.

$$\mu_X(x_{i,j}) = \frac{x_{i,j}^0 - x_{i,j}}{x_{i,j}^0 - x_{i,j}^1}, \ i = \{1, \ldots, d\} \qquad j = \{1, \ldots, N\}$$

$$\mu_Y(y_{k,i}) = \frac{y_{k,j} - y_{k,j}^1}{y_{k,j}^0 - y_{k,j}^1}, \qquad k = \{1, \ldots, q\} \quad j = \{1, \ldots, N\}$$

Step 2.

$$\text{Min } \theta_p$$

$$\text{s.t. } \theta_p(x_{i,p}^0 - (x_{i,p}^0 - x_{i,p}^1)\mu) - \sum_{j=1}^{N} \gamma_j (x_{i,j}^0 - (x_{i,j}^0 - x_{i,j}^1)\mu) \geq 0, \ i = \{1, \ldots, d\}$$

$$\sum_{j=1}^{N} \gamma_j ((y_{k,j}^0 - y_{k,j}^1)\mu + y_{k,j}^1) \geq (y_{k,j}^0 - y_{k,j}^1)\mu + y_{k,j}^1, \ k = \{1, \ldots, q\}$$

$$\theta_p \geq 0, \gamma_j \geq 0, j = \{1, \ldots, N\}$$

Step 3.
Resolve the linear program in step 2, adjusting the value of μ on a prespecified interval (for example $\mu = 0, 0.2, 0.4, \ldots, 1$).

Figure 10.1 Algorithm for fuzzy DEA [24].

and $y_{1,j}^1, \ldots, y_{q,j}^1$ and $x_{1,j}^1, \ldots, x_{d,j}^1$ are the impossible and risk-free bounds on the random output and input vectors, respectively. The functions $\mu_X(x_{i,j})$ and $\mu_Y(y_{k,i})$ are the membership functions. Triantis and Girod's step 3 is a sensitivity analysis with respect to the membership level.

The data requirements for fuzzy methods are input–output data and additional data specifying the impossible and risk-free bounds on the production variables, for Triantis and Girod's method. Other methods use different data for defining the membership function.

10.2.4   Random Production Possibility Sets

The methods in this subsection consider a random production possibility set. There are multiple related methods that fall within this category, but the relationships between these models have only been explored on a limited basis. In this group we include Banker's stochastic DEA, chance-constrained programming on both the primal and the dual DEA programs, and the order-m and order-α estimators. All of these methods relax Postulate 10.3 to allow some of the data to lie above the frontier. These methods have the similarity that a prespecified parameter either directly or indirectly determines how much of the data will lie above the frontier. However, specifying this parameter is the primary outstanding challenge for these methods.

Banker [25] recognized the relationship between the DEA formulation and the conditions for characterizing subgradients for a concave function from the wider optimization literature, and used the resulting insights to rewrite the additive DEA formulation with a goal-programming-type objective function. Specifically, the objective function was changed to include a random factor, $\sum_{j=1}^{n} \left(u_j^+ + u_j^- + cv_j \right)$, where the u_j^+ are random positive deviations, u_j^- are random negative deviations, v_j are the systematic inefficiency, and c is the prespecified weight defining the ratio of random noise to systematic inefficiency. By varying c, the least-absolute-deviation regression model and DEA can be obtained as special cases. For a more recent treatment of this model, see for example Banker *et al.* [26]. The data requirements for this method are the input–output data and specification of the parameter c.

Land, Lovell, and Thore [27] (LLT) were the first to apply chance-constrained programming to DEA and allowed the constraints in the envelopment model to be violated by a particular percentage of the observations. Cooper *et al.* [28] extended the envelopment formulation to include joint chance constraints. Alternatively, Olesen and Petersen [29] considered a DEA model in the multiplier form. Olesen [30] showed that, for the LLT model, the production function shape estimated for different levels of constraint violations did not necessarily satisfy the properties of monotonicity and convexity, depending on the covariance structure of the noise terms across observations. Olesen [30] also suggested how to integrate the LLT model with the dual formulation of Olesen and Petersen and proposed

$$\min \theta$$

$$\text{s.t.} \sum_{j=1}^{n} \lambda_j \bar{Y}_{kj} + \xi_k^+ \geq y_k \quad k = 1, \ldots, q$$

$$-\sum_{j=1}^{n} \lambda_j \bar{X}_{ij} + \xi_i^- \geq \theta x_i \quad i = 1, \ldots, d$$

$$\xi_k^+ + \frac{1}{\eta} \|\lambda\|_2 \leq 0 \quad k = 1, \ldots, q \tag{10.2}$$

$$\xi_i^- + \frac{1}{\eta} \|\lambda\|_2 \leq 0 \quad i = 1, \ldots, d$$

$$\frac{1}{\kappa} \|\lambda\|_1 + \left\| \xi_k^+, \xi_i^- \right\|_2 \geq 0$$

$$\lambda \in \mathbb{R}_+^n, \theta \in \mathbb{R}, \xi^- \in \mathbb{R}_+^d, \xi^+ \in \mathbb{R}_+^q$$

Here again, there are q outputs $Y_{1j}, \ldots, Y_{qj}$ and d inputs $X_{1j}, \ldots, X_{dj}$ for production units $j = 1, \ldots, n$, where both the inputs and the outputs are random variables. Let the mean vectors of the outputs and inputs be denoted by $\bar{Y}_{1j}, \ldots, \bar{Y}_{qj}$ and $\bar{X}_{1j}, \ldots, \bar{X}_{dj}$, respectively. Also, as in standard DEA notation, θ is the efficiency measure of the firm under evaluation, and $\lambda_1, \ldots, \lambda_n$ are the intensity weights associated with each observation and are specific to the current firm under evaluation. Let $\|x\|_1 = \sum_{j=1}^{n} x_j$ and $\|x\|_2 = \sqrt{\sum_{j=1}^{n} x_j^2}$ be the L^1 and L^2 norms, respectively. Let η^{-1} be the fractile corresponding to the chosen probability level α with which the envelopment constraints from the LLT model should hold. Let κ^{-1} be the fractile corresponding to the chosen probability level α with which the multiplier constraints from the Olesen and Petersen model should hold. The variables ξ^- and ξ^+ are referred to as contingency terms and play a role similar to the slack variables in the classic DEA model. Constraining the contingency terms more or less leads to the inner and outer approximations, respectively.

The data requirements for chance-constrained programming are the expected values of all variables (input and output) for all production units and variance–covariance matrices for each variable across production units. This information is not observable and would have to be assumed in the cross-sectional setting. Often panel data are used to construct this information. However, chance-constrained programming has similar challenges to imprecise DEA related to panel data: specifically, how to correct the data for changes over time is an open question.

Daraio and Simar [31] developed an order-m frontier, which calculates the expected minimum input among a fixed number of m potential competing firms producing more than output level y, where m must be less than or equal to n, the size of the full sample. Daraio and Simar also presented an order-α frontier, in which the probability α was selected such that with probability $(1-\alpha)$ a point is observed above the order-α frontier, rather than specifying the number of observations m. The data requirements for this method are the input–output data and specification of the parameter m or α.

Banker's stochastic DEA, chance-constrained programming, and the order-m and order-α estimators all relax Postulate 10.3 to allow some of the data to lie above the frontier. To summarize the data requirements, specifying a parameter that directly or indirectly determines how much of the data should lie above the frontier is the primary challenge for the methods in this category and largely remains an open research topic. Furthermore, the chance-constrained programming methods require extensive distributional information regarding the input–output data, including variance–covariance matrices.

10.2.5   Random Noise

Models that includes random noise are also based on a random production possibility set because the production function or the boundary of the production possibility set is not directly observed, but rather estimated via randomly observed data. However, we separate out random noise models because the data requirements and assumptions are different. Models that include random noise in DEA-type estimators often take the form of regression models and can be written in the form

$$y = f(x) - u + v \tag{10.3}$$

where $x \in X \subset \mathbb{R}^d$, $d \geq 1$ is a vector of random variables quantifying the inputs to the production process of length d, $y \in \mathbb{R}$ is a random variable quantifying the output of the production process, u is a nonnegative random variable characterizing the systematic inefficiency in the production process, and v is a random variable satisfying $E(v|x) = 0$, characterizing random noise. While all concepts in this chapter apply to the general multi-input and multi-output technology (see Kuosmanen et al. [32], for example), we restrict ourselves to the single-output case for ease of exposition. Typically, the composed error term $-u + v = \varepsilon$ is used to estimate a conditional mean function, with a second-step adjustment to shift the frontier up under a specific model for the variance of the error term. When the second-step shift is small relative to the size of the noise component, Postulate 10.3 is violated.

Some would characterize (10.3) as the stochastic frontier model [33,34]. Meeusen and van den Broeck [34] explicitly specified the use of a parametric Cobb–Douglas function in (10.3). Aigner et al. [33], Kumbhakar and Lovell [35], and Parmeter and Kumbhakar [36] used the notation $y = f(x; \beta) - u + v$, emphasizing the parametric nature of the regression function $f(\cdot)$ used in stochastic frontier analysis (SFA).

Banker and Maindiratta [37] proposed a maximum likelihood estimator for a shape-constrained production function, and residual term that is a convolution of a normally distributed noise term and a half-normally distributed inefficiency term. This objective function was generally nonlinear, which made the Banker and Maindiratta estimator difficult to compute. Kuosmanen [38], Kuosmanen and Kortelainen [39], and Kuosmanen *et al.* [32] developed stochastic nonparametric envelopment of data (StoNED) to estimate the model (10.3) using the CNLS method to estimate a shape-constrained nonparametric production function at the conditional mean of the data, followed by the method of moments [33] or pseudo-maximum likelihood [40] to create a generalization of SFA. The least squares objective placed CNLS in the class of problems P that can be solved in polynomial time, whereas Banker and Maindiratta's model is in the class NP, nondeterministic polynomial time [41]. Kuosmanen and Johnson [42] showed how to interpret DEA as a CNLS estimate with a sign constraint on the residuals.

In 2015, Kuosmanen *et al.* interpreted StoNED as a generalization of both DEA and SFA, incorporating a classical model of noise into DEA and imposing axiomatic properties (postulates of monotonicity and concavity) on the shape of the function estimated in SFA. The use of a classical noise term creates a link to the standard statistics literature [43]. Specifically, the central limit theorem states that the arithmetic mean of a sufficiently large number of independent random variables will be approximately normally distributed regardless of the underlying distributions of the random variables. The central limit theorem motivates the regression-based approaches and standard applications where a large number of modeling and measurement errors are summed together in the noise term.

Kuosmanen and Kortelainen [39] proposed two potential second-stage methods (the method of moments and the pseudo-maximum likelihood method), but both require additional distributional assumptions for both the inefficiency and the noise terms. Kuosmanen and Kortelainen [39] also proposed to use the Jondrow estimator [44] to calculate firm-specific inefficiency levels. However, Greene [45] argued that Jondrow's method results in inconsistent estimates in cross-sectional analysis. Furthermore, identification of the parameters of the inefficiency and noise distributions relies on the skewness of the residuals [46]. The two-stage method requires a separability assumption to first estimate a conditional mean and then deconvolute inefficiency. Homoskedasticity of both the inefficiency and the noise terms is a sufficient condition. Alternatively, well-defined models of heteroskedasticity, such as a multiplicative residual where $y = f(x) \exp(-u+v)$, can be estimated with the two-stage method. However, separability clearly limits the flexibility of the potential models for heteroskedasticity. Kuosmanen *et al.* [32] used a fully nonparametric kernel deconvolution estimator given by Hall and Simar [47]. Here, we assume that the inefficiency term is asymmetric. The noise term v_i has a unimodal density with a unique mode at zero. Hall and Simar [47] assume that σ_v^2 approaches zero asymptotically, which is required for proving consistency of their estimator. This allows estimation of the expected value of the inefficiency without any parametric assumptions.

The data requirements for the regression-based methods are the input–output data. To use Hall and Simar's method to estimate the average inefficiency level, several additional assumptions are needed. The unimodal density function for noise is motivated by the central limit theorem as described above. Thus, the additional assumption needed for the method described by Kuosmanen *et al.* [32] is that the inefficiency

TABLE 10.1 Additional data and assumptions needed for stochastic DEA methods.

Stochastic source (method)		Additional data	Additional assumption
Random sampling	DEA	None	Perfect model specification and data measurement
	Bias-corrected DEA	None	Bootstrap sample is to the full sample as the full sample is to the true population
Imprecise measurement of data	Imprecise DEA	Bounds on the input and output levels and bounds on the multiplier weights	Input data lie within an interval which is known, but the exact value of input or output for a particular firm is unknown
Uncertainty in the membership of observations	Fuzzy DEA	The membership function or data needed to construct the membership function	Input data lie within an interval which is known, but the exact value of input or output for a particular firm is unknown
Random production possibility set	Banker's stochastic DEA	Signal-to-noise ratio	Noise is generated from a Laplace distribution
	Chance-constrained programming	Probability of the constraints being violated; expected values and a variance–covariance matrices for all variables (input and output) across production unit	Probability of a constraint being violated is constant over the production possibility set (a variant of homoskedastic noise assumption)
	Order-m and order-α	The quantile of interest (or the frontier) in terms of the parameter α or m	Output is homoskedastic in inputs
Random noise	StoNED	None	Noise is symmetrically distributed; inefficiency either is homoskedastic or has a well-defined heteroskedastic structure; the variance of the noise approaches zero asymptotically

distribution is left-truncated and that the variance of the noise approaches zero asymptotically. The left-truncation is motivated by a nonzero density near the frontier, which is likely in competitive markets, but is debatable for public industries. Hall and Simar experimented with violations of the noise variance assumption and found that the bias introduced was small. However, further research to verify this result would be useful.

10.3 CONCLUSIONS

A variety of methods fall into the general category of stochastic DEA. The purpose of making DEA stochastic is varied and includes making efficiency estimates robust to outliers, making efficiency estimates more discriminate, modeling uncertainty, and connecting to the statistics literature, among others. This chapter has emphasized that while all stochastic DEA methods require additional information in the form of either data or assumptions, regression-based approaches that build on laws of large numbers and include a classical noise term reduce the number of arbitrary assumptions needed, in a manner consistent with the classic DEA mantra, allowing the data to speak for themselves. Table 10.1 summarizes the additional data and assumptions needed for the methods discussed.

There are relatively few theories related to production that apply across the diverse types of production that are observed throughout the economy. However, monotonicity and concavity of the production function are widely accepted in a broad number of applications. Stochastic DEA methods maintain these assumptions, which, in part, has led to their wide popularity. The StoNED framework has clarified the relationship between the two first-generation efficiency analysis techniques of DEA and stochastic frontier analysis. StoNED integrates the two analysis techniques into a framework which can both impose the postulates of monotonicity and concavity and include a classical noise term. The next generation of efficiency analysis methods should address model selection, out-of-sample performance, endogeneity, and smoothness.

REFERENCES

[1] Charnes, A., Cooper, W.W., and Rhodes, E. (1978) Measuring the efficiency of decision making units. *European Journal of Operational Research*, **2**, 429–444.

[2] Banker, R.D., Charnes, A., and Cooper, W.W. (1984) Some models for estimating technical and scale inefficiencies in data envelopment analysis. *Management Science*, **30**, 1078–1092.

[3] Varian, H.R. (1992) *Microeconomic Analysis*, 3rd edn, W.W. Norton.

[4] Färe, R., Grosskopf, S., and Lovell, C.A.K. (1994) *Production Frontiers*, Cambridge University Press, Cambridge.

[5] Olesen, O.B. and Petersen, N.C. (2016) Stochastic data envelopment analysis – A review. *European Journal of Operational Research*, **251**(1), 2–21.

[6] Sherman, H.D. and Zhu, J. (2006) Managing bank productivity, in *Service Productivity Management: Improving Service Performance Using Data Envelopment Analysis* (eds H.D. Sherman and J. Zhu), Springer, pp. 159–173.

[7] Banker, R.D. (1993) Maximum-likelihood, consistency and data envelopment analysis – a statistical foundation. *Management Science*, **39**(10), 1265–1273.

[8] Korostelev, A., Simar, L., and Tsybakov, A.B. (1995) On estimation of monotone and convex boundaries. *Publications of the Institute of Statistics University of Paris*, **39**(1), 3–18.

[9] Korostelev, A., Simar, L., and Tsybakov, A.B. (1995) Efficient estimation of monotone boundaries. *Annals of Statistics*, 476–489.

[10] Kneip, A., Park, B.U., and Simar, L. (1998) A note on the convergence of nonparametric DEA estimators for production efficiency scores. *Econometric Theory*, **14**(6), 783–793.

[11] Efron, B. and Tibshirani, R.J. (1993) *An Introduction to the Bootstrap*, Chapman & Hall, New York.

[12] Simar, L. and Wilson, P. (1998) Sensitivity analysis of efficiency scores: How to bootstrap in nonparametric models. *Management Science*, **44**, 49–61.

[13] Simar, L. and Wilson, P. (2000) A general methodology for bootstrapping in nonparametric models. *Journal of Applied Statistics*, **27**, 779–802.

[14] Ferrier, G. and Hirschberg, J. (1997) Bootstrapping confidence intervals for linear programming efficiency scores: With an illustration using Italian banking data. *Journal of Productivity Analysis*, **8**(1), 19–33.

[15] Löthgren, M. (1998) How to bootstrap DEA estimators: A Monte Carlo comparison. SSE/EFI Working Paper Series in Economics and Finance, No. 223, Stockholm School of Economics.

[16] Löthgren, M. and Tambour, M. (1999) Bootstrapping the data envelopment analysis Malmquist productivity index. *Applied Economics*, **31**(4), 417–425.

[17] Xue, M. and Harker, P.T. (1999) Overcoming the inherent dependency of DEA efficiency scores: A bootstrap Approach. Center for Financial Institutions Working Paper 99-17, Wharton School Center for Financial Institutions, University of Pennsylvania.

[18] Cooper, W.W., Park, K.S., and Yu, G. (1999) IDEA and AR-IDEA: Models for dealing with imprecise data in DEA. *Management Science*, **45**(4), 597–607.

[19] Cooper, W.W., Park, K.S., and Yu, G. (2001) An illustrative application of IDEA (imprecise data envelopment analysis) to a Korean mobile telecommunication company. *Operations Research*, **49**(6), 807–820.

[20] Hatami-Marbini, A., Emrouznejad, A., and Tavana, M. (2011) A taxonomy and review of the fuzzy data envelopment analysis literature: Two decades in the making. *European Journal of Operational Research*, **214**(3), 457–472.

[21] Sengupta, J.K. (1992) A fuzzy systems approach in data envelopment analysis. *Computers and Mathematics with Applications*, **24**(8–9), 259–266.

[22] Sengupta, J.K. (1992) Measuring efficiency by a fuzzy statistical approach. *Fuzzy Sets and Systems*, **46**(1), 73–80.

[23] Ross, T.J., Booker, J.M., and Parkinson, W.J. (2002) *Fuzzy Logic and Probability Applications: Bridging the Gap*, Society for Industrial and Applied Mathematics.

[24] Triantis, K. and Girod, O. (1998) A mathematical programming approach for measuring technical efficiency in a fuzzy environment. *Journal of Productivity Analysis*, **10**(1), 85–102.

[25] Banker, R.D. (1998) Stochastic data envelopment analysis. Carnegie Mellon University working paper.

[26] Banker, R.D., Kotarac, K., and Neralic, L. (2015) Sensitivity and stability in stochastic data envelopment analysis. *Journal of Operational Research Society*, **66**, 134–147.

[27] Land, K.C., Lovell, C.A.K., and Thore, S. (1994) Productive efficiency under capitalism and state socialism: An empirical inquiry using chance-constrained data envelopment analysis. *Technological Forecasting and Social Change*, **46**(2), 139–152.

[28] Cooper, W.W., Huang, Z., Lelas, V., Li, S.X., and Olesen, O.B. (1998) Chance constrained programming formulations for stochastic characterizations of efficiency and dominance in DEA. *Journal of Productivity Analysis*, **9**(1), 53–79.

[29] Olesen, O.B. and Petersen, N.C. (1995) Chance constrained efficiency evaluation. *Management Science*, **41**, 442–457.

[30] Olesen, O.B. (2006) Comparing and combining two approaches for chance constrained DEA. *Journal of Productivity Analysis*, **26**, 103–119.

[31] Daraio, C. and Simar, L. (2007) Conditional nonparametric frontier models for convex and nonconvex technologies: A unifying approach. *Journal of Productivity Analysis*, **28**(1–2), 13–32.

[32] Kuosmanen, T., Johnson, A.L., and Saastamoinen, A. (2015) Stochastic nonparametric approach to efficiency analysis: A unified framework, in *Data Envelopment Analysis: A Handbook of Models and Methods* (ed. J. Zhu), Vol. **2**, Springer, p. 191.

[33] Aigner, D., Lovell, C.A.K., and Schmidt, P. (1977) Formulation and estimation of stochastic frontier production function models. *Journal of Econometrics*, **6**, 21–37.

[34] Meeusen, W. and van den Broeck, J. (1977) Efficiency estimation from Cobb–Douglas production functions with composed error. *International Economic Review*, **18**(2), 435–445.

[35] Kumbhakar, S.C. and Lovell, C.A.K. (2003) *Stochastic Frontier Analysis.* Cambridge University Press, New York.

[36] Parmeter, C.F. and Kumbhakar, S.C. (2014) Efficiency analysis: A primer on recent advances. *Foundations and Trends in Econometrics*, **7**(3–4), 191–385.

[37] Banker, R.D. and Maindiratta, A. (1992) Maximum likelihood estimation of monotone and concave production frontiers. *Journal of Productivity Analysis*, **3**(4), 401–415.

[38] Kuosmanen, T. (2008) Representation theorem for convex nonparametric least squares. *Econometrics Journal*, **11**, 308–325.

[39] Kuosmanen, T. and Kortelainen, M. (2012) Stochastic non-smooth envelopment of data: Semi-parametric frontier estimation subject to shape constraints. *Journal of Productivity Analysis*, **38**(1), 11–28.

[40] Fan, Y., Li, Q., and Weersink, A. (1996) Semiparametric estimation of stochastic production frontier models. *Journal of Business and Economic Statistics*, **4**(4), 460–468.

[41] Cook, S.A. (1971) The complexity of theorem proving procedures. Proceedings of the Third Annual ACM Symposium on Theory of Computing, pp. 151–158.

[42] Kuosmanen, T. and Johnson, A.L. (2010) Data envelopment analysis as nonparametric least-squares regression. *Operations Research*, **58**, 149–160.

[43] Greene, W.H. (2012) *Econometric Analysis*, 7th edn, Prentice Hall.

[44] Jondrow, J., Lovell, C.A.K., Materov, I.S., and Schmidt, P. (1982) On the estimation of technical inefficiency in the stochastic frontier production function model. *Journal of Econometrics*, **19**(2–3), 233–238.

[45] Greene, W.H. (2008) The econometric approach to efficiency analysis, in *The Measurement of Efficiency* (eds H. Fried, C.A.K. Lovell, and S. Schmidt), Oxford University Press, Chapter 2.

[46] Waldman, D. (1982) A stationary point for the stochastic frontier likelihood. *Journal of Econometrics*, **18**(2), 275–279.

[47] Hall, P. and Simar, L. (2002) Estimating a changepoint, boundary, or frontier in the presence of observation error. *Journal of the American Statistical Association*, **97**, 523–534.

11

A COMPARATIVE STUDY OF AHP AND DEA

KAORU TONE

National Graduate Institute for Policy Studies, Tokyo, Japan

11.1 INTRODUCTION

Both the analytic hierarchy process (AHP) and data envelopment analysis (DEA) aim at the evaluation of decision-making units (DMUs) in multiple-criteria environments. AHP uses pairwise comparisons and eigenvector weightings, whereas DEA uses linear fractional programs. In this chapter, we point out some structural similarities between the two methods, by comparing the benefit/cost analysis that can be done by AHP and DEA. Also, we discuss the question of fixed versus variable weights in multiple-criteria decision making.

11.2 A GLIMPSE OF DATA ENVELOPMENT ANALYSIS

DEA was developed by Charnes *et al.* [1]. DEA estimates the relative efficiencies of DMUs that have common factors in their inputs and outputs. Let the multiple inputs to and outputs from DMU$_j$ $(j = 1, \ldots, n)$ be $\{x_{ij} : i = 1, \ldots, m\}$ and $\{y_{rj} : r = 1, \ldots, s\}$, respectively. We assume that we have $\{x_{ij}\}$ and $\{y_{rj}\}$ in the form of observations or of theoretically prescribed values and that their values are positive. Also, we assume that the data are normalized so that they satisfy

Advances in DEA Theory and Applications: With Extensions to Forecasting Models,
First Edition. Edited by Kaoru Tone.

$$\sum_{j=1}^{n} x_{ij} = 1 \ (i=1,\dots,m) \tag{11.1}$$

and

$$\sum_{j=1}^{n} y_{rj} = 1 \ (r=1,\dots,s) \tag{11.2}$$

This assumption is made for the sake of comparative study and does not influence any essential features of DEA. From the point of view of efficiency, a DMU with large outputs relative to small inputs is preferable. We define the relative efficiency of a DMU h $(h=1,\dots,n)$ by solving the following linear fractional program:

$$[FP(h)] \quad \max_{u,v} \theta_h = \left(\sum_{r=1}^{s} u_r y_{rh}\right) \Big/ \left(\sum_{i=1}^{m} v_i x_{ih}\right)$$

subject to

$$\left(\sum_{r=1}^{s} u_r y_{rj}\right) \Big/ \left(\sum_{i=1}^{m} v_i x_{ij}\right) \le 1 \ (j=1,\dots,n)$$

$$u_r \ge \varepsilon \ (r=1,\dots,s), v_i \ge \varepsilon \ (i=1,\dots,m) \tag{11.3}$$

where u_r and v_i are the weights of the rth output y_r and of the ith input x_i, respectively, and ε is a 'non-Archimedean infinitesimal' number (a positive number smaller than any positive real number; see Cooper *et al.* [2]). We define the efficiency of a DMU to be the ratio of the weighted sum of output values to the weighted sum of input values. [FP(h)] maximizes the ratio associated with DMU h, keeping the ratio of every DMU, including DMU h, not greater than 1. Let the optimal solution to [FP(h)] be u^*, v^* and θ_h^*. These values vary from one DMU to another.

Definition 11.1
If $\theta_h^* = 1$, then DMU h is *DEA-efficient*. Otherwise, if $\theta_h^* < 1$, then DMU h is *DEA-inefficient*.
Actually, this definition means the following:

(i) *Output orientation:* a DMU is inefficient if it is possible to augment any output without increasing any input and without decreasing any other output.
(ii) *Input orientation:* a DMU is inefficient if it is possible to decrease any input without augmenting any other input and without decreasing any output.

A DMU is characterized as *efficient* if, and only if, neither (i) nor (ii) obtains.
For an inefficient DMU, it is very important to find other DMUs which drive that DMU into inefficiency.

Definition 11.2

The *efficient frontier* for a DMU h is the set of DMUs

$$E(h) = \left\{ j: \left(\sum_{r=1}^{s} u_r^* y_{rj} \right) / \left(\sum_{i=1}^{m} v_i^* x_{ij} \right) = 1, j = 1, \ldots, n \right\} \tag{11.4}$$

where u^* and v^* are the optimal solutions to [FP(h)].

11.3 BENEFIT/COST ANALYSIS BY ANALYTIC HIERARCHY PROCESS

Benefit/cost (b/c) analysis by AHP consists of two processes, namely a benefit process and a cost process [3]. We estimate the benefit priority and the cost priority separately by AHP, and then their ratio gives the relative efficiency of the alternative objects. In this section, first we consider the b/c analysis in the case of a three-level perfect hierarchy structure and then show that general cases can be reduced to the three-level case.

11.3.1 Three-Level Perfect Graph Case

We will deal with a three-level hierarchy structure as depicted in Figure 11.1. We call a graph of the structure a *perfect hierarchy graph* if every node in every level is connected to every node in the succeeding level by an arc and is not connected directly to any nodes beyond the succeeding level.

We assume that we have s kinds of benefit criteria (B_1, …, B_s) in Level 2 and n kinds of alternative objects (O_1, …, O_n) in Level 3. Let y_{rj} be the priority of the object O_j associated with the criterion B_r, and let U_r be the priority of the criterion B_r. Then, the overall benefit of the object O_j is given by

$$\sum_{r=1}^{s} U_r y_{rj} \quad (j = 1, \ldots, n) \tag{11.5}$$

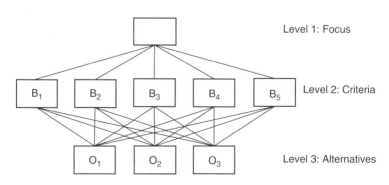

Figure 11.1 Three-level perfect hierarchy graph.

Here, y_{rj} and U_r satisfy

$$\sum_{j=1}^{n} y_{rj} = 1 \ (r = 1,\ldots,s) \text{ and } \sum_{r=1}^{s} U_r = 1 \tag{11.6}$$

Similarly, we assume that we have a perfect hierarchy cost structure with m cost criteria ($C_1, \ldots, C_m$). Let x_{ij} be the priority of the object O_j with respect to C_i and let V_i be the priority of C_i. These priorities satisfy

$$\sum_{j=1}^{n} x_{ij} = 1 \ (i = 1,\ldots,m) \text{ and } \sum_{i=1}^{m} V_i = 1 \tag{11.7}$$

Then, the overall cost of O_j is given by

$$\sum_{i=1}^{m} V_i x_{ij} \ (j = 1,\ldots,n) \tag{11.8}$$

The benefit/cost priority of the object O_h is evaluated as

$$\eta_h^* = \left(\sum_{r=1}^{s} U_r y_{rh} \right) / \left(\sum_{i=1}^{m} V_i x_{ih} \right) \tag{11.9}$$

We notice that in AHP all the elements of x, y, U and V are estimated by processes of pairwise comparisons and eigenvector weightings or by some other empirical or theoretical evaluations.

11.3.2 General Cases

We can reduce a general multilevel-structure case to a three-level problem by choosing a *key* level between the focus and the alternatives, and by aggregating the levels between them as depicted in Figure 11.2. If some arcs bypass the key level (Level 2), we introduce additional nodes in the level so that any path connecting the Level 1 node (the focus) to a Level 3 node (the alternatives) will meet a node in Level 2. Also, we

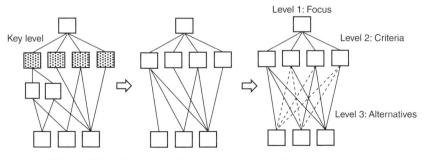

Figure 11.2 Reduction of general case to three-level perfect graph.

introduce additional dummy arcs with very small x or y values to make the three-level structure 'perfect', if necessary. It is easy to see that we can calculate the x, y, U and V values corresponding to the aggregated three-level structure from the original values. Thus, general multilevel cases can be reduced to a three-level perfect case by deliberately selecting a key level, which usually exists in AHP.

11.4   EFFICIENCIES IN AHP AND DEA

The discussion in Sections 11.2 and 11.3 shows the structural similarity between b/c analyses done by AHP and DEA. Differences exist in the way they estimate the x, y, u, v, U and V values.

11.4.1   Input x and Output y

DEA uses available numerical data for the input x and output y, while AHP creates them by processes of pairwise comparisons and eigenvector weightings. Originally, DEA was aimed at evaluating relative efficiencies of DMUs in environments where numerical or theoretically prescribed data exist. On the other hand, AHP works in a world where only subjective or psychological factors prevail in making decisions. Although the two methods stem from extremely different motivations, they exhibit a certain similarity in the presence of data, that is, an input x and an output y, and of a ratio scale of efficiency evaluations. They can trade off their inputs and outputs. AHP could benefit by using the same numerical data as DEA. DEA could expand its world by incorporating qualitative factors that AHP has exposed for the first time.

11.4.2   Weights

DEA determines the weights u and v by solving the fractional program [FP(h)] corresponding to the decision-making unit DMU$_h$. Hence, the weights differ from one DMU to another. We call this kind of weights *variable weights*. The weights are determined in such a way that they will be the most favourable for the DMU concerned. AHP uses pairwise comparisons and eigenvector weightings in determining the weighs U and V of the key-level criteria. The values are common to all alternative objects. We call this kind of weights *fixed*.

11.4.3   Efficiency

The AHP-efficiency η_h^* of an object O_h is given by the formula (11.9). The DEA-efficiency of DMU$_h$ is the optimal objective function value θ_h^* for [FP(h)],

$$\theta_h^* = \left(\sum_{r=1}^{s} u_r^* y_{rh} \right) \Big/ \left(\sum_{i=1}^{m} v_i^* x_{ih} \right) \qquad (11.10)$$

where u^* and v^* are the optimal solution to [FP(h)].

For any (U, V) in AHP, let

$$p = \max_j \left(\sum_{r=1}^{s} U_r y_{rj} \right) / \left(\sum_{i=1}^{m} V_i x_{ij} \right)$$

and   (11.11)

$$u_r = U_r / p \ (r = 1, \ldots, s) \ and \ v_i = V_i \ (i = 1, \ldots, m)$$

Then, (u, v) is feasible for [FP(h)].

Conversely, for any DEA-feasible solution (u, v), let $T = \sum_{r=1}^{s} u_r$ and $S = \sum_{i=1}^{m} v_i$, and define

$$U_r = u_r / T \ (r = 1, \ldots, s) \ and \ V_i = v_i / S \ (i = 1, \ldots, m) \qquad (11.12)$$

Then, (U, V) is an AHP-feasible solution priority. Since both transformations are scaling, they have the same priority relations in the b/c analysis.

11.4.4   Several Propositions

The above discussions lead us to several propositions [4]. Throughout this subsection, we assume x and y to be constant.

Proposition 11.1 For any AHP weight (U, V), there exists a DMU$_h$ that has the transformed (u, v) as the optimal solution to [FP(h)]. Indeed, h is the DMU that gives the maximum value to (11.11).

Proposition 11.2 DEA is the most generous method among multiple-criteria methods for evaluating the efficiency of DMUs by a ratio scale, in the sense that an efficient DMU under the latter criteria has a corresponding DEA optimal weight (u, v) which makes that DMU DEA-efficient.

Proposition 11.3 A DEA-inefficient DMU is also AHP-inefficient for any weighting of the criteria. Moreover, a DEA-inefficient DMU is inefficient under any fixed-weight multiple-criteria benefit/cost analysis.

11.5   CONCLUDING REMARKS

Both AHP and DEA have turned out to give strong impulses to the multiple-criteria decision-making community, although their origins and motivations were quite different. In this chapter, we have pointed out structural similarities between them in the case of b/c analysis and suggested their potential trade-offs. In short, AHP could be made more objective by incorporating the DEA-efficiency. AHP could exclude essentially inefficient objects by using DEA-inefficiency. Conversely, DEA could be made more subjectively oriented by incorporating some features of AHP. For example, if constraints such as $u_1 \geq u_2$ or $3v_1 \leq v_2$ were added to [FP(h)], DEA would become more intensive in judging the efficiency of the DMU concerned. Although we have been concerned mainly with a comparative study of b/c analyses by AHP and DEA, it should be noted that the usual AHP could be regarded as a special case of the AHP b/c analysis where the cost factor has only one criterion, with equal weights for each object of the alternatives. Hence, Propositions 11.1 to 11.3 remain valid in the latter case, where the corresponding [FP(h)] of DEA reduces to a linear program.

A collaboration between AHP and DEA in a group decision-making scenario is presented in Chapter 31.

REFERENCES

[1] Charnes, A., Cooper W.W. and Rhodes, E. (1978) Measuring the efficiency of decision making units. *European Journal of Operational Research*, **2**, 429–444.

[2] Cooper, W.W., Seiford, L.M. and Tone, K. (2007) *Data Envelopment Analysis: A Comprehensive Text with Models, Applications, References and DEA-Solver Software*, 2nd edn, Springer, New York.

[3] Saaty, T.L. (1980) *The Analytic Hierarchy Process, McGraw-Hill.*

[4] Tone, K. (1989) A comparative study on AHP and DEA. *International Journal on Policy and Information*, **13**, 57–63.

12

A COMPUTATIONAL METHOD FOR SOLVING DEA PROBLEMS WITH INFINITELY MANY DMUs[1]

ABRAHAM CHARNES

University of Texas at Austin, Austin, Texas, USA

KAORU TONE

National Graduate Institute for Policy Studies, Tokyo, Japan

12.1 INTRODUCTION

Usually, DEA deals with a finite set of DMUs. In this chapter, we study DEA problems with infinitely many DMUs. We assume that every DMU have common multiple input- and output- factors, and that each factor varies continuously with respect to DMUs. The problem is to find the efficient DMUs within a tolerance where *efficiency* is measured by a ratio of weighted inputs vs. weighted outputs. Section 12.2 describes Primal and Dual sides of the problem. The outline of the solution process (Discretization, Deletion and Subdivision) is explained in Section 12.3. Details of

---

[1] Reprinted from Research Report CCS 561, Center for Cybernetic Studies, The University of Texas at Austin. This article was written under the co-authorship of Professor Charnes when I (Tone) was invited to Texas in January 1987. This research was partly supported by ONA Contracts N00014-86-C-0398 and N00014-82-K-0295, and National Science Foundation Grant SES-8520806 with the Center for Cybernetic Studies, The University of Texas at Austin. Reproduction in whole or in part is permitted for any purpose of the United States Government. I have changed the original notation a little in order to make it consistent with Chapter 2 and added proofs of theorems as appendices.

Advances in DEA Theory and Applications: With Extensions to Forecasting Models,
First Edition. Edited by Kaoru Tone.
© 2017 John Wiley & Sons Ltd. Published 2017 by John Wiley & Sons Ltd.

the method when the infinite set is one-dimensional are presented in Section 12.4. Section 12.5 deals with general cases.

12.2 PROBLEM

Solve the following LP with an infinite set $\mathbf{Z}$ of DMUs for a suitable subset of elements typically designated by $\mathbf{z}^h$.

[Problem]

$$\text{(P)} \quad v_P = \max_{\mathbf{w}} \mathbf{u}^T \mathbf{y}(\mathbf{z}^h)$$

subject to

$$\mathbf{u}^T \mathbf{y}(\mathbf{z}) - \mathbf{v}^T \mathbf{x}(\mathbf{z}) \leq 0 \quad \text{for } \forall \mathbf{z} \in \mathbf{Z}$$
$$\mathbf{v}^T \mathbf{x}(\mathbf{z}^h) = 1$$
$$\mathbf{u} \geq \varepsilon \tilde{\mathbf{e}}$$
$$\mathbf{v} \geq \varepsilon \mathbf{e}$$

where

$$\mathbf{Z} : \text{a compact convex set, } \dim(\mathbf{Z}) = L$$
$$\mathbf{u} \in R^s, \tilde{\mathbf{e}} = (1,1,\ldots,1)^T \in R^s$$
$$\mathbf{v} \in R^m, \mathbf{e} = (1,1,\ldots,1)^T \in R^m$$
$$\mathbf{y}(\mathbf{z}) \in R^s \ (\text{Outputs}) : \mathbf{y}(\mathbf{z}) \geq \mathbf{0} \text{ and } C^2 \text{ on } \mathbf{Z}$$
$$\mathbf{x}(\mathbf{z}) \in R^m \ (\text{Inputs}) : \mathbf{x}(\mathbf{z}) \geq \mathbf{0} \text{ and } C^2 \text{ on } \mathbf{Z}$$
$$\varepsilon : \text{a positive infinitesimal non-Archimedian quantity.}$$

[Dual Problem]

$$\text{(D)} \quad v_D = \min\left(\theta - \varepsilon \tilde{\mathbf{e}}^T \mathbf{s}^+ - \varepsilon \mathbf{e}^T \mathbf{s}^-\right)$$

subject to

$$\sum_{\mathbf{z} \in \mathbf{Z}} \mathbf{y}(\mathbf{z}) \lambda(\mathbf{z}) - \mathbf{s}^+ = \mathbf{y}(\mathbf{z}^h)$$

$$\sum_{\mathbf{z} \in \mathbf{Z}} \mathbf{x}(\mathbf{z}) \lambda(\mathbf{z}) + \mathbf{s}^- = \theta \mathbf{x}(\mathbf{z}^h)$$

$\lambda(\mathbf{z}) \geq 0$: for every $\mathbf{z} \in \mathbf{Z}$ and $\lambda(\mathbf{z}) = 0$ except for a finite number of points

$$\mathbf{s}^+ \geq \mathbf{0}, \ \mathbf{s}^+ \in R^s$$
$$\mathbf{s}^- \geq \mathbf{0}, \ \mathbf{s}^- \in R^m.$$

12.3 OUTLINE OF THE METHOD

The method consists of three main parts: Initial discretization, deletion and subdivision. The discretized problems are solved by the simplex method throughout the iterations.

Step 0. (Discretization)
The dual pair (P)–(D) is discretized, i.e., the infinite index set $\mathbf{Z}$ is replaced by a finite set. Let the finite set be $(\mathbf{z}^1, \ldots, \mathbf{z}^n)$. We call such sets *grid*.

Solve the resulting dual pair of linear programs (P_h)–(D_h) $(h = 1, \ldots, n)$ by means of the simplex method.

$$(P_h) \quad v_P = \max_{\mathbf{u}} \mathbf{u}^T \mathbf{y}(\mathbf{z}^h)$$

subject to

$$\mathbf{u}^T \mathbf{y}(\mathbf{z}^i) - \mathbf{v}^T \mathbf{x}(\mathbf{z}^i) \leq 0 \quad \text{for } i = 1, \ldots, n$$

$$\mathbf{v}^T \mathbf{x}(\mathbf{z}^h) = 1$$

$$\mathbf{u} \geq \varepsilon \tilde{\mathbf{e}}$$

$$\mathbf{v} \geq \varepsilon \mathbf{e}$$

$$(D_h) \quad v_D = \min\left(\theta - \varepsilon \tilde{\mathbf{e}}^T \mathbf{s}^+ - \varepsilon \mathbf{e}^T \mathbf{s}^-\right)$$

subject to

$$\sum_{i=1}^{n} \mathbf{y}(\mathbf{z}^i) \lambda_i - \mathbf{s}^+ = \mathbf{y}(\mathbf{z}^h)$$

$$\sum_{i=1}^{n} \mathbf{x}(\mathbf{z}^i) \lambda_i + \mathbf{s}^- = \theta \mathbf{x}(\mathbf{z}^h)$$

$$\lambda_i \geq 0 : \text{for } i = 1, \ldots, n$$

$$\mathbf{s}^+ \geq \mathbf{0}, \ \mathbf{s}^+ \in R^s$$

$$\mathbf{s}^- \geq \mathbf{0}, \ \mathbf{s}^- \in R^m.$$

Let optimal solutions to (P_h) and (D_h) be

$$\mathbf{u} = (u_1, \ldots, u_s)^T, \mathbf{v} = (v_1, \ldots, v_m)^T$$

and

$$\theta, \ \lambda = (\lambda_1, \ldots, \lambda_n)^T, \ \mathbf{s}^+ = \left(s_1^+, \ldots, s_s^+\right)^T, \ \mathbf{s}^- = \left(s_1^-, \ldots, s_m^-\right).$$

Step 1. (Deletion)
Apply the "Deletion rule" as explained later in Sections 12.4 and 12.5 to the grid.

Step 2. (Subdivision or Bisection)
Apply the "Subdivision (bisection) rule" as explained in Sections 12.4 and 12.5 to the grid.

Step 3. (New (P_h) and (D_h))
Formulate new dual LPs (P_h)–(D_h) by deleting/augmenting constraints/variables to (P_h)–(D_h). Solve them by the simplex method.

Step 4. (Convergence Check)
Stop the process if the subdivision parameter as explained in Sections 12.4 and 12.5 becomes less than the tolerance. Otherwise go back to Step 1.

12.4 DETAILS OF THE METHOD WHEN Z IS ONE-DIMENSIONAL

In this section, we will show details of the method in case $\mathbf{Z}$ is one dimensional. Cases with dim $(\mathbf{Z}) > 1$ will be discussed in Section 12.5.

12.4.1 Initial Discretization and Subdivision Parameter

Let the set $\mathbf{Z}$ be $[a,b] \subset \mathrm{R}$ and arrange the *grid* $z_0, \ldots, z_n$ as

$$a = z_0 < z_1 < \cdots < z_n = b \tag{12.1}$$

where

$$z_i = z_0 + i(b-a)/n \quad (i = 0, \ldots, n). \tag{12.2}$$

We define the *subdivision parameter* (or mesh size) T to be

$$T = (b-a)/n. \quad \text{(the length of an interval)} \tag{12.3}$$

12.4.2 Solving (D_h)

We solve the dual program (D_h) by means of the simplex method. The reason for dealing with the dual program will be clarified later on. The optimal information related to the primal program is easily obtained from the optimal basis of (D_h).

Let the optimal solution to (P_h) and (D_h) be

$$\mathbf{u} = (u_1, \ldots, u_s)^T, \ \mathbf{v} = (v_1, \ldots, v_m)^T \tag{12.4}$$

and

$$\theta, \ \boldsymbol{\lambda} = (\lambda_0, \ldots, \lambda_n)^T, \ \mathbf{s}^+ = (s_1^+, \ldots, s_s^+)^T, \ \mathbf{s}^- = (s_1^-, \ldots, s_m^-)^T. \tag{12.5}$$

12.4.3   Deletion/Subdivision Rules

Since the optimal solutions (12.4)–(12.5) solve the discretized problems, we have at grid point z_i,

$$\mathbf{u}^T\mathbf{y}(z_i) - \mathbf{v}^T\mathbf{x}(z_i) = 0, \text{ if } \lambda_i > 0 \qquad (12.6)$$

and

$$\mathbf{u}^T\mathbf{y}(z_i) - \mathbf{v}^T\mathbf{x}(z_i) \leq 0, \text{ if } \lambda_i = 0. \qquad (12.7)$$

However, it is not certain if the relations

$$\mathbf{u}^T\mathbf{y}(z) - \mathbf{v}^T\mathbf{x}(z) \geq 0 \qquad (12.8)$$

hold for every $z \in \mathbf{Z}$.

Let

$$\psi(z) \equiv \mathbf{u}^T\mathbf{y}(z) - \mathbf{v}^T\mathbf{x}(z). \qquad (12.9)$$

The *discrepancy* $\delta(\mathbf{u}, \mathbf{v})$ of $(\mathbf{u}, \mathbf{v})$ is defined as

$$\delta(\mathbf{u},\mathbf{v}) = \max_{z \in [a,b]} \psi(z). \qquad (12.10)$$

Theorem 12.1

An upper bound to $\delta(\mathbf{u}, \mathbf{v})$ is given by

$$\Delta = \left(FMT^2\right)/8 \qquad (12.11)$$

[2], where F is an upper bound to u_r $(r = 1,\dots,s)$ and v_k $(k = 1,\dots,m)$,

$$M = \max_{z \in [a,b]} \left(\sum_{r=1}^{s} |y_r''(z)| + \sum_{k=1}^{m} |x_k''(z)| \right) < \infty \qquad (12.12)$$

and T is defined by (12.3). (See Appendix 12.A for a proof.)

Theorem 12.2

If at two successive grid points z_i and z_{i+1} we have

$$\psi(z_i) < -\Delta \text{ and } \psi(z_{i+1}) < -\Delta,$$

then it follows that

$$\psi(z) < 0 \quad \text{for every } z \in [z_i, z_{i+1}].$$

(See Appendix 12.B for a proof.)

Thus, we have the deletion rule for grids.

[Deletion Rule]

If at three successive grid points z_i, z_{i+1}, and z_{i+2}, we have

$$\psi(z_i) < -\Delta, \psi(z_{i+1}) < -\Delta \text{ and } \psi(z_{i+2}) < -\Delta, \tag{12.13}$$

then we delete z_{i+1} and hence the whole interval (z_i, z_{i+2}) from further consideration. Notice that the rule needs to be changed a little at the boundary points.

[Subdivision Rule]

We subdivide the remaining intervals by introducing a new grid at the mid-point of each interval.

Thus, we have

$$\text{new } T = T/2 \quad \text{and} \tag{12.14}$$

$$\text{new } \Delta = \Delta/4. \tag{12.15}$$

Remark 1 Usually it is not easy to determine Δ as defined by (12.11). In such a case, Δ should be taken to be a *threshold* for deleting grid points. A smaller Δ deletes more grid points. If $\psi(z)$ is well approximated by a quadratic curve at a local maximum, the relation (12.15) will generally hold after the subdivision.

12.4.4   Solving the New LP

We delete the columns corresponding to the deleted grid points from the dual tableau and introduce new columns corresponding to the new grid points to the tableau. The new columns will be priced out by using the optimal dual basis of the preceding iteration and the primal simplex method will determine the new optimal solution.

12.4.5   Convergence Check

We stop the iterations if T comes to satisfy, for some tolerance T_{tol},

$$T < T_{\text{tol}}. \tag{12.16}$$

Remark 2 A typical process of subdivision (or bisection) is sketched in Figure 12.1, where the curves represent $\psi(z)$ with z as abscissa and the tolerance $(-\Delta)$ for each iteration is given by the dashed line.

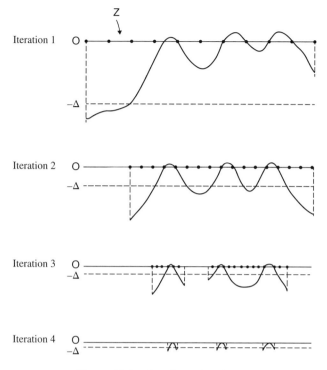

Figure 12.1   One-dimensional case.

12.5   GENERAL CASE

In this section, we will deal with the dual pair of problems (P)–(D) when $\mathbf{Z}$ is a compact convex set with $L = \dim(\mathbf{Z}) > 1$.

12.5.1   Initial Discretization

We discretize $\mathbf{Z}$ by using L-dimensional cubes with edge length T. The mesh points are the initial grid points $\{\mathbf{z}^1, \ldots, \mathbf{z}^n\}$. The grid points are used to formulate (P_h) and (D_h), which are solved by the simplex method. Let the optimal solutions be $\mathbf{u}$, $\mathbf{v}$, θ, λ, $\mathbf{s}^+$, and $\mathbf{s}^-$.

12.5.2   Deletion and Subdivision (Bisection) Rules

Every grid has at most $2L$ neighbors.

[Deleting Rule]

If the relation

$$\psi(\mathbf{z}) \equiv \mathbf{u}^T \mathbf{y}(\mathbf{z}) - \mathbf{v}^T \mathbf{x}(\mathbf{z}) < -\Delta \qquad (12.17)$$

holds at a grid and its neighbors, then we delete the center grid point and edges connecting the center with its neighbors from further consideration. $-\Delta$ is a threshold similar to (12.11) (see also Remark). For higher dimensional Ls, it would be difficult to estimate Δ by a formula such as (12.11). A practical way to estimate Δ is as follows:

After the initial LPs are solved, we estimate the discrepancy $\delta(\mathbf{u}, \mathbf{v})$ by sampling $\mathbf{z}$ from $\mathbf{Z}$. The value will be used as the initial Δ, which will be updated by dividing by 4 at each iteration.

[Subdivision Rule]

We divide the remaining edges by introducing a new grid at the midpoint of each edge. Thus we have

$$\text{new } T = T/2 \qquad (12.18)$$

and

$$\text{new } \Delta = \Delta/4 \qquad (12.19)$$

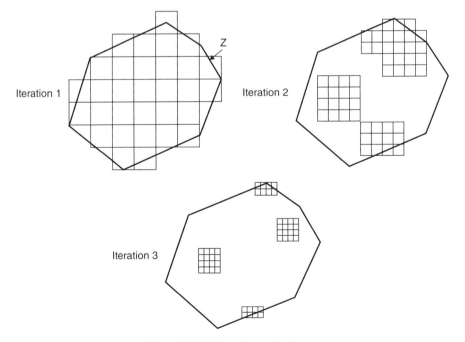

Figure 12.2 Two-dimensional case.

12.5.3   Solving New LPs and Checking Convergence

These steps are quite similar to those in the one-dimensional case as explained in Subsections 12.4.4 and 12.4.5.

Remark 3 A typical subdivision process of the two-dimensional $\mathbf{Z}$ is depicted in Figure 12.2.

12.6   CONCLUDING REMARKS (BY TONE)

As described in Section 12.2, the input function $\mathbf{x}(\mathbf{z})$ and output function $\mathbf{y}(\mathbf{z})$ must be defined on the compact set $\mathbf{Z}$ in advance. The purpose of this model is to obtain the efficient inputs or outputs or both factors in terms of $\mathbf{Z}$. Such situations may arise in the case of design problems. Although Dr. Charnes did not explicitly disclose the purpose of this study at the time, I think it was an optimal design problem.

APPENDIX 12.A   PROOF OF THEOREM 12.1

Assume that $\delta(\mathbf{u}, \mathbf{v})$ has a maximum at z^* in $z_i < z^* < z_{i+1}$ and that $\psi(z_i) \leq 0$, $\psi(z^*) > 0$, and $\psi(z_{i+1}) \leq 0$. Furthermore, without losing generality, we assume that z^* is closer to z_{i+1} than to z_i, i.e., $z_{i+1} - z^* \leq T/2$. We expand $\psi(z_{i+1})$ around $\psi(z^*)$ as

$$\psi(z_{i+1}) = \psi(z^*) + (z_{i+1} - z^*)\psi'(z^*) + \frac{(z_{i+1} - z^*)^2}{2}\psi''(z^{**})$$

where $z^* < z^{**} < z_{i+1}$.

Since $\psi(z) \in C^2$ and has a maximum at z^*, we have $\psi'(z^*) = 0$. Hence, it holds that

$$\psi(z_{i+1}) = \psi(z^*) + \frac{(z_{i+1} - z^*)^2}{2}\psi''(z^{**}) \leq 0$$

Thus, we have

$$\psi(z^*) \leq \frac{T^2}{8}|\psi''(z^{**})| = \frac{T^2}{8}|\mathbf{u}^T\mathbf{y}''(z^{**}) - \mathbf{v}^T\mathbf{x}''(z^{**})|$$

$$\leq \frac{T^2}{8}(|\mathbf{u}^T\mathbf{y}''(z^{**})| + |\mathbf{v}^T\mathbf{x}''(z^{**})|) \leq (FMT^2)/8$$

APPENDIX 12.B PROOF OF THEOREM 12.2

Assume that $\psi(z_i) < -\Delta$, $\psi(z_{i+1}) < -\Delta$, and $\max_{z_i \leq z \leq z_{i+1}} \psi(z) = \psi(z^*)$. Furthermore, without losing generality, we assume that z^* is closer to z_{i+1} than to z_i, i.e., $z_{i+1} - z^* \leq T/2$. Then, we have

$$\psi(z_{i+1}) = \psi(z^*) + \frac{(z_{i+1} - z^*)^2}{2} \psi''(z^{**}) \leq -\Delta$$

Hence, it holds that

$$\psi(z^*) < -\Delta - \frac{(z_{i+1} - z^*)^2}{2} \psi''(z^{**}) < -\Delta + \frac{T^2}{8} |\psi''(z^{**})|$$

$$\leq -\Delta + \frac{T^2 F}{8} \max_{z \in [a,b]} \left(\sum_{r=1}^{s} |y_r''(z)| + \sum_{k=1}^{m} |x_k''(z)| \right) = -\Delta + \Delta = 0$$

REFERENCE

[1] Gustafson, S.Å. and Kortanek, K.O. (1973) Numerical treatment of a class of semi-infinite programming problems. *Naval Research Logistics Quarterly*, **20**, 3, 477–507.

PART II

DEA APPLICATIONS (PAST–PRESENT SCENARIO)

13

EXAMINING THE PRODUCTIVE PERFORMANCE OF LIFE INSURANCE CORPORATION OF INDIA[1]

KAORU TONE

National Graduate Institute for Policy Studies, Tokyo, Japan

BIRESH K. SAHOO

Xavier Institute of Management, Xavier University, Bhubaneswar, India

13.1 INTRODUCTION

Though agriculture has been the main preoccupation of the bulk of the Indian population, policy planners in India want to see how important for the development of the country the significant contribution from insurance services is. According to government sources, the insurance and banking services' contribution to the country's gross domestic product (GDP) is 7%, and the funds available to the state-owned Life Insurance Corporation (LIC) for investments are approximately 8% of GDP. Our objective in this chapter is therefore to empirically examine the performance behavior of LIC as a case study.

LIC was formed in September 1, 1956 with a capital contribution of 5 crore (1 crore = 10 million) of rupees (the rupee is the Indian currency, and is denoted by "Rs.") from the Government of India. Since nationalization, the life insurance

---

[1] Part of this chapter is based upon Tone, K. & Sahoo, B.K. (2005) [1], "Evaluating Cost Efficiency and Returns to Scale in the Life Insurance Corporation of India Using Data Envelopment Analysis", *Socio-Economic Planning Sciences*, **39**(4), 261–285, with permission from Elsevier.

Advances in DEA Theory and Applications: With Extensions to Forecasting Models,
First Edition. Edited by Kaoru Tone.
© 2017 John Wiley & Sons Ltd. Published 2017 by John Wiley & Sons Ltd.

business in India has been coterminous with the state-owned LIC. LIC has played a dominant role in the economic development of the country in two ways. First, as a life insurer, it has served to pool and distribute life risks associated with the millions of deaths of earners (policyholders). Life insurance has thus served the twin purposes of an economic and a social security umbrella to millions of households, especially the rural poor and senior citizens (in terms of providing savings for old age). Second, as a major savings institution, LIC has been a dominant financial intermediary, channeling funds to the productive sectors of the economy, mostly financing government-sponsored planned development programs [2].

Since its inception, LIC has grown manyfold. LIC's new business (individual) in terms of sum assured has gone up from Rs.283.07 crore in 1957 to Rs.1,24,950.63 crore in 2000–01, and in terms of the number of individual policies it has increased from 8.16 lakh (1 lakh = 0.1 million) to 196.65 lakh for the same period, reflecting more than 15% average annual growth in the post-1980s period in terms of both sum assured (real) and number of policies. Another main indicator of growth, the individual business in force in terms of sum assured, grew from about Rs.1473 crore in 1957 to Rs.6,45,042 crore in 2000–01.

In the preliberalization period, LIC sold mostly savings plans. These were tax-efficient (with exemptions) compared with other common forms of saving. Protection business was a relatively small proportion of its total business, and riders were not popular. Before liberalization, distribution was entirely via agencies. But, in response to changing needs and requirements over time, the Corporation has been devising various products, albeit at a modest pace, to spread the message of life insurance, and this has been reflected in increased sales as seen above. Among its various products, endowment assurance (participating) and money back (participating) are the most popular, comprising 80% of the life insurance business. To further growth, LIC has recently launched "Bima Plus," the first unit-linked plan in the country; it is reducing the guarantees on its single premium product, and reportedly repricing its annuity products in a bid to improve profitability; it is linking all its branches into a computer network, thereby enabling it to establish arrangements with various internet gateways to allow the payment of premiums through the internet; it is engaging premier educational institutions in India to train its employees in areas such as human resource development, marketing, investment, and information technology; it is introducing a portfolio of riders to compete with the wide range of riders offered by new entrants; and, finally, in response to the growth of bancassurance, it is taking equity stakes and forming significant bancassurance ties with Corporation Bank and Oriental Bank of Commerce. Also, to reduce the likelihood of future competition from private insurers, LIC is in the process of bundling savings and investment plans, offering attractive returns.

Notwithstanding the phenomenal growth of LIC and its efforts to diversify its product range to spread its life insurance business, the life insurance business in India falls ways below the achievements in developed countries [3]. For instance, according to estimates reported by Swiss Reinsurance Company, insurance penetration in 1997 was 1.39%, compared with 9.42% in Japan. Insurance density in 1997 was $5.4, compared with \$3092 in Japan. Besides, the performance of LIC has come under close scrutiny with regard to its operational efficiency [4], especially in terms of its financial performance. Opening up of the insurance sector to both domestic and foreign companies has been at the center of policy debate alongside financial sector reforms as part of the macroeconomic stabilization cum structural adjustment programs initiated in 1991.

Despite this history, insurance sector reforms had to wait until the end of the year 2000 owing to a contentious and politically charged debate over the pervasive implications of privatization and foreign participation in the insurance sector. See Rao [5] and Ranade and Ahuja [6] for a detailed discussion of the likely implications of privatization and foreign participation in the life insurance sector, including regulatory-related issues. After a prolonged stalemate, following the recommendations of an official committee, the Committee on Reforms in the Insurance Sector, popularly known as Malhotra Committee 1996, which has recommended privatization and foreign participation in the insurance sector, the Insurance Regulatory and Development Authority (IRDA) had issued licenses to 11 life insurers and six nonlife insurers by the end of the year 2000.

Though there are compelling arguments in favor of both parametric and nonparametric approaches to the estimation of cost efficiency and returns to scale, we choose the latter because they do not require the specification of arbitrary functional forms and because they have the natural advantage of eliminating the effects of all productive and scale inefficiencies prior to calculating returns to scale. Recent applications of data envelopment analysis (DEA) models to the insurance sector include, among others, work on the efficiency of organizational forms and distribution systems in the US property and liability insurance industry [7, 8]. However, there has been no such study, to our knowledge, which has applied DEA to evaluating the performance of the Indian insurance sector. The current chapter, using aggregate time series data, thus utilizes DEA to evaluate LIC's performance in terms of both cost efficiency and returns to scale for the period 1982–83 to 2000–01.

The rest of this chapter unfolds as follows. Section 13.2 first discusses the various measures of scale elasticity in the DEA literature, then points out their limitations, and finally introduces a new variant of the DEA model to circumvent these limitations. The dataset for LIC operations is discussed in Section 13.3. Section 13.4 deals with results and provides a discussion, followed by concluding remarks in Section 13.5.

13.2 NONPARAMETRIC APPROACH TO MEASURING SCALE ELASTICITY

Throughout, we deal with a number n of firms; each uses m inputs to produce s outputs. For firm h, we denote the input and output vectors by $x_h \in R^m$ and $y_h \in R^s$, respectively. The input and output matrices are defined by $\mathbf{X} = (x_1,\ldots,x_n) \in R^{m \times n}$ and $\mathbf{Y} = (y_1,\ldots,y_n) \in R^{s \times n}$. We assume that $\mathbf{X} > 0$ and $\mathbf{Y} > 0$.

13.2.1   Technology and Returns to Scale

The standard neoclassical characterization of the production function for multiple outputs and multiple inputs is the transformation function $\psi(x, y)$, which satisfies the following properties:

$$\psi(x,y) = 0, \frac{\partial \psi(x,y)}{\partial y_r} < 0 \ (\forall r) \text{ and } \frac{\partial \psi(x,y)}{\partial x_i} > 0 \ (\forall i) \tag{13.1}$$

The returns to scale (RTS), or scale elasticity in production (ρ_p), degree of scale economies (DSE), or Passus coefficient, is defined as the ratio of the maximum proportional expansion (β) of outputs to a given proportional expansion (μ) of inputs. So, differentiating the transformation function $\Psi(\mu x, \beta y) = 0$ with respect to the scaling factor μ and then equating it to zero yields the following local scale elasticity measure:

$$\rho_p(x,y) \equiv - \sum_{i=1}^{m} x_i \frac{\partial \psi}{\partial x_i} \bigg/ \sum_{r=1}^{s} y_r \frac{\partial \psi}{\partial y_r} \tag{13.2}$$

See Hanoch [9], Starrett [10], Panzar and Willig [11] and Baumol et al. [12] for a detailed discussion. For a discussion of the evolution of the concept of scale and the computational procedure for it in DEA, see, for example, Sahoo et al. [13], Tone and Sahoo [1, 14–16], Sengupta and Sahoo [17], Podinovski et al. [18], Podinovski and Førsund [19], Sahoo and Tone [20, 21], Sahoo and Sengupta [22], and Podinovski et al. [23], among others.

However, in the case of a single-input, single-output technology, ρ_p is simply expressed as the ratio of marginal product (MP) to average product (AP), that is,

$$\rho_p(x,y) \equiv \frac{\text{MP}}{\text{AP}} = \frac{dy/dx}{y/x} \tag{13.3}$$

For a neoclassical "S-shaped" production function (or *regular ultra Passum law* (RUPL) in the words of Frisch [24]), ρ_p takes on values ranging from "greater than one" for suboptimal output levels, through "one" at the optimal scale level, to values "less than one" at superoptimal output levels. So, the production function satisfies RUPL if $\partial \rho_p / \partial y < 0$ and $\partial \rho_p / \partial x < 0$ [25]. The RTS are increasing, constant, and decreasing if $\rho_p > 1$, $\rho_p = 1$, and $\rho_p < 1$, respectively.

Following Baumol et al. [12], the dual measure of the production elasticity, called the cost elasticity (ρ_c), is defined in a multiple-input and multiple-output environment as

$$\rho_c \equiv C(y;w) \bigg/ \sum_{r=1}^{s} y_r \frac{\partial C(y;w)}{\partial y_r} \tag{13.4}$$

where $C(y; w)$ is the minimum cost of producing the output vector y when the input price vector is w. However, ρ_c can be expressed as the ratio of average cost to marginal cost in the case of a single output. The RTS are increasing, constant, or decreasing depending upon whether $\rho_c > 1$, $\rho_c = 1$, or $\rho_c < 1$, respectively.

13.2.2 Qualitative Information on Returns to Scale

The CCR input-oriented model [26], which is based on the assumption of constant returns to scale (CRS), is used to qualitatively describe the local RTS for firm h:

$$[\text{CCR}] \quad \min \theta$$

$$\text{subject to} \quad -\sum_{j=1}^{n} x_{ij}\lambda_j + \theta x_h \geq 0 \ (\forall i), \ \sum_{j=1}^{n} y_{rj}\lambda_j \geq y_h \ (\forall r), \ \lambda_j \geq 0 \ (\forall j) \tag{13.5}$$

If $\sum_{j=1}^{n} \lambda_j^* = 1$ for any alternate optima, then CRS prevails for firm h; if $\sum_{j=1}^{n} \lambda_j^* < 1$ for all alternate optima, then increasing returns to scale (IRS) prevails; and if $\sum_{j=1}^{n} \lambda_j^* < 1$ for all alternate optima, then decreasing returns to scale (DRS) prevails.

The dual of the BCC model [27], which is based on the assumption of variable returns to scale (VRS), is also used to obtain qualitative information about the local RTS for firm h:

$$[\text{BCC}] \quad \max \varphi = \sum_{i=1}^{m} u_r y_{rh} + u_o$$

$$\text{subject to} \quad \sum_{r=1}^{s} u_r y_{rj} - \sum_{i=1}^{m} v_i x_{ij} + u_o \leq 0, \ (\forall j), \ \sum_{r=1}^{s} v_i x_{ih} = 1, \ u_r, \ v_i \geq 0, \text{ and } u_o : \text{free}.$$

$$\tag{13.6}$$

If $u_o^* = 0$ (where $*$ represents the optimal value) in any alternate optima, then CRS prevails for firm h; if $u_o^* > 0$ for all alternate optima, then IRS prevails; and if $u_o^* < 0$ for all alternate optima, then DRS prevails for firm h.

Färe et al. [28] introduced the following "scale efficiency index" (SEI) method, which is based on nonincreasing returns to scale (NIRS), to determine the nature of the local RTS for firm h as follows:

$$[\text{SEI}] \quad \min f$$

$$\text{subject to} \quad -\sum_{j=1}^{n} x_{ij}\lambda_j + f x_h \geq 0 \ (\forall i), \ \sum_{j=1}^{n} y_{rj}\lambda_j \geq y_h \ (\forall r), \ \sum_{j=1}^{n} \lambda_j \leq 1, \ \lambda_j \geq 0 \ (\forall j) \tag{13.7}$$

If $\theta^* = \phi^*$, then firm h exhibits CRS; otherwise, if $\theta^* < \phi^*$, then firm h exhibits IRS iff $\phi^* > f^*$, and firm h exhibits DRS iff $\phi^* = f^*$.

These three different RTS methods are equivalent to estimating the RTS parameter [29,30]. In empirical applications, however, one finds that the CCR and BCC RTS methods may fail when DEA models have alternate optima. However, the scale efficiency index method does not suffer from the above problem, and hence is found to be robust.

In the light of all possible multiple-optima problems in the CCR and BCC methods, Banker and Thrall [31] generalized the structure by introducing new variables u_o^+ and u_o^-, which represent optimal solutions obtained by solving the dual of the output-oriented BCC model. In Banker and Thrall's approach, the constraint $\sum u_r y_{ro} + u_o = 1$ was added, while the objective function was replaced by either $u_o^+ = \max u_o$ or $u_o^- = \min u_o$. It was shown that IRS operates iff $u_o^+ \geq u_o^- > 0$, DRS operates iff $0 > u_o^+ \geq u_o^-$, and CRS operates iff $u_o^+ \geq 0 \geq u_o^-$.

Banker et al. [32] pointed out that the concept of RTS is unambiguous only at points on the efficient facets of the production technology. So the RTS for inefficient units may depend upon whether the efficiency estimation is done in an input-oriented or output-oriented manner. A detailed method for doing so can be found in the studies of Banker et al. [29], Tone [33], and Cooper et al. [34].

13.2.3 Quantitative Information on Returns to Scale

We will discuss the quantitative evaluation of both production and cost elasticity, then point out their limitations, and then suggest an alternative measure to get rid of such limitations.

13.2.3.1 Production Elasticity If firm h is efficient in [BCC], then it holds that

$$\sum_{r=1}^{s} u_r^* y_{rh} - \sum_{i=1}^{m} v_i^* x_{ih} + u_o^* = 0$$

In order to unify multiple outputs and multiple inputs, let us define a scalar output y and a scalar input x as $y = \sum_{r=1}^{s} u_r^* y_{rh}$ and $x = \sum_{i=1}^{m} v_i^* x_{ih}$, respectively Then, we have a relationship between the output (y) and the input (x) $y = x - u_o^*$. From this equation, we find $MP = dy/dx = 1$ and $AP = y/x = 1 - u_o^*$, since $x = \sum_{r=1}^{s} v_i^* x_{ih} = 1$. Now, the production elasticity (ρ_p) is defined as

$$\rho_p = \frac{MP}{AP} = \frac{1}{1 - u_o^*} \tag{13.8}$$

However, if firm h is inefficient, then ρ_p equals $\varphi^*/(\varphi^* - u_o^*)$. The RTS are increasing, constant, and decreasing if $u_o^* > 0$, $u_o^* = 0$, and $u_o^* < 0$, respectively.

Note here that, as pointed out by Førsund and Hjalmarsson [25], the production elasticity ρ_p does not satisfy fully the requirement of RUPL, as

$$\frac{\partial \rho_p(x,y)}{\partial y_{rh}} = -\frac{u_o^*(\partial\varphi/\partial y_{rh})}{(\varphi^* - u_o^*)^2} = -u_o^* u_{rh} \left/ \left(\sum_{r=1}^{s} u_r y_{rh}\right)^2 \right. \quad (\forall r)$$

IRS ($u_o^* > 0$) implies decreasing production elasticity, which is in accordance with RUPL, while DRS ($u_o^* < 0$) implies an increasing ρ_p, thus violating the law.

The evaluation of production elasticity has been extended to network DEA models [35, 36].

13.2.3.2 *Cost Elasticity* Sueyoshi [37, 38] used the following dual of the VRS cost DEA model:

$$[\text{COST}] \quad \gamma^* = \max \sum_{r=1}^{s} u_r y_{ro} + \omega_o$$

$$-\sum_{i=1}^{m} v_i x_{ij} + \sum_{r=1}^{s} u_r y_{rj} + \omega_o \leq 0,\ (\forall j),\ v_i \leq w_i,\ (\forall i),\ u_r,\ v_i \geq 0,\ (\forall r,\ i),\ \omega_o : \text{free}$$

$$\tag{13.9}$$

to compute the cost elasticity for firm h (where $*$ represents the optimal value). Following Baumol *et al.* [12], he computed the cost elasticity (ρ_c) at (w_h, y_h) as

$$\rho_c = \frac{\gamma^*}{\displaystyle\sum_{r=1}^{s} u_r^* y_{rh}} \tag{13.10}$$

and showed the equivalence of IRS to $\rho_c > 1$, CRS to $\rho_c = 0$, and DRS to $\rho_c < 1$.

It should be noted here that under the assumption of a unique optimal solution, the production elasticity (ρ_p) in the BCC model and the cost elasticity (ρ_c) in the VRS cost model are same when $\phi^* = 1$ and $v_o^* = \omega_o^*/(\omega_o^* - \gamma^*)$. Otherwise,

$$\frac{\rho_c}{\rho_p} = \frac{1 - \dfrac{\omega_o^*}{\omega_o^* - \gamma^*}}{1 - \dfrac{1}{\varphi^*} v_o^*} \tag{13.11}$$

The details of the duality relationship between ρ_p and ρ_c can be found in Cooper *et al.* [39] and Sueyoshi [38].

13.2.4 An Alternative Measure of Scale Elasticity

The DEA model [COST] (13.9) may be of limited use in actual applications, as this model is based on a number of simplifying assumptions. First, not only are factor inputs homogeneous, but also their prices are exogenous. As a result, the scale elasticities in both the production and the cost environments are equal, thus giving the illusion that RTS and economies of scale are the one and same.

With an expansion in production, firms experience *changes* in the organization of their processes or in the characteristics of their inputs that are economically more attractive than the replicated alternatives of those already in use. Therefore, the technique and inputs used at higher scale are very different from those used at lower scale. Hence, the inputs are heterogeneous and, as a result of this, their prices may vary across firms. Since the input resources vary in their quality, the construction of the *technology* in (13.9) becomes problematic.

Input prices are also not exogenous, but instead vary according to the actions of firms. Firms often face *ex ante* price uncertainty when making production decisions. Economic theory suggests that firms enjoying some degree of monopoly power should charge different prices if there is productivity heterogeneity in their inputs. This is empirically valid, since most firms face an upward-sloping supply curve in their input purchase decisions. This observation also suggests that the assumption of common prices for firms, that is, the law of one price, which has long been maintained as a necessary and sufficient condition for Pareto efficiency in competitive markets, is not at all justified when one is aiming to reveal the proper scale economy behavior of firms when market imperfections exist in any form.

Second, the factor-based technology employed in (13.9) is convex. Convexity, as argued by Farrell [40], assumes away some important technological features such as *indivisible* production activities, *economies of scale*, and *economies of specialization*, which all in fact result from *concavities* in production.

Third, the [COST] model (13.9) may also be of limited value in actual applications even when the inputs are homogeneous. This is because, as pointed out by several scholars [20–22, 41–44], the cost efficiency (CE) reflects only input inefficiencies (i.e., *technical inefficiency* and/or *allocative efficiency*) and not *market* (price) inefficiencies. Therefore, these authors suggested a very comprehensive scheme to measure CE that can be attributed to both inputs and market inefficiencies.

Note that when market imperfections exist, ρ_h is not very comprehensive, as it involves the cost effects of output expansion only. To make it comprehensive, one needs to link ρ_h with further cost reductions due to other sources such as pecuniary economies. Therefore, when the inputs are heterogeneous, in order to account for varying input prices, the alternative CE model of Tone [45] should be used, where the technology is defined in cost–output space so as to account for varying input heterogeneity.

Let us describe Tone [45]'s cost DEA model. The cost-based technology T_c is

$$T_c = \left\{ (\bar{x}, y) : \bar{x} \geq \bar{X}\lambda, \, y \leq Y\lambda, \, \lambda \geq 0 \right\} \tag{13.12}$$

where $\bar{\mathbf{x}} = (\bar{x}_1, \ldots, \bar{x}_n)$ with $\bar{\mathbf{x}}_j = \left(w_{1j}x_{1j}, \ldots, w_{mj}x_{mj} \right)^T$. Based on this new production possibility set T_c, a new technical and scale efficiency (NTSE), $\bar{\theta}^*$, is obtained as the optimal solution of the following LP problem:

$$[\text{NTech}_{\text{crs}}] \quad \bar{\theta}^* = \min \bar{\theta}$$

$$\text{subject to} \quad \bar{\theta}\bar{x}_h \geq \bar{X}\lambda, \ y_h \leq Y\lambda, \ \lambda \geq 0 \tag{13.13}$$

Similarly, a new technical efficiency of firm h is computed from the $[\text{NTech}_{\text{vrs}}]$ model, which is obtained by imposing a convexity constraint ($\mathbf{e}\lambda = 1$) in $[\text{NTech}_{\text{crs}}]$, where $\mathbf{e} \in R^n$ is a row vector with each of its elements equal to one.

The new overall scale efficiency (NOSE), $\bar{\gamma}^*$, is defined as $\bar{\gamma}^* = e\bar{x}_h^* / e\bar{x}_h$, where $\bar{x}_h^*$ is the optimal solution of the LP given below:

$$[\text{NCost}_{\text{crs}}] \quad \min e\bar{x}$$

$$\text{subject to} \quad \bar{x} \geq \bar{X}\lambda, \ y_h \leq Y\lambda, \ \lambda \geq 0 \tag{13.14}$$

The new allocative scale efficiency (NASE), $\bar{\alpha}^*$, is then defined as the ratio of $\bar{\gamma}^*$ to $\bar{\theta}^*$, that is, NASE $(\bar{\alpha}^*) = $ NOSE $(\bar{\gamma}^*)$/NTSE $(\bar{\theta}^*)$. Similarly, the $[\text{NCost}_{\text{vrs}}]$ model can be introduced by adding a convexity constraint ($\mathbf{e}\lambda = 1$) in $[\text{NCost}_{\text{crs}}]$, where the new allocative efficiency is obtained as the ratio of the new overall efficiency to the new technical efficiency. It should be noted here that the NOSE is not greater than the NTSE, and these new efficiency measures are all units invariant.

The dual of the $[\text{NCost}_{\text{vrs}}]$ model can be represented by the following LP problem:

$$[\text{NCost}_{\text{vrs}}] \quad \delta = \max \ \sum_{r=1}^{s} u_r y_{rh} + \sigma_1 - \sigma_2$$

$$\text{subject to} \ -\sum_{i=1}^{m} v_i \bar{x}_{ij} + \sum_{r=1}^{s} u_r y_{rj} + \sigma_1 - \sigma_2 \leq 0, \ (\forall j), \ v_i = 1 \ (\forall i), \ u_r \geq 0 \ (\forall r), \ \sigma_1 \geq 0, \ \sigma_2 \geq 0$$

$$\tag{13.15}$$

The primal and dual of $[\text{NCost}_{\text{vrs}}]$ can be considered as special forms of the assurance region (AR) DEA model of Thompson *et al.* [46, 47] and the cone ratio (CR) model of Charnes *et al.* [48, 49], respectively, where the availability of the reasonable price vectors enters as input weights in the general DEA model. See also Schaffnit *et al.* [50] for a detailed discussion.

If firm h is efficient, then it holds that $-\sum_{i=1}^{m} v_i \bar{x}_{ih} + \sum_{r=1}^{s} u_r y_{rh} + \sigma_1 - \sigma_2 = 0$. Unifying the total cost (c) as $\sum_{i=1}^{m} v_i \bar{x}_{ih} = \sum_{i=1}^{m} \bar{x}_{ih}$, and the total output ($y$) as $\sum_{r=1}^{s} u_r y_{rh}$, the cost–output relationship is represented as $c = y + \sigma_1 - \sigma_2$. From this, we derive the marginal cost (MC) as $dc/dy = 1$ and the average cost (AC) as $c/y = \delta^* / \sum_{r=1}^{s} u_r y_{rh}$.

Now we define the cost elasticity (ρ_c) as

$$\rho_c = \frac{\text{AC}}{\text{MC}} = \frac{\text{AC}}{1} = \delta^* / \left(\sum_{r=1}^{s} u_r^* y_{rh} \right) = \delta^* / u^* y_h \qquad (13.16)$$

The degree of scale economies cannot be uniquely determined at $(w_h = e, y_h)$ only when there is a problem of degeneracy, that is, when there are multiple supporting hyperplanes. The upper and lower bounds of ω then need to be identified from the following LP model:

$$\max/\min \sigma_1 - \sigma_2$$

$$\text{s.t.} -\sum_{i=1}^{m} v_i \bar{x}_{ij} + \sum_{r=1}^{s} u_r y_{rj} + \sigma_1 - \sigma_2 \leq 0 \ (\forall j),$$

$$v_i = 1 \ (\forall i), \ \sum_{r=1}^{s} u_r y_{rh} = \delta^*, \ u_r \geq 0 \ (\forall r), \ \sigma_1 \geq 0, \ \sigma_2 \geq 0 \qquad (13.17)$$

The problem of degeneracy in the unique determination of returns to scale in production-based DEA models has been discussed extensively by Banker and Thrall [31], Banker *et al.* [29, 32], and Tone [33].

Note that the measure of RTS defined in (13.16) is very different from the standard measure discussed by Aly *et al.* [51]. While the former is derived from the cost–output-based technology set, the latter is from the input–output-based technology set.

13.3   THE DATASET FOR LIC OPERATIONS

As with all service sectors, the measurement of output in the insurance sector is an insurmountable problem [52]. Therefore, insurance, being essentially a service industry, thus requires a distinct set of criteria for carrying out such an exercise. Let us first briefly discuss why conventional financial ratios are not meaningful output measures for financial intermediaries. The principal reason is that such intermediaries do not exist to produce financial ratios; rather, they seek to produce financial services. Output measures should thus be a proxy for the volume of financial services provided. For example, in the case of CRS, if the inputs are increased by 10%, then the outputs should increase by 10%, which does not necessarily occur with a financial ratio, since a larger ratio is not necessarily better than a smaller one. For many ratios, such as capital-to-assets or the liquid assets ratio, there is likely to be some optimal value for the ratio, such that the firm is worse off if the ratio is much lower or higher than the optimum. For example, investing more in liquid assets is fine, up to a point, beyond which the firm would begin to encounter operating constraints due to underinvestment in nonliquid (capital) assets such as computers.

Another problem with the use of financial ratios is that there are many such ratios used by financial analysts and regulators in judging the financial health of an institution. No one ratio necessarily dominates any others as a measure of financial stability; and, again, virtually none of these ratios has a monotonic relationship to input or output quantities. The main problem is that financial ratios are quality variables, not output proxies. Quality variables certainly have a role to play in evaluating a firm, but they should not be used to represent output quantities.

The question then becomes what measure or measures to use as a proxy for the volume of financial services. As suggested in the literature, for a proxy for outputs in the financial sector, one should seek a measure or measures that are highly correlated with the volume of financial services provided. Premiums might seem to be a logical measure of output volume, but this is not necessarily the case. As Yuengert [52] pointed out, premiums equal price times quantity, whereas output volumes should represent only quantity. This suggests a loss-based measure, which has been used in the majority of existing studies of insurance efficiency (see, e.g., Cummins and Weiss [53]). This could be losses incurred or, in life insurance, benefits incurred plus addition to reserves. Losses are an appropriate measure because the purpose of insurance is to pool the experience of all policyholders and pay claims to those who suffer loss during a given period. Losses are also highly correlated with other services provided by insurers, such as financial planning.

It has also been argued that losses might not be appropriate because insurers can sustain unusually high losses owing to random fluctuations and that paying for these higher-than-expected losses does not represent output. This argument is incorrect, however, because one of the important financial services provided by an insurer is the payment of losses even when they are higher than expected. This is called the residual risk-bearing function in the literature.

Considering these difficulties, and to overcome them, a modified version of the value added approach to measuring life insurance output was adopted in our study. The value added approach counts as important outputs those that represent significant value added, as judged using operating cost allocations [54]. We follow the recent insurance efficiency literature by defining insurance output as the present value of real losses incurred (e.g., Berger *et al.* [55] and Cummins *et al.* [8]). We have taken the losses as the claims settled during the year, including claims written back (y_1). Losses were deflated to a base of 1994–95 using the Consumer Price Index (CPI). The CPI data were taken from the *International Financial Statistics Year Book*, 1999.

Following the study of Brockett *et al.* [7], the ratio of liquid assets to liabilities (y_2) was considered for use as a second output in our study. This ratio reflects a company's claims-paying ability, and is an important objective of an insurance firm, with improvement in claims-paying ability contributing to the likelihood of attracting and retaining customers. Despite its importance, however, this second output was completely dropped from our analysis because this output, along with the first output, was tested and found not to be meaningful. The reasons behind the occurrence of such results can be viewed from two angles. First, as already discussed above, financial intermediaries do not exist to produce financial ratios; rather, they exist to produce financial services. Second, without knowing whether it is a ratio or a volume measure, a closer look at the data reveals that there is little variation in this ratio, indicating the company's constant claims-paying ability. So taking this ratio as another output is just like taking a constant output term for all firms in any DEA model, which has, in essence, no effect on the efficiency scores.

Insurance inputs can be classified into four groups: business services (x_1), labor (x_2), debt capital (x_3), and equity capital (x_4). The business services were taken as commission to agents, which is material input, and this was deflated by the CPI. The input price for business services (w_1) was calculated by dividing the total deflated commission to agents by the total number of active agents. The labor variable was taken as the total number of employees. The price per unit of labor (w_2) was calculated by dividing the total deflated salary and other benefits to employees by the total number of employees.

The debt capital of insurers consists of funds borrowed from policyholders. These funds were measured in real terms as the life insurance fund deflated using the CPI. The price of the policyholder-supplied debt capital (w_3) was the rate of interest realized on the mean life insurance fund. Equity capital is an input for the risk-pooling function because it provides assurance that the company can pay claims even if there are larger than expected losses. The equity capital was taken as the sum of shareholders' paid-up capital, a general reserve, a reserve for bad and doubtful debts, loans, a reserve for house property, and an investment reserve. This value of equity capital deflated by the CPI was considered an input category. Following Gutfinger and Meyers [56], the price of equity capital (w_4) was taken as 9% + rate of inflation. To summarize, we used four inputs: business services, labor, policyholder-supplied debt capital, and equity capital. The dataset, related to LIC's operations in 19 annual periods, is summarized in Table 13.1.

Our primary data source was the annual statements of LIC for the period from 1982–83 to 2000–01. Though LIC has several branches all over India, the relevant data are not available for each of these branches. The annual statement of LIC is the only database which compiles aggregate figures of the necessary operational and financial data for all its branches. In the spirit of earlier studies by Boussofiane *et al.* [57], Ray and Kim [58], and Sueyoshi [37, 38], we have treated each of LIC's 19 years of operation as a distinct firm.

13.4   RESULTS AND DISCUSSION

The analysis of efficiency on the input side rather than the output side is becoming common in DEA applications for a variety of reasons. First, real-world managers are never given a bundle of inputs and told to produce the maximum output from it. Instead, they are given output targets and told to produce them in the most efficient way possible, that is, with the minimum inputs. Second, profitability in any business hinges on the efficiency of operations. But if the business involves a commodity, then what depends on efficient operations is survival. When prices are beyond a company's control, what remain are costs of inputs. This reflects companies' emphasis on the input dimensions of policies. On a tentative basis, it has been suggested in the literature that costs (or inputs) are generally more predictable than outputs, giving cost targets a greater credibility than those for outputs. Sengupta [59] has argued that: "… data variations may arise in practical situations … when the output measures have large and uncertain measurement errors which are much more significant than in the input measures (p. 2,290). For example in school efficiency studies, the input costs, such as teachers' salaries, administrative expenses, etc., may have low measurement errors whereas the performance test scores of students may contain large errors of measurement of true student quality." This argument is most compelling where measurement errors are large relative to true random fluctuations in the production process.

The efficiency estimates were calculated using the assumption of CRS for the reference technology. As pointed out by Färe et al. [60], this technology has some useful features in that it captures the notion of maximal average product (consistent with the minimum point on a long-run U-shaped average cost curve), which provides a very nice benchmark for identifying the optimal scale.

Three-way analysis was done from our efficiency/RTS estimates, the first two elements being the production-and cost-based analysis and the third one being the RTS. The production-based results are reported in Table 13.1.

TABLE 13.1   Production-based efficiency scores.

Year	TSE	S-SBM	AR	NTSE	NS-SBM
1982–83	0.851	0.730	0.722	1	1.004
1983–84	0.915	0.856	0.747	1	1.014
1984–85	1	1.026	0.825	1	1.048
1985–86	0.991	0.967	0.831	0.980	0.950
1986–87	0.994	0.949	0.828	0.967	0.957
1987–88	0.952	0.784	0.860	0.978	0.837
1988–89	0.921	0.733	0.891	0.939	0.808
1989–90	0.896	0.702	0.879	0.929	0.775
1990–91	0.907	0.721	0.888	0.923	0.780
1991–92	0.910	0.749	0.893	0.909	0.824
1992–93	0.978	0.865	0.962	1	1.006
1993–94	0.994	0.911	0.980	0.988	0.946
1994–95	1	1.052	1	1	1.037
1995–96	0.951	0.862	0.925	0.897	0.852
1996–97	1	1.005	0.980	0.961	0.946
1997–98	0.996	0.987	0.962	0.997	0.890
1998–99	0.990	0.967	0.926	1	1.016
1999–00	0.991	0.971	0.943	1	1.001
2000–01	1	1.081	1	1	1.064

13.4.1 Production-Based Analysis

It can be seen from Table 13.1 that though the technical and scale efficiency scores exhibit a slightly upward trend, the efficiency scores were consistently high (around one) since 1994–95. In order to differentiate the efficient units, we have reported the super-slack-based measure (S-SBM) efficiency scores introduced by Tone [61, 62]. Since input price data were available, we have also used the assurance region model to calculate an AR efficiency score, where the weight ratios (v_i/v_j), bounded between $\min(w_i/w_j)$ and $\max(w_i/w_j)$ for all $i < j$, were as follows:

$\min(w_i/w_j)$ (v_i/v_j) $\max(w_i/w_j)$

$$0.172344 \leq v_1/v_2 \leq 0.407958$$
$$0.000153 \leq v_1/v_3 \leq 0.000266$$
$$0.000083 \leq v_1/v_4 \leq 0.000209$$
$$0.000638 \leq v_2/v_3 \leq 0.001271$$
$$0.000279 \leq v_2/v_4 \leq 0.001123$$
$$0.427670 \leq v_3/v_4 \leq 0.972358$$

Use of this AR model serves two purposes. First, it addresses the issue of the degrees-of-freedom problem (our data are for 19 years only), and second, it protects against the frequent occurrence of zero weights for some of the inputs. We see here that although the AR scores rise, they do so only up to 1994–95. They then decline until 1998–99, after which they rise to 2000–01. However, if we consider the new efficiency scores (NTSE) obtained from the cost-based production technology, the overall trend remains more or less constant. Nevertheless, the year-specific score suggests that full efficiency was maintained for the first three years, followed by a declining trend until 1991–92. Scores then remained high, approaching unity after 1997–98.

The improvement in technical efficiency, particularly after 1997–98, can be claimed to arise from two phenomena. First, LIC has of late geared itself up to face future competition. It has devised a more tailor-made, diversified product range, bundling savings and investment plans offering attractive returns. It is also going through the process of overhauling itself, with significant decentralization in the management and organizational structure so as to make itself more efficient. But, what is more important is the changing macroeconomic environment in India. After an initial stock market boom, especially in the information and technology sector, which started with liberalization and gained momentum around 1993–94, households swayed by the speculative stock market boom diverted a significant proportion of their financial savings into investing in the stock market. But, after the collapse of stock prices as the information and technology stock price boom changed to bust, households lost confidence in the stock market and resorted to secured forms of savings such as banks and insurance. See Rao [63] for a discussion of household financial savings behavior and macroeconomic dynamics in India. Thus, a combination of both an improvement in the efficiency of LIC and the macroeconomic environment appears to have a definite bearing on technical efficiency.

Leibenstein [64] maintains that the theoretical basis for claims that exposure to competition will generate improvement in efficiency is the notion of X-efficiency (or technical efficiency). He argues that enterprises exposed to the bracing atmosphere of competition will respond by eliminating internal inefficiency and seeking out opportunities for innovation. He refers to the productivity gains arising from this process as improvement in X-efficiency. To Stigler [65], this X-efficiency gain is nothing but simply an increase in the intensity of labor or, equivalently, a reduction in on-the-job leisure. Ganley and Grahl [66] pointed out that where labor productivity has increased owing to such competition, there is evidence of increased work intensity.

A closer look at our dataset reveals that labor productivity shows a monotonically increasing trend, confirming the above-mentioned claim of increased work intensity. Further, LIC has recently adopted information technology; for instance, it has used UnixWare 7 to link over 2000 branches throughout India and to serve approximately 11.6 million customers. UnixWare 7 links LIC's local area networks, metropolitan area networks, wide area networks, and interactive voice response system, and LIC has also adopted other labor-saving technologies. This allows each branch office to act as a stand-alone entity with mutual access to all transactions, information, and computer support for all policyholders. This adds further support to our finding of LIC running efficiently in terms of technical and scale dimensions.

13.4.2   Cost-Based Analysis

Since the cost-based efficiency scores obtained using the earlier cost–DEA model seem to be misleading, we have decided not to report them. Rather, we report here our new cost and new allocative efficiency scores in Table 13.2.

We see here that, contrary to our AR trend, the NOSE trend is of decline up to 1991–92, with an abrupt rise in year 1992–93, after which the trend is again of decline up to 1999–2000. The year 2000–01 again sees a marked increase in efficiency. The declining trend in the NOSE scores up to 1991–92 is due principally to the fall in the NTSE scores, whereas the declining trend after 1992–93 is due to a fall in the new allocative efficiency.

Since LIC has pursued computerization vigorously in recent times, it has incurred substantial costs for such modernization. Therefore, it is not surprising to see that the cost efficiency has either shown a fluctuating trend or declined from 1992–93 to 1999–00, as it will take a substantial amount of time for any organization to internalize the initial high fixed cost incurred in modernization of its operations. However, again as expected, the cost efficiency has shown a significant increase from 1999–00, and hopefully LIC will continue to show this cost efficiency accrual.

13.4.3   Returns-to-Scale Issue

Table 13.3 presents the estimated minimum cost, and the infimum, supremum, and average of the scale elasticity (represented by Inf ρ_c, Sup ρ_c, and Avg. ρ_c, respectively), and the RTS in our new VRS cost model. We find here that LIC operates under IRS for the first two years, followed by CRS in 1984–85, after which DRS applies.

TABLE 13.2   Cost-based efficiency scores.

Year	NOSE	NASE
1982–83	0.979	0.979
1983–84	0.959	0.959
1984–85	1	1
1985–86	0.973	0.994
1986–87	0.937	0.969
1987–88	0.958	0.979
1988–89	0.933	0.994
1989–90	0.921	0.991
1990–91	0.913	0.988
1991–92	0.892	0.981
1992–93	0.988	0.988
1993–94	0.953	0.964
1994–95	0.975	0.975
1995–96	0.879	0.979
1996–97	0.915	0.952
1997–98	0.903	0.906
1998–99	0.882	0.882
1999–00	0.873	0.873
2000–01	0.960	0.960

TABLE 13.3   Scale elasticity and RTS.

Year	[Ncost$_{vrs}$]	Inf ρ_c	Sup ρ_c	Avg. ρ_c	RTS
1982–83	25.297	1.106	∞	∞	IRS
1983–84	27.86	1.096	1.096	1.096	IRS
1984–85	31.8	0.977	1.083	1.03	CRS
1985–86	34.244	0.978	0.978	0.978	DRS
1986–87	37.349	0.98	0.98	0.98	DRS
1987–88	39.928	0.981	0.981	0.981	DRS
1988–89	46.452	0.984	0.984	0.984	DRS
1989–90	51.586	0.985	0.985	0.985	DRS
1990–91	56.049	0.987	0.987	0.987	DRS
1991–92	62.847	0.988	0.988	0.988	DRS
1992–93	67.783	0.951	0.951	0.951	DRS
1993–94	78.478	0.957	0.957	0.957	DRS
1994–95	84.512	0.96	0.96	0.96	DRS
1995–96	95.676	0.965	0.965	0.965	DRS
1996–97	107.688	0.969	0.969	0.969	DRS
1997–98	113.097	0.97	0.97	0.97	DRS
1998–99	127.337	0.973	0.973	0.973	DRS
1999–00	143.954	0.976	0.976	0.976	DRS
2000–01	155.139	0.000	0.978	0.489	DRS

13.4.4   Sensitivity Analysis

Since the selection of outputs is problematic in the insurance literature, it is worth testing sensitivity, using premiums or transaction-based variables such as the number of policies or the number of claims settled as output variables. As we know, the use of time series data for one firm over 19 years and retaining a technology specification with more than one output at a time, along with four inputs, might lead to objections from a methodological viewpoint. The small number of observations and the detailed specification of the technology can lead to dimensionality problems. High dimensionality generates statistical problems in the convergence of DEA estimators [67] and in the form of model misspecification [68]. With only 19 observations, the pertinence of an analysis in a six-or-more-dimensional space might be questionable. For example, the lack of possible comparisons may explain most of the NTSE scores being one. This problem is also highlighted by quasi-systematic zero shadow prices of the second, third, and fourth outputs in some of these years (not shown here). We thus decided to consider each of these outputs separately. The sensitivity analysis was carried out using the number of policies, premiums, and number of claims separately as output variables along with the four inputs. The overall and scale efficiency scores are plotted in Figure 13.1.

The sensitivity analysis shows that two measures (NOSE_1 and NOSE_2) follow the pattern of NOSE without throwing up any dramatic changes except for a

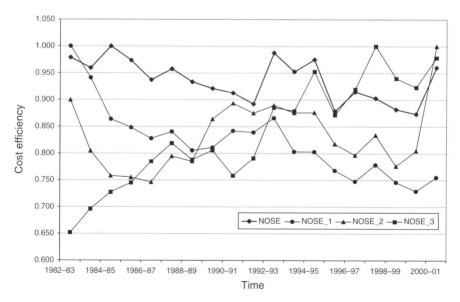

Figure 13.1   Sensitivity analysis of cost efficiency. NOSE: overall and scale efficiency when output is "real loss incurred." NOSE_1: overall and scale efficiency when output is "number of policies." NOSE_2: overall and scale efficiency when output is "premium income." NOSE_3: overall and scale efficiency when output is "claims settled including claims written back."

continuous improvement in claim settlement (NOSE_3). This is due to the fact that LIC has had a good reputation throughout the last two decades for being efficient in settling claims in the shortest possible time [62].

13.5 CONCLUDING REMARKS

The results on the performance trends of Life Insurance Corporation of India suggest a significant heterogeneity in overall and scale efficiencies over the 19-year study period. More importantly, there has been a downward trend in performance, measured in terms of cost efficiency, since 1994–95. This decline in performance is due to the huge initial fixed cost of modernizing the corporation's operations. A significant increase in cost efficiency in 2000–01 suggests that LIC may be beginning to benefit from such modernization, which will stand it in good stead in terms of future competition.

Future extensions of this research study include, first, the development of a non-linear DEA model accounting for the relationship between input price and input quantity, as cost has a linkage with production changes (e.g., a bulk purchase), and second, the development of new concepts concerning technical, cost, and allocative efficiencies by formulating a time series DEA cost model.

REFERENCES

[1] Tone, K. and Sahoo, B.K. (2005) Evaluating cost efficiency and returns to scale in the Life Insurance Corporation of India using data envelopment analysis. *Socio-Economic Planning Sciences*, **39**, 261–285.

[2] Rao, D.T. (1998) Operational efficiency of Life Insurance Corporation of India. *Journal of Indian School of Political Economy*, **10**, 473–489.

[3] Rao, D.T. (2000) Privatisation and foreign participation in (life) insurance sector. *Economic and Political Weekly*, **35**, 1107–1120.

[4] Malhotra, R.N. (1994) Report of the Committee on Reforms in the Insurance Sector. Government of India, Ministry of Finance.

[5] Rao, D.T. (1999) Life insurance business in India: Analysis of performance. *Economic and Political Weekly*, **34**, 2174–2181.

[6] Ranade, A. and Ahuja, R. (2000) Issues in regulation of insurance. *Economic and Political Weekly*, **35**, 331–338.

[7] Brockett, P.L., Cooper, W.W., Golden, L.L., Rousseau, J.J., and Wang, Y. (1998) DEA evaluations of the efficiency of organizational forms and distribution systems in the US property and liability insurance industry. *International Journal of Systems Science*, **29**, 1235–1247.

[8] Cummins, J.D., Weiss, M.A., and Zi, H. (1999) Organizational form and efficiency: The coexistence of stock and mutual property-liability insurers. *Management Science*, **45**, 1254–1269.

[9] Hanoch, G. (1970) Homotheticity in joint production. *Journal of Economics Theory*, **2**, 423–426.

[10] Starrett, D.A. (1977) Measuring returns to scale in the aggregate, and the scale effect of public goods. *Econometrica*, **45**, 1439–1455.

[11] Panzar, J.C. and Willig, R.D. (1997) Economies of scale in multi-output production. *Quarterly Journal of Economics*, **41**, 481–493.

[12] Baumol, W.J., Panzar, J.C., and Willig, R.D. (1982) *Contestable Markets and the Theory of Industry Structure,* Harcourt Brace Jovanovich, New York.

[13] Sahoo, B.K., Mohapatra, P.K.J., and Trivedi, M.L. (1999) A comparative application of data envelopment analysis and frontier translog production function for estimating returns to scale and efficiencies. *International Journal of Systems Science*, **30**, 379–394.

[14] Tone, K. and Sahoo, B.K. (2003) Scale, indivisibilities and production function in data envelopment analysis. *International Journal of Production Economics*, **84**, 165–192.

[15] Tone, K. and Sahoo, B.K. (2004) Degree of scale economies and congestion: A unified DEA approach. *European Journal of Operational Research*, **158**, 755–772.

[16] Tone, K. and Sahoo, B.K. (2006) Re-examining scale elasticity in DEA. *Annals of Operations Research*, **145**, 69–87.

[17] Sengupta, J.K. and Sahoo, B.K. (2006) *Efficiency Models in Data Envelopment Analysis: Techniques of Evaluation of Productivity of Firms in a Growing Economy,* Palgrave Macmillan, London.

[18] Podinovski, V.V., Førsund F.R., and Krivonozhko,V.E. (2009) A simple derivation of scale elasticity in data envelopment analysis. *European Journal of Operations Research*, **197**, 149–153.

[19] Podinovski, V.V. and Førsund, F.R. (2010) Differential characteristics of efficient frontiers in data envelopment analysis. *Operations Research*, **58**, 1743–1754.

[20] Sahoo, B.K. and Tone, K. (2013) Non-parametric measurement of economies of scale and scope in non-competitive environment with price uncertainty. *Omega*, **41**, 97–111.

[21] Sahoo, B.K. and Tone, K. (2015) Scale elasticity in non-parametric DEA approach, in *Data Envelopment Analysis: A Handbook of Models and Methods* (ed. J. Zhu), Springer, New York, pp. 269–290.

[22] Sahoo, B.K. and Sengupta, J.K. (2014) Neoclassical characterization of returns to scale in nonparametric production analysis. *Journal of Quantitative Economics*, **12**, 78–86.

[23] Podinovski, V.V., Chambers, R.G., Atici, K.B., and Deineko, I.D. (2016) Marginal values and returns to scale for nonparametric production frontiers. *Operations Research*, DOI: http://dx.doi.org/10.1287/opre.2015.1457.

[24] Frisch, R. (1965) *Theory of Production,* Reidel, Dordrecht.

[25] Førsund, F.R. and Hjalmarsson, L. (2004) Are all scales optimal in DEA? Theory and empirical evidence. *Journal of Productivity Analysis*, **21**, 25–48.

[26] Charnes, A., Cooper, W.W., and Rhodes, E. (1978): Measuring the efficiency of decision making units. *European Journal of Operational Research*, **2**, 429–444.

[27] Banker, R.D., Charnes, A., and Cooper, W.W. (1984) Some models for estimating technical and scale inefficiencies in data envelopment analysis. *Management Science*, **30**, 1078–1092.

[28] Färe, R., Grosskopf, S., and Lovell, C.A.K. (1985) *The Measurement of Efficiency of Production,* Kluwer Nijhoff, Boston, MA.

[29] Banker, R.D., Bardhan, I., and Cooper, W.W. (1996) A note on returns to scale in DEA. *European Journal of Operational Research*, **88**, 583–585.

[30] Färe, R., Grosskopf, S., and Lovell, C.A.K. (1994) *Production Frontiers,* Cambridge University Press, Cambridge.

[31] Banker, R.D. and Thrall, R.M. (1992) Estimation of returns to scale using data envelopment analysis. *European Journal of Operational Research*, **62**, 74–84.

[32] Banker, R.D., Chang, H., and Cooper, W.W. (1996): Equivalence and implementation of alternative methods for determining returns to scale in data envelopment analysis. *European Journal of Operational Research*, **89**, 473–481.

[33] Tone, K. (1996) A simple characterization of returns to scale in DEA. *Journal of the Operations Research Society of Japan*, **39**, 604–613.

[34] Cooper, W.W., Seiford, L.M., and Tone, K. (2000) *Data Envelopment Analysis: A Comprehensive Text with Models, Applications, References and DEA-Solver Software,* Kluwer Academic, Boston, MA.

[35] Sahoo, B.K., Zhu, J., Tone, K., and Klemen, B.M. (2014) Decomposing technical efficiency and scale elasticity in two-stage network DEA. *European Journal of Operational Research*, **233**, 584–594.

[36] Sahoo, B.K., Zhu, J. and Tone, K. (2014) Decomposing efficiency and returns to scale in two-stage network systems, in *Data Envelopment Analysis: A Handbook of Modeling Internal Structure and Network* (eds W.D. Cook and J. Zhu), Springer, New York, pp. 137–164.

[37] Sueyoshi, T. (1997) Measuring efficiencies and returns to scale of Nippon Telegraph & Telephone in production and cost analyses. *Management Science*, **43**, 779–796.

[38] Sueyoshi, T. (1999) DEA duality on returns to scale (RTS) in production and cost analyses: An occurrence of multiple solutions and differences between production-based and cost-based RTS estimates. *Management Science*, **45**, 1593–1608.

[39] Cooper, W.W., Thompson, R.G., and Thrall, R.M. (1996) Introduction: Extensions and new developments in DEA. *Annals of Operations Research*, **66**, 3–46.

[40] Farrell, M.J. (1959) Convexity assumptions in theory of competitive markets. *Journal of Political Economy*, **67**, 377–391.

[41] Camanho, A.S. and Dyson, R.G. (2005) Cost efficiency measurement with price uncertainty: A DEA application to bank branch assessments. *European Journal of Operational Research*, **161**, 432–446.

[42] Camanho, A.S. and Dyson, R.G. (2008) A generalization of the Farrell cost efficiency measure applicable to non-fully competitive settings. *Omega*, **36**, 147–162.

[43] Sahoo, B.K., Kerstens, K., and Tone, K. (2012) Returns to growth in a non-parametric DEA approach. *International Transactions in Operational Research*, **19**, 463–486.

[44] Sahoo, B.K., Mehdiloozad, M., and Tone, K. (2014) Cost, revenue and profit efficiency measurement in DEA: A directional distance function approach. *European Journal of Operational Research*, **237**, 921–931.

[45] Tone, K. (2002) A strange case of the cost and allocative efficiency in DEA. *Journal of Operational Research Society*, **53**, 1225–1231.

[46] Thompson, R.G., Langemeier, L.N., Lee, C.T., and Thrall, R.M. (1990) The role of multiplier bounds in efficiency analysis with application to Kansas farming. *Journal of Econometrics*, **46**, 93–108.

[47] Thompson, R.G., Singleton, F.D., Thrall, R.M., and Smith, B.A. (1986) Comparative site evaluations for locating high-energy lab in Texas. *Interfaces*, **16**, 1380–1395.

[48] Charnes, A., Cooper, W.W., Wei, Q.L., and Huang, Z.M. (1989) Cone ratio data envelopment analysis and multi-objective programming. *International Journal of Systems Science*, **20**, 1099–1118.

[49] Charnes, A., Cooper, W.W., Huang, Z.M., and Sun, B. (1990) Polyhedral cone-ratio DEA models with an illustrative application to large commercial banks. *Journal of Econometrics*, **46**, 71–91.

[50] Schaffnit, C., Rosen, D., and Paradi, J.C. (1997) Best practice analysis of bank branches: An application of DEA in a large Canadian bank. *European Journal of Operational Research*, **98**, 269–289.

[51] Aly, H.Y., Grabowski, R., Pasurka, C., and Rangan, N. (1990) Technical, scale, and allocative efficiencies in US banking: An empirical investigation. *Review of Economics and Statistics*, **72**, 211–218.

[52] Yuengert, A.M. (1993) The measurement of efficiency in life insurance: Estimates of a mixed normal-gamma error model. *Journal of Banking and Finance*, **17**, 483–496.

[53] Cummins, J.D. and Weiss, M.A. (2000) Analyzing firm performance in the insurance industry using frontier efficiency and productivity methods, in *Handbook of Insurance (ed. G. Dionne), Kluwer Academic, Boston, MA.*

[54] Berger, A.N. and Humphrey, D.V. (1992) Measurement and efficiency issues in commercial banking, in *Output Measurement in the Services Sector* (ed. Z. Griliches), University of Chicago Press, Chicago, IL.

[55] Berger, A.N., Cummins, J.D., and Weiss, M.A. (1997) The coexistence of multiple distribution systems for financial services. *Journal of Business*, **70**, 515–546.

[56] Gutfinger, M. and Meyers, S. (2000) Embedded value – Part 2 – Implementation issues. 2nd Global Conference of Actuaries, Actuarial Society of India, Delhi Chapter.

[57] Boussofiane, A., Dyson, R.G., and Thanassoulis, E. (1991) Applied data envelopment analysis. *European Journal of Operational Research*, **52**, 1–15.

[58] Ray, S.C. and Kim, H.L. (1995) Cost efficiency in the U.S. steel industry: A nonparametric analysis using data envelopment analysis. *European Journal of Operational Research*, **80**, 654–671.

[59] Sengupta, J.K. (1987) Efficiency measurement in non-market systems through data envelopment analysis. *International Journal of Systems Science*, **18**, 2279–2304.

[60] Färe, R., Grosskopf, S., and Norris, M. (1997) Productivity growth, technical progress and efficiency change in industrialized countries: reply. *American Economic Review*, **87**, 1040–1043.

[61] Tone, K. (2001) A slacks-based measure of efficiency in data envelopment analysis. *European Journal of Operational Research*, **130**, 498–509.

[62] Tone, K. (2002) A slacks-based measure of super-efficiency in data envelopment analysis. *European Journal of Operational Research*, **143**, 32–41.

[63] Rao, D.T. (2001) Economic reforms, anatomy of behavioural agents and macroeconomic outcomes: A critical review of the performance of Indian economy. *Journal of Indian School of Political Economy*, **13**, 401–427.

[64] Leibenstein, H. (1966) Allocative efficiency vs. X-efficiency. *American Economic Review*, **56**, 392–45.

[65] Stigler, G.J. (1976) The existence of X-efficiency. *American Economic Review*, **66**, 213–216.

[66] Ganley, J. and Grahl, J. (1988) Competitive tendering and efficiency in refuse collection: A critical comment. *Fiscal Studies*, **9**, 80–85.

[67] Korostelev, A.P., Simar, L., and Tsybakov, A.B. (1995) Efficient estimation of monotone boundaries. *Annals of Statistics*, **23**, 476–489.

[68] Olesen, O.B. and Petersen, N.C. (1996) Indicators of ill-conditioned data sets and model misspecification in data envelopment analysis: An extended facet approach. *Management Science*, **42**, 205–219.

14

AN ACCOUNT OF DEA-BASED CONTRIBUTIONS IN THE BANKING SECTOR

JAMAL OUENNICHE

Business School, University of Edinburgh, Edinburgh, UK

SKARLETH CARRALES

Business School, University of Edinburgh, Edinburgh, UK

KAORU TONE

National Graduate Institute for Policy Studies, Tokyo, Japan

HIROFUMI FUKUYAMA

Faculty of Commerce, Fukuoka University, Fukuoka, Japan

14.1 INTRODUCTION

The banking sector plays a crucial socio-economic role at the regional, national and international levels. Banks are at the heart of financial systems; in fact, they act as financial intermediaries. To be more specific, banks borrow money by accepting deposits and issuing debt securities, and lend money both directly to their customers and indirectly through capital markets by investing in debt securities. Banks play an important role in money supply and the efficient allocation of financial resources in an economy. They make profits in exchange for their services, including risk

Advances in DEA Theory and Applications: With Extensions to Forecasting Models,
First Edition. Edited by Kaoru Tone.

management. Nowadays, banks have a diversified portfolio of activities that range from personal, corporate and investment banking to trading of currency, commodities and financial securities on stock markets. Because of the crucial importance of banking systems to the economy and the financial risks they face, banks are required to comply with both national and international regulations, and their performance is constantly monitored by both regulatory bodies and investors. In fact, poor performance often leads to distress, which might lead to bankruptcy under some circumstances along with substantial undesirable financial, economic and social consequences. In this chapter, we shall report on the current state-of-the-art research on data envelopment analysis (DEA) in the banking sector, with an emphasis on static DEA methodologies.

DEA is a data-driven, non-parametric, frontier-based methodology originally designed for the evaluation of the relative performance of a set of entities commonly referred to as decision-making units (DMUs). Within a DEA framework, benchmarking is done with respect to the best or the worst peers rather than the average performers, which is the case for other methodologies such as stochastic frontier analysis. Since the publication of the seminal paper by Charnes, Cooper and Rhodes in 1978 [1], DEA has witnessed growing popularity amongst academics and practitioners, as suggested by the relatively large number of both methodological and application-oriented publications [2–4]. In banking, DEA typically addresses two types of problems, namely, performance evaluation problems and risk assessment problems. With respect to performance evaluation problems, the DEA literature on banking can be further divided into two categories depending on whether one is concerned with assessing the relative performance of banks or the relative performance of the branches of a given bank. As to risk assessment problems in the banking sector, the DEA literature could also be further divided into several categories depending on whether one is concerned with distress and bankruptcy of banks, or distress and default of a bank's customers. In this chapter, the focus is on assessing the relative performance of commercial banks.

The remainder of this chapter is organized as follows. In Section 14.2, we provide a detailed account of the literature on the performance evaluation of banks using static DEA methodologies. In Section 14.3, we provide a summary of the current state of the art. Finally, Section 14.4 concludes this chapter.

14.2   PERFORMANCE EVALUATION OF BANKS: A DETAILED ACCOUNT

In this section, we report in detail on the literature on the relative performance evaluation of banks using static DEA methodologies in chronological order. As early as 1938, empirical studies investigated the performance of banks and their risk of failure either directly or indirectly [5,6].

The first use of DEA in banking can be traced back to Rangan *et al.* [7], who investigated a sample of 215 US banks with data from 1986. They used the CCR model [1] to compute an *overall technical efficiency* index and the BCC model [8] to compute a *pure technical efficiency* index. These indexes or scores were computed using three inputs (i.e. labour, capital and purchased funds) and five outputs (i.e. real estate loans, commercial and industrial loans, consumer loans, demand deposits, and time and savings deposits). *Scale efficiency* was then computed as the ratio of the CCR score to the BCC score. The empirical results revealed that, on average, the banks in their sample had an overall technical efficiency index of 70% and that the source of inefficiency was mainly technical, as their scale efficiency index was 97%. In addition, after linearly regressing the overall technical efficiency and the pure technical efficiency against the bank size, the level of product diversity and the extent to which bank branching was allowed, the empirical results revealed that the technical efficiency of the banks was positively related to size, negatively related to product diversity, and not related to the extent to which branch banking was allowed.

In 1990, Ferrier and Lovell [9] used an input-oriented variable-returns-to-scale (VRS) model with both categorical and continuous environmental variables – an approach first proposed by Banker and Morey [10] – to assess the pure technical efficiency of a sample of 575 US banks with data from 1984. This model was fed with three inputs (i.e. labour, occupancy costs and expenditure on furniture and equipment, and expenditure on materials), five outputs (i.e. number of demand deposit accounts, number of time deposit accounts, number of real estate loans, number of instalment loans and number of commercial loans) and 12 environmental variables (i.e. average size of demand deposit account, average size of time deposit account, average size of real estate loan, average size of instalment loan, average size of commercial loan, location in unit or branch, number of branches operated, membership of a multibank holding company, and institutional type (non-commercial, savings and loan, mutual savings, and credit union)). They also used an input-oriented VRS cost allocation model with both categorical and continuous environmental variables to investigate the *cost efficiency* of banks by decomposing the amount by which cost is increased into technical and allocative inefficiencies, where their cost allocation model minimized the cost-weighted sum of inputs under a set of constraints similar to the above-mentioned VRS model with environmental variables. Their empirical results revealed that the banks in their sample exhibited a relatively high technical inefficiency and modest allocative inefficiency relative to a technology that exhibits increasing returns to scale, where the most efficient banks belonged to the smallest size class, and this efficiency advantage enabled them to compete despite the potential cost disadvantage they suffered owing to the structure of the efficient technology.

In the same year, Elyasiani and Mehdian [11] investigated the *rate of technological change* (RTC) of a sample of 191 US banks between 1980 and 1985, where the RTC was defined as $1 - \theta_{CCR-IO}^{1980;1985} / \theta_{CCR-IO}^{1980}$; $\theta_{CCR-IO}^{1980;1985}$ was the overall technical efficiency index computed by solving an input-oriented CCR model (CCR-I) using 1980 and 1985 data, and θ_{CCR-IO}^{1980} was the overall technical efficiency index computed by solving a CCR-I model using 1980 data only. Both of the CCR-IO models used four inputs (i.e. deposits, total demand deposits, capital and labour) and four outputs (i.e. investment, real estate loans, commercial and industrial loans, and other loans), where the choice of these inputs and outputs was motivated by an intermediation perspective on banks, where the intermediation approach or perspective considers banks as intermediation agents that collect funds and provide loans and other assets. In addition, RTCs were linearly regressed against the intensities of inputs and outputs obtained from the solution of CCR-IO models. The first-stage empirical results suggested that had the banks included in the sample been fully efficient in 1980, on average, they could have produced the same level of output with 89.55% of the inputs they actually used. Also, Elyasiani and Mehdian found that the efficiency frontier shifted inward between 1980 and 1985, reflecting a high pace of technological advancement achieved by the banks in the sample. The pace, however, varied significantly across the banks, with some banks even regressing over time. In the second-stage analysis, linear regression analysis revealed that technological change, over the sample period, was non-neutral and essentially labour biased.

At the same time, Aly *et al.* [12] investigated the overall technical, pure technical, scale, cost and allocative efficiencies of a sample of 322 independent US banks with data from 1986. The overall and pure technical efficiency measures were computed by solving a CCR-IO model and a BCC-IO model, respectively. Then, the scale efficiency measure was computed as the ratio of the CCR-IO score to the BCC-IO score. The cost efficiency measure – also known as the overall efficiency measure – was computed as the ratio of minimum cost to actual cost, where the minimum cost was determined by solving a cost allocation model under the constant returns-to-scale regime. Finally, the measure of *allocative efficiency* was computed as the ratio of cost efficiency to technical efficiency. The CCR-IO, BCC-IO and cost allocation models used three inputs (i.e. labour, capital and loanable funds) and five outputs (i.e. demand deposits, real estate loans, commercial and industrial loans, consumer loans, and other loans), and the costs used in the allocation model were the price of labour, as measured by the ratio of total expenditure on employees to the total number of employees, a proxy for the price of capital, as measured by the ratio of total expenditure on premises and fixed assets to book value, and the price of loanable funds, as measured by the ratio of the sum of interest expenses on time deposits and other loanable funds to loanable funds. The empirical results suggested a low level of overall efficiency, which was mainly technical in nature rather than allocative. In addition, it was found that the distributions of efficiency measures for branching and non-branching banks were not significantly different.

Charnes *et al.* [13] were the first to propose a cone-ratio (CR) CCR-IO model, which they used, with data from 1980 to 1985, to assess the relative performance of 48 US commercial banks drawn from the top 300 banks headquartered in America which were also members of Federal Deposit Insurance Corporation (FDIC). The CR-CCR-IO model was fed with four inputs (i.e. total operating expenses, total non-interest expenses, provision for loan losses and actual loan losses) and four outputs (i.e. total operating income, total interest income, total non-interest income and total net loans). The empirical results remain illustrative of DEA analysis.

Several studies revealed that minority-owned banks (MOBs) charged higher loan rates, paid lower deposit rates and yet consistently failed to achieve profitability ratios comparable to those of the non-minority-owned banks (NMOBs) – see, for example, Fukuyama *et al.* [14]. Elyasiani and Mehdian [15] looked into whether this phenomenon was due to technical, scale, cost and/or allocative inefficiencies or whether it was caused by factors outside the control of the MOB management (e.g. limited portfolio choices due to deposit instability, scarcity of profitable lending opportunities, higher operating costs due to neighbourhood location, and higher loan losses and information-gathering costs due to the particular clientele that MOBs serve), by investigating the relationship between bank ownership and efficiency for a sample of 160 US banks with data from 1988. Their CCR-IO, BCC-IO and cost allocation models were fed with four inputs (i.e. certificates of deposit and time and savings deposits; demand deposits; labour; and capital) and four outputs (i.e. commercial and industrial loans, real estate loans, other loans and investment securities), and the costs used in the allocation model were measured by the sum of interest on deposits, wages, and expenses on premises, machinery and equipment. The findings supported the hypothesis that, when the regional, regulatory, size and maturity characteristics of banks were abstracted, the efficiency differentials between MOBs and NMOBs were not statistically significant.

Yue [16] assessed the management of 60 US commercial banks for the period ranging from 1984 to 1990 using CCR-IO and weighted additive models with four inputs (i.e. interest expenses, non-interest expenses, transaction deposits and non-transaction deposits) and three outputs (i.e. interest income, non-interest income and total loans), where bank deposits were disaggregated into transaction and non-transaction deposits because they had different turnover and cost structures. The additive model was first proposed by Charnes *et al.* [17]. The weighting scheme used by the weighted additive model consisted of the inverses of the absolute values of the inputs and outputs. The efficiency score, however, was computed as follows:

$$\left(\sum_{i=1}^{m} x_{i,j}^* + \sum_{r=1}^{s} y_{r,j}^* \right) \Big/ \left(\sum_{i=1}^{m} x_{i,j} + \sum_{r=1}^{s} y_{r,j} + \sum_{r=1}^{s} s_{r,j}^+ \right)$$

where $x_{i,j}^*$ and $y_{r,j}^*$ denote the inputs and outputs, respectively, of the projection of DMU_j on the efficiency frontier. In addition, Yue also performed a window analysis to find out about the evolution of DEA efficiency scores and to identify the most stable and the most variable banks in terms of their seven-year average DEA scores. This paper has been included in our survey because of the quality of its pedagogical exposition of DEA. The empirical results remain illustrative of DEA analysis.

Some studies revealed that the *quality and efficiency of bank management* was a leading cause of failure [18–23], either by analysing financial indicators of non-failed and failed banks using statistical tests or by using modelling and prediction frameworks such as linear regression analysis, logistic regression analysis and discriminant analysis. Barr *et al.* [24] made use of a DEA model, namely, the CCR-IO model of Charnes, Cooper and Rhodes [1], to assess the managerial efficiency of banks for a sample of 930 US banks over a period ranging from December 1984 to December 1998.

They chose six inputs (i.e. full-time equivalent employees, salary expenses, premises and fixed assets, other non-interest expenses, total interest expenses, and purchased funds) and three outputs (i.e. core deposits, earning assets and total interest income) to capture the importance of management to a bank's survival – these variables were used as proxies to reflect the quality of management in making decisions related to input allocation and the product mix needed to attract deposits and make loans and investments. The empirical results revealed statistically significant differences in management quality scores between surviving and failing banks, which tended to increase as the failure date approached, suggesting that a DEA analysis could prove a valuable tool in detecting signs of distress before failure takes place. In 1994, the same authors [25], using the same sample of banks, compared the performance of two probit models with and without CCR scores as proxies for management quality, along with some financial ratios as proxies for the remaining dimensions of the CAMEL scoring system (i.e. equity capital/total loans as a proxy for capital adequacy, non-performing loans/total assets as a proxy for asset quality, net income/total assets as a proxy for earnings ability, and large deposits/total assets as a proxy for liquidity) and a proxy for the local economic climate (i.e. percentage of change in residential construction), in predicting bank failure with logit and probit models from the literature, and reported that the CCR-IO scores enhanced the classification accuracy of the model significantly. Then, in 1997, Barr and Siems performed an additional analysis with the same methodological choices as made by Barr *et al.* [25] and a sample of 1010 US banks to assess the sensitivity of the results to misclassification of costs, and reported similar findings [26].

Grabowski *et al.* [27] investigated the relative performance of two organizational forms, namely, branch banking and a bank holding company, by comparing the overall, allocative, technical, pure technical and scale efficiencies of a sample of 522 US banks affiliated to multibank holding companies and 407 US banks with branches, with data from 1989. The CCR-IO, BCC-IO and allocation models were fed with three inputs (i.e. labour, capital and loanable funds) and five outputs (i.e. real estate loans, commercial and industrial loans, consumer loans, demand deposits, and investment securities), and the costs used in the allocation model were the price of labour, as measured by the ratio of annual salaries plus employee benefits to the number of full-time equivalent employees on the payroll at the end of the year; the price of capital, as measured by the ratio of annual expenses for premises and fixed assets to the book value of the premises and fixed assets at the end of the year; and the price of loanable funds, as measured by the ratio of annual interest and expenses on time deposits and other borrowed funds to the dollar value of the end-of-the-year time deposits and the other borrowed funds. The empirical findings suggested that branch banking was a more efficient organizational form than a bank holding company.

Fukuyama [28] studied the performance of a sample of 143 Japanese commercial banks with data from 1991 by comparing their overall technical, pure technical, and scale efficiencies. The CCR-IO and BCC-IO models with VRS and non-increasing returns to scale (NIRS) used in this study were fed with three inputs (i.e. labour, capital and funds from customers) and two outputs (i.e. revenue from loans and revenue from other business activities) under the assumption that interest rates were the same for any loan type across banks. He also investigated the relationship between bank size (as measured by total assets, on the one hand, and total revenue, on the other hand) and returns to scale. Finally, he looked into whether the form of organization (i.e. city banks, regional banks or former *sogo* banks) implied different levels of efficiency, using non-parametric tests (i.e. the median test, Kruskal–Wallis test, van der Waerden test and Savage test) and analysis of variance. His empirical results suggested that the major cause of overall technical inefficiency was pure technical inefficiency, not scale inefficiency. Nonetheless, there still existed some degree of scale inefficiency. The scale inefficiency for pooled data was found to be mainly due to increasing returns to scale. When commercial banks were divided into three organizational forms – city banks, regional banks and former *sogo* banks – similar statements could be made for regional and former *sogo* banks, but not for city banks. With respect to both asset and revenue size definitions, scale efficiency was weakly associated with bank size, while a relationship of bank size to pure technical efficiency and to overall technical efficiency was not clearly indicated.

Favero and Papi [29] investigated the efficiency of a sample of 174 Italian banks with data from 1991 using a two-stage analysis framework. To be more specific, in the first stage, they analysed the technical and scale efficiencies of commercial banks using CCR-IO and BCC-IO scores derived under two different perspectives, namely, the asset approach and the intermediation approach. Under the asset approach, these models were fed with five inputs (i.e. labour, capital, financial capital available for investment, loanable funds (i.e. current accounts and savings deposits), certificates of deposit (CDs), and net funds borrowed by other banks) and three outputs (i.e. loans to other banks and non-financial institutions, investment in securities and bonds, and non-interest income). Under the intermediation approach, the same inputs and outputs were used except that current accounts and savings deposits were shifted from being inputs to being outputs. In the second stage, Favero and Papi linearly regressed the BCC-IO scores against size (measured by a categorical variable reflecting major, large, medium, small and minor sizes, which were defined with reference to deposits, capital and managed external funds), productive specialization (measured by the ratio of the profit from banking services to the total intermediation margin, where the latter was defined as the sum of profit from banking services, profit from non-banking services and interest margin), ownership (measured by a categorical variable, where POP = banche popolan, CR = Casse di Risparmio, BIN = banche di interesse nazionale, BCO = banche di credito ordinario and ICDP = istituti di credito di diritto pubblico), market structure (measured by the difference between the regional interest rate on loans and the average national interest rate on loans, weighted to take 'bad credit' into account), and localization (measured by two indicators, where the first indicator took account of the size of the population of the area of localization and whether that area was industrial or rural, and the second indicator was a categorical variable reflecting the region, namely, Northern Italy, Central Italy or Southern Italy). The empirical results suggested that, for the sample under consideration, Italian banks in 1991 operated on average at 88% of their potential overall technical efficiency and achieved about 97% of scale efficiency under the intermediation approach. These figures, however, were lower by 10% or so under the asset approach. The second-stage analysis revealed that specialization was the only variable that seemed to consistently explain the efficiency.

Zaim [30] investigated the effect of the 1980 financial liberalization of the banking sector in Turkey on the efficiency of a sample of 95 commercial banks by performing pre- and post-financial-liberalization analyses and comparing the overall, allocative, technical, pure technical and scale efficiencies of banks. The measures of these efficiencies were computed directly or indirectly by solving input-oriented CRS, VRS, IRS, NIRS and cost allocation models with both categorical and uncontrollable continuous environmental variables. These models were fed with four inputs (i.e. total number of employees, total interest expenditure, depreciation expenditure and expenditure on materials), four outputs (i.e. total balance of demand deposits, total balance of time deposits, total balance of short-term loans and total balance of long-term loans), and four environmental variables. Two of the latter were considered as uncontrollable inputs (i.e. number of branches and institutional type (1 for national banks and 0 for foreign banks)) and the other two as uncontrollable outputs (i.e. average size of demand deposit accounts and average size of time deposit accounts). In the cost allocation model, the price of labour was measured by the ratio of total expenditure on salaries and fringe benefits to the total number of employees; however, the prices of the remaining inputs were set to 1 on the assumption that all banks faced the same input prices. The empirical results, based on averages of DEA scores, suggested that the financial reform had succeeded in stimulating the commercial banks to take measures that would enhance both their technical and their allocative efficiencies. In addition, this study revealed that state banks were more efficient than their private counterparts, which for the Turkish banking industry contradicted the hypothesis that public ownership is inherently less inefficient. Furthermore, banks seemed to have gone through a considerable scale adjustment and were successful in achieving the optimal scale. Last but not least, the effects of allocative and technical inefficiencies on cost increases were different for private and state banks; to be more specific, while state banks were more vulnerable to allocative inefficiency, the effect of technical inefficiency on cost increases was more dominant for private banks.

Miller and Noulas [31] investigated the efficiency of a sample of 201 US large commercial banks with data from 1984 to 1990 using a two-stage analysis framework. In the first stage, they analysed the technical and scale efficiencies of banks using CCR-IO and BCC-IO scores. The models were fed with four inputs (i.e. total transaction deposits, total non-transaction deposits, total interest expenses and total non-interest expenses) and six outputs (i.e. commercial and industrial loans, consumer loans, real estate loans, investments, total interest income, and total non-interest income). In the second stage, Miller and Noulas linearly regressed the overall technical efficiency scores against bank size (measured by total assets), profitability (measured by the ratio of net operating income to total assets), market power (the ratio of bank deposits to the total deposits in the state within which the bank operated) and location (measured by several different dummy variables for location – one that reflected the degree of metropolitanization and two that captured regional aspects of the US). The empirical results suggested, on one hand, that the average inefficiency, including both pure technical and scale inefficiency, across all 201 banks was small at just over 5%, which was due to the stiffer competition for markets and market share in the late 1980s that forced more efficiency on bank operations, and that the majority of banks were too large and experienced decreasing returns to scale. On the other hand, larger and more profitable banks had higher pure technical efficiency. Market power did not seem to have significantly affected efficiency. Finally, if bank size and profitability effects were held constant, banks in the Mideast (or Northeast) had significantly higher pure technical efficiency in the latter half of the 1980s.

Thompson *et al.* [32] investigated the efficiency of a sample of 48 US large commercial banks with data from 1980 to 1990 using CCR-IO, assurance region (AR) CCR-IO, linked-cone (LC) CCR-IO and allocative LC-CCR-IO (i.e. maximum profit ratio and minimum profit ratio) models fed with five inputs (i.e. total labour in terms of number of employees; total physical capital in terms of book value of bank premises, furniture and equipment; total purchased funds, including federal funds purchased, large (> $100 k) CDs, foreign deposits and other liabilities for borrowed money; total number of branches, including the main office; and total deposits, including demand deposits, time and savings deposits, and small CDs) and two outputs (i.e. total loans, including commercial/industrial, instalment and real estate loans, and total non-interest income), where the space of admissible multipliers was specified by imposing bounding constraints on the relative magnitude of the multipliers that take account of the range of values of inputs and outputs. The empirical results revealed that maximum profit ratios were relatively low across the 48 banks in each year analysed, which suggests that all 48 banks analysed were assured of losses. The authors of the study claimed that their results were in accordance with the low actual profit ratios observed.

Bhattacharyya *et al.* [33] investigated the impact of liberalization of the banking sector in India on performance using a sample of 70 commercial banks with data from 1986 to 1991 and a two-stage analysis framework. In the first stage, pure technical efficiency and scale efficiency scores were computed by solving output-oriented CCR and BCC models (CCR-O and BCC-O) fed with two inputs (i.e. interest expenses and operating expenses) and three outputs (i.e. advances, investments and deposits). Then, in the second stage, the pure technical efficiency scores were regressed against six bank-specific exogenous variables that took account of the expansion of the banking sector into suburban and rural areas as well as national and international regulatory requirements (i.e. number of branches in rural areas, number of branches in suburban areas, number of branches in urban areas, number of branches in metropolitan areas, ratio of priority sector lending to total advances, and capital adequacy ratio), along with time dummies to model the evolution of bank performance through time relative to performance in 1986, and ownership-type dummies. The regression framework was based on stochastic frontier analysis, which allows one to decompose variations in pure technical efficiency scores into three components related to time, ownership and random noise. Once the stochastic frontier analysis model (without ownership-type dummies) was estimated, the authors of the study estimated an index of efficiency change as the difference between time dummy coefficients in two consecutive periods, following the lead of Baltagi and Griffin [34]. The empirical findings suggested that publicly owned Indian banks were the most efficient, followed by foreign-owned banks and privately owned Indian banks. In addition, out of the 43 banks that turned out to be on the efficiency frontier, 33 displayed decreasing returns to scale. Furthermore, only foreign-owned frontier banks showed any tendency towards increasing or constant returns to scale. However, an analysis of the index of efficiency change by bank category suggested that publicly owned Indian banks experienced a decline in performance, foreign-owned banks experienced an improvement in performance and privately owned Indian banks did not experience any trend in their performance. Finally, the authors found that, on average, across all three ownership forms and throughout the sample period, only 5.7% of calculated efficiency variation remained unexplained by interaction between temporal and ownership form effects.

Pastor *et al.* [35] investigated the efficiency, differences in technology, and productivity of the Spanish banking system and performed a comparison with six European countries and the US for the year 1992. The sample details can be summarized as follows: 168 US banks, 45 Austrian banks, 59 Spanish banks, 22 German banks, 18 UK banks, 31 Italian banks, 17 Belgian banks and 67 French banks. To be more specific, CCR-IO and BCC-IO models were used to investigate efficiency and differences in technology, whereas Malmquist indices computed under the constant-returns-to-scale assumption were used to investigate productivity change. The choice of Malmquist indices – instead of the productivity change indices of Fisher [36] and Törnqvist [37] – was motivated by the fact that Malmquist indices are decomposable into technical efficiency (catching up) and technical change (frontier shifts). The CCR-IO and BCC-IO models were fed with two inputs (i.e. non-interest expenses other than personnel expenses, and personnel expenses) and three outputs (i.e. loans, other productive assets and deposits). Note that the efficiency scores were obtained by solving these models so that each bank was compared with its own banking system, whereas the productivity indices were obtained by solving CCR-IO so that a bank was compared with a frontier composed of other banking systems as well. The empirical findings suggested that French, Spanish and Belgian banks were the most efficient ones, whereas UK, Austrian and German banks were the least efficient. In addition, some evidence of scale inefficiencies in Austrian, German and US banks was found, and almost no trace of scale inefficiency was found in the French and UK samples. On the other hand, with respect to productivity, the empirical results revealed that Austrian, Italian, German and Belgian banks were more productive than US, UK, French and Spanish ones. Furthermore, the decomposition of the Malmquist index into catching up and distance from the efficiency frontier revealed that different banks operated under different combinations of the two factors; for example, banks in countries such as Spain and France showed relatively high efficiency and a relatively low level of technology simultaneously, whereas other banks in countries such as Austria and Germany combined a very productive technology with a low level of efficiency.

Taylor *et al.* [38] investigated the efficiency and profitability of 13 Mexican commercial banks with data from 1989 to 1991 using the CCR-IO model, the BCC-IO model, the cone-ratio assurance region (CR-AR-IO) model under CRS and the LC-AR profit model [39]. These models were fed with two inputs (i.e. total deposits and total non-interest expenses) and one output (i.e. total income). The main finding lay in the fact that DEA-inefficient banks could have higher profits than DEA-efficient banks. Thus, although LC-profitability and DEA-efficiency are different concepts, they can complement each other in an empirical analysis.

Chen [40] investigated the impact of liberalization on the performance of Taiwanese commercial banks using a sample of seven publicly owned and 27 privately owned banks with data from 1996 and a two-stage analysis framework. In the first stage, overall technical, pure technical and scale efficiency scores were computed using CCR-IO and BCC-IO models fed with three inputs (i.e. labour, assets and interest expenses) and four outputs (i.e. loans services, investments, interest income and non-interest income). Chen compared the overall technical efficiency scores of this set-up with seven other set-ups where different measures of different criteria were used (e.g. deposits as an alternative to interest expenses, and business loans and individual loans as an alternative to loan services) to assess the impact of the choice of measures on the efficiency scores, on the one hand, and considered additional inputs or outputs (e.g. number of branches), on the other hand. In the second stage, the efficiency scores were linearly regressed against ownership (as measured by a dummy variable representing public and private ownership) and bank size (as measured by assets, staff or deposit balances). The empirical findings suggested that the whole-sample mean of the overall technical efficiency was quite high (0.969); that is, Taiwanese commercial banks could have produced the same level of output by using 96.9% of the input actually used. In addition, publicly owned banks (with an average overall technical efficiency of 0.923) were relatively less efficient than the privately owned ones (with an average overall technical efficiency of 0.979). The decomposition of overall technical efficiency into pure technical efficiency and scale efficiency revealed that, on average, these scores were very close; however, publicly owned banks were less scale efficient than they were pure technically efficient. On the other hand, ownership seemed to be the main driver of the differences in efficiency scores.

Chu and Lim [41] investigated the relationship between the share prices of six local Singapore-listed groups of banks and their efficiency using a two-stage analysis framework, with data from 1992 to 1996. In the first stage, overall technical, pure technical and scale efficiencies were computed by solving CCR-OO and BCC-OO models fed with three inputs (i.e. shareholders' fund, interest expenses, and operating expenses including provisions) and two outputs (i.e. annual increase in average assets, and total income or profit, depending on the perspective from which one looks at banks). In the second stage, annual stock returns (adjusted for capitalization changes) were linearly regressed against percentage changes in efficiency scores, where the super-efficiency model of Andersen and Petersen [42] was used instead of the CCR model to compute these scores, which allowed the authors of the study to break the ties between banks on the efficiency frontier and thus enhance the statistical fit.

The empirical findings suggested that all banks within the sample under consideration had higher overall and pure technical efficiency scores when computed using total income – rather than total profit – as an output. In addition, larger banks were in general more efficient than smaller ones, regardless of the type of efficiency. On the other hand, the second-stage results suggested that the percentage changes in share prices were better explained by percentage changes in the super-efficiency scores computed with total profit – rather than total income – as an output, which could be explained by the fact that shareholders are more concerned with their profits/dividends than with the banks' income.

Pastor [43] investigated the efficiency of four European banking systems (i.e. commercial banks in Spain, Italy, France and Germany, with data from 1988 to 1994), adjusted for credit risk and environment using a three-phase methodology, where credit risk was measured by bad loans and decomposed into internal and external components. To be more specific, in the first phase, an indicator of *risk management efficiency* was computed using one of three methodologies (i.e. a single-stage, two-stage or three-stage input-oriented methodology), where the proportion of bad loans attributable to bad risk management (as measured by the provision for loans losses, PLL), the volume of loans, and economic-cycle-related environmental variables (i.e. the coefficient of variation of the nominal GDP for the period, the growth rate of the nominal GDP for the period and the cumulative annual growth rate in the last five years) were taken into account. In the second phase, an *efficiency measure adjusted for credit risk due to internal factors* was computed using a BCC-IO model fed with three inputs (i.e. personnel expenses, operating costs and proportion of PLL due to internal factors) and three outputs (i.e. loans, deposits and other earning assets). Finally, in the third phase, an *efficiency measure adjusted for both credit risk due to internal factors and the environment* was computed using an input-oriented VRS model with environmental variables, fed with the three inputs used in phase 2 adjusted for slacks, along with the economic-cycle-related environmental variables mentioned above, as well as efficiency-related environmental variables which were structural (i.e. per capita wages, density of deposits, national income per branch and capital adequacy ratio), used as inputs or outputs depending on whether they were to be maximized or minimized. The empirical results suggested that the ranking of countries changed substantially when credit risk was considered in the performance evaluation of banks. However, environmental variables did not seem to have a marked effect on efficiency. Finally, increased competition generated by the deregulation of the EU banking system did not seem to have pushed banks into riskier business and/or behaviour.

Drake *et al.* [44] investigated the impact of macroeconomic and regulatory factors on the efficiency of the Hong Kong banking system using a three-stage analysis framework. The sample details can be summarized as follows: 59 banks (1995), 66 banks (1996), 52 banks (1997), 66 banks (1998), 62 banks (1999), 61 banks (2000) and 47 banks (2001). The first stage of the analysis used BCC-IO and SBM-IO models to compute efficiency scores and slacks. In the second stage, the radial and non-radial slacks were regressed against environmental variables – divided into macroeconomic continuous variables and regulatory categorical variables – and the inputs were adjusted by the difference between the predicted maximum slack and the predicted slack. These adjusted inputs were then used in the third stage to compute new efficiency scores using BCC-IO and SBM-IO models, respectively. This three-stage analysis framework was implemented under both the profit-oriented approach and the intermediation approach. Under the profit-oriented approach, both the BCC-IO and the SBM-IO models were fed with three inputs (i.e. employee expenses, other non-interest expenses and loan loss provisions) and three outputs (i.e. net interest income, net commission income and total other income). On the other hand, under the intermediation approach, both the BCC-IO and the SBM-IO models were fed with four inputs (i.e. personnel expenses, total deposits + total money market funds + total other funding, total fixed assets, and loan loss provisions and other provisions) and three outputs (i.e. total customer loans + total other lending, total other earning assets, and other non-interest income). The empirical results suggested that Hong Kong banks, on average, exhibited a relatively high degree of inefficiency regardless of whether BCC or SBM scores were used. Such high levels of inefficiency are common in bank efficiency studies which do not incorporate environmental factors. In addition, the dominant external influence on efficiency in the Hong Kong banking system is the macroeconomic cycle. Furthermore, the authors of the study found, as expected, that not incorporating environmental factors would lead to biased efficiency scores. Also, they found that the efficiency scores were generally higher under the intermediation approach than under the profit approach. Finally, the authors reported that once environmental factors were taken into account, the intermediation approach offered little scope for discriminating between bank categories, compared with the profit-oriented approach, which produced a much greater diversity in relative efficiency scores, both across different asset size groups and across different categories of banks.

Liu and Tone [45] investigated the efficiency of the Japanese banking sector by performing a three-stage analysis on a sample of Japanese commercial banks. The details of the sample can be summarized as follows: 138 banks (1997), 134 banks (1998), 133 banks (1999), 129 banks (2000) and 126 banks (2001). In the first stage, Liu and Tone solved output-oriented weighted SBM (WSBM-OO) models [46] to compute efficiency scores and slacks, where the WSBM-OO model was fed with three inputs (i.e. interest expenses, credit costs, and general and administrative expenses) and two outputs (i.e. interest-accruing loans and lending revenues). In the second stage, they regressed the normalized slacks obtained in the first stage against environmental variables using a doubly heteroscedastic stochastic frontier analysis framework to allow control for the impacts of both environmental factors and statistical noise, along with a mechanism to adjust the outputs to an ideal level where there was an absence of environmental influences and random shocks. Within the doubly heteroscedastic stochastic frontier analysis framework, the authors of the study used three categories of environmental variables, namely, environmental variables used within the log-linear Cobb–Douglas function (i.e. monetary aggregate to GDP ratio, bank lending to GDP ratio, short-term risk spread, long-term risk spread, Japan premium, real land price index, real GDP growth index, real stock price index and real bankrupt debt per case), environmental variables used in the heteroscedastic model of the technical efficiency term (i.e. residuals in the non-performing loan ratio and residuals in the capital adequacy ratio) and environmental variables used in the heteroscedastic model of the noise or random shock term (i.e. bank heterogeneity in the non-performing loan ratio and bank heterogeneity in the capital adequacy ratio). Finally, in the third and last stage, these adjusted outputs were used alongside the original inputs to compute efficiency scores using WSBM-OO. The empirical results revealed that the mean efficiency scores had a volatile pattern when the characteristics of the operating environment of the banks and random noise were not controlled for, which hid the learning process of bankers. However, after controlling for the impacts of environmental factors and statistical noise, the mean efficiency scores exhibited a stable upward trend, while the standard deviation narrowed over time, suggesting that Japanese bankers were in fact learning from past experience.

In the next section, we shall analyse the literature surveyed above and provide the big picture on the current state of the art of static DEA in banking.

14.3   CURRENT STATE OF THE ART SUMMARIZED

So far, the overall technical efficiency, pure technical efficiency, scale efficiency, and cost and allocative efficiencies of banks have been investigated by a variety of studies – see the previous section for details. In terms of the DEA-based methodologies used in these investigations, they fall into three main categories, namely, single-stage, two-stage and three-stage methodologies.

The single-stage methodologies consist of using a DEA model with or without environmental variables to compute the efficiency scores of banks. To be more specific, a typical single-stage methodology uses one or several classical DEA models (e.g. the CCR, BCC, SBM, assurance region, cone ratio, linked-cone and allocative models) with or without environmental variables to compute relevant efficiency scores (e.g. overall technical, pure technical, scale, cost and allocative efficiency scores), as well as slacks. Although single-stage methodologies have been and are still very popular, in practice they are not without limitations. In fact, in many practical settings, the choice of inputs and outputs is often not subject to scrutiny, which might lead to biased performance profiles due to over- or under-estimated efficiency scores. One way to overcome this issue is to double-check whether the inputs and outputs are actually responsible for the performance figures. A simple approach to addressing this issue is to regress the efficiency scores against the inputs and outputs and reconsider the choice of those inputs and outputs accordingly. In sum, this issue can be overcome by using an iterative two-stage methodology, which can be summarized as follows:

- *Stage 1.* Given a specific choice of inputs and outputs, compute the efficiency scores most relevant for the analysis under consideration, as well as slacks, using the appropriate DEA models.
- *Stage 2.* Regress the efficiency scores computed in Stage 1 against the inputs and outputs chosen in Stage 1 using a linear regression framework, reconsider the choice of those inputs and outputs accordingly, and go to Stage 1 if necessary.

On the other hand, when environmental variables are taken into account in a relative performance evaluation exercise, the efficiency scores obtained with a single-stage methodology are environmentally biased in that the environment of a bank might advantage or disadvantage that bank relative to others and therefore lead to an unfair comparison. This issue can be overcome by using a two-stage methodology, which can be summarized as follows:

- *Stage 1*. Compute the efficiency scores most relevant for the analysis under consideration, as well as slacks, using the appropriate classical DEA models fed with the relevant environment-independent inputs and outputs (e.g. financial information).
- *Stage 2*. Regress the efficiency scores computed in Stage 1 against environmental variables using a linear regression framework or a non-linear one (e.g. tobit or logit) to find whether or not the efficiency is environment-related, and estimate new efficiency scores that control for the environment if necessary.

Note, however, that the efficiency scores obtained by this two-stage process will still be environmentally biased because the inputs and outputs used in Stage 1 are not adjusted for the environment. In order to properly control for the environmental variables, one can use a three-stage methodology, which can be summarized as follows:

- *Stage 1*. Compute the efficiency scores most relevant for the analysis under consideration, as well as slacks, using the appropriate classical DEA models fed with the relevant environment-independent inputs and outputs (e.g. financial information). It would be unfair to use the efficiency scores obtained at this stage for an evaluation of the relative performance of banks, since these operate in different environments, which could advantage or disadvantage them.
- *Stage 2*. Filter the slacks computed in Stage 1 for the influence of environmental variables using a DEA framework. To be more specific, if the DEA analysis is input-oriented, then the inputs are the slacks computed in Stage 1 and the environmental variables amongst those under consideration which are to be minimized, whereas the outputs are the environmental variables amongst those under consideration which are to be maximized. On the other hand, if the DEA analysis is output-oriented, then the outputs are the slacks computed in Stage 1 and the environmental variables amongst those under consideration which are to be maximized, whereas the inputs are the environmental variables amongst those under consideration which are to be minimized. Finally, if the DEA analysis is non-oriented, the input surpluses computed in Stage 1 (i.e. input-related slacks) and the environmental variables amongst those under consideration which are to be minimized are used as inputs, whereas the output shortfalls computed in Stage 1 (i.e. output-related slacks) and the environmental variables amongst those under consideration which are to be maximized are used as outputs. The resulting filtered slacks are then used to adjust the inputs, outputs or both depending on the orientation of the DEA model.

- *Stage 3*. Compute the efficiency scores most relevant for the analysis under consideration, as well as slacks, using the appropriate DEA models fed with the adjusted inputs and outputs computed in Stage 2. The efficiency scores thus obtained are environment-independent and therefore more appropriate for an evaluation of the relative performance of banks.

The reader is referred to Table 14.1 for a snapshot of the literature on DEA-based methodologies or analyses and the underlying models, and to Table 14.2 for a summary of the response and explanatory variables used in multistage analyses. As to the inputs and outputs with which the DEA models used in the above-mentioned methodologies are fed, their choice is typically driven by the perspective from which banks are assessed, namely, the intermediation approach, the asset approach, the production approach – sometimes referred to as the profit approach – and the value added approach. The intermediation approach or perspective considers banks as intermediation agents that collect funds and provide loans and other assets. The asset approach is a variant of the intermediation approach which considers banks as financial intermediaries between liability holders and those who receive bank funds. The production approach considers banks as production units that transform inputs into outputs, or producers of deposit accounts and loan services. In the literature, the production approach is sometimes referred to as the profit approach – although we believe there is a distinction between these two approaches because, under the profit approach, profit should guide the choice of inputs and outputs. Finally, under the value added approach, the share of value added guides the choice of inputs and outputs. We refer the reader to Table 14.3 for a snapshot of the literature on the choice of inputs and outputs under each of these approaches and to Table 14.4 for a summary of the measures of inputs and outputs and other variables used in analyses of banks' performance (when not properly reflected in the definition). For a summary of the environmental variables used in DEA analyses, we refer the reader to Table 14.5. Also, Table 14.6 provides a summary of the data used in assessing the performance of banks, the period of analysis, and the data provider or database. Since the empirical results and related findings of any DEA analysis are sample-dependent, it would be inappropriate to make any attempt to draw any general conclusions – for the main findings of different studies, the reader is referred to the previous section. However, to conclude this section, we would like to provide the reader with a snapshot of the main types of empirical investigations covered in our survey, summarized in the following bullet points:

- Investigation of the *relationship between type of ownership and efficiency*. For example, Elyasiani and Mehdian [15] considered minority-owned and non-minority-owned US banks, Bhattacharyya *et al.* [33] considered publicly owned Indian banks, privately owned Indian banks and foreign-owned banks, and Chen [40] considered publicly owned and privately owned Taiwanese banks.
- Investigation of the *relationship between type of organizational form and efficiency*. For example, Aly *et al.* [12] considered unit banking and branch banking in the US, Grabowski *et al.* [27] considered branch banking and bank holding

TABLE 14.1 **Summary of analyses and underlying models for assessing the performance of banks.**

Reference	First-stage models	Second-stage models	Third-stage models
Single-stage analysis			
Ferrier and Lovell [9]	Input-oriented VRS and VRS cost allocation models with both categorical and continuous environmental variables	N/A	N/A
Charnes *et al.* [13]	CR-CCR-IO	N/A	N/A
Elyasiani and Mehdian [15]	CCR-IO; BCC-IO; cost allocation model	N/A	N/A
Yue [16]	CCR-IO; weighted ADD; window analysis	N/A	N/A
Grabowski *et al.* [27]	CCR-IO; BCC-IO; cost allocation model	N/A	N/A
Barr *et al.* [24]	CCR-IO	N/A	N/A
Fukuyama [28]	CCR-IO; BCC-IO with VRS and NIRS	N/A	N/A
Zaim [30]	Input-oriented CRS, VRS, IRS, NIRS and cost allocation models with both categorical and uncontrollable continuous environmental variables	N/A	N/A
Pastor *et al.* [35]	Input-oriented CRS and VRS models with both categorical and continuous environmental variables; Malmquist indices	N/A	N/A
Taylor *et al.* [38]	CCR-IO; BCC-IO; CRS-CR-AR-IO; LC-AR-based profit model	N/A	N/A
Two-stage analysis			
Rangan *et al.* [7]	CCR; BCC	Linear regression analysis	N/A
Elyasiani and Mehdian [11]	CCR-IO; rate of technological change (RTC)	Linear regression analysis	N/A

(continued overleaf)

TABLE 14.1   (*continued*)

Reference	First-stage models	Second-stage models	Third-stage models
Aly *et al.* [12]	CCR-IO; BCC-IO; cost allocation model	Linear regression analysis	N/A
Favero and Papi [29]	CCR-IO; BCC-IO	Linear regression analysis	N/A
Miller and Noulas [31]	CCR-IO; BCC-IO	Linear regression analysis	N/A
Bhattacharyya *et al.* [33]	CCR-OO; BCC-OO	Stochastic frontier analysis	N/A
Chen [40]	CCR-IO; BCC-IO	Linear regression analysis	N/A
Chu and Lim [41]	CCR-OO; BCC-OO	Linear regression analysis	N/A
Barr *et al.* [25]	CCR-IO	Logit and probit analyses	N/A
Barr and Siems [26]	CCR-IO	Logit and probit analyses	N/A
Three-stage analysis			
Pastor [43]	1. Input-oriented VRS with environmental variables; 2. BCC-IO and regression with environmental variables; 3. BCC-IO, input-oriented VRS with environmental variables and BCC-IO	BCC-IO	Input-oriented VRS with environmental variables
Drake *et al.* [44]	BCC-IO; SBM-IO	Tobit analysis with both categorical and continuous environmental variables	BCC-IO; SBM-IO with inputs adjusted for slacks
Liu and Tone [45]	WSBM-OO	Doubly heteroscedastic stochastic frontier analysis with environmental variables	WSBM-OO with outputs adjusted for slacks

TABLE 14.2 Summary of response and explanatory variables used in second-stage models for assessing the performance of banks.

Reference	Response/dependent variable	Explanatory variables
Rangan et al. [7]	Overall technical efficiency; pure technical efficiency	Bank size (+); level of product diversity (−); extent to which bank branching is allowed (no relationship)
Elyasiani and Mehdian [11]	Rate of technological change (RTC)	Intensities (λ_j) of deposits, total demand deposit, capital and labour obtained from the solution to CCR-IO model
Aly et al. [12]	Efficiency measures	Bank size; bank product diversity; degree of urbanization that characterizes a bank's environment
Favero and Papi [29]	Pure technical efficiency	Bank size; productive specialization; ownership; market structure; localization
Miller and Noulas [31]	Pure technical efficiency	Bank size; profitability; market power; location
Bhattacharyya et al. [33]	Pure technical efficiency	Number of branches in rural areas; number of branches in suburban areas; number of branches in urban areas; number of branches in metropolitan areas; ratio of priority sector lending to total advances; capital adequacy ratio; time dummies show how bank performance evolves through time relative to performance in 1986; ownership dummies corresponding to the three ownership forms
Chen [40]	Overall technical efficiency; pure technical efficiency; scale efficiency	Ownership; size; other bank characteristics
Chu and Lim [41]	Annual stock returns (adjusted for capitalization changes)	Percentage changes in super-efficiency scores
Pastor [43]	Risk management efficiency without correcting for environmental variables	Economic-cycle-related environmental variables, i.e. coefficient of variation of the nominal GDP for the period, growth rate of nominal GDP for the period and cumulative annual growth rate in the last five years
Drake et al. [44]	Radial and non-radial slacks	Macroeconomic variables: private consumption expenditure; government expenditure; gross fixed capital formation; net export of goods; net export of services;

(continued overleaf)

TABLE 14.2   (*continued*)

Reference	Response/dependent variable	Explanatory variables
		discount window base rate; unemployment; retail sales values; expenditure on housing; and the current account balance.
		Regulatory variables: dummy variable for the Hong Kong property crash/ Asian financial crisis; dummy variable for handover to the People's Republic of China; dummy variable for 1999 (Hong Kong Monetary Authority agreed to phase out the remaining interest rate controls, i.e. caps); and a dummy variable for 2001 (remaining interest rate controls removed).
		Environmental variables used within the log-linear Cobb–Douglas function: monetary aggregate to GDP ratio; bank lending to GDP ratio; short-term risk spread; long-term risk spread, Japan premium; real land price index; real GDP growth index; real stock price index; real bankrupt debt per case.
Liu and Tone [45]	Normalized slacks obtained in the first stage	Environmental variables used in the heteroscedastic model of the technical efficiency term: residuals in non-performing loan ratio; residuals in capital adequacy ratio.
		Environmental variables used in the heteroscedastic model of the noise or random shock term: bank heterogeneity in non-performing loan ratio; bank heterogeneity in capital adequacy ratio.

companies in the US, Fukuyama [28] considered city banks, regional banks and former *sogo* banks in Japan, and Zaim [30] considered state banks and private banks in Turkey.

- Investigation of the *relationship between some measure of efficiency and one or several endogenous or exogenous variables*. For example, Aly *et al.* [12] considered size, extent of product diversity and level of urbanization; Fukuyama [28] considered bank size; Favero and Papi [29] considered bank size, productive specialization, ownership, market structure and localization; Miller and Noulas

TABLE 14.3   Summary of inputs and outputs used in DEA models for assessing the performance of banks.

Reference	Inputs	Outputs
Intermediation approach		
Rangan *et al.* [7]	Labour; capital; purchased funds	Real estate loans; commercial and industrial loans; consumer loans; demand deposits; time and savings deposits
Ferrier and Lovell [9]	Total number of employees; occupancy costs and expenditure on furniture and equipment; expenditure on materials	Number of demand deposit accounts; number of time deposit accounts; number of real estate loans; number of instalment loans; number of commercial loans
Charnes *et al.* [13]	Total operating expenses; total non-interest expenses; provision for loan losses; actual loan losses	Total operating income; total interest income; total non-interest income; total net loans
Elyasiani and Mehdian [11]	Labour; capital; deposits; total demand deposits	Investment; real estate loans; commercial and industrial loans; other loans
Aly *et al.* [12]	Labour; capital; loanable funds	Demand deposits; real estate loans; commercial and industrial loans; consumer loans; other loans
Elyasiani and Mehdian [15]	Labour; capital; certificates of deposit; time and savings deposits; demand deposits	Commercial and industrial loans; real estate loans; other loans; investment securities
Yue [16]	Interest expenses; non-interest expenses; transaction deposits; non-transaction deposits	Interest income; non-interest income; total loans
Grabowski *et al.* [27]	Labour; capital; loanable funds	Real estate loans; commercial and industrial loans; consumer loans; demand deposits; investment securities
Fukuyama [28]	Labour; capital; funds from customers	Revenue from loans; revenue from other business activities
Zaim [30]	Total number of employees; total interest expenditure; depreciation expenditure; expenditure on materials	Total balance of demand deposits; total balance of time deposits; total balance of short-term loans; total balance of long-term loans
Favero and Papi [29]	Labour; capital; financial capital available for investment; loanable funds (i.e. CDs); net funds borrowed by other banks	Current accounts and savings deposits; loans to other banks and non-financial Institutions; investment in securities and bonds; non-interest income
Miller and Noulas [31]	Total transaction deposits; total non-transaction deposits; total interest expenses; total non-interest expenses	Commercial and industrial loans; consumer loans; real estate loans; investments; total interest income; total non-interest income

(continued overleaf)

TABLE 14.3   (*continued*)

Reference	Inputs	Outputs
Taylor *et al.* [38]	Total deposits; total non-interest expenses	Total income
Chen [40]	Labour; assets; interest expenses	Loan services; investments; interest income; non-interest income
Drake *et al.* [44]	Personnel expenses; total deposits + total money market funds + total other funding; total fixed assets; loan loss provisions and other provisions	Total customer loans + total other lending; total other earning assets; other non-interest income
	Asset approach	
Favero and Papi [29]	Labour; capital; financial capital available for investment; loanable funds (i.e. current accounts and savings deposits); CDs; net funds borrowed by other banks	Loans to other banks and non-financial institutions; investment in securities and bonds; non-interest income
	Value added approach	
Bhattacharyya *et al.* [33]	Interest expenses; operating expenses	Advances to priority sector activities; investments; deposits
Pastor *et al.* [35]	Non-interest expenses other than personnel expenses; personnel expenses	Loans; other productive assets, including all existing deposits with banks, short-term investments, other investments and equity investments; deposits, including customer and short-term funding, which is the sum of demand, savings, time, interbank and other deposits
Chu and Lim [41]	Shareholders' fund; interest expenses; operating expenses (including provisions)	Annual increase in average assets as a proxy for future income or future profit; total income or profit depending on whether X-efficiency or P-efficiency is evaluated
Pastor [43]	Personnel expenses; operating costs, excluding personnel expenses and including financial costs; proportion of provision for loan losses due to internal factors; all inputs adjusted for slacks (for third phase); structural environmental variables: per capita wages; density of deposits; national income per branch; capital adequacy ratio; economic-	Loans; deposits; other earning assets; economic-cycle environmental variables: coefficient of variation of the nominal GDP for the period

TABLE 14.3 (*continued*)

Reference	Inputs	Outputs
	cycle environmental variables: growth rate of nominal GDP of the period; cumulative annual growth rate in the last five years	
	Production/profit-oriented approach	
Drake *et al.* [44]	Employee expenses; other non-interest expenses; loan loss provisions	Net interest income; net commission income; total other income
Liu and Tone [45]	Interest expenses; credit costs; general and administrative expenses	Interest-accruing loans; lending revenues

[31] considered bank size, profitability, market power and location; Bhattachar-yya *et al.* [33] considered six bank-specific exogenous variables that take account of the expansion of the banking sector into suburban and rural areas, as well as national and international regulatory requirements (i.e. number of branches in rural areas, number of branches in suburban areas, number of branches in urban areas, number of branches in metropolitan areas, ratio of priority sector lending to total advances, and capital adequacy ratio), along with ownership type; and Chen [40] considered ownership and bank size.

- Investigation of the *effect of an event on the efficiency of banks*. For example, Zaim [30] considered the effect of post-1980 financial liberalization policies on the economic efficiency of Turkish commercial banks, and Drake *et al.* [44] considered the impact of macroeconomic and regulatory factors on the efficiency of the Hong Kong banking system.

14.4 CONCLUSION

In this chapter, we have provided a detailed account of DEA-based contributions in the banking sector, with emphasis on static conventional DEA models, often referred to as black box models. Our account starts from the first paper on DEA in banking, published in 1988, and covers all major contributions to date. Apart from assessing the efficiency profiles of banks, the authors of these contributions have investigated the relationship between the type of ownership and efficiency, the relationship between the type of organizational form and efficiency, the relationship between some measure of efficiency and one or several endogenous or exogenous variables, and the effect of an event (e.g. deregulation) on the efficiency of banks. For those researchers who are unfamiliar with this field, we have summarized the literature into tables that provide snapshots of the landscape of this research area. These snapshots could also serve as an 'aide-memoire' for readers who are familiar with DEA and its applications in banking.

TABLE 14.4 Summary of measures of inputs and outputs and other variables used in analyses of bank performance.

Variable	Measure and reference
Labour	Number of full-time employees on the payroll [7,11,12,15,27–29,40,43]; employee expenses [44]
Capital	Book value of premises and fixed assets [7,11,12,15,27,29]; bank premises and equipment; suspense payments for constitutions unfinished and surety deposits and intangibles [28]
Purchased funds	Certificates of deposit greater than $100,000; notes and debentures; other borrowed funds [7]
Deposits	Savings and time deposits – including large ($100,000 or more) negotiable CDs – and total demand deposits [11,15]; transaction deposits and non-transaction deposits [16]; customer and short-term funding, which is the sum of demand, savings, time, interbank and other deposits [35,43]
Total loans	Loans and leases net of unearned income [16]; business and individual loans [40]
Loanable funds	Sum of time deposits and other borrowed funds [27]
Funds from customers	Part of the liabilities in the balance sheet, including deposits, CDs, call money, bills sold, borrowed money, foreign exchange and others [28]
Shareholders' fund	Capital provided by bank's shareholders [41]
Interest expenses	Expenses for Federal funds, purchase and sale of securities, and interest on demand notes and other borrowed money [16]; interest on deposit (savings, fixed or time, and current or checking) accounts [41]; external financial cost [45]
Non-interest expenses	Salaries; expenses associated with premises and fixed assets, taxes and other expenses [16]; non-interest expenses other than personnel expenses [35,44]
Operating expenses	Operating expenses, including provisions [41]
General and administrative expenses	Cost of information production, in an economic sense [45]
Credit cost	Credit cost covers unexpected, expected and realized losses due to credit risk exposures and is calculated as transfer to reserve for possible loan losses + net provision of specific reserve for possible loan losses + write-off claims + losses in sale of claims – recoveries of written-off claims [45]
Interest income	Interest and fee income on loans, income from lease-financing receivables, interest and dividend income on securities, and other income [16]; net interest income [44]

TABLE 14.4   (*continued*)

Variable	Measure and reference
Non-interest income	Service charges on deposit accounts, income from fiduciary activities and other non-interest income [16]
Interest-accruing loans	Loans and bills discounted + 0.5 × customers' liabilities for acceptances and guarantees − loans to borrowers in legal bankruptcy + past due loans in arrears by 6 months or more [45]. In Japan, banks are required to stop accruing interest on a loan that is past due for 6 months or more.
Investments	Government securities and shares and securities in public and private enterprises [40]
Revenue from loans	Interest on loans and discounts and interest on bills bought – these are the traditional primary business activities of banks [28]; lending revenue computed as net interest income + net fees and commission income [45]
Bad loans attributable to bad risk management	Provision for loan losses [43,44]
Revenue from other business activities	Total operating income minus any other operating income, after deducting gains on foreign exchange and trading account securities transactions, as well as gains on sales and redemption of bonds minus revenue from loans [28]
Bank size	Total deposits [7,12]; number of branches [12]; assets, staff or deposits [40]
Level of product diversity	Minus the logarithm of the sum over products of the squared proportion of a bank's total dollar revenue or sales accounted for by a product [7,12]
Extent to which bank branching is allowed	Categorical variable that takes values of 0, 1 or 2 depending on whether no branch banking is allowed by the state, limited branch banking is allowed or unlimited branch banking is allowed [7]
Degree of urbanization that characterizes a bank's environment	Measured by two dummy variables. The first takes a value of one if the bank operates in a standard metropolitan statistical area (SMSA), but not in a consolidated metropolitan statistical area (CMSA), and zero otherwise. The second dummy variable takes a value of one if the bank operates in an SMSA that is also part of a CMSA, and zero otherwise [12].

TABLE 14.5 Summary of environmental variables used in DEA analyses for assessing the performance of banks.

Reference	Inputs	Outputs
	Intermediation approach	
Ferrier and Lovell [45]	*Categorical environmental variables*: institutional type (non-commercial; savings and loan; mutual savings; credit union); membership of a multibank holding company; location in unit or branch	
	Number of branches operated	Average size of demand deposit account; average size of time deposit account; average size of real estate loan; average size of instalment loan; average size of commercial loan
Zaim [30]	*Categorical environmental variables*: institutional type (national bank; foreign bank)	
	Number of branches as uncontrollable input	Average size of demand deposit accounts; average size of time deposit accounts as uncontrollable outputs
	Value added approach	
Pastor [43]	*Economic environmental variables*: coefficient of variation of the nominal GDP of the period	*Economic environmental variables*: growth rate of nominal GDP for the period; cumulative annual growth rate in the last five years; per capita wages
	Efficiency-related/structural environmental variables: capital adequacy ratio	*Efficiency-related/structural environmental variables*: density of deposits; national income per branch
	Profit-oriented approach	
Drake et al. [44]	*Regulatory variables*: dummy variable for the Hong Kong property crash/Asian financial crisis; dummy variable for handover to the People's Republic of China; dummy variable for 1999 (Hong Kong Monetary Authority agreed to phase out the remaining interest rate controls, i.e. caps); dummy variable for 2001 (remaining interest rate controls removed)	
	Macroeconomic variables: private consumption expenditure; government expenditure; gross fixed capital formation; net export of goods; net export of services; discount window base rate; unemployment; retail sales values; expenditure on housing; current account balance	
Liu and Tone [45]	Monetary aggregate to GDP ratio; bank lending to GDP ratio; short-term risk spread; long-term risk spread; Japan premium; real land price index; real GDP growth index; real stock price index; real bankrupt debt per case	

TABLE 14.6   **Summary of data, period of analysis and its source used in assessing the performance of banks.**

Reference	Data/DMUs	Period of analysis	Source of data/data provider
Rangan *et al.* [7]	215 US banks	1986	Federal Deposit Insurance Corporation (FDIC)
Ferrier and Lovell [45]	575 US banks	1984	The Federal Reserve System's Functional Cost Analysis Program
Charnes *et al.* [13]	48 US commercial banks drawn from the top 300 banks headquartered in America which are also members of the FDIC	1980 to 1985	FDIC
Elyasiani and Mehdian [11]	191 US banks	1980; 1985	Call and income report tapes published by the National Technical Information Service (NTIS) of the Department of Commerce
Aly *et al.* [12]	322 independent US banks	1986	FDIC; tapes on the Reports of Condition and Reports of Income (call reports)
Elyasiani and Mehdian [15]	160 minority-owned and non-minority-owned US banks selected to be from the same state, county, SMSA, CMSA and Federal Reserve district to control for geographical factors and regulatory environment	1988	1988 call and income report tapes
Yue [16]	60 of the largest US commercial banks located in Missouri	1984 to 1990	Not provided
Grabowski *et al.* [27]	522 US banks affiliated to multibank holding companies and 407 US banks with branches	1989	FDIC files on the Report of Income and Condition (call report)
Fukuyama [28]	143 Japanese commercial banks	1991	Analysis of financial statements of all banks from the Federation of Bankers Associations of Japan
Barr *et al.* [24]	930 US banks	December 1984 to December 1989	Not provided

(*continued overleaf*)

TABLE 14.6   (*continued*)

Reference	Data/DMUs	Period of analysis	Source of data/data provider
Zaim [30]	95 Turkish commercial banks	1981 (39 banks) and 1990 (56 banks)	Banks Association of Turkey
Favero and Papi [29]	174 Italian commercial banks	1991	Centrale dei Bilanci-ABI data set
Miller and Noulas [31]	201 US large commercial banks	1984 to 1990	Call report data – reports of condition and income
Thompson *et al.* (1996)	48 US large commercial banks	1980 to 1990	FDIC reports
Bhattacharyya *et al.* [33]	70 Indian commercial banks	1986 to 1991	Indian Banks' Association
Pastor *et al.* [35]	168 US banks, 45 Austrian banks, 59 Spanish banks, 22 German banks, 18 UK banks, 31 Italian banks, 17 Belgian banks, 67 French banks	1992	International Bank Credit Analysis Ltd
Taylor *et al.* [38]	13 Mexican commercial banks	1989 to 1991	Comision Nacional Bancaria (National Banking Commission)
Chen [40]	7 publicly owned and 27 privately owned Taiwanese commercial banks	1996	Not provided
Chu and Lim [41]	6 local Singapore-listed groups of banks	1992 to 1996	End-of-the-year stock prices, duly adjusted for capitalization changes, obtained from Dbank financial database, maintained at the National University of Singapore
Pastor [43]	Commercial banks in Spain, Italy, France and Germany, resulting in 2598 bank-year observations	1988 to 1994	IBCA Ltd, an international rating agency which homogenizes information and classifies firms in terms of specialization, so that accounting uniformity is guaranteed. Data on environmental variables were taken from the Economic Bulletin of the Bank of Spain, Bank

TABLE 14.6   (*continued*)

Reference	Data/DMUs	Period of analysis	Source of data/data provider
			Profitability, Eurostat and the National Statistical Institute of Spain (INE).
Drake *et al.* [44]	Hong Kong banks: 59 (1995), 66 (1996), 52 (1997), 66 (1998), 62 (1999), 61 (2000), 47 (2001)	1995 to 2001	Bank-scope
Liu and Tone [45]	Japanese commercial banks: 138 (1997), 134 (1998), 133 (1999), 129 (2000), 126 (2001).	1997 to 2001	Multiple data sources: Japanese Bankers Association; Bank of Japan; Government of Japan; Japanese Ministry of Land, Infrastructure and Transport; Tokyo Commercial & Industrial Research

REFERENCES

[1] Charnes, A., Cooper, W.W. and Rhodes, E. (1978) Measuring the efficiency of decision making units. *European Journal of Operational Research*, **2**, 429–444.

[2] Seiford, L. (1996) Data envelopment analysis: The evolution of the state of the art (1978–1995). *Journal of Productivity Analysis*, **7**, 99–137.

[3] Emrouznejad, A., Parker, B.R. and Tavares, G. (2008) Evaluation of research in efficiency and productivity: A survey and analysis of the first 30 years of scholarly literature in DEA. *Socio-Economic Planning Sciences*, **42**, 151–157.

[4] Liu, J.S., Lu, L.Y.Y., Lu, W.-M. and Lin, B.J.Y. (2013) A survey of DEA applications. *Omega*, **41**, 893–902.

[5] Secrist, H. (1938) *National Bank Failures and Non-failures: An Autopsy and Diagnosis, Principia Press.*

[6] Kumar, P.R. and Ravi, V. (2007) Bankruptcy prediction in banks and firms via statistical and intelligent techniques – A review. *European Journal of Operational Research*, **180**, 1–28.

[7] Rangan, N., Grabowski, R., Aly, H.Y. and Pasurka, C. (1988) The technical efficiency of US banks. *Economics Letters*, **28**, 169–175.

[8] Banker, R.D., Charnes, A. and Cooper, W.W. (1984) Some models for estimating technical and scale inefficiencies in data envelopment analysis. *Management Science*, **30**, 1078–1092.

[9] Ferrier, G.D. and Lovell, C.K. (1990) Measuring cost efficiency in banking: Econometric and linear programming evidence. *Journal of Econometrics*, **46**, 229–245.

[10] Banker, R.D. and Morey, R.C. (1986) Efficiency analysis for exogenously fixed inputs and outputs. *Operations Research*, **34**, 513–521.

[11] Elyasiani, E. and Mehdian, S.M. (1990) A nonparametric approach to measurement of efficiency and technological change: The case of large US commercial banks. *Journal of Financial Services Research*, **4**, 157–168.

[12] Aly, H.Y., Grabowski, R., Pasurka, C. and Rangan, N. (1990) Technical, scale, and allocative efficiencies in US banking: An empirical investigation. *Review of Economics and Statistics*, 211–218.

[13] Charnes, A., Cooper, W.W., Hung, Z.M. and Sun, D.B. (1990) Polyhedral cone-ratio DEA models with an illustrative application to large commercial banks. *Journal of Econometrics*, **46**, 73–91.

[14] Fukuyama, H., Guerra, R. and Weber, W.L. (1999) Efficiency and ownership: Evidence from Japanese credit cooperatives. *Journal of Economics and Business*, **51**(6), 473–487.

[15] Elyasiani, E. and Mehdian, S. (1992) Productive efficiency performance of minority and nonminority-owned banks: A nonparametric approach. *Journal of Banking and Finance*, **16**, 933–948.

[16] Yue, P. (1992) Data envelopment analysis and commercial bank performance: A primer with applications to Missouri banks. *Federal Reserve Bank of St. Louis Review*, **74**, 31–45.

[17] Charnes, A., Cooper, W.W., Golany, B., Seiford, L. and Stutz, J. (1985) Foundations of data envelopment analysis for Pareto–Koopmans efficient empirical production functions. *Journal of Econometrics*, **30**, 91–107.

[18] Meyer, P.A. and Pifer, H.W. (1970) Prediction of bank failures. *Journal of Finance*, **25**, 853–868.

[19] Sinkey, J.F. (1975) A multivariate statistical analysis of the characteristics of problem banks. *Journal of Finance*, **30**, 21–36.

[20] Fraser, D.R. (1976) The determinants of bank profits: An analysis of extremes. *Financial Review*, **11**, 69–87.

[21] Martin, D. (1977) Early warning of bank failure: A logit regression approach. *Journal of Banking & Finance*, **1**, 249–276.

[22] Pantalone, C.C. and Platt, M.B. (1987) Predicting commercial bank failure since deregulation. *New England Economic Review*, 37–47.

[23] Seballos, L.D. and Thomson, J.B. (1990) Underlying causes of commercial bank failures in the 1980s. *Economic Commentary*, Sep., Federal Reserve Bank of Cleveland, ISSN 0428-1276.

[24] Barr, R.S., Seiford, L.M. and Siems, T.F. (1993) An envelopment-analysis approach to measuring the managerial efficiency of banks. *Annals of Operations Research*, **45**, 1–19.

[25] Barr, R.S., Seiford, L.M. and Siems, T.F. (1994) Forecasting bank failure: A nonparametric frontier estimation approach. *Recherches Économiques de Louvain/Louvain Economic Review*, 417–429.

[26] Barr, R.S. and Siems, T.F. (1997) Bank failure prediction using DEA to measure management quality, in *Interfaces in Computer Science and Operations Research (eds R.S. Barr, R.V. Helgason and J.L. Kennington), Springer, pp. 341–365.*

[27] Grabowski, R., Rangan, N. and Rezvanian, R. (1993) Organizational forms in banking: An empirical investigation of cost efficiency. *Journal of Banking & Finance*, **17**, 531–538.

[28] Fukuyama, H. (1993) Technical and scale efficiency of Japanese commercial banks: A non-parametric approach. *Applied Economics*, **25**, 1101–1112.

[29] Favero, C.A. and Papi, L. (1995) Technical efficiency and scale efficiency in the Italian banking sector: A non-parametric approach. *Applied Economics*, **27**, 385–395.

[30] Zaim, O. (1995) The effect of financial liberalization on the efficiency of Turkish commercial banks. *Applied Financial Economics*, **5**, 257–264.

[31] Miller, S.M. and Noulas, A.G. (1996) The technical efficiency of large bank production. *Journal of Banking & Finance*, **20**, 495–509.

[32] Thompson, R., Dharmapala, P., Humphrey, D., Taylor, W. and Thrall, R. (1996) Computing DEA/AR efficiency and profit ratio measures with an illustrative bank application. *Annals of Operations Research*, **68**, 301–327.

[33] Bhattacharyya, A., Lovell, C.K. and Sahay, P. (1997) The impact of liberalization on the productive efficiency of Indian commercial banks. *European Journal of Operational Research*, **98**, 332–345.

[34] Baltagi, B.H. and Griffin, J.M. (1988) A general index of technical change. *Journal of Political Economy*, 20–41.

[35] Pastor, J., Perez, F. and Quesada, J. (1997) Efficiency analysis in banking firms: An international comparison. *European Journal of Operational Research*, **98**, 395–407.

[36] Fisher, I. (1922) *The Making of Index Numbers: A Study of Their Varieties, Tests, and Reliability, Houghton Mifflin.*

[37] Törnqvist, L. (1936) *Consumption price index, Bank of Finland.*

[38] Taylor, W.M., Thompson, R.G., Thrall, R.M. and Dharmapala, P. (1997) DEA/AR efficiency and profitability of Mexican banks: A total income model. *European Journal of Operational Research*, **98**, 346–363.

[39] Thompson, R.G. and Thrall, R.M. (1994) Polyhedral assurance regions with linked constraints, in *New Directions in Computational Economics (eds W.W. Cooper and A.B. Whinston), Springer, pp. 121–133.*

[40] Chen, T.-Y. (1998) A study of bank efficiency and ownership in Taiwan. *Applied Economics Letters*, **5**, 613–616.

[41] Chu, S.F. and Lim, G.H. (1998) Share performance and profit efficiency of banks in an oligopolistic market: Evidence from Singapore. *Journal of Multinational Financial Management*, **8**, 155–168.

[42] Andersen, P. and Petersen, N.C. (1993) A procedure for ranking efficient units in data envelopment analysis. *Management Science*, **39**, 1261–1264.

[43] Pastor, J.M. (2002) Credit risk and efficiency in the European banking system: A three-stage analysis. *Applied Financial Economics*, **12**, 895–911.

[44] Drake, L., Hall, M.J. and Simper, R. (2006) The impact of macroeconomic and regulatory factors on bank efficiency: A non-parametric analysis of Hong Kong's banking system. *Journal of Banking & Finance*, **30**, 1443–1466.

[45] Liu, J. and Tone, K. (2008) A multistage method to measure efficiency and its application to Japanese banking industry. *Socio-Economic Planning Sciences*, **42**, 75–91.

[46] Cooper, W.W., Seiford, L.M. and Tone, K. (2006) *Introduction to Data Envelopment Analysis and Its Uses: With DEA-Solver Software and References, Springer Science & Business Media.*

15

DEA IN THE HEALTHCARE SECTOR

HIROYUKI KAWAGUCHI

Economics Faculty, Seijo University, Setagaya-ku, Tokyo, Japan

KAORU TONE

National Graduate Institute for Policy Studies, Tokyo, Japan

MIKI TSUTSUI

Central Research Institute of Electric Power Industry, Tokyo, Japan

15.1 INTRODUCTION

Japanese municipal hospitals have experienced financial crises throughout the last few decades. There are 9000 hospitals in Japan, half of which are owned by private not-for-profit organizations, and the remainder of which are run by public organizations. One thousand public hospitals are owned and operated by municipal governments, and most of these hospitals have been losing money for a long time.

As the Japanese government has huge cumulative deficits, it is important that the municipal hospitals have sound financial foundations. The municipal hospitals depend financially on a subsidy from central government through local government. The master plan for the reform of Japan's municipal hospitals included five steps from fiscal year 2007 to fiscal year 2014, as described below. First, the central government designed guidelines regarding proposed reforms and a timeframe for those reforms in fiscal year 2007. The government ordered the reform of all municipal hospitals according to those guidelines. Therefore, all reform of municipal hospitals was to start in fiscal year 2007.

Advances in DEA Theory and Applications: With Extensions to Forecasting Models,
First Edition. Edited by Kaoru Tone.
© 2017 John Wiley & Sons Ltd. Published 2017 by John Wiley & Sons Ltd.

Second, the central government ordered individual municipal hospitals to formulate a reform plan, including performance indicators for the evaluation of the reform, within fiscal year 2008. The contents of the reform plan had some range of autonomy, and municipal hospitals could freely select countermeasures.

The guidelines illustrated several countermeasures that could be used in the reform of municipal hospitals. These countermeasures can be grouped into four categories. The first is the introduction of private business management systems. For example, the guidelines recommend outsourcing to private companies and the adoption of a 'private finance initiative'. The second category is the restructuring and consolidation of the hospital organization. For example, the guidelines recommend the merging of several hospitals and the conversion of hospitals into long-term care facilities. The third category refers to a reduction in hospitals' operating costs. For example, the guidelines propose a revision of wage systems and reductions in the purchase prices of medical materials. The fourth category is an increase in revenue. For example, the guidelines recommend increasing occupancy rates and unit values per inpatient (nearly equal to 'unit revenue per inpatient per day').

Municipal hospitals could choose countermeasures from the examples in the guidelines and could include their own reform countermeasures. Individual reform plans proposed the recruitment of highly skilled professionals, further education for healthcare professionals and a revision of the range of medical services. Thus, each hospital formulated its own reform plan and then self-evaluated the results.

Third, municipal hospitals were required to report the results of the reform plan annually to central government. The first report was submitted in fiscal year 2009.

Fourth, municipal hospitals were required to submit intermediate reports on the results of efficiency promotion from fiscal year 2007 to fiscal year 2010 at the end of fiscal year 2011.

Fifth, municipal hospitals were required to submit a final report on the results of their individual reform plans at the end of fiscal year 2014. Fiscal year 2014 was the deadline for the reform. If the reform was not effective, then central government would request that the municipal hospital shut down, or sell the operation of the hospital.

As explained above, the guidelines for the reform of municipal hospitals mainly targeted hospital administration, because the main objective was to reduce the amount of subsidies that were covering the deficit of the hospitals. Therefore, the central government was more interested in the financial situation of the hospital than in the quality of medical services.

Harris [1] pointed out that a hospital can be considered as two separate firms. There are two heterogeneous internal organizations: a medical-examination division and an administration division. The administration division carries out business management activities to contain medical expenses within medical revenue. The medical-examination division provides various medical care services directly. This unique characteristic of hospitals is particularly strong among municipal hospitals. In a municipal hospital, executive managers in the administration division are ordinarily dispatched from the municipal government. They are reshuffled every few years in the same way as other municipal officials. These two organizations can be described as internal mutual exchange services. The administration division provides medical beds to the medical-examination division, and the medical-examination division repays the revenue through the use of medical beds for inpatient services.

For local residents, many of the problems with municipal hospitals arise through the curtailment of certain medical services. Thus, some residents are no longer able to receive specialized care at those hospitals. The chief medical officer in charge of the medical-examination division is typically the target of criticism from the stakeholders of the hospital. The administration division tends to operate from behind closed doors and avoids blame for any failures. However, previous research has not compared efficiency improvements between the two divisions. Japanese hospitals have acute beds and long-term care beds in various ratios. The larger hospitals tend to concentrate on acute care services. In addition, physicians and surgeons are hired and paid by hospitals, as in the case of National Health Service hospitals in the United Kingdom. These physicians provide services not only to inpatients but also to outpatients at the same hospitals.

The purpose of this study is to evaluate the policy effects of the reform of municipal hospitals in Japan. We have estimated efficiency scores from 2007FY to 2012FY not only for each hospital as a whole but also for the two divisions. In addition, we consider further policy implications to address the financial problems of Japanese municipal hospitals.

The structure of this chapter is as follows. The background and purpose of our study have been discussed in this first section. The methods and data are discussed in the second section. After the estimation of efficiency, we report the efficiency scores. The results of the analyses are presented in the third section. The last section includes a discussion of the results and future challenges.

15.2 METHOD AND DATA

15.2.1 Previous Literature

Data envelopment analysis (DEA) is a popular method with which to estimate the efficiency of hospitals [2]. DEA is a non-parametric method used in operations research to evaluate the efficiency performance of decision-making units (DMUs). The traditional DEA model is often considered a 'black-box' model, because it does not take account of the internal structure of DMUs. Several previous studies have evaluated the efficiency of Japanese hospitals using the method [3–6]. These studies used cross-sectional data from Japanese public hospitals and adopted largely traditional DEA approaches. Average efficiency scores ranged from 0.8869 to 0.9456 in terms of revenue efficiency [3, 4]; for technical efficiency, the scores ranged from 0.8585 to 0.90008 [5, 6].

As an extension of the above traditional DEA model, the 'network DEA model' accounts for divisional efficiencies as well as overall efficiency in a unified framework. Through the network DEA model, we can observe not only the efficiency of DMUs but also divisional efficiencies as its components. Network DEA models were first introduced by Färe and Grosskopf [7–9]. These models have been extended by several authors. The network DEA model proposed by Lewis and Sexton [10] has a multistage structure, as an extension of the two-stage DEA model proposed by Sexton and Lewis [11]. That study solved the DEA model for each node independently. Prieto and Zofio [12] applied a network efficiency analysis within an input–output model, as initiated by Koopmans [13]. Löthgren and Tambour [14] applied a network DEA model to a sample of Swedish pharmacies with organizational objectives that necessitated the monitoring of efficiency, productivity and customer satisfaction. They compared the results of the network DEA models with those of traditional DEA models. Tone and Tsutsui [15] developed this model using a slacks-based measure called the network slacks-based measure (NSBM). The NSBM approach is a non-radial method and is suitable for measuring efficiencies when inputs and outputs may change non-proportionally.

In contrast, the dynamic DEA model can measure the efficiency score obtained from long-term optimization using carry-over variables. The traditional DEA model focuses only on a single period, and therefore the measurement of intertemporal efficiency change has long been a subject of concern in DEA. The window analysis of Klopp [16] was the first approach to account for intertemporal efficiency change. Based on Malmquist [17], Färe et al. [9] developed the Malmquist index in the DEA framework. The dynamic DEA model proposed by Färe and Grosskopf [8] was the first innovative scheme to formally deal with interconnecting activities. Tone and Tsutsui [15] extended their model within the slacks-based measurement framework proposed by Tone [18] and Pastor et al. [19]. Hence, this model is non-radial and can deal with inputs and outputs individually, which enables us to obtain non-uniform input/output factor efficiencies. This is in contrast to radial approaches, which assume proportional changes in inputs or outputs and provide only uniform input/output factor efficiency.

The dynamic network DEA (DN DEA) model takes into account the internal heterogeneous organizations of DMUs, where divisions are mutually connected by link variables and trade internal products with each other. This DN DEA model can evaluate (i) the overall efficiency over the entire observed term, (ii) dynamic changes in the period efficiency and (iii) dynamic changes in the divisional efficiency. In addition, each DMU has carry-over variables that take into account a positive or negative factor for the previous period. We have employed a dynamic DEA model involving the network structure proposed by Tone and Tsutsui [20]. This DN DEA model has advantages of being able to evaluate a policy effect on the individual divisions of each DMU. Tone and Tsutsui [20] provided detailed information about the notation for the DN DEA model. Recently, researchers have started to apply the DN DEA model to the banking sector [21–23] and the hospital sector [24].

The study by Kawaguchi et al. [24] was the first application to a Japanese hospital. The present study expanded that of Kawaguchi et al. [24] in three ways. First, we doubled the observation time from three years to six years. Secondly, we added a new input variable as a proxy for expensive medical equipment. Thirdly, we calculated a Malmquist index score from the DN DEA model. Therefore, this study should provide a more precise evaluation in terms of efficiency change induced by policy intervention.

15.2.2 Formulas for Efficiency Estimation by DN DEA Model

We deal with n DMUs ($j = 1, ..., n$), which consist of K divisions ($k = 1, ..., K$), over T time periods ($t = 1, ..., T$). Let m_k and r_k be, respectively, the numbers of inputs to and outputs from division k. We denote the link leading from division k to division h by $(k, h)_l$ and the set of links by L_{kh}. The observed data are as follows:

- $x_{ijk}^t \in R_+$ ($i = 1, ..., m_k; j = 1, ..., n; k = 1, ..., K; t = 1, ..., T$) is the input resource i for DMU$_j$ for division k in period t, and
- $y_{rjk}^t \in R_+$ ($r = 1, ..., r_k; j = 1, ..., n; k = 1, ..., K; t = 1, ..., T$) is the output product r from DMU$_j$ for division k in period t.

If some outputs are undesirable, we treat them as inputs to division k.

- $z_{j(kh)_l}^t \in R_+$ ($j = 1, ..., n; l = 1, ..., L_{kh}; t = 1, ..., T$) represents the linking intermediate products of DMUj from division k to division h in period t, where L_{kh} is the number of items in the links from k to h.
- $z_{jk_l}^{(t, t+1)} \in R_+$ ($j = 1, ..., n; l = 1, ..., L_k; k = 1, ..., K, t = 1, ..., T-1$) is the carry-over of DMUj for division k from period t to period $t + 1$, where L_k is the number of items in the carry-over from division k.

DMU$_o$ ($o = 1, ..., n$) $\in P^t$ can be expressed as follows. The input and output constraints are

$$
\begin{aligned}
\mathbf{x}_{ok}^t &= \mathbf{X}_k^t \boldsymbol{\lambda}_k^t + \mathbf{s}_{ko}^{t-} \quad (\forall k, \forall t) \\
\mathbf{y}_{ok}^t &= \mathbf{Y}_k^t \boldsymbol{\lambda}_k^t - \mathbf{s}_{ko}^{t+} \quad (\forall k, \forall t) \\
\mathbf{e}\boldsymbol{\lambda}_k^t &= 1 \quad (\forall k, \forall t) \\
\boldsymbol{\lambda}_k^t &\geq \mathbf{0}, \ \mathbf{s}_{ko}^{t-} \geq \mathbf{0}, \ \mathbf{s}_{ko}^{t+} \geq \mathbf{0}, \quad (\forall k, \forall t)
\end{aligned}
\tag{15.1}
$$

where $\mathbf{X}_k^t = \left(\mathbf{x}_{1k}^t, ..., \mathbf{x}_{nk}^t \right) \in R^{m_k \times n}$ and $\mathbf{Y}_k^t = \left(\mathbf{y}_{1k}^t, ..., \mathbf{y}_{nk}^t \right) \in R^{r_k \times n}$ signify the input and output matrices and $\mathbf{s}_{ko}^{t-}$ and $\mathbf{s}_{ko}^{t+}$ are the input and output slacks, respectively.

With regard to the linking constraints, there are several options, for which we present four possible cases. There are, for example, 'as input' and 'as output' link value cases.

In the 'as input' link value case, the linking activities are treated as an input to the succeeding division, and excesses are accounted for in the input inefficiency:

$$\mathbf{z}^t_{o(kh)\text{in}} = \mathbf{Z}^t_{(kh)\text{in}}\boldsymbol{\lambda}^t_k + \mathbf{s}^t_{o(kh)\text{in}} \quad ((kh)\text{in} = 1, \ldots, linkin_k) \tag{15.2}$$

where $\mathbf{s}^t_{o(kh)\text{in}} \in R^{L_{(kh)\text{in}}}$ represents slacks and is non-negative, and $linkin_k$ is the number of 'as input' links from division k.

In the 'as output' link value case, the linking activities are treated as an output from the preceding division, and shortages are accounted for in the output inefficiency:

$$\mathbf{z}^t_{o(kh)\text{out}} = \mathbf{Z}^t_{(kh)\text{out}}\boldsymbol{\lambda}^t_k - \mathbf{s}^t_{o(kh)\text{out}} \quad ((kh)\text{out} = 1, \ldots, linkout_k) \tag{15.3}$$

where $\mathbf{s}^t_{o(kh)\text{out}} \in R^{L_{(kh)\text{out}}}$ represents slacks and is non-negative, and $linkout_k$ is the number of 'as output' links from division k.

We classify carry-over activities into four categories as follows. Corresponding to each category of carry-over, we derive the following equations:

$$
\begin{aligned}
z^{(t,\,t+1)}_{ok_l\text{good}} &= \sum_{j=1}^{n} z^{(t,\,t+1)}_{jk_l\text{good}}\lambda^t_{jk} - s^{(t,\,t+1)}_{ok_l\text{good}} \quad (k_l = 1, \ldots, ngood_k; \forall k; \forall t)\\
z^{(t,\,t+1)}_{ok_l\text{bad}} &= \sum_{j=1}^{n} z^{(t,\,t+1)}_{jk_l\text{bad}}\lambda^t_{jk} + s^{(t,\,t+1)}_{ok_l\text{bad}} \quad (k_l = 1, \ldots, nbad_k; \forall k; \forall t)\\
z^{(t,\,t+1)}_{ok_l\text{free}} &= \sum_{j=1}^{n} z^{(t,\,t+1)}_{jk_l\text{free}}\lambda^t_{jk} + s^{(t,\,t+1)}_{ok_l\text{free}} \quad (k_l = 1, \ldots, free_k; \forall k; \forall t)\\
z^{(t,\,t+1)}_{ok_l\text{fix}} &= \sum_{j=1}^{n} z^{(t,\,t+1)}_{jk_l\text{fix}}\lambda^t_{jk} \quad (k_l = 1, \ldots, \text{fix}_k; \forall k; \forall t)\\
s^{(t,\,t+1)}_{ok_l\text{good}} &\geq 0, \; s^{(t,\,t+1)}_{ok_l\text{bad}} \geq 0 \text{ and } s^{(t,\,t+1)}_{ok_l\text{free}} : \text{free} \quad (\forall k_l; \forall t)
\end{aligned}
\tag{15.4}
$$

where $s^{(t,\,t+1)}_{ok_l\text{good}}$, $s^{(t,\,t+1)}_{ok_l\text{bad}}$ and $s^{(t,\,t+1)}_{ok_l\text{free}}$ represent slacks denoting carry-over shortfall, carry-over excess and carry-over deviation, respectively, and $ngood_k$, $nbad_k$ and $nfree_k$ indicate the numbers of desirable (good), undesirable (bad) and free carry-overs, respectively, for each division k.

The overall efficiency is evaluated by the following program:

$$\theta^*_o = \min \frac{\sum_{t=1}^{T} W^t \left[\sum_{k=1}^{K} w^k \left[1 - \dfrac{1}{m_k + linkin_k + nbad_k}\left(\sum_{i=1}^{m_k} \dfrac{s^{t-}_{iok}}{x^t_{iok}} + \sum_{(kh)_i=1}^{linkin_k} \dfrac{s^t_{o(kh)_i\text{in}}}{z^t_{o(kh)_i\text{in}}} + \sum_{k_l=1}^{nbad_k} \dfrac{s^{(t,t+1)}_{ok_l\text{bad}}}{z^{(t,t+1)}_{ok_l\text{bad}}}\right)\right]\right]}{\sum_{t=1}^{T} W^t \left[\sum_{k=1}^{K} w^k \left[1 + \dfrac{1}{r_k + linkout_k + ngood_k}\left(\sum_{r=1}^{r_k} \dfrac{s^{t+}_{rok}}{y^t_{rok}} + \sum_{(kh)_i=1}^{linkout_k} \dfrac{s^t_{o(kh)_i\text{out}}}{z^t_{o(kh)_i\text{out}}} + \sum_{k_l=1}^{ngood_k} \dfrac{s^{(t,t+1)}_{ok_l\text{good}}}{z^{(t,t+1)}_{ok_l\text{good}}}\right)\right]\right]} \tag{15.5}$$

subject to (15.1)–(15.4), where W^t ($\forall t$) is the weight of period t and w^k ($\forall k$) is the weight of division k. These weights satisfy the condition $\sum_{t=1}^{T} W^t = 1,$

$\sum_{k=1}^{K} w^k = 1, W^t \geq 0(\forall t), w^k \geq 0 \ (\forall k)$ They are supplied exogenously. The numerator includes terms associated with the relative slacks of inputs ('as-input' links and bad carry-overs), whereas the denominator includes the relative slacks of outputs ('as-output' links and good carry-overs). These terms are weighted by the divisional weight w_k and further by the period weight W^t, and they result in the overall efficiency θ_o^*. This objective function is a generalization of the slacks-based measure developed by Tone [19]. The divisional weights indicate the importance of the division, for example in terms of cost and manpower, whereas the period weights reflect, for example, the discount rate by period. $\theta_o^* \leq 1$ and $\theta_o^* = 1$ hold if and only if all slacks are zero. The input- and output-oriented models can be defined by dealing with the numerator and denominator, respectively, of the above objective function. Utilizing the optimal slacks obtained by solving the program (15.5), we define the period and divisional efficiencies as follows. The period efficiency is defined by

$$
\tau_o^{t*} = \frac{\sum_{k=1}^{K} w^k \left[1 - \frac{1}{m_k + linkin_k + nbad_k} \left(\sum_{i=1}^{m_k} \frac{s_{iok}^{t-}}{x_{iok}^t} + \sum_{(kh)_l=1}^{linkin_k} \frac{s_{o(kh)_l \text{in}}^t}{z_{o(kh)_l \text{in}}^t} + \sum_{k_l=1}^{nbad_k} \frac{s_{ok_l \text{bad}}^{(t,t+1)}}{z_{ok_l \text{bad}}^{(t,t+1)}} \right) \right]}{\sum_{k=1}^{K} w^k \left[1 + \frac{1}{r_k + linkout_k + ngood_k} \left(\sum_{r=1}^{r_k} \frac{s_{rok}^{t+}}{y_{rok}^t} + \sum_{(kh)_l=1}^{linkout_k} \frac{s_{o(kh)_l \text{out}}^t}{z_{o(kh)_l \text{out}}^t} + \sum_{k_l=1}^{ngood_k} \frac{s_{ok_l \text{good}}^{(t,t+1)}}{z_{ok_l \text{good}}^{(t,t+1)}} \right) \right]} \quad (\forall t)
$$

(15.6)

where the variables on the right-hand side indicate optimal values for the overall efficiency θ_o^*. The divisional efficiency is defined by

$$
\delta_{ok}^* = \frac{\sum_{t=1}^{T} W^t \left[1 - \frac{1}{m_k + linkin_k + nbad_k} \left(\sum_{i=1}^{m_k} \frac{s_{iok}^{t-}}{x_{iok}^t} + \sum_{(kh)_l=1}^{linkin_k} \frac{s_{o(kh)_l \text{in}}^t}{z_{o(kh)_l \text{in}}^t} + \sum_{k_l=1}^{nbad_k} \frac{s_{ok_l \text{bad}}^{(t,t+1)}}{z_{ok_l \text{bad}}^{(t,t+1)}} \right) \right]}{\sum_{t=1}^{T} W^t \left[1 + \frac{1}{r_k + linkout_k + ngood_k} \left(\sum_{r=1}^{r_k} \frac{s_{rok}^{t+}}{y_{rok}^t} + \sum_{(kh)_l=1}^{linkout_k} \frac{s_{o(kh)_l \text{out}}^t}{z_{o(kh)_l \text{out}}^t} + \sum_{k_l=1}^{ngood_k} \frac{s_{ok_l \text{good}}^{(t,t+1)}}{z_{ok_l \text{good}}^{(t,t+1)}} \right) \right]} \quad (\forall t)
$$

(15.7)

Finally, the period-divisional efficiency is defined by

$$
\rho_{ok}^{t*} = \frac{1 - \frac{1}{m_k + linkin_k + nbad_k} \left(\sum_{i=1}^{m_k} \frac{s_{iok}^{t-}}{x_{iok}^t} + \sum_{(kh)_l=1}^{linkin_k} \frac{s_{o(kh)_l \text{in}}^t}{z_{o(kh)_l \text{in}}^t} + \sum_{k_l=1}^{nbad_k} \frac{s_{ok_l \text{bad}}^{(t,t+1)}}{z_{ok_l \text{bad}}^{(t,t+1)}} \right)}{1 + \frac{1}{r_k + linkout_k + ngood_k} \left(\sum_{r=1}^{r_k} \frac{s_{rok}^{t+}}{y_{rok}^t} + \sum_{(kh)_l=1}^{linkout_k} \frac{s_{o(kh)_l \text{out}}^t}{z_{o(kh)_l \text{out}}^t} + \sum_{k_l=1}^{ngood_k} \frac{s_{ok_l \text{good}}^{(t,t+1)}}{z_{ok_l \text{good}}^{(t,t+1)}} \right)} \quad (\forall k; \forall t)
$$

(15.8)

In the input- and output-oriented models, the numerator and denominator, respectively, of the above formulas are applied.

15.2.3 Formulas for Malmquist Index by DN DEA Model

Based on the period-divisional efficiency score, we define a new Malmquist index as follows. We define the divisional catch-up index as the ratio of the period-divisional efficiencies between t and $t+1$ as follows:

$$\gamma_{ok}^{t\to t+1} = \frac{\rho_{ok}^{t+1^*}}{\rho_{ok}^{t^*}} \quad (t=1,\ldots,T-1; k=1,\ldots,K; o=1,\ldots,n) \tag{15.9}$$

We define the position effect from t to $t+1$ as

$$\sigma_{ok}^{t\to t+1} = \sqrt{\varphi_1\varphi_2} \quad (t=1,\ldots,T-1; k=1,\ldots,K; o=1,\ldots,n)$$

$$\text{where } \varphi_1 = \frac{\rho_{ok}^{t^*}}{\frac{1}{n}\sum_{j=1}^{n}\rho_{jk}^{t^*}} \text{ and } \varphi_2 = \frac{\rho_{ok}^{t+1^*}}{\frac{1}{n}\sum_{j=1}^{n}\rho_{jk}^{t+1^*}} \tag{15.10}$$

Using the above catch-up index and position effect, we define the divisional Malmquist index by their geometric mean:

$$\mu_{ok}^{t\to t+1} = \gamma_{ok}^{t\to t+1}\sigma_{ok}^{t\to t+1} \quad (t=1,\ldots,T-1; k=1,\ldots,K; o=1,\ldots,n) \tag{15.11}$$

15.2.4 Empirical Data

The data used in this empirical investigation concerned 74 municipal hospitals from 2007FY to 2012FY in a balanced panel. There are approximately 1000 municipal hospitals in Japan and there is large heterogeneity among them. We selected municipal hospitals with more than 300 beds. Therefore, this sample may represent larger acute hospitals owned by Japanese municipalities. The data were collected from the *Annual Databook of Local Public Enterprise* published by the Ministry of Internal Affairs and Communications. It is a legal requirement that the local chief executive of each municipal government submits audited financial statements to the ministry. Therefore, the data should be accurate. Accuracy is required for DEA because it cannot take into account measurement errors in the data. DEA also implicitly assumes a correct model specification and the correct specification of inputs, outputs and other variables.

The objective of the administration division is to realize a sound financial situation through labour inputs and capital inputs. The objective of the medical-examination division is to provide a certain amount of medical services using hospital beds that are maintained by the administration division at the same hospital. The DN DEA model makes it possible to have a two-stage production structure in one hospital, that is, both an administration division and a medical-examination division. The administration division raises funds for and maintains medical beds and expensive medical equipment. The medical-examination division uses the medical beds and provides medical services.

Furthermore, the medical-examination division earns medical revenue in return for medical services and the administration division collects the revenue from the medical-examination division and manages financial matters. Previous literature that adopted traditional DEA models in the study of Japanese hospitals did not consider intermediate products in a hospital. In the case of the DN DEA model, we can use link variables as intermediate products for both divisions. This benefit of the DN DEA model (compared with the traditional DEA model) is that it makes it possible to reflect the actual situation. We adopted three link variables in our model.

In addition, if we were to add variables related to the administration division in the traditional DEA model, we would suffer from inadequate correspondence between inputs and outputs. For example, the administration staff do not directly engage in the production of medical services. In the case of the traditional DEA model, the input from the administration staff may correspond to the number of inpatients as an output. However, the relationship between the administration staff and the number of inpatients would cause an undesirable bias in the efficiency estimation. Therefore, the DN DEA model conceptually reduces bias (compared with the traditional DEA model) in the estimation of efficiency both by considering the multiple-step production structure and by excluding inadequate interactions between inputs and outputs. However, we did not consider more detailed divisions in this study. For example, we did not consider pharmaceutical or clinical laboratory divisions.

Many previous studies that have adopted traditional DEA models to examine Japanese hospitals have focused on the activities of the medical-examination division. These studies typically adopt the numbers of doctors and nurses as inputs and the numbers of inpatients and outpatients as outputs. Therefore, such studies do not contain the activities of the administration division, by way of either an input variable or an output variable. However, the DN DEA model enables us to consider activities in both divisions. We can observe the activities of the administration division separately from the medical-examination division.

The inputs, outputs, links and carry-overs of the DN DEA model are described in Figure 15.1. For Division 1, (the administration division), we adopted two labour inputs and three capital inputs. The administration division does not directly provide a medical service to patients. The division is in charge of providing medical beds to the medical-examination division and maintains a sound financial situation for the hospital. Therefore, administration staff should manage the financial situation of the hospital. They also receive subsidies from the municipal government and manage the reimbursement of hospital bonds issued. Maintenance staff maintain all the hospital buildings for hospital activities. As labour inputs, we used both the number of administration officers and the number of maintenance officers. All labour inputs were full-time equivalents (FTEs). However, we did not consider differences in productivity and wage levels of staff. As capital inputs, we used the interest cost for financial arrangements and the municipal subsidies to cover deficits.

For the output of Division 1, we intended to adopt the 'balance ratio of medical income to medical expenses'; the break-even point has a value of 1 and a surplus has a value exceeding 1. However, using a ratio as an input or output makes the

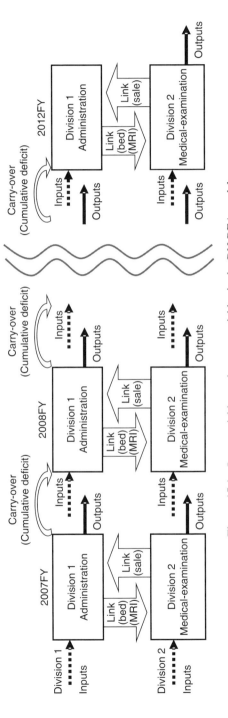

Figure 15.1 Input variables and output variables in the DN DEA model.

convexity issue of DEA problematic [25]. Emrouznejad and Amin [26] recommended not using constant returns to scale when there is a ratio in the input/output variables. Therefore, we decomposed the ratio into medical income and medical expenses: the numerator was used as an output for Division 1 and the denominator as an input for Division 1.

For Division 2 (the medical-examination division), we adopted four labour inputs: the number of doctors, number of nurses, number of assistant nurses and number of medical technologists. All labour inputs were the corresponding value of FTEs. For the outputs of Division 2, we adopted the number of inpatients per operation day, the number of outpatients per operation day and the number of beds in emergency units. In Japan, there is no gatekeeping system involving general practitioners. Therefore, hospitals accept a large number of outpatients to attract potential inpatients. In this study, the number of beds in emergency units was used as a surrogate variable for the emergency care service because we could not distinguish between emergency care patients and ordinary patients from the data source.

Previous studies regarding efficiency estimations of Japanese hospitals did not include emergency medical services. A core function of public hospitals in Japan is to ensure a quick response for emergency patients. However, some municipal hospitals have closed their emergency units to reduce costs, despite the increasing need for emergency medical services. Although we consider that the evaluation of the efficiency of municipal hospitals should include the number of emergency patients, we were not able to obtain numbers from the available data. Therefore, we adopted the number of emergency beds as a proxy for the number of emergency patients. This proxy variable has limitations because it did not control for differences in severity of emergency patients, the quality of the emergency medical service and the occupancy rate of the emergency beds.

The existence of a link variable is one of the key characteristics of the DN DEA model. The link variable is an intermediate product that acts simultaneously as an output from Division 1 and an input to Division 2. Using an intermediate product, we can evaluate multiple production steps among divisions in one DMU. Tone and Tsutsui [20] presented four possible scenarios. For example, the 'fixed link value case' means that the linking activities are unchanged.

There are peculiar characteristics in the case of municipal hospitals compared with private hospitals in Japan. Because of a soft budget constraint and sometimes the election tactics of governors, municipal hospitals tend to overinvest in both beds and expensive medical equipment. Decisions about capital investment are ordinarily made by the administration division. We need to take account of the effect of capital investment by the administration division in the medical-examination division in evaluating the efficiency of municipal hospitals.

We used the both 'number of beds' and 'number of tesla of magnetic resonance imaging [MRI] scanners' as link variables from Division 1 to Division 2. We assumed that Division 1 was in charge of the funding and maintenance of medical beds and expensive medical devices. Division 1 supplied these beds and devices to Division 2. Division 2 used the medical beds and devices for delivering medical care services to patients. We adopted a non-discretionary 'fixed' link, where the linking activity remains constant. The reason for this is that it would be unusual for the medical-examination division to negotiate with the administration division to change the number of beds. The administration division also has an incentive to generate sufficient medical revenue (to offset the medical expenses) and to use all available beds. MRI scanners are expensive medical devices and are very popular in Japanese hospitals. The word 'tesla' is a unit of magnetic field strength, and the latter is related to the fineness of diagnostic imaging. When there were two MRI scanners in one hospital, we summed the numbers of tesla of the two MRI scanners. Therefore, we used the number of tesla of MRI scanners as a proxy variable for both the quality and the quantity of the service provided by MRI scanners.

We used the 'average revenue per inpatient per day' as a link variable from Division 2 to Division 1. We assumed that the average revenue was the consideration to be paid to Division 1 for the beds from Division 2. The average revenue per inpatient may represent the density of medical care services. We adopted an 'as-output' link, where the linking activity is treated as an output from Division 1. The reason for this is that this matter was not negotiable between the two divisions. Division 1 should be efficient enough to provide higher-density medical services under the given resource constraints.

There are other peculiar characteristics in the case of municipal hospitals compared with private hospitals in Japan. Because of the soft budget constraint, municipal hospitals can have a huge cumulative deficit. We need to consider the negative effect of the deficit in our evaluation.

The carry-over variable is one of the benefits of using the DN DEA model compared with the traditional DEA model. A DMU ordinarily continues its activities over several terms. Furthermore, intertemporal factors can affect its efficiency. The carry-over variable makes it possible to account for the effect of connecting activities between terms. The carry-over variable has four characteristics, according to Tone and Tsutsui [20]. For example, 'desirable (good) carry-over' variables are treated as outputs, and a comparative shortage of carry-overs is seen as inefficiency.

We used the 'balance account of the public enterprise bond' (hospital bond) as an undesirable (bad) carry-over. The hospital bond was chosen as the carry-over because municipal hospitals issue these bonds to raise funds for capital investment in hospital beds. The municipal hospital gradually redeems the bond from any revenue surplus. We adopted the 'undesirable (bad)' carry-over; thus, the connecting activity from Period 1 to Period 2 was treated as an input. The reason for this is that newly built hospitals are more attractive to patients but represent a heavier fiscal burden in terms of repaying the principal. Therefore, treating the public enterprise bond as a carry-over reflects accurately the competitive condition of the market in which patients can freely access any hospital. However, we did not consider either the average life or the interest rate of hospital bonds (Figure 15.1).

According to the first principle that a public hospital is expected to accomplish a policy goal with a minimum budget, we selected an input-oriented model. We adopted both a constant-returns-to-scale (CRS) and a variable-returns-to-scale (VRS) model in the analysis. We also employed a Malmquist productivity index approach and decomposed the Malmquist index (MI) into technological change (frontier shift) and the efficiency change of non-best-practice DMUs (catch-up). In the case of the MI, we selected the CRS model according to the results of Grifell-Tatje and Lovell [27]. Descriptive statistics of all variables in the analysis are provided in Table 15.1.

Before we move on to efficiency estimation, we should check the three main outputs to grasp the time trend of the management of the municipal hospitals during the observation period. The number of inpatients per operation day decreased by 3 percentage points from 2007FY to 2012FY. The number of outpatients per operation day decreased by 6 percentage points from 2007FY to 2012FY. The balance ratio of medical expenses to medical income improved by 5 percentage points (but was still in the red) from 2007FY to 2012FY (Figure 15.2).

15.3 RESULTS

15.3.1 Estimated Efficiency Scores

Table 15.2 presents the key statistics of the estimated efficiency scores obtained by the DN DEA model. In Table 15.2, the first set of rows shows the efficiency scores of the overall hospital organization as determined by the DN DEA model. The second set of rows shows the efficiency scores of the administration divisions of the sample hospitals. The third set of rows shows the efficiency scores of the medical-examination divisions of the sample hospitals. From the results of the DN DEA model, we obtained four key findings.

First, the average overall efficiency obtained by the DN DEA model was 0.912 (VRS model) for 2007FY. The average efficiency score estimated by the DN DEA model was almost at the same level as the average efficiency level estimated in previous studies of Japanese municipal hospitals [3–6].

Second, the average level of relative efficiency in 2012FY was slightly less than for 2007FY overall. The average efficiency score was 0.912 for 2007FY and 0.895 for 2012FY (VRS model).

Third, because of the advantages of the network structure in the DN DEA model, we can observe the efficiency changes separately for different internal organizations. The average level of the estimated period-divisional efficiency of the administration division decreased from 0.901 in 2007FY to 0.881 in 2012FY (VRS model). The average period-divisional efficiency of the medical-examination division also decreased from 0.922 in 2007FY to 0.909 in 2012FY (VRS model). On average, there was no significant efficiency improvement in the two divisions for the 6-year period.

15.3.2 Estimated Malmquist Index Scores

The Malmquist productivity index is suitable for evaluating the dynamic change in efficiency of the samples. We estimated the MI of both the administration division and the medical-examination division separately from 2007FY to 2012FY.

TABLE 15.1 Descriptive statistics of all variables in the DN DEA model.

	Variable Names	Average	S.D.	Max	Min	units
Division 1 Input	Number of administration officers	35.48	16.40	92.00	10.00	person
	Number of maintenance officers	13.39	15.52	99.00	0.00	person
	Interest cost per year	237	234	1,120	5	million Yen
	Subsidy from municipality	1,847	1,423	7,195	345	million Yen
	Medical expense	13,036	6,123	30,582	4,035	minion Yen
Output	Medical income	12,368	6,010	29,151	3,107	million Yen
Link(Div1 → Div2)	Number of beds	504.8	169.3	1063.0	300.0	unit
Link(Div1 → Div2)	Number of MRI scanners	2.7	1.2	6.0	1.0	tesla
Division2 Input	Number of doctors	85.45	39.31	182.00	19.00	person
	Number of nurses	420.06	170.58	991.00	138.00	person
	Number of assistant nurses	1.77	3.31	21.00	0.00	person
	Number of medical technologists	84.34	31.13	156.00	33.00	person
Output	Number of inpatients per operation day	402.00	161.99	850.00	119.00	person
	Number of outpatients per operation day	850.14	403.44	1884.00	17.00	person
	Number of beds for emergency units	17.56	13.97	50.00	0.00	unit
Link(Div2 → Div1)	Inpatient revenue	8,499	4,121	21,205	1,866	million Yen
Carry over	Cumulative deficit	4,678	5,380	21,355	0	million Yen

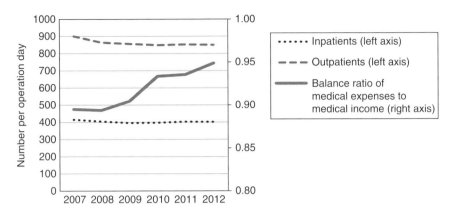

Figure 15.2 Main outputs of sample hospitals from 2007FY to 2012 FY.

TABLE 15.2 Estimation results from DN DEA model.

Division	Model	Fiscal year	2007	2008	2009	2010	2011	2012
Overall	CRS-I	Average	0.829	0.841	0.841	0.831	0.832	0.803
		SD	0.166	0.164	0.164	0.161	0.160	0.170
		Maximum	1.000	1.000	1.000	1.000	1.000	1.000
		Minimum	0.382	0.346	0.474	0.506	0.518	0.481
	VRS-I	Average	0.912	0.917	0.915	0.902	0.905	0.895
		SD	0.139	0.133	0.139	0.140	0.136	0.148
		Maximum	1.000	1.000	1.000	1.000	1.000	1.000
		Minimum	0.542	0.524	0.540	0.531	0.534	0.522
Division1	CRS-I	Average	0.800	0.809	0.809	0.796	0.791	0.765
(admin)		SD	0.214	0.212	0.214	0.215	0.219	0.221
		Maximum	1.000	1.000	1.000	1.000	1.000	1.000
		Minimum	0.301	0.217	0.268	0.318	0.296	0.343
	VRS-I	Average	0.901	0.908	0.907	0.891	0.882	0.881
		SD	0.169	0.163	0.168	0.178	0.180	0.193
		Maximum	1.000	1.000	1.000	1.000	1.000	1.000
		Minimum	0.385	0.452	0.421	0.394	0.371	0.364
Division	2CRS-	Average	0.859	0.873	0.873	0.866	0.873	0.840
(medical)	I	SD	0.149	0.147	0.142	0.149	0.135	0.156
		Maximum	1.000	1.000	1.000	1.000	1.000	1.000
		Minimum	0.463	0.476	0.548	0.528	0.557	0.495
	VRS-I	Average	0.922	0.926	0.923	0.914	0.927	0.909
		SD	0.130	0.130	0.129	0.135	0.120	0.139
		Maximum	1.000	1.000	1.000	1.000	1.000	1.000
		Minimum	0.539	0.591	0.576	0.541	0.561	0.551

We can observe the efficiency change separately for different internal organizations. The average level of the estimated MI of the administration division increased from an initial value of 1.058 from 2007FY to 2008FY, and from an initial value of 1.049 from 2011FY to 2012FY (CRS model). In contrast, the average level of the MI of the medical-examination division was almost unchanged from an initial value of 0.968 in 2007FY to 2008FY, and an initial value of 1.002 from 2011FY to 2012FY (CRS model) (Table 15.3).

The MI of the administration division improved year by year by about 6%. In contrast, the MI of the medical-examination division seems to have shown no change during 2007FY–2012FY. To investigate the reason for the improvement of the administration division, we decomposed the MI into a frontier shift effect and catch-up effect. The improvement in the MI of the administration division may come from a 'frontier shift' rather than a 'catch-up' (Figure 15.3). There may be 'technological change' in the administration division in the form of a frontier shift effect.

TABLE 15.3 Estimation results from Malmquist productivity index scores.

Division	Model	Year	07 → 08	08 → 09	09 → 10	10 → 11	11 → 12
Division1	CRS-I	Average	1.058	1.070	1.107	1.085	1.049
(admin)		SD	0.169	0.137	0.136	0.124	0.132
		Maximum	1.608	1.516	1.669	1.632	1.465
		Minimum	0.714	0.736	0.833	0.820	0.520
Division 2	CRS-I	Average	0.968	0.980	1.001	0.972	1.002
(medical)		SD	0.074	0.106	0.113	0.090	0.118
		Maximum	1.275	1.494	1.385	1.478	1.554
		Minimum	0.761	0.641	0.693	0.750	0.786

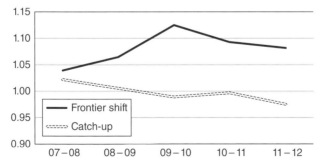

Figure 15.3 Decomposition of the Malmquist productivity index of the administration division.

15.4 DISCUSSION

15.4.1 Estimation Results and Policy Implications

Japanese municipal hospitals, which are about a thousand in number, have experienced financial crises throughout the last few decades. The Japanese central government established a new reform policy aimed at restructuring hospital operations to reduce the debt of municipal hospitals that is sustained by the subsidies that it provides. This would be one of the most extensive reform policies for hospitals in the world.

The planning for the reform involved several steps from 2007FY to 2014FY. Recently, the Japanese central government announced that the financial situation of the municipal hospitals had improved because of the intensive policy campaign during 2009FY–2011FY. However, this announcement did not include any analysis of the efficiency improvement of these municipal hospitals. On the contrary, the Ministry of Internal Affairs and Communication [28] established the following two points in a sample survey of annual reports of municipal hospitals for 2009FY. First, almost all such hospitals did not achieve the targets that had been set for that year. Second, in an interim evaluation, the ministry concluded that reform of municipal hospitals should be considered and effective measures implemented. Thus, the purpose of the present study was to evaluate the policy effect in terms of efficiency improvement.

To evaluate the policy effect, we separately estimated the efficiency change in both the medical-examination division and the administration division, which are heterogeneous internal organizations of a hospital. The administration division conducts business management, while the medical-examination division provides medical care services directly. Furthermore, the administration division provides medical beds to the medical-examination division as interim products, in exchange for the medical revenue from the medical-examination division obtained by using medical beds. We believed that both efficiency scores and the countermeasures to improve efficiency would be different in each division.

We employed a DN DEA model to perform the evaluation. This model makes it possible to simultaneously estimate both the efficiencies of the individual organizations and the dynamic changes in the efficiencies. We have already published a preliminary evaluation of the policy during 2007FY–2009FY [24]. We extended the observation period from three years (2007FY–2009FY) to six years (2007FY–2012FY) with a focus on 2009FY–2011FY, the period of an intensive policy campaign for efficiency improvement as a part of the reform policy.

We also estimated the Malmquist productivity index of both divisions to evaluate the dynamic change in productivity of the municipal hospitals. The MI of the administration division improved year by year about 6% during 2007FY–2012FY. In contrast, the MI of the medical-examination division seems to have shown no change. The improvement in the MI of the administration division may come from a 'frontier shift' effect rather than a 'catch-up' effect. The cause of the improvement could be that accounting standards had been temporarily loosened in terms of the prerequisites for municipal hospitals issuing hospital bonds with financial support from central government. If so, the improvement in the financial situation of municipal hospitals may be based on some kind of passing the burden on to the future rather than efficiency improvement. This would cause a further fiscal burden on the Japanese government. As a conclusion, we cannot state that there is a positive policy effect of the reform of municipal hospitals in terms of efficiency improvement.

15.4.2   Further Research Questions

This study may be one of the earliest empirical applications of the DN DEA model, and thus there are some limitations that need to be addressed. For example, we were unable to use variables regarding the 'quality' of medical services and the 'severity' of the patients' condition. Therefore, we assumed that the sample hospitals would be homogeneous in terms of quality of service and severity of patients' condition. We could, however, narrow the range of samples according to the number of hospital beds to ensure homogeneity of the sample hospitals on some level. We used the number of beds in emergency units as an output and used the total number of medical beds as the link variable. This double counting of medical beds could affect the results in some way.

There are other limitations regarding the control of several factors which could influence the estimated efficiencies. For example, the price of medical services, which is covered by public health insurance, changes every two years in Japan. The rate of reimbursement by public health insurance for medical services changed several times during the observation time. However, the rate of change was relatively small and did not exceed 1% in total. Therefore, the relative efficiency scores should show only a small influence. Many Japanese acute hospitals decided to voluntarily change their reimbursement system from 'fee for service' to 'per diem based on diagnosis groups' (DPC). This study did not fully consider these external environmental changes in the Japanese hospital market.

Regarding policy implications, we did not consider either the relative costs of the two divisions or the relative costs of improving efficiency in each division. For example, one division may be less efficient on average, but the other may be more costly so that a given efficiency improvement is more beneficial. If we are to play an active role in policy implementation, we need to consider relative costs in addition to efficiency scores. Future studies will require a larger sample set and a more complex model.

ACKNOWLEDGEMENTS

We would like to thank the attendees of the DEA Symposium 2012 for their helpful comments on our earlier study. We also appreciate the valuable comments from the attendees of the 10th World Congress 2014 of the International Health Economics Association (iHEA). This study was supported by a Grant-in-Aid for Scientific Research (KAKENHI, grant number 22320092) and a Grant-in-Aid for Scientific Research (KAKENHI, grant number 24243039).

REFERENCES

[1] Harris, J.E. (1977) The internal organization of hospitals: Some economic implications. *Bell Journal of Economics*, **8**, 467–482.

[2] Hollingsworth, B. (2008) The measurement of efficiency and productivity of health care delivery. *Health Economics*, **17**, 1107–1128.

[3] Num, S., Ishikawa, K. (1994) An application of DEA for labor efficiency of Japanese hospitals. *Journal of the Operations Research Society of Japan*, **39**, 292–296.

[4] Num, S., Gunji, A. (1994) A study on managerial efficiency in medical facilities: evaluating human resource efficiency in municipal hospitals using data envelopment analysis. *Journal of the Japan Society of Health Administration*, **l31**, 33–39.

[5] Nakayama, N. (2003) A comparison of parametric and non-parametric distance functions: A case study of Japanese public hospitals. *Iryo to Shakai*, **13**, 83–95.

[6] Nakayama, N. (2004) Technical efficiency and subsidies in Japanese public hospitals. *Iryo to Shakai*, **14**, 69–79.

[7] Färe, R., Grosskopf, S. (2000) Network DEA. *Socio-Economic Planning Science*, **34**, 35–49.

[8] Färe, R., Grosskopf, S. (1996) *Intertemporal Production Frontiers: With Dynamic DEA. Kluwer Norwell.*

[9] Färe, R., Grosskopf, S., Norris, S., Zhang, Z. (1994) Productivity growth, technical progress, and efficiency change in industrialized countries. *American Economic Review*, **84**(1), 66–83.

[10] Lewis, H.F., Sexton, T.R. (2004) Network DEA: Efficiency analysis of organisations with complex internal structure. *Computers and Operations Research*, **31**, 1365–1410.

[11] Sexton, T.R., Lewis, H.F. (2003) Two-stage DEA: An application to major league baseball. *Journal of Productivity Analysis*, **19**, 227–249.

[12] Prieto, A.M., Zofio, J.L. (2007) Network DEA efficiency in input–output models: With an application to OECD countries. *European Journal of Operational Research*, **178**, 292–304.

[13] Koopmans, T. (1951) Analysis of production as an efficient combination of activities, in *Activity Analysis of Production and Allocation (ed. T. Koopmans), Cowles Commission for Research in Economics Monograph, vol.13. John Wiley and Sons, New York, p. 33.*

[14] Löthgren, M., Tambour, M. (1999) Productivity and customer satisfaction in Swedish pharmacies: A DEA network model. *European Journal of Operational Research*, **115**, 449–458.

[15] Tone, K., Tsutsui, M. (2009) Network DEA: A slacks based measurement approach. *European Journal of Operational Research*, **197**, 243–252.

[16] Klopp, G.A. (1985) The analysis of the efficiency of production system with multiple inputs and outputs. PhD dissertation. Industrial and System Engineering College, University of Illinois at Chicago.

[17] Malmquist, S. (1953) Index numbers and indifference surfaces. *Trabajos de Estadistica*, **4**(2), 209–242.

[18] Tone, K. (1999) A slacks-based measure of efficiency in data envelopment analysis. *European Journal of Operational Research*, **130**(3), 498–509.

[19] Pastor, J.T., Ruiz, J.L., Sirvent, I. (1999) An enhanced DEA Russell graph efficiency measure. *European Journal of Operational Research*, **115**(3), 596–607.

[20] Tone, K., Tsutsui, M. (2014) Dynamic DEA with network structure: A slacks-based measure approach. *Omega*, **42**(1), 124–131.

[21] Fukuyama, H., Weber, W.H. (2014) Measuring Japanese bank performance: A dynamic network DEA approach. *Journal of Productivity Analysis First Online*, 1–16.

[22] Avkiran, N.K. (2015) An illustration of dynamic network DEA in commercial banking including robustness tests. *Omega*, **55**, 141–150.

[23] Lu, W.M., Kweh, Q.L., Nourani, M., Wang, W.K. (2014) The effects of intellectual capital on dynamic network bank performance. Proceedings of the International Conference on Contemporary Economic Issues 2014, pp. 71–78.

[24] Kawaguchi, H., Tone, K., Tsutsui, M. (2014) Estimation of the efficiency of Japanese hospitals using a dynamic and network data envelopment analysis model. *Health Care Management Science*, **17**(2), 101–112.

[25] Hollingsworth, B., Smith, P.C. (2003) The use of ratios in data envelopment analysis. *Applied Economics Letters*, **10**, 733–735.

[26] Emrouznejad, A., Amin, G.R. (2009) DEA models for ratio data: Convexity consideration. *Applied Mathematical Modelling*, **33**(1), 486–498.

[27] Grifell-Tatje, E., Lovell, C.A.K. (1995) A note on the Malmquist productivity index. *Economics Letters*, **4**, 169–175.

[28] Ministry of Internal Affairs and Communication (2011) Results from a survey of implementation of the reformation plans of municipal hospitals.

16

DEA IN THE TRANSPORT SECTOR[1]

MING-MIIN YU

Department of Transportation Science, National Taiwan Ocean University, Keelung, Taiwan

LI-HSUEH CHEN

Department of Transportation Science, National Taiwan Ocean University, Keelung, Taiwan

16.1 INTRODUCTION

The evaluation of operational performance has become a critical indicator for the management of transport services. Traditionally, partial indicators are used to measure the operational performance of transport organizations (e.g. vehicle-miles per vehicle, passengers per revenue vehicle hour, and revenue vehicle hours per dollar operating cost). However, partial indicators only focus on single operational factors or parts of them. They may lead to misleading results, because the operations of transport organizations are characterized by multiple inputs and multiproduct capability. Therefore, more advanced techniques are needed to reflect the multidimensional nature of transport services. To date, the literature has developed some methods to assess the operational performance of transport organizations, including data envelopment analysis (DEA) (e.g. [1–11]), stochastic frontier analysis (SFA) (e.g. [12–14]), multiple linear regression (e.g. [15]), total factor analysis (TFA) (e.g. [16–[18]), the free disposal hull (FDH) method (e.g. [19,20]) and multiple-criteria decision making

[1] Part of the material in this chapter is adapted from Yu, M.M., Chen, L.H. and Hsiao, B., 2016, 'Dynamic performance assessment of bus transit with the multi-activity network structure', *Omega*, **60**, 15–25, and Yu, M.M., Hsiao, B., Hsu, S.H. and Li, S.Y., 2012, 'Measuring harbour management, stevedoring and warehousing performance of Taiwanese container ports using the multi-activity network DEA model', *Journal of International Logistics and Trade*, **10**(2), 77–115, with permission from Elsevier Science and the Jungseok Research Institute of International Logistics and Trade.

Advances in DEA Theory and Applications: With Extensions to Forecasting Models,
First Edition. Edited by Kaoru Tone.
© 2017 John Wiley & Sons Ltd. Published 2017 by John Wiley & Sons Ltd.

(MCDM) (e.g. [21,22]). Among these methods, DEA is considered to be one of the best approaches for organizing and analysing data, owing to its simple framework. It applies a mathematical programming approach to set up an overall measurement indicator, where input and output variables are used to calculate the relative efficiency of individual decision-making units (DMUs) [23]. In addition, it allows efficiency to evolve over time and requires no prior assumptions for the specification of the best-practice frontier. There is extensive literature on DEA and it has been applied to a wide diversity of economic topics.

However, conventional DEA models treat the operational process as a black box, and use aggregate data to evaluate efficiency, without considering the linking items in a series. Services provided by transport organizations are unstorable and must be consumed immediately. If they are not consumed, they will disappear [24]. The quantities of service consumed may be a proportion of the quantities of service produced. Hence, in general, the operation of a transport organization involves two processes: the production process and the service process, and these two processes are interdependent. The capacities produced in the production process are treated as inputs to generate service outputs in the service process. In order to reflect the actual operational situations, Färe and Grosskopf [25,26] proposed a network DEA (NDEA) model to explore the divisional correlations in the evaluation of operational efficiency. Afterwards, various models were proposed to measure the efficiencies of individual processes (see Kao [27] for a review).

The operation of a transport organization is not independent between periods. When operators plan operationally, they will consider the interrelationship between consecutive terms, and reserve a proportion of outputs or revenue to the next period. Hence, in the consideration of long-term planning and investment, a single-period optimization model is not favourable. Since multiperiod benchmarking can identify the best industry practices over time, it can grasp long-term business variations. Wu *et al.* [28] argued that there are three advantages of multiperiod benchmarking. First, since multiperiod benchmarking can identify the industry leaders over time, it can provide suitable models for industry followers. Second, since some industries, such as transport, have seasonal fluctuations, multiperiod benchmarking can obtain more reliable results based on monthly data. Third, multiperiod benchmarking can specify the potential effects of lagged-productive or carry-over items. Although window analysis and the Malmquist index have been used to account for intertemporal efficiency, they ignore the effects of carry-over items. In response to the interrelationship between consecutive terms, Färe and Grosskopf [25] introduced a dynamic DEA model, which connected storable inputs and carry-over outputs from individual periods, to study dynamical and historical systems. Since then, various dynamic DEA models have been proposed to overcome the problem of intertemporal input–output dependence (e.g. [29–32]).

In addition, in the transport industry, it occurs often that undesirable outputs, such as pollution and noise, are produced jointly with desirable outputs. They are unwillingly but inevitably generated. Undesirable outputs appear to have a harmful impact on the service of a DMU. Without considering the undesirable outputs in the evaluation, efficiency evaluation methods may produce misleading results [33]. When evaluating the performance of transport organizations, the trade-off between the utilization of desirable outputs and the control of undesirable outputs should be considered.

In response to these operational characteristics of transport organizations, this chapter develops a dynamic NDEA (DNDEA) model to explore their operational performance. Then, we extend the DNDEA model to investigate the situation in which the operation of a transport organization includes multiple activities. In addition, this chapter provides an application to illustrate the performance of bus transit firms by applying a multi-activity DNDEA (MDNDEA) model. The concept described in this chapter can be developed to solve more complex problems.

This chapter is structured as follows. Following the introduction, the DNDEA model for performance evaluation in transport is formulated. Then, the DNDEA model is extended to consider the effect of interdependence among activities. Afterwards, a related application to the transport industry is provided to investigate the applicability of a multiprocess, multiperiod and multi-activity framework. Finally, conclusions are drawn.

16.2   DNDEA IN TRANSPORT

Conventional DEA models treat the operational process as a black box, without examining the structure of the processes in a DMU's operation. However, the structure of a transport organization is complex. It includes mainly two processes: the production process and the service process. In addition, carry-over items exist in the transport industry, because the operation of a transport organization in one period is not independent of that in the next one. In response to these operational characteristics of transport organizations, the DNDEA model, which considers the effects of interrelationships among processes and the impacts of carry-over items between two consecutive terms, has been designed to improve on the weaknesses of conventional DEA models. In addition, the outputs of transport services may include undesirable outputs. In order to deal with problems where some outputs (desirable outputs) are expected to be maximized and some outputs (undesirable outputs) are expected to be minimized, the directional distance function proposed by Luenberger [34] is a more adequate tool. This permits simultaneous expansion of desirable outputs and contraction of undesirable outputs. Hence, we build the performance measurement model by using the DNDEA method and the directional distance function.

Figure 16.1 outlines the structure of our model. For each DMU, the operations of transport services are assessed for the two processes. The production process transfers the original inputs while maintaining their capacities, and the production efficiency (PE) is examined. In addition, some outputs in the production process in the current period are transferred into the next period. The second process, known as the service process, uses its previous process capacities as inputs in order to produce service outputs, including both undesirable and desirable outputs, and the service efficiency (SE)

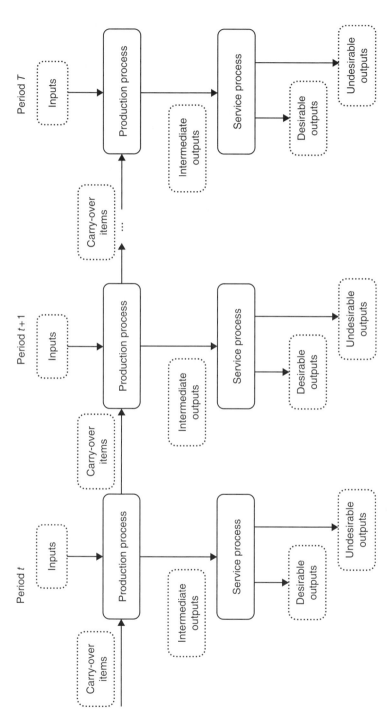

Figure 16.1 The operational structure of a transport organization.

is then examined. In the evaluation of the production process, it can be understood that if input resources are used inadequately, this will lead to waste. Alternatively, service inefficiency can be measured as the utilization of production capacity. Finally, the overall operational efficiency (OE) is determined by mixing the PE and SE.

Suppose that there are n DMUs in period t ($t = 1, \ldots, T$), and that each DMU engages in production and service processes. Let $X_{i_P jP}^t = \left(x_{1jP}^t, \ldots, x_{m_P jP}^t \right)$ denote the input vector associated with the production process in period t. For the production process, in period t, each DMU produces the intermediate output vector $Z_{j(PS)_{l_{PS}}}^t = \left(z_{j(PS)_1}^t, \ldots, z_{j(PS)_{L_{PS}}}^t \right)$, which flows into the service process, and carries the carry-over item vector $Z_{jP_{l_P}}^{(t, t+1)} = \left(z_{jP_1}^{(t, t+1)}, \ldots, z_{jP_{L_P}}^{(t, t+1)} \right)$ to period $t + 1$. For the service process, the desirable output vector $Y_{r_d jS}^t = \left(y_{1jS}^t, \ldots, x_{s_d jS}^t \right)$ and the undesirable output vector $Y_{r_{nd} jS}^t = \left(y_{1jS}^t, \ldots, y_{s_{nd} jS}^t \right)$ are jointly produced in period t.

Microeconomic theory indicates that one of a firm's objectives is to produce the level of outputs where constant returns to scale (CRS) exist. Although firms may operate under variable returns to scale (VRS) in the short run, they will adjust their scale of operations to move towards CRS in the long run [35]. In addition, Månsson [36] and Färe *et al.* [37] argued that CRS captured the long-run results, while VRS was suitable for the short run. Hence, in a multiperiod context, it is reasonable to calculate efficiency estimates under the assumption of CRS. Accordingly, there are two production technologies, T_P^t and T_S^t, in our DNDEA model.

16.2.1 The Production Technology for the Production Process

The production technology T_P^t for the production process under the assumption of CRS is written as follows:

$$
T_P^t = \Bigg\{ \left(x^t, z^t, z^{(t, t+1)} \right) : \sum_{j=1}^n \lambda_{jP}^t x_{i_P jP}^t \leq x_{i_P kP}^t, \quad i_P = 1, \ldots, m_P,
$$

$$
\sum_{j=1}^n \lambda_{jP}^t z_{j(PS)_{l_{PS}}}^t \geq z_{k(PS)_{l_{PS}}}^t, \quad l_{PS} = 1, \ldots, L_{PS},
$$

$$
\sum_{j=1}^n \lambda_{jP}^t z_{jP_{l_P}}^{(t, t+1)} \geq z_{kP_{l_P}}^{(t, t+1)}, \quad l_P = 1, \ldots, L_P,
$$

$$
\sum_{j=1}^n \lambda_{jP}^t z_{jP_{l_P}}^{(t-1, t)} \leq z_{kP_{l_P}}^{(t-1, t)}, \quad l_P = 1, \ldots, L_P,
$$

$$
\lambda_{jP}^t \geq 0, \quad j = 1, \ldots, n \Bigg\} (t = 1, \ldots, T) \tag{16.1}
$$

However, if $t = 1$, $\sum_{j=1}^{n} \lambda_{jp}^1 z_{jP_{l_p}}^{(0,1)} \leq z_{kP_{l_p}}^{(0,1)}$ is substituted for $\sum_{j=1}^{n} \lambda_{jp}^1 z_{jP_{l_p}}^{(0,1)} = z_{kP_{l_p}}^{(0,1)}$; if $t = T$,

$\sum_{j=1}^{n} \lambda_{jP_l}^T z_{jP_{l_p}}^{(T,T+1)} \geq z_{kP_{l_p}}^{(T,T+1)}$ is removed.

16.2.2 The Production Technology for the Service Process

Since undesirable outputs are produced together with desirable outputs in the service process, we model the service process technology by imposing null-jointness between desirable and undesirable outputs, as well as weak disposability. Then, T_S^t is an output set, as $\left(y_{r_dS}^t, y_{r_{nd}S}^t \right) \in T_S^t$ and $0 \leq \theta \leq 1$ imply $\left(\theta y_{r_dS}^t, \theta y_{r_{nd}S}^t \right) \in T_S^t$. In other words, this means that a reduction in undesirable outputs is feasible only if desirable outputs are simultaneously reduced, given a fixed level of inputs. In addition, we assume that the desirable outputs are freely disposable, as $\left(y_{r_dS}^t, y_{r_{nd}S}^t \right) \in T_S^t$ and $y_{r_dS}^{t\,'} \leq y_{r_dS}^t$ imply $\left(y_{r_dS}^{t\,'}, y_{r_{nd}S}^t \right) \in T_S^t$. The notion that the desirable outputs are jointly produced with the undesirable outputs is modelled by stating that if $\left(y_{r_dS}^t, y_{r_{nd}S}^t \right) \in T_S^t$ and $y_{r_{nd}S}^t = 0$ then $y_{r_dS}^t = 0$. This means that if a desirable output is produced in a positive amount, some undesirable outputs must also be produced [38].

Then, the production technology T_S^t for the service process under the assumption of CRS is constructed as follows:

$$T_S^t = \left\{ (z^t, y^t) : \sum_{j=1}^{n} \lambda_{jS}^t z_{j(PS)_{l_{PS}}}^t \leq z_{k(PS)_{l_{PS}}}^t, \quad l_{PS} = 1, \ldots, L_{PS}, \right.$$

$$\sum_{j=1}^{n} \lambda_{jS}^t y_{r_djS}^t \geq y_{r_dkS}^t, \quad r_d = 1, \ldots, s_d,$$

$$\sum_{j=1}^{n} \lambda_{jS}^t y_{r_{nd}jS}^t = y_{r_{nd}kS}^t, \quad r_{nd} = 1, \ldots, s_{nd},$$

$$\left. \lambda_{jS}^t \geq 0, \quad j = 1, \ldots, n \right\} (t = 1, \ldots, T) \qquad (16.2)$$

where λ_P^t and λ_S^t are intensity variables associated with the production process and service process, respectively, in period t.

Based on manipulation of the directional distance function, in period t, the kth DMU's production inefficiency score β_{kP}^t can be represented as the directional distance function defined by the technology T_P^t, and its service inefficiency score β_{kS}^t can be represented as the directional distance function defined by the technology T_S^t.

Then the overall operational ineffectiveness for DMU_k can be estimated by solving the following DNDEA model:

$$\max \beta_k = \sum_{t=1}^{T} W^t \left(w^P \cdot \beta_{kP}^t + w^S \cdot \beta_{kS}^t \right) \tag{16.3}$$

s.t.

(Production process)

$$\sum_{j=1}^{n} \lambda_{jP}^t x_{i_P jP}^t \leq \left(1 - \beta_{kP}^t\right) x_{i_P kP}^t, \quad i_P = 1, \ldots, m_P, \; t = 1, \ldots, T \tag{16.4}$$

$$\sum_{j=1}^{n} \lambda_{jP}^t z_{j(PS)_{l_{PS}}}^t = z_{k(PS)_{l_{PS}}}^t, \quad l_{PS} = 1, \ldots, L_{PS}, \; t = 1, \ldots, T \tag{16.5}$$

$$\sum_{j=1}^{n} \lambda_{jP}^t z_{jP_{l_P}}^{(t,t+1)} = \sum_{j=1}^{n} \lambda_{jP}^{t+1} z_{jP_{l_P}}^{(t,t+1)}, \quad l_P = 1, \ldots, L_P, t = 1, \ldots, T-1 \tag{16.6}$$

$$\sum_{j=1}^{n} \lambda_{jP}^t z_{jP_{l_P}}^{(t,t+1)} = z_{kP_{l_P}}^{(t,t+1)}, \quad l_P = 1, \ldots, L_P, t = 1, \ldots, T-1 \tag{16.7}$$

(Service process)

$$\sum_{j=1}^{n} \lambda_{jS}^t z_{j(PS)_{l_{PS}}}^t = z_{k(PS)_{l_{PS}}}^t, \quad l_{PS} = 1, \ldots, L_{PS}, \; t = 1, \ldots, T \tag{16.8}$$

$$\sum_{j=1}^{n} \lambda_{jS}^t y_{r_d jS}^t \geq \left(1 + \beta_{kS}^t\right) y_{r_d kS}^t, \quad r_d = 1, \ldots, s_d, t = 1, \ldots, T \tag{16.9}$$

$$\sum_{j=1}^{n} \lambda_{jS}^t y_{r_{nd} jS}^t = \left(1 - \beta_{kS}^t\right) y_{r_{nd} kS}^t, \quad r_{nd} = 1, \ldots, s_{nd}, t = 1, \ldots, T \tag{16.10}$$

(Initial condition)

$$\sum_{j=1}^{n} \lambda_{jP}^1 z_{jP_{l_P}}^{(0,1)} = z_{kP_{l_P}}^{(0,1)}, \quad l_P = 1, \ldots, L_P \tag{16.11}$$

(Additional conditions)

$$\sum_{t=1}^{T} W^t = 1 \tag{16.12}$$

$$w^P + w^S = 1 \tag{16.13}$$

$$\lambda_{jP}, \; \lambda_{jS}, \; W^t, \; w^P, \; w^S \geq 0, \quad j = 1,\ldots,n, \; t = 1,\ldots,T \qquad (16.14)$$

where β_{kP}^t, β_{kS}^t, λ_{jP}^t, λ_{jS}^t and $s_{kP_{l_p}}^{(t,t+1),\,\text{free}}$, $t = 1,\ldots,T$, $j,k = 1,\ldots,n, l_P = 1,\ldots,L_P$ are variables of this model. W^t, w^P and w^S are the weights of period t, the production process and the service process, respectively, and represent the relative importance of these periods and processes. It is assumed that the linking items between the production and service processes are fixed by the constraints (16.5) and (16.8), and the carry-over items in the production process act as the non-discretionary link because of the constraints (16.6) and (16.7). The constraint (16.6) imposes a continuity condition between consecutive periods. In addition, the initial conditions can be accounted for by the constraint (16.11), and are given and fixed. Based on the above DNDEA model, various efficiencies can be defined as follows:

$$\text{Period-production efficiency (PPE)} = 1 - \beta_{kP}^t$$

$$\text{Period-service efficiency (PSE)} = 1 - \beta_{kS}^t$$

$$\text{Period-operational efficiency (POE)} = 1 - \left(w^P \cdot \beta_{kP}^t + w^S \cdot \beta_{kS}^t \right)$$

$$\text{PE} = 1 - \sum_{t=1}^{T} W^t \cdot \beta_{kP}^t$$

$$\text{SE} = 1 - \sum_{t=1}^{T} W^t \cdot \beta_{kS}^t$$

$$\text{OE} = 1 - \beta_k$$

β_k is equal to zero if and only if the DMU is operationally efficient and $\beta_{kP}^t = \beta_{kS}^t = 0$, $t = 1,\ldots,T$.

With regard to the constraints on linking items (intermediate outputs) and carry-over items, there are several options. Referring to Tone and Tsutsui [39], we present two possible cases for linking items and four cases for carry-over items. In terms of linking items, there are fixed and free link value cases. Equations (16.5) and (16.8) represent the fixed linking constraints, meaning that the linking items are kept unchanged. If the linking items are freely adjustable, the fixed linking constraints (16.5) and (16.8) can be replaced with the constraint (16.15):

$$\sum_{j=1}^{n} \lambda_{jP}^t z_{j(PS)_{l_{PS}}}^t = \sum_{j=1}^{n} \lambda_{jS}^t z_{j(PS)_{l_{PS}}}^t, \quad l_{PS} = 1,\ldots,L_{PS}, \; t = 1,\ldots,T \qquad (16.15)$$

In terms of carry-over items, there are desirable, undesirable, discretionary and non-discretionary link value cases. Equation (16.7) represents the non-discretionary linking constraint, meaning that the values of the carry-over items are unchanged. If the carry-over items are desirable, they are treated as outputs, and the target values cannot be less than the observed values. The non-discretionary linking constraint (16.7) can be substituted by the constraint (16.16):

$$\sum_{j=1}^{n} \lambda_{jP}^{t} z_{jP_{l_P}}^{(t,t+1)} \geq z_{kP_{l_P}}^{(t,t+1)}, \quad l_P = 1,\dots,L_P, t = 1,\dots,T-1 \qquad (16.16)$$

In contrast to desirable links, undesirable links are considered as inputs, and the target values cannot be greater than the observed values. Hence, the non-discretionary linking constraint (16.7) can be replaced with the constraint (16.17):

$$\sum_{j=1}^{n} \lambda_{jP}^{t} z_{jP_{l_P}}^{(t,t+1)} \leq z_{kP_{l_P}}^{(t,t+1)}, \quad l_P = 1,\dots,L_P, t = 1,\dots,T-1 \qquad (16.17)$$

Finally, if the carry-over items are discretionary, their values can be freely increased or decreased. The constraint (16.7) can be substituted by the constraint (16.18):

$$\sum_{j=1}^{n} \lambda_{jP}^{t} z_{jP_{l_P}}^{(t,t+1)} = z_{kP_{l_P}}^{(t,t+1)} - s_{kP_{l_P}}^{(t,t+1),\text{free}}, \quad l_P = 1,\dots,L_P, t = 1,\dots,T-1 \qquad (16.18)$$

where $s_{kP_{l_P}}^{(t,t+1),\text{free}}$, $k = 1,\dots,n$, $l_P = 1,\dots,L_P$, $t = 1,\dots,T-1$ are slack variables denoting link deviation.

16.3   EXTENSION

In the above section, we considered the interrelationships between processes and the effects of carry-over items. However, a transport organization may consist of several identifiable activities [40]. Take the operations of a container port as an example. A container port has harbour management, stevedoring and warehousing activities. Since a transport organization with efficiency in one activity may not be efficient in other activities, different efficiency ratings for different activities should be distinguished. When a DMU jointly carries out various activities and processes which cannot be assumed to be technologically identical, these activities and processes are separated into different technologies in a multi-activity DEA model [41]. In addition, parts of the resources are unseparated, and are shared among different activities and/or processes. For example, the straddle carriers of a container port work for both harbour management and warehousing activities. In order to understand more deeply the operational performance of a transport organization, the multi-activity structure and the allocation of common inputs also need to be taken into account. Hence, in this section, we develop an MDNDEA model to evaluate the performance of transport organizations. Since the operational characteristics of different transport organizations are different, we construct an MDNDEA model by taking bus transit firms, which provide highway bus (HB) and urban bus (UB) services in the production process, as an example.

The structure of the multi-activity model for a specific bus transit firm is depicted in Figure 16.2. Specifically, inputs are divided into two parts. One part consists of

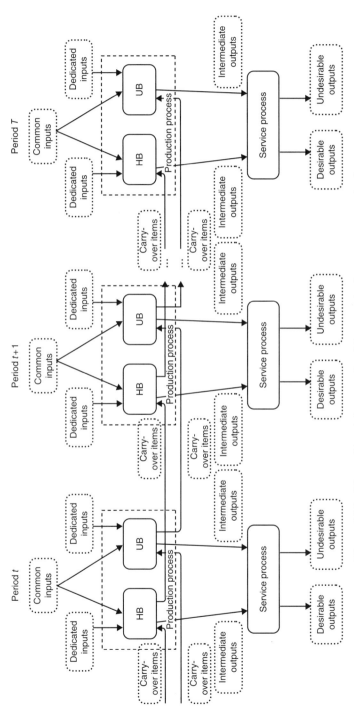

Figure 16.2 The multi-activity operational framework of a transport organization.

dedicated inputs that contribute to the specific activity or process, and the other consists of common inputs that are shared between the HB and UB production activities. The production capacities of these two activities are utilized as the inputs in the service process. Similarly, some outputs of the HB and UB activities in the current period will be transferred into the next period.

Similarly, suppose that there are n DMUs in period t ($t = 1, \ldots, T$), and that each DMU engages in HB and UB production activities as well as a service process. Let $X_{i_H jH}^t = \left(x_{1jH}^t, \ldots, x_{m_H jH}^t \right)$ and $X_{i_U jU}^t = \left(x_{1jU}^t, \ldots, x_{m_U jU}^t \right)$ denote the dedicated input vectors associated with the HB and UB production activities, respectively, in period t, and let $X_{i_{HU} jHU}^t = \left(x_{1jHU}^t, \ldots, x_{m_{HU} jHU}^t \right)$ be a common input vector shared by the HB and UB production activities in period t. It is assumed that, in period t, DMU$_j$ allocates some portion $\alpha_{i_{HU} jH}^t$ of the common input quantities $x_{i_{HU} jHU}^t$ to the HB production activity, and the remaining $\left(1 - \alpha_{i_{HU} jH}^t \right)$ to the UB production activity. For the HB production process, in period t, each DMU produces the intermediate output vector $Z_{j(HS)_{l_{HS}}}^t = \left(z_{j(HS)_1}^t, \ldots, z_{j(HS)_{L_{HS}}}^t \right)$, which flows into the service process, and carries the carry-over item vector $Z_{jH_{l_H}}^{(t, t+1)} = \left(z_{jH_1}^{(t, t+1)}, \ldots, z_{jH_{L_H}}^{(t, t+1)} \right)$ to period $t + 1$. For the UB production process, in period t, each DMU produces the intermediate output vector $Z_{j(US)_{l_{US}}}^t = \left(z_{j(US)_1}^t, \ldots, z_{j(US)_{L_{UC}}}^t \right)$, which flows into the service process, and carries the carry-over item vector $Z_{jU_{l_U}}^{(t, t+1)} = \left(z_{jU_1}^{(t, t+1)}, \ldots, z_{jU_{L_U}}^{(t, t+1)} \right)$ to period $t + 1$. For the service process, the desirable output vector $Y_{r_d jS}^t = \left(y_{1jS}^t, \ldots, x_{s_d jS}^t \right)$ and the undesirable output vector $Y_{r_{nd} jS}^t = \left(y_{1jS}^t, \ldots, y_{s_{nd} jS}^t \right)$ are jointly produced in period t.

Accordingly, there are three production technologies, T_H^t, T_U^t and T_S^t, in our MDNDEA model.

16.3.1   The Production Technology for HB Activity

The production technology T_H^t with CRS for the HB activity is defined as follows:

$$T_H^t = \left\{ \left(x^t, z^t, z^{(t,\ t+1)} \right) : \sum_{j=1}^n \lambda_{jH}^t x_{i_H jH}^t \le x_{i_H kH}^t, \quad i_H = 1, \ldots, m_H, \right.$$

$$\sum_{j=1}^n \alpha_{i_{HU} jH}^t \lambda_{jH}^t x_{i_{HU} jHU}^t \le \alpha_{i_{HU} kH}^t x_{i_{HU} kHU}^t, \quad i_{HU} = 1, \ldots, m_{HU},$$

$$0 < \alpha_{i_{HU} jH}^t < 1, \quad i_{HU} = 1, \ldots, m_{HU},$$

$$\sum_{j=1}^n \lambda_{jH}^t z_{j(HS)_{l_{HS}}}^t \ge z_{k(HS)_{l_{HS}}}^t, \quad l_{HS} = 1, \ldots, L_{HS},$$

$$\sum_{j=1}^{n} \lambda_{jH}^{t} z_{jH_{l_H}}^{(t,t+1)} \geq z_{kH_{l_H}}^{(t,t+1)}, \quad l_H = 1,\ldots,L_H,$$

$$\sum_{j=1}^{n} \lambda_{jH}^{t} z_{jH_{l_H}}^{(t-1,t)} \leq z_{kH_{l_H}}^{(t-1,t)}, \quad l_H = 1,\ldots,L_H,$$

$$\lambda_{jH}^{t} \geq 0, \quad j=1,\ldots,n \Big\} (t=1,\ldots,T) \qquad (16.19)$$

If $t=1$, $\displaystyle\sum_{j=1}^{n} \lambda_{jH}^{1} z_{jH_{l_H}}^{(0,1)} \leq z_{kH_{l_H}}^{(0,1)}$ is substituted for $\displaystyle\sum_{j=1}^{n} \lambda_{jH}^{1} z_{jH_{l_H}}^{(0,1)} = z_{kH_{l_H}}^{(0,1)}$; if $t=T$,

$\displaystyle\sum_{j=1}^{n} \lambda_{jH}^{T} z_{jH_{l_H}}^{(T,T+1)} \geq z_{kH_{l_H}}^{(T,T+1)}$ is removed.

16.3.2   The Production Technology for UB Activity

The production technology T_U^t with CRS for the UB activity is expressed as follows:

$$T_U^t = \Bigg\{ \left(x^t,\ z^t,\ z^{(t,\ t+1)} \right) : \sum_{j=1}^{n} \lambda_{jU}^{t} x_{i_U jU}^{t} \leq x_{i_U kU}^{t}, \quad i_U = 1,\ldots,m_U,$$

$$\sum_{j=1}^{n} \left(1-\alpha_{i_{HU} jH}^{t}\right) \lambda_{jU}^{t} x_{i_{HU} jHU}^{t} \leq \left(1-\alpha_{i_{HU} kH}^{t}\right) x_{i_{HU} kHU}^{t}, \quad i_{HU} = 1,\ldots,m_{HU},$$

$$0 < \alpha_{i_{HU} jH}^{t} < 1, \quad i_{HU} = 1,\ldots,m_{HU},$$

$$\sum_{j=1}^{n} \lambda_{jU}^{t} z_{j(US)_{l_{US}}}^{t} \geq z_{k(US)_{l_{US}}}^{t}, \quad l_{US} = 1,\ldots,L_{US},$$

$$\sum_{j=1}^{n} \lambda_{jU}^{t} z_{jU_{l_U}}^{(t,t+1)} \geq z_{kU_{l_U}}^{(t,t+1)}, \quad l_U = 1,\ldots,L_U,$$

$$\sum_{j=1}^{n} \lambda_{jU}^{t} z_{jU_{l_U}}^{(t-1,t)} \leq z_{kU_{l_U}}^{(t-1,t)}, \quad l_U = 1,\ldots,L_U,$$

$$\lambda_{jU}^{t} \geq 0, \quad j=1,\ldots,n \Big\} (t=1,\ldots,T) \qquad (16.20)$$

Similarly, if $t=1$, $\displaystyle\sum_{j=1}^{n} \lambda_{jU}^{1} z_{jU_{l_U}}^{(0,1)} \leq z_{kU_{l_U}}^{(0,1)}$ is substituted for $\displaystyle\sum_{j=1}^{n} \lambda_{jU}^{1} z_{jU_{l_U}}^{(0,1)} = z_{kU_{l_U}}^{(0,1)}$; if $t=T$,

$\displaystyle\sum_{j=1}^{n} \lambda_{jU}^{T} z_{jU_{l_U}}^{(T,T+1)} \geq z_{kU_{l_U}}^{(T,T+1)}$ is removed.

16.3.3   The Production Technology for the Service Process

The production technology T_S^t with CRS for the service process is written as follows:

$$
T_S^t = \left\{ (z^t, y^t) : \sum_{j=1}^{n} \lambda_{jS}^t z_{j(HS)_{l_{HS}}}^t \leq z_{k(HS)_{l_{HS}}}^t, \quad l_{HS} = 1,\ldots,L_{HS}, \right.
$$

$$
\sum_{j=1}^{n} \lambda_{jS}^t z_{j(US)_{l_{US}}}^t \leq z_{k(US)_{l_{US}}}^t, \quad l_{US} = 1,\ldots,L_{US},
$$

$$
\sum_{j=1}^{n} \lambda_{jS}^t y_{r_d jS}^t \geq y_{r_d kS}^t, \quad r_d = 1,\ldots,s_d,
$$

$$
\sum_{j=1}^{n} \lambda_{jS}^t y_{r_{nd} jS}^t = y_{r_{nd} kS}^t, \quad r_{nd} = 1,\ldots,s_{nd},
$$

$$
\left. \lambda_{jS}^t \geq 0, \quad j = 1,\ldots,n \right\} (t = 1,\ldots,T) \tag{16.21}
$$

where λ_H^t, λ_U^t and λ_S^t are intensity variables associated with the HB production activity, UB production activity and service process, respectively, in period t.

Based on manipulation of the directional distance function, in period t, the kth DMU's inefficiency score for the HB activity φ_{kH}^t can be represented as the directional distance function defined by the technology T_H^t, its inefficiency score for the UB activity φ_{kU}^t can be represented as the directional distance function defined by the technology T_U^t, and its service inefficiency score φ_{kS}^t can be represented as the directional distance function defined by the technology T_S^t. Then the operational inefficiency for DMU_k can be estimated by solving the following MDNDEA model:

$$
\max \; \varphi_k = \sum_{t=1}^{T} W^t \left[w^P \left(w^H \cdot \varphi_{kH}^t + w^U \cdot \varphi_{kU}^t \right) + w^S \cdot \varphi_{kS}^t \right] \tag{16.22}
$$

s.t.

(HB production activity)

$$
\sum_{j=1}^{n} \lambda_{jH}^t x_{i_H jH}^t \leq \left(1 - \varphi_{kH}^t\right) x_{i_H kH}^t, \quad i_H = 1,\ldots,m_H, \; t = 1,\ldots,T \tag{16.23}
$$

$$
\sum_{j=1}^{n} \lambda_{jH}^t z_{j(HS)_{l_{HS}}}^t = z_{k(HS)_{l_{HS}}}^t, \quad l_{HS} = 1,\ldots,L_{HS}, \; t = 1,\ldots,T \tag{16.24}
$$

$$
\sum_{j=1}^{n} \lambda_{jH}^t z_{jH_{l_H}}^{(t,t+1)} = \sum_{j=1}^{n} \lambda_{jH}^{t+1} z_{jH_{l_H}}^{(t,t+1)}, \quad l_H = 1,\ldots,L_H, t = 1,\ldots,T-1 \tag{16.25}
$$

$$\sum_{j=1}^{n}\lambda_{jH}^{t}z_{jH_{l_{H}}}^{(t,t+1)}=z_{kH_{l_{H}}}^{(t,t+1)}, \quad l_{H}=1,\ldots,L_{H}, t=1,\ldots,T-1 \tag{16.26}$$

(UB production activity)

$$\sum_{j=1}^{n}\lambda_{jU}^{t}x_{i_{U}jU}^{t}\leq\left(1-\varphi_{kU}^{t}\right)x_{i_{U}kU}^{t}, \quad i_{U}=1,\ldots,m_{U}, \ t=1,\ldots, T \tag{16.27}$$

$$\sum_{j=1}^{n}\lambda_{jU}^{t}z_{j(US)_{l_{US}}}^{t}=z_{k(US)_{l_{US}}}^{t}, \quad l_{US}=1,\ldots,L_{US}, \ t=1,\ldots,T \tag{16.28}$$

$$\sum_{j=1}^{n}\lambda_{jU}^{t}z_{jU_{l_{U}}}^{(t,t+1)}=\sum_{j=1}^{n}\lambda_{jU}^{t+1}z_{jU_{l_{U}}}^{(t,t+1)}, \quad l_{U}=1,\ldots,L_{U}, t=1,\ldots,T-1 \tag{16.29}$$

$$\sum_{j=1}^{n}\lambda_{jU}^{t}z_{jU_{l_{U}}}^{(t,t+1)}=z_{kU_{l_{U}}}^{(t,t+1)}, \quad l_{U}=1,\ldots,L_{U}, t=1,\ldots,T-1 \tag{16.30}$$

(Service process)

$$\sum_{j=1}^{n}\lambda_{jS}^{t}z_{j(HS)_{l_{HS}}}^{t}=z_{k(HS)_{l_{HS}}}^{t}, \quad l_{HS}=1,\ldots,L_{HS}, \ t=1,\ldots,T \tag{16.31}$$

$$\sum_{j=1}^{n}\lambda_{jS}^{t}z_{j(US)_{l_{US}}}^{t}=z_{k(US)_{l_{US}}}^{t}, \quad l_{US}=1,\ldots,L_{US}, \ t=1,\ldots,T \tag{16.32}$$

$$\sum_{j=1}^{n}\lambda_{jS}^{t}y_{r_{d}jS}^{t}\geq\left(1+\varphi_{kS}^{t}\right)y_{r_{d}kS}^{t}, \quad r_{d}=1,\ldots,s_{d}, t=1,\ldots, T \tag{16.33}$$

$$\sum_{j=1}^{n}\lambda_{jS}^{t}y_{r_{nd}jS}^{t}=\left(1-\varphi_{kS}^{t}\right)y_{r_{nd}kS}^{t}, \quad r_{nd}=1,\ldots,s_{nd}, t=1,\ldots,T \tag{16.34}$$

(Shared inputs)

$$\sum_{j=1}^{n}\alpha_{i_{HU}jH}^{t}\lambda_{jH}^{t}x_{i_{HU}jHU}^{t}\leq\left(1-\varphi_{kH}^{t}\right)\alpha_{i_{HU}kH}^{t}x_{i_{HU}kHU}^{t},$$
$$i_{HU}=1,\ldots,m_{HU}, t=1,\ldots,T \tag{16.35}$$

$$\sum_{j=1}^{n}\left(1-\alpha_{i_{HU}jH}^{t}\right)\lambda_{jU}^{t}x_{i_{HU}jHU}^{t}\leq\left(1-\varphi_{kU}^{t}\right)\left(1-\alpha_{i_{HU}kH}^{t}\right)x_{i_{HU}kHU}^{t},$$
$$i_{HU}=1,\ldots,m_{HU}, t=1,\ldots,T \tag{16.36}$$

$$L^t_{i_{HU}H} < \alpha^t_{i_{HU}H} < U^t_{i_{HU}H}, \quad i_{HU} = 1,\ldots,m_{HU}, t = 1,\ldots,T \qquad (16.37)$$

(Initial conditions)

$$\sum_{j=1}^{n} \lambda^1_{jH} z^{(0,1)}_{jH_{l_H}} = \sum_{j=1}^{n} \lambda^1_{jH} z^{(0,1)}_{jH_{l_H}}, \quad l_H = 1,\ldots,L_H \qquad (16.38)$$

$$\sum_{j=1}^{n} \lambda^1_{jU} z^{(0,1)}_{jU_{l_U}} = \sum_{j=1}^{n} \lambda^1_{jU} z^{(0,1)}_{jU_{l_U}}, \quad l_U = 1,\ldots,L_U \qquad (16.39)$$

(Additional conditions)

$$\sum_{t=1}^{T} W^t = 1 \qquad (16.40)$$

$$w^H + w^U = 1 \qquad (16.41)$$

$$w^P + w^S = 1 \qquad (16.42)$$

$$\lambda_{jH}, \ \lambda_{jU}, \ \lambda_{jS}, \ W^t, \ w^H, \ w^U, \ w^P, \ w^S \geq 0,$$
$$j = 1,\ldots,n, \ t = 1,\ldots,T \qquad (16.43)$$

where $\varphi^t_{kH}, \varphi^t_{kU}, \varphi^t_{kS}, \lambda^t_{jH}, \lambda^t_{jU}, \lambda^t_{jS}$, and $\alpha^t_{i_{HU}jH}, t = 1,\ldots,T, \ j,k = 1,\ldots,n, \ i_{HU} = 1,\ldots,m_{HU}$ are variables of this model. L and U are the lower and upper bounds placed on the various shared inputs. W^t, w^H, w^U, w^P and w^S are the weights of period t, the HB production activity, the UB production activity, the production process and the service process, respectively. The constraints (16.24) and (16.31), as well as the constraints (16.28) and (16.32), show that the linking items between the HB production activity and the service process, as well as between the UB production activity and the service process, are fixed. The constraints (16.25) and (16.26) and the constraints (16.29) and (16.30) indicate that the carry-over items in the HB and UB production activities act as non-discretionary links. The constraints (16.25) and (16.29) impose a continuity condition between two consecutive periods. Note that the linking and carry-over items have several forms, based on the characteristics of these items. The related constraints of these forms have been shown in (16.15)–(16.18). In addition, the initial conditions can be accounted for by the constraints (16.38) and (16.39). With the proposed MDNDEA model, the individual efficiencies can be defined as follows:

Period-production efficiency in the HB activity (PHBPE) $= 1 - \varphi^t_{kH}$ $\qquad (16.44)$

Period-production efficiency in the UB activity (PUBPE) $= 1 - \varphi^t_{kU}$ $\qquad (16.45)$

$$\text{PSE} = 1 - \varphi^t_{kS} \qquad (16.46)$$

$$\text{PPE} = 1 - \left(w^H \cdot \varphi_{kH}^t + w^U \cdot \varphi_{kU}^t \right) \tag{16.47}$$

$$\text{POE} = 1 - \left[w^P \left(w^H \cdot \varphi_{kH}^t + w^U \cdot \varphi_{kU}^t \right) + w^S \cdot \varphi_{kS}^t \right] \tag{16.48}$$

$$\text{Production efficiency of the HB activity (HBPE)} = 1 - \sum_{t=1}^{T} W^t \cdot \varphi_{kH}^t \tag{16.49}$$

$$\text{Production efficiency of the UB activity (UBPE)} = 1 - \sum_{t=1}^{T} W^t \cdot \varphi_{kU}^t \tag{16.50}$$

$$\text{PE} = 1 - \sum_{t=1}^{T} W^t \left(w^H \cdot \varphi_{kH}^t + w^U \cdot \varphi_{kU}^t \right) \tag{16.51}$$

$$\text{SE} = 1 - \sum_{t=1}^{T} W^t \cdot \varphi_{kS}^t \tag{16.52}$$

$$\text{OE} = 1 - \varphi_k \tag{16.53}$$

φ_k is equal to zero if and only if the bus transit firm is operationally efficient and $\varphi_{kH}^t = \varphi_{kU}^t = \varphi_{kS}^t = 0$, $t = 1, \ldots, T$. Since the model combines the measures of PHBPE, PUBPE and PSE to compute the OE measure, the results can provide further insight into the sources of OE.

16.4   APPLICATION[2]

This section provides an example based on 20 bus transit firms in Taiwan for the period 2004–2012 to investigate performance issues by an MDNDEA model.

16.4.1   Input and Output Variables

The operational framework of a bus transit firm in the MDNDEA model is represented in Figure 16.3. The input and output variables of the bus transit firm that are adopted in this example are illustrated as follows:

1. *Dedicated inputs*: (i) HB service: the number of drivers (DRIVER), the number of vehicles (VEHICLE) and the number of litres of fuel (FUEL). (ii) UB service: DRIVER, VEHICLE and FUEL. (iii) Consumption service: the number of ticket agents (TICKET).
2. *Common inputs*: (i) Shared between HB and UB services: the number of technicians (TEC). (ii) Shared among HB, UB and consumption services: the number of management staff (MGT).

[2] Adapted from Yu *et al.* [42].

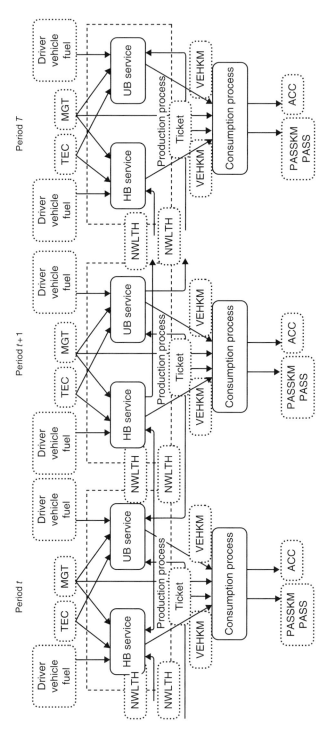

Figure 16.3 The operational framework of a bus transit firm.

3. *Intermediate outputs*: (i) HB service: vehicle-kilometres (VEHKM). (ii) UB service: VEHKM.

4. *Final desirable outputs*: Passenger-kilometres (PASSKM) and the number of passengers (PASS).

5. *Final undesirable output*: The number of accidents (ACC).

6. *Carry-over items*: (i) HB service: network length (NWLTH). (ii) UB service: NWLTH.

However, the MDNDEA model does not allow these common inputs to take weights of 0 or 1. The weights for these common inputs need to be limited. We considered the proportion of TEC shared with the HB service in period t to range from 0.3 to 0.7, while the proportions of MGT shared with the HB and UB services in period t ranged from 0.2 to 0.8. Finally, since the weights of periods, the production process and the consumption process are exogenously pre-assigned scalars, we assumed for simplicity that the weight of each period was 0.1111, the weights of the HB and UB services were equal to 0.5, and the weights of the production and consumption processes were equal to 0.5.

16.4.2 Empirical Results

Table 16.1 displays the average results for the OE scores and its components for the bus transit firms obtained by use of the MDNDEA model. Looking first at the average OE score of the bus transit firms, its value was 0.8540, with a range from 0.7507 to 0.9871. This indicates that, on average, there was room for bus transit firms to enhance their performance by 14.6% in the study period. Since the OE is defined as the weighted-average performance of the production and consumption processes, we can explore the contributions of these two processes. As can be seen from Table 16.1, the PE (0.7960) was worse than the SE (0.9121), implying that the operational inefficiency came mainly from the production process. For the production process, the PE score was determined by the weighted average of the HBPE and UBPE scores. We can investigate further where the production inefficiency comes from. As shown in Table 16.1, the HBPE (0.7904) was slightly lower than the UBPE (0.8015). This means that inefficiencies in both the HB and the UB activities lead to production inefficiency.

For the individual bus transit firms, the results show that no bus transit firm was efficient in terms of the OE. Since the OE score is equal to unity if and only if all production and consumption processes are simultaneously efficient in each period, this result signifies that none of the bus transit firms performed efficiently in terms of all their three services in each period. All bus transit firms could enhance their performance in at least one of these three services. In terms of individual activities, four bus transit firms (CitiAir, Hualien, Fengyuan and Chiayi) were efficient in the HB activity, six firms (Sanchung, Taipei, Kuang-hua, Tansui, Chungli and Chiayi) were efficient in the UB activity, and eight firms (Sanchung, Capital, Taipei, Chih-nan,

TABLE 16.1   Operational efficiency and its components for individual bus transit firms.[a]

Firm	OE	PE	HBPE	UBPE	SE
Sanchung	0.8729 (9)	0.7459 (14)	0.4917 (19)	1.0000 (1)	1.0000 (1)
Capital	0.8816 (8)	0.7633 (13)	0.5770 (17)	0.9496 (8)	1.0000 (1)
Taipei	0.8469 (12)	0.6937 (17)	0.3874 (20)	1.0000 (1)	1.0000 (1)
Chih-nan	0.8919 (5)	0.7837 (10)	0.5875 (16)	0.9799 (7)	1.0000 (1)
CitiAir	0.8180 (15)	0.9715 (3)	1.0000 (1)	0.9431 (10)	0.6644 (19)
Chung-shing	0.7566 (18)	0.6880 (18)	0.8069 (12)	0.5692 (17)	0.8252 (16)
Kuang-hua	0.7930 (16)	0.9229 (5)	0.8458 (9)	1.0000 (1)	0.6631 (20)
Tansui	0.9347 (4)	0.9358 (4)	0.8717 (8)	1.0000 (1)	0.9336 (13)
Chungli	0.9444 (2)	0.8966 (6)	0.7932 (13)	1.0000 (1)	0.9923 (10)
Taoyuan	0.7776 (17)	0.5553 (20)	0.4997 (18)	0.6108 (15)	1.0000 (1)
Hsinchu	0.8519 (11)	0.7038 (16)	0.9372 (6)	0.4704 (19)	1.0000 (1)
Hualien	0.9871 (1)	0.9743 (2)	1.0000 (1)	0.9485 (9)	1.0000 (1)
Fengyuan	0.8217 (14)	0.7650 (12)	1.0000 (1)	0.5300 (18)	0.8783 (15)
Taichung	0.8846 (6)	0.7961 (9)	0.7897 (14)	0.8025 (12)	0.9731 (11)
Changhua	0.8818 (7)	0.7679 (11)	0.9628 (5)	0.5729 (16)	0.9958 (9)
Ubus	0.8618 (10)	0.7236 (15)	0.8306 (10)	0.6166 (14)	1.0000 (1)
Geya	0.7507 (20)	0.5564 (19)	0.8114 (11)	0.3013 (20)	0.9451 (12)
Kaohsiung	0.7511 (19)	0.8145 (8)	0.6884 (15)	0.9406 (11)	0.6877 (18)
Pingtung	0.8284 (13)	0.8609 (7)	0.9269 (7)	0.7949 (13)	0.7959 (17)
Chiayi	0.9440 (3)	1.0000 (1)	1.0000 (1)	1.0000 (1)	0.8880 (14)
Average	0.8540	0.7960	0.7904	0.8015	0.9121
Std. dev.	0.0681	0.1280	0.1907	0.2250	0.1207

[a] Rankings are provided in parentheses.

Taoyuan, Hsinchu, Hualien and Ubus) were efficient in the consumption process. However, some firms with efficiency in one dimension were relatively inefficient in others. For example, Taoyuan was efficient in the consumption process, but had a lower PE in both production activities. Taoyuan could improve its resource utilization. The ranking of the bus transit firms is also listed in Table 16.1. Hualien was the best among all in terms of the OE, with the first ranking in HBPE and SE, and the ninth ranking in UBPE. On the other hand, Geya had the lowest OE. The main reason for this is that Geya had an extraordinarily low UBPE (0.3013). Although Kuang-hua and Taoyuan ranked first in terms of UBPE and SE, respectively, they were 16th and 17th in OE. This result indicates that the ranking of bus transit firms in different dimensions of the performance measures is inconsistent. In other words, the sources of operational inefficiency of the bus transit firms are different. Hence, compared with the conventional DEA model, the proposed MDNDEA model can reveal inefficiency in individual activities and processes, and provide operators with more information about where to improve performance.

One of the merits of the MDNDEA model is that it can measure the period performance in a unified model, so that it can provide an overall trend of performance change. This is why the MDNDEA model is superior to the static multi-activity

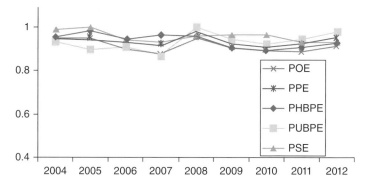

Figure 16.4 Period performance.

network DEA model. Hence, we can investigate further the average trend of performance change over the period 2004–2012. Figure 16.4 indicates that the average POE scores maintained a stable variance over the sample period. The POE can be decomposed into PPE and PSE. From Figure 16.4, it can be seen that the PPE scores were lower than the PSE scores during 2004–2011, while the PSE score was worse than the PPE score in 2012. It is worth noting that the average PSE scores showed higher levels over the sample period. This implies that these bus transit firms performed well in the consumption process over the sample period. We can explore the PPE further between the HB and UB activities. As can be seen in Figure 16.4, the PHBPE scores were better than the PUBPE scores during 2004–2007, while the PUBPE scores were greater than the PHBPE scores during 2008–2012. However, PHBPE and PUBPE appeared to have similar patterns over the sample period.

In addition, the proposed MDNDEA model considers the impacts of carry-over items. Hence, it can help bus transit firms to modify their long-term planning and investments by investigating changes in carry-over items. Table 16.2 shows by what amount the average of the network lengths of the HB and UB services can be reduced or expanded during 2004–2012.[3] Only one bus transit firm (Chiayi) shows no changes in both carry-over items. Two firms (Sanchung and Kuang-hua) should increase the network length of their highway bus service, while two firms (Taipei and Chungli) should decrease the network length of their highway bus service. Three firms (CitiAir, Hualien and Fengyuan) require a reduction of the network length of their urban bus service. Four firms (Taoyuan, Hsinchu, Taichung and Kaohsiung) need to expand both carry-over items, while three firms (Chih-nan, Changhua and Geya) need to reduce both carry-over items. Five firms (Capital, Chung-shing, Tansui, Ubus and Pingtung) should enlarge the network length of their highway bus service, but reduce the network length of their urban bus service.

---

[3] Since the network lengths of the HB and UB services have been defined as a discretionary link in this application, they can be freely increased or decreased.

TABLE 16.2   Average slack values of network lengths of HB and UB services during 2004–2012.

Firm	Network length of HB service	Network length of UB service
Sanchung	1976.51	0
Capital	400.56	−18.83
Taipei	−24.78	0
Chih-nan	−49.53	−46.54
CitiAir	0	−1.95
Chung-shing	624.24	−2.67
Kuang-hua	662.66	0
Tansui	44.06	−0.60
Chungli	−38.14	0
Taoyuan	2320.81	172.87
Hsinchu	1217.62	11.89
Hualien	0	−3.40
Fengyuan	0	−15.16
Taichung	1437.29	77.26
Changhua	−5.98	−14.58
Ubus	27.97	−3.73
Geya	−11.50	−4.31
Kaohsiung	769.82	6.61
Pingtung	345.25	−2.85
Chiayi	0	0
Average	484.84	7.20
Std. dev.	719.95	44.83

16.5   CONCLUSIONS

This chapter has provided a more comprehensive analysis to reflect the operational characteristics of transport organizations in an efficiency evaluation, and has presented the construction of a DNDEA model that illustrates a network and dynamic structure with undesirable outputs. The DNDEA model is designed to evaluate the performance achieved by transport organizations which have several operational processes and carry-over items between two consecutive terms. In order to provide more accurate performance measurement in the transport sector, we have extended the DNDEA model further to consider a multi-activity framework. Based on these models, the sources of inefficiency within a transport organization can be identified. In addition, we have chosen one related application to transport organizations to illustrate the selection of inputs and outputs, and to investigate the applicability of a multi-process, multiperiod and multi-activity framework.

REFERENCES

[1] Chang, K.P. and Kao, P.H. (1992) The relative efficiency of public versus private municipal bus firms: An application of data envelopment analysis. *Journal of Productivity Analysis*, **3**(1), 67–84.

[2] Roll, Y. and Hayuth, Y. (1993) Port performance comparison applying data envelopment analysis (DEA). *Maritime Policy and Management*, **20**(2), 153–161.

[3] Viton, P. (1997) Technical efficiency in multimode bus transit: A production frontier analysis. *Transport Research Part B*, **31**, 23–39.

[4] Viton, P. (1998) Changes in multimode bus transit efficiency. *Transport*, **25**, 1–21.

[5] Nolan, J.F., Ritchie, P.C. and Rowcroft, J.R. (2001) Measuring efficiency in the public sector using nonparametric frontier estimators: A study of transit agencies in the USA. *Applied Economics*, **33**, 913–922.

[6] Coelli, T., Grifell-Tatjé, E. and Perelman, S. (2002) Capacity utilization and profitability: A decomposition of short-run profit efficiency. *International Journal of Production Economics*, **79**(3), 261–278.

[7] Barros, C.P. and Athanassiou, M. (2004) Efficiency in European seaports with DEA: Evidence from Greece and Portugal. *Maritime Economics & Logistics*, **6**, 122–140.

[8] Odeck, J. (2006) Congestion, ownership, region of operation, and scale: Their impact on bus operator performance in Norway. *Socio-Economic Planning Sciences*, **40**, 52–69.

[9] Färe, R., Grosskopf, S. and Sickles, R.C. (2007) Productivity? of US airlines after deregulation. *Journal of Transport Economics and Policy*, **41**(1), 1–21.

[10] Assaf, G.A. and Jossiassen, A. (2009) The operational performance of UK airlines: 2002–2007. *Journal of Economic Studies*, **38**, 5–16.

[11] Assaf, G.A. and Jossiassen, A. (2011) European vs. U.S. airlines: Performance comparison in a dynamic market. *Tourism Management*, **3**(2), 317–326.

[12] Liu, Z. (1995) The comparative performance of public and private enterprises: The case of British ports. *Journal of Transport Economics and Policy*, **29**(3), 263–274.

[13] Cullinane, K., Song, D.W. and Gray, R. (2002) A stochastic frontier model of the efficiency of major container terminals in Asia: Assessing the influence of administrative and ownership structures. *Transport Research Part A*, **36**, 743–762.

[14] Cullinane, K. and Song, D.W. (2003) A stochastic frontier model of the productive efficiency of Korean container terminals. *Applied Economics*, **35**, 251–267.

[15] Tongzon, J. (1995) Determinants of port performance and efficiency. *Transport Research*, **29A**(3), 245–252.

[16] Oum, T.H. and Yu, C. (1995) A productivity comparison of the world's major airlines. *Journal of Air Transport Management*, **2**(3–4), 181–195.

[17] Estache, A., Tovar de la Fé, B. and Trujillo, L. (2004) Sources of efficiency gains in port reform: A DEA decomposition of a Malmquist index for Mexico. *Utility Policy*, **30**(4), 221–230.

[18] Barbot, C., Costa, A. and Sochirca, E. (2008) Airlines performance in the new market context: A comparative productivity and efficiency analysis. *Journal of Air Transport Management*, **14**(5), 270–274.

[19] Wang, T., Cullinane, K. and Song, D.W. (2003) Container port production efficiency: A comparative study of DEA and FDH approach. *Journal of the Eastern Asia Society for Transport Studies*, **5**, 698–713.

[20] Cullinane, K., Song, D.W. and Wang, T.F. (2005) The application of mathematical programming approaches to estimating container port production efficiency. *Journal of Productivity Analysis*, **24**, 73–92.

[21] Lee, H.S., Chu, C.W., Chen, K.K. and Chou, M.T. (2005) A fuzzy multiple criteria decision making model for airline competitiveness evaluation. *Eastern Asia Society for Transport Studies*, **5**, 507–519.

[22] Wang, Y.J. and Lee, H.S. (2007) Generalizing TOPSIS for fuzzy multiple-criteria group decision-making. *Computers and Mathematics with Applications*, **53**, 1762–1772.

[23] Gillen, D. and Lall, A. (1997) Developing measures of airport productivity and performance: An application of data envelopment analysis. *Transport Research Part E*, **4**, 261–273.

[24] Tomazinis, A.R. (1975) *Productivity, Efficiency, and Quality in Urban Transport Systems,* D.C. Heath and Company, Lexington, MA.

[25] Färe, R. and Grosskopf, S. (1996) Productivity and intermediate products: A frontier approach. *Economics Letters*, **50**(1), 65–70.

[26] Färe, R. and Grosskopf, S. (2000) Network DEA. *Social-Economics Planning Science*, **34**, 35–49.

[27] Kao, C. (2014) Network data envelopment analysis: A review. *European Journal of Operational Research*, **239**, 1–16.

[28] Wu, W.W., Lan, L.W. and Lee, Y.T. (2013) Benchmarking hotel industry in a multi-period context with DEA approaches: A case study. *Benchmarking: An International Journal*, **20**, 152–168.

[29] Nemoto, J. and Goto, M. (1999) Dynamic data envelopment analysis: Modeling intertemporal behaviour of a firm in the presence of productive inefficiencies. *Economics Letters*, **64**, 51–56.

[30] Emrouznejad, A. and Thanassoulis, E. (2005) A mathematical model for dynamic efficiency using data envelopment analysis. *Applied Mathematics and Computation*, **160**, 363–378.

[31] Tone, K. and Tsutsui, M. (2010) Dynamic DEA: A slacks-based measure approach. *Omega*, **38**, 145–156.

[32] Kao, C. and Liu, S.T. (2014) Multi-period efficiency measurement in data envelopment analysis: The case of Taiwanese commercial banks. *Omega*, **47**, 101–112.

[33] Lovell, C.A.K., Pastor, J.T. and Turner, J.A. (1995) Measuring macroeconomic performance in the OECD: A comparison of European and non-European countries. *European Journal of Operational Research*, **87**, 507–518.

[34] Luenberger, D.G. (1992) Benefit function and duality. *Journal of Mathematical Economics*, **21**, 461–481.

[35] Cummins, J.D. and Xie, X. (2013) Efficiency, productivity, and scale economies in the U.S. property-liability insurance industry. *Journal of Productivity Analysis*, **39**(2), 141–164.

[36] Månsson, J. (1996) Technical efficiency and ownership: The case of booking centres in the Swedish taxi market. *Journal of Transport Economics and Policy*, **30**, 83–93.

[37] Färe, R., Grosskopf, S. and Norris, M. (1997) Productivity growth, technical progress, and efficiency change in industrialized countries: Reply. *American Economic Review*, **87**(5), 1040–1044.

[38] Chung, Y.H., Färe, R. and Grosskopf, S. (1997) Productivity and undesirable outputs: A directional distance function approach. *Journal of Environmental Management*, **51**, 229–240.

[39] Tone, K. and Tsutsui, M. (2014) Dynamic DEA with network structure: A slacks-based measure approach. *Omega*, **42**(1), 124–131.

[40] Beasley, J.E. (2003) Allocating fixed costs and resources via data envelopment analysis. *European Journal of Operational Research*, **147**(1), 198–216.

[41] Mar Molinero, C. (1996) On the joint determination of efficiencies in a data envelopment analysis context. *Journal of the Operational Research Society*, **47**(10), 1279–1279.

[42] Yu, M.M., Chen, L.H. and Hsiao, B. (2016) Dynamic performance assessment of bus transit with the multi-activity network structure. *Omega*, **60**, 15–25.

17

DYNAMIC NETWORK EFFICIENCY OF JAPANESE PREFECTURES

HIROFUMI FUKUYAMA

Faculty of Commerce, Fukuoka University, Fukuoka, Japan

ATSUO HASHIMOTO

Fukuoka Girls' Commercial High School, Chikushi-gun, Fukuoka, Japan

KAORU TONE

National Graduate Institute for Policy Studies, Tokyo, Japan

WILLIAM L. WEBER

Southeast Missouri State University, Cape Girardeau, USA

17.1 INTRODUCTION

In this chapter, we develop a multiperiod dynamic network DEA (data envelopment analysis) model and apply it to production in Japanese prefectures during 2007–2009. Our method assumes that a human capital sector, a private physical capital sector, and a social overhead capital sector jointly produce a final output. Private physical capital and social overhead capital from a preceding period affect current production possibilities and, in turn, both types of capital can be carried over to a subsequent period.

Advances in DEA Theory and Applications: With Extensions to Forecasting Models,
First Edition. Edited by Kaoru Tone.
© 2017 John Wiley & Sons Ltd. Published 2017 by John Wiley & Sons Ltd.

The objective function of our model seeks to maximize the size of the technology sets for a decision-making unit (DMU), in this case a prefecture, over all periods by choosing the amounts of private physical and social overhead capital to be used in the current period and the amounts to be carried over to the subsequent period. Resources can be reallocated between periods as long as the decline in output or increase in input in one period is more than offset by an increase in output or decrease in input in a subsequent period.

We first provide a general model for incorporating a network structure and dynamics. Färe and Grosskopf [1] presented a dynamic methodology that comprises a sequence of technologies which are connected by storable inputs and carry-over outputs from period to period. Färe and Grosskopf [2] proposed a DEA technique, which they called network DEA, to measure the efficiency of a DMU with a network production structure. Following and building upon the foundation laid by Färe and Grosskopf [1,2], various authors have incorporated quasi-fixed inputs [3], assessed dynamic efficiency and examined the correspondence between carry-over products [4], and incorporated lagged effects of input consumption using DEA [5,6]. Tone and Tsutsui [7] developed a dynamic slacks-based measure of performance and classified carry-over activities as either good (enhancing production), bad, free, or fixed. Tone and Tsutsui [8] presented a slacks-based dynamic DEA model with a network structure by combining their previous studies.

We build further on the dynamic network foundation and develop a weighted dynamic network (WDN) model similar to the slacks-based form of Tone and Tsutsui [7,8]. Our WDN model accounts for the slacks in the exogenous inputs and final outputs but does not include the slacks from various divisions or subprocesses in the objective function of the optimization problem.[1] Furthermore, our WDN model allows for joint outputs to be produced by more than one division and incorporates the effects of lagged outputs/inputs on current production.

17.2 MULTIPERIOD DYNAMIC MULTIPROCESS NETWORK

In this section, we define a dynamic network technology that can be represented using DEA. We assume there are $j = 1, \ldots, J$ DMUs that use various inputs to produce outputs in $t = 1, \ldots, T$ periods. Each DMU consists of $k = 1, \ldots, h, \ldots, g, \ldots, K$ subprocesses or divisions. Each division is endowed with $n = 1, \ldots, N$ exogenous inputs that must be used by that division contemporaneously. The divisions produce $m = 1, \ldots, M$ final outputs and/or $q = 1, \ldots, Q$ intermediate products that can be used as inputs by other divisions. Each division has access to $\bar{r} = 1, \ldots, \bar{R}$ unused inputs that have been carried

---

[1] Fukuyama and Mirdehghan [9] discussed how to identify divisional efficiency.

over from a previous period and, in turn, each division can forgo current production and carry over some inputs $r = 1,\ldots,R$ for use in subsequent periods. For DMU_j, we define the following:

x_{nj}^{kt}: exogenous input n consumed by division k in period t;

$z_{qj}^{(kt,ht)}$: intermediate product (input) q produced by division k in period t and consumed by division h in period t;

y_{mj}^{kt}: final output m produced by process k in period t;

$c_{rj}^{(kt,g\tau)}$: carry-over product r produced by division k in period t and consumed by division g in period $\tau > t$; and

$\bar{c}_{\bar{r}j}^{(h\bar{\tau},kt)}$: lagged carry-over product $\bar{r}$ coming from division h in period $\bar{\tau} < t$ and entering division k in period t.

Here, t, τ, and $\bar{\tau}$ are the index sets of the relevant time periods. For the intermediate products (z), the first superscript in parentheses corresponds to the division that produces the intermediate product in t and the second superscript corresponds to the division that uses the intermediate product as an input in t. We denote the set of intermediate inputs q entering division k in time t from division h in time t $\left(z_q^{(ht,kt)}\right)$ by $\underline{L}$. Similarly, we denote the set of intermediate products produced by division k and used by division h in period t $\left(z_q^{(kt,ht)}\right)$ by $\overline{\overline{L}}$. For the carry-over products (c), the first superscript indicates the division which generates the carry-overs in t and the second superscript indicates the division that receives those carry-overs for use in period $\tau > t$. The set of outflows from k in t to h in $\tau > t \left(c_r^{kt,h\tau}\right)$ is represented by $\bar{F}$. Finally, for the lagged carry-over products ($\bar{c}$), the first superscript indicates the division which generated the lagged carry-over in period $\bar{\tau} < t$ and the second superscript indicates the division which uses the lagged carry-over product in period t. Thus, $\bar{c}_{\bar{r}j}^{(h\bar{\tau},kt)}$ represents the lagged carry-over from period $\bar{\tau} < t$ generated by division h and used by division k in period t.

We make the following assumptions about our dynamic network technology.

- *Assumption 1.* The objective function of our framework includes slacks of exogenous inputs but does not include slacks associated with intermediate products and carry-overs.
- *Assumption 2.* Lagged carry-over products constrain a division's production possibilities but are independent of which division they came from and can affect production possibilities in only a finite number of future periods.
- *Assumption 3.* Final outputs are jointly produced by several divisions.

Assumption 1 is consistent with the two-stage procedures of Kao and Hwang [10], Chen *et al.* [11], and Fukuyama and Weber [12] in that slacks associated with intermediate products are not included in the objective function. Assumption 2 means that carry-overs do not depreciate or spoil as they are moved across time. In addition, the effects of lagged carry-overs are finite [6]. For example, bad loans were used as a carry-over output in a bank efficiency context by Akther *et al.* [13] and Fukuyama and Weber [14], where carry-over inputs negatively affect production in later periods. Furthermore, we note that these carry-over products might also be summed over several past periods as in Fukuyama *et al.* [15]. Regarding Assumption 3, some final outputs can be produced by combining several subtechnologies or subprocesses.

The production possibility set for division k in period t consists of the exogenous inputs, the intermediate products (inputs) from other divisions, the sum of all past carry-overs that have not yet been used that can produce carry-overs to future periods, and contemporaneous intermediate products and final outputs. In set notation, the production possibility set is represented as

$$
T^{kt} = \left\{
\begin{array}{l}
\left(x_n^{kt}, \bar{c}_{\bar{r}}^{(\bar{\tau},kt)},\; z_q^{(ht,kt)}, z_q^{(kt,ht)},\quad c_r^{(kt,g\tau)},\; y_m^{kt} \right) \text{ such that} \\[2mm]
\left(x_n^{kt}, \bar{c}_{\bar{r}}^{(\bar{\tau},kt)},\; z_q^{(ht,kt)} \right) \text{ can produce} \left(z_q^{(kt,ht)},\quad c_r^{(kt,g\tau)},\; y_m^{kt} \right)
\end{array}
\right\}
\tag{17.1}
$$

In DEA, the following equations represent the feasible inputs and outputs (left-hand side) that can be produced by the technology formed by taking linear combinations of inputs and outputs (on the right-hand side):

$$
\text{Inputs} \begin{cases}
\bar{c}_{\bar{r}}^{(\bar{\tau},kt)} \geq \sum_{j=1}^{J} \bar{c}_{\bar{r}j}^{(\bar{\tau},k)} \lambda_j^{kt}, \quad \bar{r}=1,\ldots,\bar{R}^k \\[4mm]
x_n^{kt} \geq \sum_{j=1}^{J} x_{nj}^{kt} \lambda_j^{kt}, \qquad n=1,\ldots,N^k \\[4mm]
z_q^{(ht,kt)} \geq \sum_{j=1}^{J} z_q^{(ht,kt)} \lambda_j^{kt}, \quad (ht,kt) \in \overline{\mathbf{L}}, \forall ht;\;\; q=1,\ldots,\bar{Q}^k
\end{cases}
$$

$$
\text{Outputs} \begin{cases}
y_m^{kt} \leq \sum_{j=1}^{J} y_{mj}^{kt} \lambda_j^{kt}, \qquad m=1,\ldots,M^k \\[4mm]
z_q^{(kt,ht)} \leq \sum_{j=1}^{J} z_q^{(kt,ht)} \lambda_j^{kt}, \quad (kt,ht) \in \overline{\overline{\mathbf{L}}}, \forall ht;\;\; q=1,\ldots,\bar{\bar{Q}}^k \\[4mm]
c_r^{(kt,g\tau)} \leq \sum_{j=1}^{J} c_{rj}^{(kt,g\tau)} \lambda_j^{kt}, \qquad (kt,g\tau) \in \bar{\mathbf{F}};\;\; r=1,\ldots,R^k
\end{cases}
$$

$$
\lambda_j^{kt} \geq 0, \quad j=1,\ldots,J, k=1,\ldots,K, t=1,\ldots,T
$$

Using (17.1), we define a network production possibility set by

$$NT^t = \left\{ \left(\underbrace{\bar{c}_{\bar{r}}^{(h\bar{\tau},\,kt)}, x_n^{kt}}_{\text{inputs}},\; \underbrace{z_q^{(ht,\,kt)}, z_{\tilde{q}}^{(kt,\,ht)},\; c_r^{(kt,\,g\tau)},\; y_m^{kt}}_{\text{outputs}} \right) \in T^{kt}\; (\forall k) \right\} \qquad (17.2)$$

The internal structure of the network technology is illustrated in Figure 17.1 for a single DMU that has three divisions, k, h, and h'. A general multiperiod dynamic multiprocess network technology takes the form

$$\left\{ \left(\bar{\mathbf{c}}^t,\; \mathbf{x}^t,\; \mathbf{z}^t(\bar{\mathbf{L}}), \mathbf{z}^t(\bar{\bar{\mathbf{L}}}),\; \mathbf{c}^t(\bar{\mathbf{F}}),\; \mathbf{y}^t \right) \in NT^t \quad \forall t = 1,\ldots,T \right\} \qquad (17.3)$$

where

$$\bar{\mathbf{c}}^t = \left(\bar{c}_{\bar{r}}^{(h\bar{\tau},\,1t)},\ldots,\bar{c}_{\bar{r}}^{(h\bar{\tau},\,Kt)} \right), \qquad \bar{r} = 1,\ldots,\bar{R},\; h = 1,\ldots,K,$$

$$\mathbf{x}^t = \left(x_n^{1t},\ldots,x_n^{Kt} \right), \qquad\qquad k = 1,\ldots,K,$$

$$\mathbf{z}^t\left(\bar{\mathbf{L}}\right) = \left(z_q^{(ht,\,1t)},\ldots,z_q^{(ht,\,Kt)} \right), \qquad q = 1,\ldots,Q,\; (ht,kt) \in \bar{\mathbf{L}},$$

$$\mathbf{z}^t\left(\bar{\bar{\mathbf{L}}}\right) = \left(z_{\tilde{q}}^{(1t,\,ht)},\ldots,z_{\tilde{q}}^{(Kt,\,ht)} \right), \qquad \tilde{q} = 1,\ldots,\tilde{Q},\; (kt,ht) \in \bar{\bar{\mathbf{L}}},$$

$$\mathbf{c}^t\left(\bar{\mathbf{F}}\right) = \left(c_r^{(1t,\,g\tau)},\ldots,c_r^{(Kt,\,g\tau)} \right), \qquad r = 1,\ldots,R,\; (kt,g\tau) \in \bar{\mathbf{F}}$$

$$\mathbf{y}^t = \left(y_m^{1t},\ldots,y_m^{Kt} \right), \qquad\qquad m = 1,\ldots,M$$

(17.4)

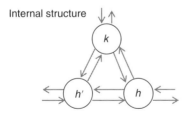

Figure 17.1   Dynamic network structure.

The dynamic structure is represented at the bottom of Figure 17.1, where divisions g', k, and g send and receive carry-over products from each other across periods.

17.3 EFFICIENCY/PRODUCTIVITY MEASUREMENT

Using (17.3) as the dynamic technology, we define a weighted multiperiod dynamic multidivision network (WDN) model for DMU o as

$$
\theta_{o,\text{WDN}} = \min \frac{\displaystyle\sum_{t=1}^{T} W^t \sum_{k=1}^{K} w^k \left\{ 1 - \frac{1}{N^k} \times \left(\sum_{n=1}^{N^k} \frac{s_n^{kt-}}{x_{no}^{kt}} \right) \right\}}{\displaystyle\sum_{t=1}^{T} W^t \sum_{k=1}^{K} w^k \left\{ 1 + \frac{1}{M^k} \times \left(\sum_{m=1}^{M^k} \frac{s_m^{kt+}}{y_{mo}^{kt}} \right) \right\}}
\tag{17.5}
$$

subject to:

$$
c_{\bar{r}o}^{(h\bar{\tau},kt)} \geq \sum_{j=1}^{J} c_{\bar{r}j}^{(h\bar{\tau},kt)} \lambda_j^{kt}, \quad \bar{r} = 1,\ldots,\bar{R}^k
$$

$$
x_{no}^{kt} - s_n^{kt-} = \sum_{j=1}^{J} x_{nj}^{kt} \lambda_j^{kt}, \quad k = 1,\ldots,K\,; n = 1,\ldots,N^k
$$

$$
z_q^{(ht,kt)} \geq \sum_{j=1}^{J} z_{qj}^{(ht,kt)} \lambda_j^{kt}, \quad (ht,kt) \in \overline{\mathbf{L}}\,; q = 1,\ldots,\bar{Q}^k
$$

$$
y_{mo}^{kt} + s_m^{kt+} = \sum_{j=1}^{J} y_{mj}^{kt} \lambda_j^{kt}, \quad k = 1,\ldots,K\,; m = 1,\ldots,M^k
$$

$$
z\tilde{q}^{(kt,ht)} \leq \sum_{j=1}^{J} z_{\tilde{q}j}^{(kt,ht)} \lambda_j^{ht}, \quad (kt,ht) \in \overline{\overline{\mathbf{L}}}\,; \tilde{q} = 1,\ldots,\tilde{Q}^k
\tag{17.6}
$$

$$
c_{ro}^{(kt,h\tau)} = \sum_{j=1}^{J} c_{rj}^{(kt,g\tau)} \lambda_j^{kt}, \quad (kt,g\tau) \in \bar{\mathbf{F}}\,; r = 1,\ldots,R^k
$$

$$
s_n^{kt-} \geq 0 \;\; (\forall n,k,t); \;\; s_m^{kt-} \geq 0 \;\; (\forall m,k,t); \;\; \lambda_j^{kt} \geq 0 \;\; (\forall m,k,t)
$$

where[2] w^k $(\forall k)$ are exogenous weights attached to division k and W^t $(\forall t)$ are exogenous weights associated with time t. The time weights might correspond to discount rates for a given rate of time preference. The input efficiency measure is defined as

---

[2] Note that w^k and W^t can be endogenized.

$$\theta_{o,\text{inp}} = \min \sum_{t=1}^{T} W^t \sum_{k=1}^{K} w^k \left\{ 1 - \frac{1}{N^k} \times \left(\sum_{n=1}^{N^k} \frac{s_n^{kt-}}{x_{no}^{kt}} \right) \right\} \qquad (17.7)$$

and the output efficiency measure is defined as

$$\theta_{o,\text{out}} = \min \frac{1}{\sum_{t=1}^{T} W^t \sum_{k=1}^{K} w^k \left\{ 1 + \frac{1}{M^k} \times \left(\sum_{m=1}^{M^k} \frac{s_m^{kt+}}{y_{mo}^{kt}} \right) \right\}} \qquad (17.8)$$

Each of the efficiency measures ($\theta_{o,\text{WDN}}$, $\theta_{o,\text{inp}}$, and $\theta_{o,\text{out}}$) is efficient if it is equal to one; it is inefficient if it is less than one. Each ratio-form programming problem (after applying the Charnes–Cooper transformation) is solved J times, once for each DMU in the sample.

17.4   EMPIRICAL APPLICATION

17.4.1   Prefectural Production and Data

We follow Fukuyama *et al.* [16], who proposed a prefectural production model where the input sectors jointly produce the prefectural gross domestic product (GDP). Based on conventional economic growth theory, we assume that labor and two types of physical capital are transformed into a single product of prefectural GDP. The prefectural technology comprises three input sectors, called the human capital (HC) sector, the private physical capital (PPC) sector, and the social overhead capital (SOC) sector. We assume an internal parallel structure with the final output being jointly produced by the three sectors. Furthermore, in order to implement a dynamic structure in our model, we assume that the PPC and SOC sectors receive and send carry-overs from the preceding period and to the subsequent period. Figure 17.2 shows the prefectural dynamic network structure.

The HC sector consists of general education and training investments, which are important factors for economic growth. We measure the quantity of human capital (number of employees) and also the quality of human capital using Fukao and Yue's [17] human resource quality index method at the prefectural level. Fukao and Yue [17] estimated the index for 1955 to 1995, and we have extended their method and estimated the index for the period 2007 to 2009. Multiplying the human capital quality index by the number of employees yields our measure of human capital input. The total fixed capital assets possessed by private firms equals the PPC. The public SOC equals the total amount of fixed assets owned by the prefecture. Social overhead capital enhances worker productivity and contributes to GDP by enhancing communication and transportation. We distinguish between PPC and SOC because their effects on the GDP can be different.

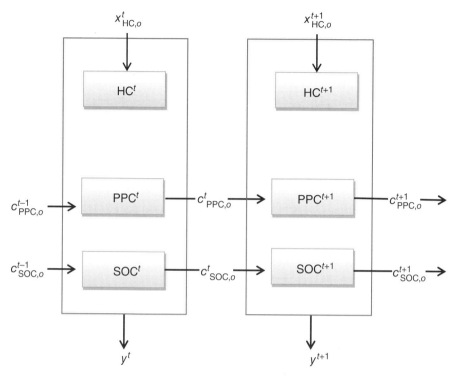

Figure 17.2 Prefectural production structure.

The final product equals the prefecture's contribution to GDP and is jointly produced by the three sectors. The exogenous inputs, carry-overs, and output for prefecture o are defined as follows:

$x^t_{HC,o}$ = human capital in t (Fukao and Yue's human capital index multiplied by number of workers)

$c^t_{PPC,o}$ = private capital stock in t

$c^t_{SOC,o}$ = public capital stock in t

y^t_o = prefectural GDP in t.

In our empirical analysis, we used the output efficiency measure defined in (17.8) and estimated it for the two year sequences 2007–2008 and 2008–2009. For prefecture o, we solve the following:

$$\theta_{\text{overall}} = \min \frac{1}{W^t\left\{1 + \frac{1}{M} \times \left(\sum_{m=1}^{M} \frac{s^{t+}_m}{y^t_{mo}}\right)\right\} + W^{t+1}\left\{1 + \frac{1}{M} \times \left(\sum_{m=1}^{M} \frac{s^{(t+1)+}_m}{y^{t+1}_{mo}}\right)\right\}} \quad (17.9)$$

subject to the constraints
 (year t)

$$x_{\text{HC},no}^t \geq \sum_{j=1}^{J} x_{\text{HC},nj}^t \lambda_{\text{HC},j}^t, \qquad n=1,\ldots,N_{\text{HC}}^t \tag{17.10}$$

$$y_{mo}^t + s_m^{t+} = \sum_{j=1}^{J} y_{mj}^t \lambda_{\text{HC},j}^t, \qquad m=1,\ldots,M^t \tag{17.11}$$

$$y_{mo}^t + s_m^{t+} = \sum_{j=1}^{J} y_{mj}^t \lambda_{\text{PPC},j}^t, \qquad m=1,\ldots,M^t \tag{17.12}$$

$$c_{\text{PPC},ro}^{t-1} \geq \sum_{j=1}^{J} c_{\text{PPC},rj}^{t-1} \lambda_{\text{PPC},j}^t, \qquad r=1,\ldots,R_{\text{PPC}}^{t-1} \tag{17.13}$$

$$c_{\text{PPC},ro}^t \leq \sum_{j=1}^{J} c_{\text{PPC},rj}^t \lambda_{\text{PPC},j}^t, \quad r=1,\ldots,R_{\text{PPC}}^t \tag{17.14}$$

$$y_{mo}^t + s_m^{t+} = \sum_{j=1}^{J} y_{mj}^t \lambda_{\text{SOC},j}^t, \qquad m=1,\ldots,M^t \tag{17.15}$$

$$y_{mo}^t + s_m^{t+} \geq \sum_{j=1}^{J} c_{\text{SOC},rj}^{t-1} \lambda_{\text{SOC},j}^t, \qquad r=1,\ldots,R_{\text{SOC}}^{t-1} \tag{17.16}$$

$$c_{\text{SOC},ro}^t \leq \sum_{j=1}^{J} c_{\text{SOC},rj}^t \lambda_{\text{SOC},j}^t, \quad r=1,\ldots,R_{\text{SOC}}^t \tag{17.17}$$

 (year $t+1$)

$$x_{\text{HC},no}^{t+1} \geq \sum_{j=1}^{J} x_{\text{HC},nj}^{t+1} \lambda_{\text{HC},j}^{t+1}, \qquad n=1,\ldots,N_{\text{HC}}^{t+1} \tag{17.18}$$

$$y_{mo}^{t+1} + s_m^{(t+1)+} = \sum_{j=1}^{J} y_{mj}^{t+1} \lambda_{\text{HC},j}^{t+1}, \qquad m=1,\ldots,M^{t+1} \tag{17.19}$$

$$y_{mo}^{t+1} + s_m^{(t+1)+} = \sum_{j=1}^{J} y_{mj}^{t+1} \lambda_{\text{PPC},j}^{t+1}, \qquad m=1,\ldots,M^{t+1} \tag{17.20}$$

$$c_{\text{PPC},ro}^t \geq \sum_{j=1}^{J} c_{\text{PPC},rj}^t \lambda_{\text{PPC},j}^{t+1}, \qquad r=1,\ldots,R_{\text{PPC}}^t \tag{17.21}$$

$$c_{\text{PPC},ro}^{t+1} \leq \sum_{j=1}^{J} c_{\text{PPC},rj}^{t+1} \lambda_{\text{PPC},j}^{t+1}, \qquad r=1,\ldots,R_{\text{PPC}}^{t+1} \tag{17.22}$$

$$y_{mo}^{t+1} + s_m^{(t+1)+} = \sum_{j=1}^{J} y_{mj}^{t+1} \lambda_{\text{SOC},j}^{t+1}, \quad m = 1,\ldots,M^{t+1} \tag{17.23}$$

$$c_{\text{SOC},ro}^{t} \geq \sum_{j=1}^{J} c_{\text{SOC},rj}^{t} \lambda_{\text{SOC},j}^{t+1}, \quad r = 1,\ldots,R_{\text{SOC}}^{t} \tag{17.24}$$

$$c_{\text{SOC},ro}^{t+1} \leq \sum_{j=1}^{J} c_{\text{SOC},rj}^{t+1} \lambda_{\text{SOC},j}^{t+1}, \quad r = 1,\ldots,R_{\text{SOC}}^{t+1} \tag{17.25}$$

$$s_m^{t+} \geq 0, \; s_m^{(t+1)+} \geq 0, \; (\forall m)$$

$$\lambda_{\text{HC},j}^{t} \geq 0, \; \lambda_{\text{PPC},j}^{t} \geq 0, \; \lambda_{\text{SOC},j}^{t} \geq 0, \; (\forall j)$$

$$\lambda_{\text{HC},j}^{t+1} \geq 0, \; \lambda_{\text{PPC},j}^{t+1} \geq 0, \; \lambda_{\text{SOC},j}^{t+1} \geq 0, \; (\forall j)$$

We choose equal weights for each two-period sequence, with $W^t = W^{t+1} = 0.5$. Since the three sectors jointly produce the prefectural GDP, we add the following restrictions on the intensity variables:

$$\sum_{j=1}^{J} y_{mj}^{t} \lambda_{\text{HC},j}^{t} = \sum_{j=1}^{J} y_{mj}^{t} \lambda_{\text{PPC},j}^{t} = \sum_{j=1}^{J} y_{mj}^{t} \lambda_{\text{SOC},j}^{t} = y_{mo}^{t} + s_m^{t+} \; (m = 1,\ldots,M) \tag{17.26}$$

Equation (17.26) indicates that, while we observe one GDP value for a prefecture each year, the intensity variables $\lambda_{\text{HC},j}$ $(j = 1,\ldots,J)$ in (17.11) provide information about the human capital subtechnology consisting of (17.10) and (17.11), the intensity variables $\lambda_{\text{PPC},j}$ $(j = 1,\ldots,J)$ provide information about the physical capital formation sector consisting of (17.12)–(17.14), and the intensity variables $\lambda_{\text{SOC},j}$ $(j = 1,\ldots,J)$ provide information on the social overhead capital subtechnology consisting of (17.15)–(17.17). Similar restrictions on the intensity variables for $t + 1$ are imposed:

$$\sum_{j=1}^{J} y_{mj}^{t+1} \lambda_{\text{HC},j}^{t+1} = \sum_{j=1}^{J} y_{mj}^{t+1} \lambda_{\text{PPC},j}^{t+1} = \sum_{j=1}^{J} y_{mj}^{t+1} \lambda_{\text{SOC},j}^{t+1} = y_{mo}^{t+1} + s_m^{(t+1)+} \; (m = 1,\ldots,M)$$

$$\tag{17.27}$$

Table 17.1 provides summary statistics of the inputs, output, and carry-overs. The monetary values have been deflated by the 2005 GDP deflator.

17.4.2 Efficiency Estimates and Their Determinants

Table 17.2 shows efficiency estimates for the fiscal years 2007–2008 and 2008–2009, where the fiscal year starts on April 1 and ends on March 31. The estimates indicate

TABLE 17.1 Data description (inputs, output, carry-overs).[a]

Period		Human capital	Previous period $C^{t-1}_{PPC,j}$ (million yen)	Previous period $C^{t-1}_{SOC,j}$ (million yen)	Current period $C^{t}_{PPC,j}$ (million yen)	Current period $C^{t}_{SOC,j}$ (million yen)	GDP (million yen)
2007–2009	Mean	1,659,470	22,068,681	9,761,130	22,171,369	9,705,514	11,227,607
	Std. dev.	2,062,459	20,125,147	7,046,991	20,339,269	7,017,393	15,285,715
	Max	13,628,005	138,496,761	35,214,924	139,490,220	35,015,750	100,061,637
	Min	344,775	10,079,838	3,787,431	10,087,270	3,784,257	2,027,794
2007	Mean	1,594,230	21,898,680	9,822,438	22,197,761	9,770,545	11,588,971
	Std. dev.	1,920,465	19,802,785	7,084,482	20,325,682	7,050,147	15,774,189
	Max	11,818,594	135,011,740	35,214,924	138,496,761	35,006,161	100,061,637
	Min	344,775	10,079,838	3,825,600	10,124,231	3,810,468	2,160,115
2008	Mean	1,657,262	22,197,761	9,770,545	22,109,601	9,690,407	11,280,426
	Std. dev.	2,064,125	20,325,682	7,050,147	20,241,880	7,005,493	15,364,942
	Max	12,797,412	138,496,761	35,006,161	138,054,626	34,837,462	97,840,393
	Min	351,277	10,124,231	3,810,468	10,091,808	3,787,431	2,092,722
2009	Mean	1,726,917	22,109,601	9,690,407	22,206,745	9,655,590	10,813,422
	Std. dev.	2,191,712	20,241,880	7,005,493	20,449,565	6,995,925	14,688,356
	Max	13,628,005	138,054,626	34,837,462	139,490,220	35,015,750	93,842,542
	Min	365,369	10,091,808	3,787,431	10,087,270	3,784,257	2,027,794

[a] PPC = private physical capital; SOC = social overhead capital.

TABLE 17.2   Two-period efficiency estimates.[a]

Prefecture		2007–2008	2008–2009	All years
All	Mean	0.812	0.823	0.818
	Std. dev.	0.052	0.060	0.056
	Max	0.907	1	1
	Min	0.721	0.722	0.721
Hokkaido-Tohoku	Mean	0.828	0.852	0.840
	Std. dev.	0.057	0.055	0.057
Kanto	Mean	0.797	0.808	0.803
	Std. dev.	0.043	0.047	0.046
Hokuriku-Tokai	Mean	0.825	0.839	0.832
	Std. dev.	0.033	0.042	0.039
Kansai	Mean	0.819	0.813	0.816
	Std. dev.	0.074	0.067	0.070
Chugoku-Shikoku	Mean	0.800	0.809	0.804
	Std. dev.	0.049	0.051	0.050
Kyushu-Okinawa	Mean	0.812	0.823	0.817
	Std. dev.	0.050	0.078	0.066
Urbanized industrial	Mean	0.854	0.839	0.847
prefectures (Tokyo,	Std. dev.	0.034	0.029	0.030
Kanagawa, Osaka, Aichi)				
Other prefectures	Mean	0.808	0.822	0.815
	Std. dev.	0.052	0.061	0.057

[a] The Kanto region consists of Tokyo, Kanagawa, Saitama, Gunma, Tochigi, Ibaraki, and Chiba, as well as the two Koshin-area prefectures of Yamanashi and Nagano.

that only Okinawa prefecture was efficient for both periods. Lehman Brothers went bankrupt in the fall of 2008 at the start of the financial crisis in the US. The economic downturn precipitated by the bankruptcy seems to have quickly affected the Japanese prefectures except for Okinawa, Japan's southernmost island prefecture.

This efficiency estimate for Okinawa is consistent with the evidence that most Japanese prefectures suffered negative growth in 2008 but Okinawa's growth rate was 1.21%, which was the highest among all prefectures.[3] Okinawa's industrial characteristics in 2008 showed that the tertiary sector, providing services for consumers or businesses, constituted 86.1% of prefectural GDP, an amount much larger than Japan's average of 70.0%.The large tertiary sector is due to the relative importance of tourism in Okinawa. The average efficiency for the 47 prefectures was 0.818.

The last column of Table 17.2 reports the efficiency for all years. These estimates show that the urban prefectures of Tokyo, Kanagawa, Aichi, and Osaka, where the four largest cities are located, have higher efficiency scores than nonurban prefectures. Therefore, we conjecture that there is a positive relationship between efficiency and

[3] The second and third highest growth rates were observed for Shimane (0.33%) and Nagasaki (0.25%).

agglomeration economies, where firms gain when activities external to the firm are clustered near the firm. Moreover, the economic downturn arising from the Lehman Brothers bankruptcy caused a decrease in industrial production[4] (production in the secondary sector of industry). Therefore, increasing industrial production might have possibly enhanced prefectural efficiencies. To examine these possibilities, we estimated the following regression in a second stage using the efficiency estimates from (17.9) as the first stage. The regression employed was as follows:

$$\text{Overall efficiency} = f(\text{DEN, MA, REG}) \tag{17.28}$$

We measured potential agglomeration economies by using the population density as a proxy. Otsuka et al. [18] stated that the higher the population, the more industrial and service production are enhanced. Next, we considered how market access might also result in agglomeration economies. We measured market access in the industrial structure by the ratio of the secondary sector's production to the whole prefectural production. The larger the ratio, the greater is the market access in the prefecture.

Finally, we included regional dummies (REG) because Japan is an island nation, comprising an archipelago extending along the Asian–Pacific coast with different climates. Since the explanatory variable contains two-period estimates, we used arithmetic average values for DEN and MA between the two periods. To control for the possibility of correlation between the efficiency estimates in the first phase and the explanatory variables in the second regression phase, we used bootstrap regression analysis. Our actual regression model was

$$\hat{\theta}_{\text{overall}} = \alpha + \alpha_1 \cdot \text{DEN} + \alpha_2 \cdot \text{MA} + \sum_i \beta_i \cdot \text{dummy}_i + \varepsilon \tag{17.29}$$

where $\alpha, \alpha_1, \alpha_2, \beta_i$ are parameters to be estimated. We specified six regions and dropped the Kyushu–Okinawa region dummy to avoid an exact linear dependence between the regional dummies.

Table 17.3 reports the regression estimates. Our bootstrap regression analysis indicates that the coefficient DEN is significantly positive at the 1% level. This result is consistent with Otsuka et al.'s argument [18] for agglomeration economies. The estimated coefficient MA is also positively significant at the 5% level. The regional dummy variables had no effect on efficiency except for the Kanto region, which had significantly less efficiency than the other prefectures. In summary, the regression analysis indicates that prefectures with greater population density and greater market access benefit from agglomeration economies.

---

[4] For example, in 2007 secondary sector production in Aichi and Tokyo was 15.2 and 12.2 trillion yen, respectively, in 2007. By 2008, secondary sector production in Aichi and Tokyo had fallen to 11.7 and 12.2 trillion yen, respectively. In 2009, Aichi's production was 11.1 trillion yen and Tokyo's production was 11.5 trillion yen.

TABLE 17.3   Bootstrap regression analysis.

	Coefficient		Std.error	Z	Prob. $z > Z*$		95% confidence interval
α	0.2180	***	0.007	31.05	0.000	0.204	0.232
DEN	0.0029	***	0.001	3.12	0.003	0.001	0.005
MA	0.0587	**	0.025	2.39	0.017	0.011	0.107
Hokkaido-Tohoku	0.0072		0.006	1.30	0.192	−0.004	0.018
Kanto	−0.0139	***	0.006	−2.87	0.004	−0.023	−0.004
Hokuriku-Tokai	−0.0031		0.006	−0.67	0.503	−0.012	0.006
Kansai	−0.0090		0.007	−1.25	0.212	−0.023	0.005
Chugoku-Shikoku	−0.0066		0.005	−−1.23	0.218	−0.017	0.004

***, **, * Significance at 1%, 5%, and 10% levels.

17.5   CONCLUSIONS

In real-world production technologies, various divisions within a firm or economic entity often produce intermediate products and receive them from other divisions. In addition, inputs are often saved or carried over from one period to another so as to optimize production plans for the entire firm. In this chapter, we have developed a dynamic network model that can account for various internal structures within a firm and for the fact that current production plans may be influenced by past production decisions as well as affect future production possibilities. Panel data can be used to estimate variants of our DEA model for various kinds of producers. We have offered one illustrative example of our method using data from 2007–2009 for 47 Japanese prefectures. Prefectural output is jointly produced by three internal input sectors: a human capital sector, a private physical capital sector, and a social overhead capital sector. The average efficiency was 81.8%, with Okinawa being the most efficient prefecture. We also found that prefectures with a greater population density and greater market access were more efficient. Although we focused on a single desirable final output (prefectural GDP), future studies that control for undesirable outputs such as carbon dioxide emissions might also yield insights that can be used to inform policy-makers.

REFERENCES

[1] Färe, R. and Grosskopf, S. (1996) *Intertemporal Production Frontiers: With Dynamic DEA*, Kluwer Academic, Boston, MA.

[2] Färe, R. and Grosskopf, S. (2000) Network DEA. *Socio-Economic Planning Sciences*, **34**, 35–49.

[3] Nemoto, J. and Goto, M. (2003) Measurement of dynamic efficiency in production: An application of data envelopment analysis to Japanese electric utilities. *Journal of Productivity Analysis*, **19**, 191–210.

[4] Emrouznejad, A. and Thanassoulis, E. (2005) A mathematical model for dynamic efficiency using data envelopment analysis. *Applied Mathematics and Computation*, **160**(2), 363–378.

[5] Bogetoft, P., Färe, R., Grosskopf, S., *et al.* (2009) Dynamic network DEA: An illustration. *Journal of the Operations Research Society of Japan*, **52**, 147–162.

[6] Chen, C.M. and van Dalen, J. (2010) Measuring dynamic efficiency: Theories and an integrated methodology. *European Journal of Operational Research*, **203**, 749–760.

[7] Tone, K. and Tsutsui, M. (2010) Dynamic DEA: A slacks-based measure approach. *Omega*, **38**, 145–156.

[8] Tone, K. and Tsutsui, M. (2014) Dynamic DEA with network structure: A slacks-based measure approach. *Omega*, **42**(1), 124–131.

[9] Fukuyama, H. and Mirdehghan, S.M. (2012) Identifying the efficiency status in network DEA. *European Journal of Operational Research*, **220**, 85–92.

[10] Kao, C. and Hwang, S.N. (2008) Efficiency decomposition in two-stage data envelopment analysis: An application to non-life insurance companies in Taiwan. *European Journal of Operational Research*, **185**, 418–429.

[11] Chen, Y., Liang, L., and Zhu, J. (2009) Equivalence in two-stage DEA approaches. *European Journal of Operational Research*, **193** (2), 600–604.

[12] Fukuyama, H. and Weber, W.L. (2010) A slacks-based inefficiency measure for a two-stage system with bad outputs. *Omega*, **38**(5), 239–410.

[13] Akther, S., Fukuyama, H., and Weber, W.L. (2013) Estimating two-stage network slacks-based inefficiency: An application to Bangladesh banking. *Omega*, **41**(1), 88–96.

[14] Fukuyama, H. and Weber, W.L. (2013) A dynamic network DEA model with an application to Japanese cooperative Shinkin banks, in *Efficiency and Productivity Growth: Modelling in the Financial Services Industry* (ed. F. Pasiouras), John Wiley & Sons Ltd, Chichester, pp. 193–214.

[15] Fukuyama, H., Weber, W.L., and Xia, Y. (2016) Time substitution and network effects with an application to nanobiotechnology policy for US universities. *Omega*, **60**, 34–44.

[16] Fukuyama, H., Hashimoto, A., Tone, K., and Weber, W.L. (2015) Does human capital or physical capital constrain output in Japanese prefectures? *Empirical Economics*, in press.

[17] Fukao, K. and Yue, X. (2000) Regional factor inputs and convergence in Japan: How much can we apply closed economy neoclassical growth models? [In Japanese.] *Economic Review (Keizai Kenkyu)*, **51**, 136–151.

[18] Otsuka, A., Goto, M., and Sueyoshi, T. (2010) Industrial agglomeration effects in Japan: Productive efficiency, market access, and public fiscal transfer. *Regional Science*, **89**(4), 819–840.

18

A QUANTITATIVE ANALYSIS OF MARKET UTILIZATION IN ELECTRIC POWER COMPANIES

MIKI TSUTSUI

Central Research Institute of Electric Power Industry, Tokyo, Japan

KAORU TONE

National Graduate Institute for Policy Studies, Tokyo, Japan

18.1 INTRODUCTION

The data envelopment analysis (DEA) method is very popular in the energy industry. A considerable number of studies have evaluated the efficiency performance of energy companies in many countries after deregulation of the industry [1–3]. In addition, energy regulators in some countries in Europe, where incentive-based regulation in electricity networks has been introduced, have officially applied DEA methods for efficiency benchmarking [4–7]. DEA is also used for the benchmarking of European transmission system operators [8].

Contrary to such benchmarking purposes, we have applied DEA to evaluate the effect of energy trading in the market. In European countries, the wholesale power markets are well developed enough to be utilized by many electric power companies. These companies usually have a trading unit which handles intensively all of the transactions with fuel and power markets, standing between the generation and retail divisions, even if they were vertically integrated before liberalization. Although some

Advances in DEA Theory and Applications: With Extensions to Forecasting Models,
First Edition. Edited by Kaoru Tone.
© 2017 John Wiley & Sons Ltd. Published 2017 by John Wiley & Sons Ltd.

companies have such trading units as a department, and others have a subsidiary company for trading, their basic functions are the same. In this chapter, we refer to organizations that have a trading function in the company as a 'trading division' or 'TD'.

This study quantitatively evaluates the effects of the potential use of market opportunities through TDs, and compares them under different conditions and constraints using DEA. Then we clarify the problem of what price conditions the trading function will work effectively under in the future.

In Japan, system reform in the electricity industry is now under way; for instance, the retail electricity markets for domestic customers will be opened up in April 2016. The government expects that the reform will promote new market entries, resulting in revitalization of the competition in the electricity retail market. In addition, it is also expected that the wholesale power market will be revitalized, even though liquidity has been very limited since it started to operate in 2005.

The incumbent Japanese power companies have been vertically integrated, similarly to those in European countries before liberalization. In these companies, the generation division (GD) sends most of the electricity generated to the retail division (RD) directly as a matter of course. However, this will change in accordance with the increase in market liquidity in the wholesale power market, just as in European countries. In fact, some incumbent Japanese power companies are attempting to establish TDs in preparation for the effective use of market opportunities. On the other hand, others are sceptical about the utilization of market mechanisms and the effects of TDs. Therefore, the quantitative analysis in this study will help these companies to consider the introduction of TDs.

This chapter proceeds as follows. In Section 18.2, we summarize how the internal transaction system for electricity in power companies in Europe changed from before to after liberalization. In Section 18.3, the framework of the quantitative analysis is explained in order to clarify the effect of the trading function. The results are shown in Section 18.4, and some remarks follow in the last section.

18.2   THE FUNCTIONS OF THE TRADING DIVISION

Before liberalization of the electric power industry in many countries, including Japan, typical electric power companies were vertically integrated, where several functions existed inside one company, such as generation, transmission, distribution and retail functions. It was quite common for these companies to transmit generated electricity internally to the retail division and then to customers (Figure 18.1(a)).[1] The electricity tariffs for the final customers were generally under cost-based regulation. In this case, the GD (or a fuel procurement division) procured fossil fuels for power

---

[1] In European countries, network businesses such as transmission and distribution businesses are still regulated and are required to be independent from competitive businesses such as generation and retail businesses. In Figure 18.1, we focus only on competitive businesses; therefore, network businesses are not depicted in the figure, even if the parent company owns them.

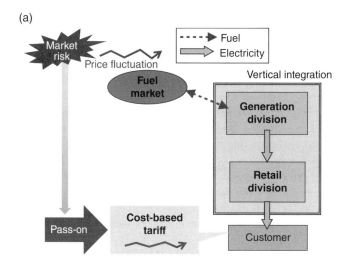

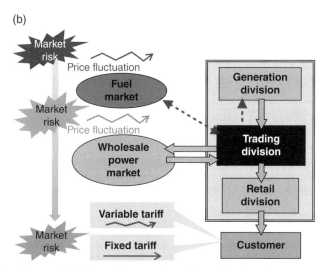

Figure 18.1 Change in the internal transaction of electricity in power companies: (a) before liberalization; (b) after liberalization.

plants from the fuel market, and was therefore exposed to market risk (price fluctuation risk). However, a power company could pass on the risk to the customers via a cost-based tariff.

However, as wholesale power markets were gradually developed in several countries in Europe after liberalization, the representative power companies established TDs in order to sell and buy electricity on the market, effectively on behalf of GDs and RDs (Figure 18.1(b)). This enables the whole (parent) company to concentrate various types of market risk management in the TD. In particular, after liberalization, customers became able to freely choose electricity tariffs, which are not regulated. If many customers choose fixed tariffs, a power company cannot pass on market risks to customers any more. Therefore, it became very important for companies to control market risk effectively. A TD controls all the market risks of a company intensively.

Moreover, TDs procure fossil fuels for generation and determine the economic dispatch by referring to market prices[2] to optimize the operation of all power plants through fuel and power trading based on profit maximization. This means that the GD produces electricity only when it can make a profit. For instance, when fuel prices are high and the price of power is relatively low, the TD will decide to purchase electricity from the market to cover the final demand in the RD instead of ordering the GD to produce electricity at its own power plants.

In this market-oriented (MO) system, there is no direct transaction between the GD and the RD; this is completely different from the vertically integrated (VI) system, which depended heavily on internal transactions before liberalization. It should be noted that the company in the MO system described in Figure 18.1(b) is actually called a 'vertically integrated company', because the whole management (or the parent company) owns both the GD and the RD functions. However, in the MO system, these functions are operated as independent businesses with independent licences.

We assume that the difference between the two systems is attributable to the volume and price constraints on the internal transactions as follows:

- *Volume constraint.* In the VI system, all of the electricity demand in the RD is covered by the electricity generated at power plants in the GD. In other words, the volume of the internal transactions is strictly constrained, while it is completely free in the MO system; that is, the TD can freely choose sources of electricity from the power market and/or internal transactions in order to cover the final demand in the RD.
- *Price constraint.* In the traditional VI system, the transfer price of the internal transactions is based on generation cost, and therefore retail tariffs for customers are also cost-based. On the other hand, in the MO system, tariffs are decided based on market prices. In the market mechanism, prices depend on supply and demand and are not based on cost. Therefore, in this situation, the cost-based price setting in the VI system can be regarded as a strict constraint.

This study compares these two systems and clarifies the effects of the trading function from three points of view as follows:

- *Profit (return).* If a power company effectively utilizes the fuel and power markets in the MO system, the total profit of the company will be maximized rather

---

[2] In Europe, TDs refer not only to fuel and power market prices, but also to CO_2 prices. Emissions trading is not as active in Japan and, therefore, to simplify the model, we do not consider CO_2 prices.

than depending on the internal transaction in the VI system. In other words, strict volume and price constraints may inhibit the profit maximization of the company.

- *Stability of profit (risk).* However, the company will be exposed to market risks in the MO system.
- *Competitiveness.* If the company utilizes the market price for the internal transfer price, profit will be optimized, but the competitiveness in the retail market may be reduced, because it cannot differentiate its retail prices from its competitors. In the VI system, the cost-based internal price may have an advantage over the MO system, especially in the case where the company owns inexpensive power plants such as hydro power plants. In this study, we employ the retail price level as a competitive index. We assume that a lower retail price can enhance competitiveness in the retail market of the company.

In general, the high-risk case could bring high return, and higher competitiveness (lower retail price level) could result in lower profit. In other words, the three factors listed above would result in different evaluations even under the same conditions. In such a case, DEA is a very powerful method for conducting a comprehensive evaluation based on multiple factors. Therefore, in this study, we have applied the slacks-based measure (SBM)-max model (Chapter 22) to evaluate VI and MO systems under several market price conditions.

18.3 MEASURING THE EFFECT OF ENERGY TRADING

In this section, we explain the framework for how the effect of the trading function under different conditions was measured in this study.

18.3.1 Definition of Transaction Volumes and Prices

Figure 18.2 summarizes the electricity transactions in a typical power company after liberalization, where the notation in parentheses indicates electricity volume and price.

18.3.1.1 Generation Division We postulate that the GD owns gas-fired, coal-fired and hydro power plants.[3] G_{it}^e is the fuel actually consumed, measured by the kilowatt hours (kWh) used in period t ($t = 1,\ldots,T$),[4] where i indicates the type of power plant (i = gas, coal or hyd). Each power plant cannot generate electricity above its capacity $\bar{G}_i^e$:

---

[3] We do not include nuclear power plants in this study; however, they could be treated in the same manner as hydro power plants in our model, as nuclear fuel is not a commodity that is in general freely traded on the market and its fuel cost is much lower than that of fossil fuel power plants.

[4] The units of the time period t may be minutes, hours, days and so on.

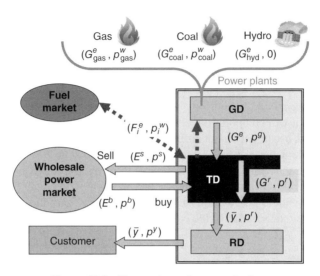

Figure 18.2   Transaction volumes and prices.

[Capacity constraint]

$$G_{it}^e \le \bar{G}_i^e \tag{18.1}$$

The fuel price is denoted by p_{it}^w, which in this study is defined as a market price fluctuating on a moment-to-moment basis for gas and coal, while the price for hydro power is zero. The total generated power (G_t^e) and the average generation (fuel) cost (p_t^w) are measured based on the volume actually consumed as

$$G_t^e = \sum_i G_{it}^e, p_t^w = \frac{\sum_i p_{it}^w G_{it}^e}{G_t^e} \tag{18.2}$$

The decision to generate electricity or not in the GD is made by the TD at different times by referring to the market prices of fuel and power.

18.3.1.2 Internal Transactions (GD → TD)   All of the generated power (G_t^e) is transmitted from the GD to the TD at the internal transfer price (p^g), which is defined as an average of p_t^w during T periods as

$$p^g = \sum_t \frac{p_t^w}{T} \tag{18.3}$$

18.3.1.3 Trading Division   The TD sells electricity generated at plant i to the power market (E_{it}^s) at the market sell price (p_t^s) and/or sends it to the RD (G_{it}^r) at the transfer price (p_t^r), whose definition will appear later in (18.6):

[Flow constraint]

$$G_t^e = \sum_i E_{it}^s + \sum_i G_{it}^r = E_t^s + G_t^r \qquad (18.4)$$

The flow constraint indicates that electric power cannot be stored, and therefore all electricity generated has to be sold to anywhere available.

The TD also has to procure electricity to cover all of the retail demand in the RD ($\bar{y}$). The TD decides the volume to be purchased from the market (E_t^b) and the volume to be generated at the power plants (G_t^r) based on the market prices of fuel and power:

[Demand constraint]

$$\bar{y} = E_t^b + \sum_i G_{it}^r = E_t^b + G_t^r \qquad (18.5)$$

For instance, when the transfer price (p_t^r) is higher than the market buy price (p_t^b), the TD will procure electricity from the market.

18.3.1.4 Internal Transactions (TD → RD)

All of the electricity demand in the RD ($\bar{y}$) is procured by the TD and transmitted to the RD at the transfer price (p_t^r), which is defined as a weighted sum between the generation cost (p_t^w) and the market buy price (p_t^b):

$$p_t^r = \beta p_t^w + (1 - \beta) p_t^b \qquad (18.6)$$

where β is a parameter that will be explained in the next subsection.

18.3.1.5 Retail Division

The RD sells electricity received from the TD to customers, adding γ % retail margin; therefore, the retail price (p_t^y) is

$$p_t^y = (1 + \gamma) p_t^r \qquad (18.7)$$

Obviously, a company can earn more profit if it sets a large margin rate. However, in reality it is difficult to set a large γ to survive against competition in a competitive retail market.

18.3.2 Constraints on Internal Transactions

In order to compare the VI and MO systems, we assume two parameters for the volume and price constraints on internal transactions.

- *Volume constraint: α.* We postulate that the TD has to use electricity from the GD, which is generated at the company's own power plants, to cover at least $\alpha \times 100\%$ of the retail demand as follows:

$$\sum_i G_{it}^r \geq \alpha \bar{y} (0 \leq \alpha \leq 1) \qquad (18.8)$$

TABLE 18.1 Simulation of constraints on internal transactions.

	(MO system)				(VI system)
	Free	⟵	Constraint	⟶	Strict
	$a1$	$a2$	$a3$	$a4$	$a5$
$\alpha=$	0	0.25	0.5	0.75	1
	$b1$	$b2$	$b3$	$b4$	$b5$
$\beta=$	0	0.25	0.5	0.75	1

$\alpha = 0$ (MO system): the TD can decide the volume to generate at power plants, to sell to the market and to buy from the market, based only on the market mechanism without any constraints.

$\alpha = 1$ (VI system): the TD has to cover all of the retail demand with electricity generated at its own plants in the GD, regardless of the market price level.

- *Price constraint: β.* As shown in (18.6), the internal transfer price (p_t^r) from the TD to the RD is defined based on the generation cost (p_t^w) and the market buy price (p_t^b), weighted by β.

$\beta = 0$ (MO system): the retail price (p_t^y) is defined based on only the market price.

$\beta = 1$ (VI system): the retail price (p_t^y) is defined based on only the generation cost of the company's own power plants in the GD regardless of the market price level.

In this study, we simulated five levels of constraints, as listed in Table 18.1.

18.3.3 Profit Maximization

Theoretically speaking, the GD wants to sell generated electricity at a higher price, while the RD wants to procure it at a lower price, which suggests a possibility of internal conflict. The TD can resolve this by mediating between the two and aiming at overall profit maximization. Divisional and overall profits are calculated as follows:
Generation division:

$$\text{Revenue}: \text{REV}_t^{\text{GD}} = p_t^g G_t^e$$

$$\text{Cost}: \text{COS}_t^{\text{GD}} = \sum_i p_{it}^w G_{it}^e = p_t^w G_t^e \qquad (18.9)$$

$$\text{Profit}: \text{PRO}_t^{\text{GD}} = \left(p_t^g - p_t^w\right) G_t^e$$

Retail division:

$$\text{Revenue}: \text{REV}_t^{\text{RD}} = p_t^y \bar{y}$$

$$\text{Cost}: \text{COS}_t^{\text{RD}} = p_t^r \bar{y} \tag{18.10}$$

$$\text{Profit}: \text{PRO}_t^{\text{RD}} = \left(p_t^y - p_t^r\right)\bar{y}$$

Trading division:

$$\text{Revenue}: \text{REV}^{\text{TD}} = p_t^s \sum_i E_{it}^s + p_t^r \bar{y}$$

$$\text{Cost}: \text{COS}^{\text{TD}} = p_t^g \sum_i G_{it}^e + p_t^b E_t^b \tag{18.11}$$

$$\text{Profit}: \text{PRO}^{\text{TD}} = p_t^s \sum_i E_{it}^s + p_t^r \bar{y} - p_t^g \sum_i G_{it}^e - p_t^b E_t^b$$

Whole company:

$$\text{Profit}: \text{PRO}_t = \text{PRO}_t^{\text{GD}} + \text{PRO}_t^{\text{RD}} + \text{PRO}_t^{\text{TD}}$$

$$= p_t^s \sum_i E_{it}^s + p_t^y \bar{y} - \sum_i p_{it}^w G_{it}^e - p_t^b E_t^b \tag{18.12}$$

$$= \sum_i \left(p_t^s - p_{it}^w\right) E_{it}^s + \sum_i \left(p_t^b - p_{it}^w\right) G_{it}^r + \left(p_t^y - p_t^b\right)\bar{y}$$

We find that the overall profit of the company consists of three types of price spreads multiplied by electricity volumes.

Then, the overall profit maximization model is formulated as

$$\max \ \text{PRO}_t = \sum_i \left(p_t^s - p_{it}^w\right) E_{it}^s + \sum_i \left(p_t^b - p_{it}^w\right) G_{it}^r + \left(p_t^y - p_t^b\right)\bar{y} \tag{18.13}$$

s.t. (18.1), (18.2), (18.4), (18.5), (18.6), (18.7) and (18.8)

In this model, the unknown variables are E_{it}^s, G_{it}^r and p_t^y.

It should be noted that the above profit maximization model is a non-linear problem because (18.2) includes unknown variables G_{it}^e $(= E_{it}^s + G_{it}^r)$ in both the numerator and the denominator. To solve the problem as a linear problem (LP), we substitute G_{it}^e in the denominator for $\bar{G}_i^e$, which can maintain the scale of the variables. By solving the substituted LP model, we can obtain the optimal values of E_{it}^{s*} and G_{it}^{r*}, and then calculate G_{it}^{e*} as $E_{it}^{s*} + G_{it}^{r*}$. The optimal average fuel cost (p_t^{w*}) can also be measured, based on G_{it}^{e*}, as

$$p_t^{w*} = \frac{\sum_i p_{it}^w G_{it}^{e*}}{G_t^{e*}} \tag{18.14}$$

18.3.4   Exogenous Variables

In (18.13), the market prices of fuel and power ($p^w_{\text{coal},t}$, $p^w_{\text{gas},t}$, p^s_t and p^b_t) are exogenous variables, and in this study, we generated data randomly for T points under several conditions.

- *Fluctuation (two cases).* We assumed two different conditions for the market price fluctuation for T periods: a stable and a volatile case. The average gas and coal prices were defined by referring to the actual market prices and converted into the units of electric energy at ¥8.789/kWh and ¥3.264/kWh, respectively. The average sell and buy prices in the power market were both defined as ¥9.229/kWh, which is 5% higher than the average gas price. The market sell and buy prices were independently generated under the same conditions (average and variance):
  - ○ *Case 1.* Stable: the variance of the coefficient was 0.05.
  - ○ *Case 2.* Volatile case: the variance of the coefficient was 0.2.
- *Trend (three cases).* We assumed three conditions for the market price trend for T periods: an increasing, a decreasing and a flat case:
  - ○ *Case 1.* Up: an increasing rate of +0.2%.
  - ○ *Case 2.* Down: a decreasing rate of −0.2%.
  - ○ *Case 3.* Flat: a rate of 0%.

As a result, we generated six (=2 × 3) price series for coal, gas and power prices, and then we had 216 (=6 × 6 × 6) combinations of cases (scenarios). Figure 18.3 shows only the generated data for the gas and coal price series. The sell and buy prices for power follow a similar trend to the gas prices.

In addition, there were 25 cases of combinations of constraints (α, β), resulting in 5400 scenarios in total.

For each of these scenarios, we solved (18.13), and then obtained profit, stability and competitiveness indices. It should be noted that in order to obtain the stability index, we needed to calculate a standard deviation of profits; therefore, repeated calculation was done using randomly generated price series (for T periods) under the same conditions for each case.[5] We then obtained an average profit, a standard deviation of profits for the stability index and an average retail price for the competitiveness index, for 5400 scenarios.

However, in the study presented in this chapter, we fixed the coal price series as the volatile/flat combination, because the three indices were very similar, even if we changed the conditions for the coal prices. Finally, we used 900 scenarios in the DEA calculation.

---

[5] In this chapter, we show results for $T = 30$.

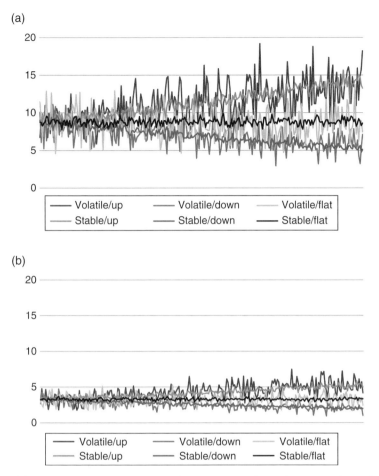

Figure 18.3   Market price setting (250 points): (a) gas prices (six cases); (b) coal prices (six cases).

TABLE 18.2   Fuel mix settings.

	Gas	Coal	Hydro	Total
Mix 1	8 000	2 000	2 000	12 000
Mix 2	2 000	8 000	2 000	12 000
Mix 3	2 000	2 000	8 000	12 000

Furthermore, we postulated three types of fuel mix. Table 18.2 shows the capacity settings for the power plants ($\bar{G}_i^e$) in the GD by type. The total retail demand ($\bar{y}$) was also an exogenous variable, which was defined as 10 000 MW in this study. The retail margin rate γ was defined as 5%, which is a typical value for UK power companies.

18.4 DEA CALCULATION

As mentioned above, the profit, stability and competitiveness indices for each scenario may be evaluated differently; for example, the profit may be large, while the competitiveness may be small. In order to obtain a comprehensive evaluation, we applied DEA.

In this study, we used the SBM-max model (as presented in Chapter 22), which refers to the nearest point of the efficiency frontier in the SBM model, whereas the original SBM model [9] refers to the farthest point of the frontier, denoted by SBM-min in Figure 18.4. It can be said that the efficiency score in the SBM-max model is measured under the best conditions for the target decision-making unit (DMU).

The input-oriented SMB-min model is formulated as follows:

$$\theta_o^* = \min_{\lambda, s^-, s^+} 1 - \sum_{i=1}^{m} \frac{s_i^-}{x_{io}}$$

subject to

$$x_{io} = \sum_{j=1}^{n} x_{ij}\lambda_j + s_i^- \quad (i = 1, \ldots, m) \tag{18.15}$$

$$y_{ro} = \sum_{j=1}^{n} y_{rj}\lambda_j - s_r^+ \quad (r = 1, \ldots, s)$$

$$\lambda_j \geq 0 \ (\forall j), \ s_i^- \geq 0 \ (\forall i), s_r^+ \geq 0 \ (\forall r)$$

where x_{ij} and y_{rj} denote input i and output r for DMU$_j$, and s_i^-, s_r^+ and λ_j are the input and output slacks and the intensity variables, respectively. θ_o^* is an efficiency score referring to the farthest point of the frontier.

In contrast to this, the model to find the nearest point on the efficiency frontier is much more complicated, with several steps, including (18.15) as the first step. Details of the procedure are explained in Chapter 22.

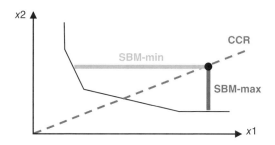

Figure 18.4 Comparison of three DEA models.

In this study, profit was regarded as an output, while stability (standard deviation of profits) and competitiveness (retail price level) were regarded as inputs, because they have better evaluations when their scores are small. The DMUs were the 900 scenarios for each fuel mix.

18.5   EMPIRICAL RESULTS

18.5.1   Results of Profit Maximization

Figure 18.5 plots the results for the three indices obtained with (18.13) for all three fuel mix cases; there are 2700 (=900 scenarios × 3 fuel mixes) dots in each part of the

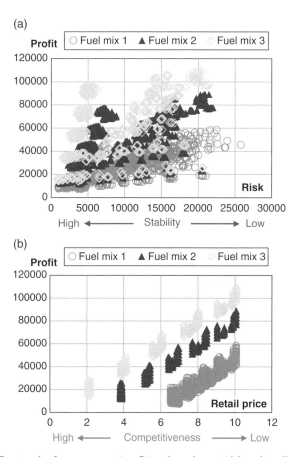

Figure 18.5   Scatter plot for one output (profit) and two inputs (risk and retail price) for all fuel mix cases: (a) Input 1 (risk) versus output (profit); (b) Input 2 (retail price) versus output (profit); (c) Input 1 (risk) versus Input 2 (retail price); (d) Input 1/output versus Input 2/output.

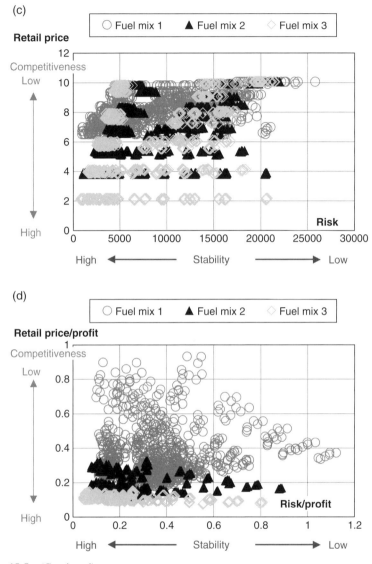

Figure 18.5 (Continued)

figure. In addition, both of the inputs (stability and competitiveness indices) are divided by the output (profit) in Figure 18.5(d). As the fuel price for hydro power is defined as 0, all indices for Fuel Mix 3 are better than those for the others (i.e. relatively larger profits, lower risks and lower retail prices), and vice versa for Fuel Mix 1, because of the strong dependence on gas power plants.

Figure 18.6 shows only the values for Fuel Mix 1, where the GD owns a large gas power plant capacity. The 900 dots shown in this figure are coloured differently based on the level of the constraints on internal transactions, $a1$ to $a5$ and $b1$ to $b5$. Intuitively, we find that the dark-coloured dots ($b1$) are relatively efficient.

Figure 18.7 shows the average of the three indices for the case of Fuel Mix 1 by level of constraint. To adjust the levels of the three indices, every result has been divided by the total average of all the constraints.

According to these figures, profit is larger under fewer constraints, while risk is lower (more stable) under stricter constraints; this is especially remarkable for the price constraints β. In addition, the volume constraints α have no influence on

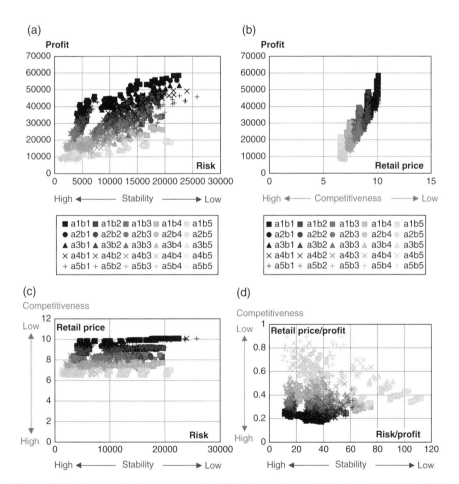

Figure 18.6 Scatter plot for one output (profit) and two inputs (risk and retail price) for Fuel Mix 1: (a) Input 1 (risk) versus output (profit); (b) Input 2 (retail price) versus output (profit); (c) Input 1 (risk) versus Input 2 (retail price); (d) Input 1/output versus Input 2/output.

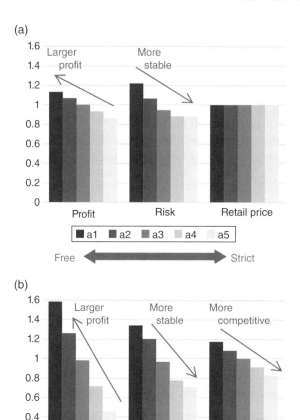

Figure 18.7   Average scores for each constraint for Fuel Mix 1: (a) average for each volume constraint; (b) average for each price constraint.

competitiveness, while strong price constraints β lead to higher competitiveness (lower price level).

As we assumed, the evaluations of these three indices are different for each case, and therefore the DEA method can help us to comprehensively evaluate them.

18.5.2   Results of DEA

Using the three indices (profit, stability and competitiveness), we solved the SBM-max model taking the 900 scenarios as DMUs. Figure 18.8 presents the average

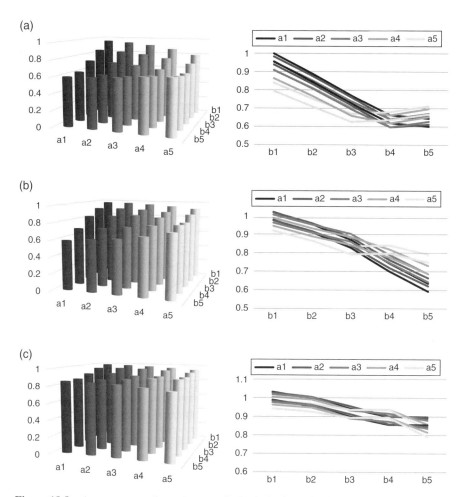

Figure 18.8   Average scores for each constraint by fuel mix: (a) Fuel Mix 1 (the capacity of the gas power plant is large); (b) Fuel Mix 2 (the capacity of the coal power plant is large); (c) Fuel Mix 3 (the capacity of the hydro power plant is large).

efficiency scores for each constraint level and each fuel mix. We find relatively little difference among the constraints in the case of Fuel Mix 3. This is attributable to the low generation cost for hydro power plants; therefore, they can generate electricity regardless of the market situation. This implies that the trading function will work more effectively in a company that owns many fossil-fuelled power plants.

In addition, it can be said that the constraint-free case ($a1$ and $b1$), which is just the case of the pure MO system, is the most efficient of all, and the system becomes less efficient as the constraints become stricter. However, in the strictest case ($a5$ and $b5$), which is the case of the pure VI system, the efficiency score can be better than that of several other combinations with fewer constraints for Fuel Mix 1 and 2.

(a)

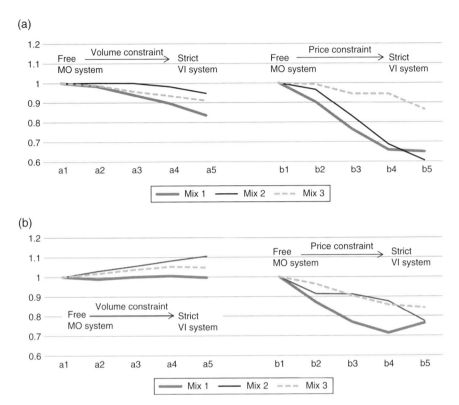

Figure 18.9 Comparison of efficiency scores for different constraints: (a) volatile case; (b) stable case.

Figure 18.9 focuses on the cases of volatility and stability in the market price of power, where the scores have been divided by the scores in the no-constraint case ($a1$ or $b1$) in order to compare the results under different constraints for all fuel mixes. In the volatile case, the MO system, with fewer constraints, performs better for every fuel mix. Therefore, if a company predicts that the price of power in the market will be volatile in the future, the MO system will be suitable, especially for a company that owns a large capacity of gas power plants.

18.6   CONCLUDING REMARKS

In this study, we have examined a quantitative analysis to evaluate the effects of the potential use of market opportunities through a TD. The results show that energy trading in an MO system without constraints on internal transactions will lead to a larger profit, but less stability of profits and less competitiveness in the retail market. We then applied a DEA model to obtain a comprehensive evaluation. According to the results, a TD will work effectively in a company that owns large fuel power plants and expects market prices to be volatile in the future.

In Japan, many incumbent companies depend heavily on fossil-fuelled power plants, as all of the nuclear power plants were shut down after the Great East Japan Earthquake in 2011. Even if many of them are restarted, fossil-fuelled power plants will continue to be an important energy source. The wholesale power market in Japan has not been particularly active so far; however, it may be revitalized in the near future in a similar way to the power markets in Europe. It will be a good option for Japanese power companies to establish TDs in order to act effectively in the fuel and power markets.

In future work, we will attempt to generate price data in a more realistic way, for instance by considering covariance between fuel and power prices. Furthermore, we will examine several different settings to obtain more robust results.

REFERENCES

[1] Jamasb, T. and Pollitt, M. (2000) Benchmarking and regulation: International electricity experience. *Utility Policy*, **9**(3), 107–130.

[2] Jamasb, T. and Pollitt, M. (2003) International benchmarking and regulation: An application to European electricity distribution utilities. *Energy Policy*, **31**(15), 1609–1622.

[3] Jamasb, T. and Nepal, R. (2014) Incentive regulation and utility benchmarking for electricity network security. Cambridge Working Papers in Economics 1434.

[4] Schweinsberg, A., Stronzik, M. and Wissner, M. (2011) Cost benchmarking in energy regulation in European countries. WIK-Consult, Final Report.

[5] Frontier Economics (2012) Trends in electricity distribution network regulation in North West Europe. Report prepared for Energy Norway, August 2012.

[6] Agrell, P. and Bogetoft, P. (2007) Development of benchmarking models for German electricity and gas distribution. Final report on the efficiency benchmarking model for electricity and gas distribution operators in Germany, for the regulator Bundesnetzagentur.

[7] Bjørndal, E., Bjørndal, M. and Camanho, A. (2009) Weight restrictions in the DEA benchmarking model for Norwegian electricity distribution companies: Size and structural variables. SNF Report 22/09, NHH Brage.

[8] Agrell, P. and Bogetoft, P. (2009) International benchmarking of electricity transmission system operators. e^{3}GRID Project, Final Report.

[9] Tone, K. (2001) A slacks-based measure of efficiency in data envelopment analysis. *European Journal of Operational Research*, **130**, 498–509.

19

DEA IN RESOURCE ALLOCATION[1]

MING-MIIN YU

Department of Transportation Science, National Taiwan Ocean University, Keelung, Taiwan

LI-HSUEH CHEN

Department of Transportation Science, National Taiwan Ocean University, Keelung, Taiwan

19.1 INTRODUCTION

As a part of resource management, resource allocation is based on a strategic plan for efficiently allocating available resources among various units to achieve future goals. These units may belong to the same organization and operate under a central decision-maker who has the power to control the allocation of resources to these units. For example, the central authorities of liner shipping companies control the allocation of human resources among dedicated container terminals in major international harbours. Previous studies have employed individual resource perspectives when analysing the operational efficiency of firms. Such techniques suggest identifying the number of resources to be reduced (or increased) for their particular input (or output) as a method to compare best practices. However, such models are not suitable for evaluating firms operated either by headquarters or by a central decision-making

---

[1] Part of the material in this chapter is adapted from Chang, S.M., Wang, J.S., Yu, M.M., Shang, K.C., Lin, S.H. and Hsiao, B., 2015, 'An application of centralized data envelopment analysis in resource allocation in container terminal operations', *Maritime Policy & Management*, **42**(8), 776–788, and Yu, M.M., Chern, C. C. and Hsiao, B., 2013, 'Human resource rightsizing using centralized data envelopment analysis: Evidence from Taiwan's airports', *Omega*, **41**(1), 119–130, with permission from Taylor & Francis and Elsevier Science.

Advances in DEA Theory and Applications: With Extensions to Forecasting Models,
First Edition. Edited by Kaoru Tone.
© 2017 John Wiley & Sons Ltd. Published 2017 by John Wiley & Sons Ltd.

controller. This kind of allocation follows the 'first-order-change' method.[2] More precisely, these studies focus on allocating resources to individual units, not on reallocating each unit from a centralized perspective. On a larger scale, enterprises should be able to focus using a 'centralized perspective' instead of an 'individual perspective' to allocate available resources and maximize organizational performance; this emphasizes the need to use a second-order-change method.[3] As such, if the goal is to achieve maximum output levels, information on best practices using an individual perspective cannot fully fit the resource reallocation requirements of specific firms.

To date, most studies on resource allocation efficiency have focused mainly on goal programming [1], dynamic programming [2], heuristic approaches [3], grey relation analysis [4], linear programming [5], data envelopment analysis (DEA) [6–10] and multiple-criteria decision making [11–13]. DEA is considered a suitable approach for organizing and analysing data, because it allows the improvement of efficiency over time and requires no prior assumptions about the specification of the best-practice frontier. The DEA method not only estimates performance, but also helps decision-making units (DMUs) remove other sources of inefficiency from the observations. This capability distinguishes DEA from other decision-making techniques. However, conventional DEA models project each DMU separately onto the efficient frontier. In a centralized decision-making environment, the aim of the central decision-maker is to optimize resource utilization by all DMUs in an organization rather than consider the resource consumption by each DMU separately. Hence, it is more reasonable to project all DMUs onto the efficient frontier by solving one model. In order to deal with such a situation, centralized DEA (CDEA) models have been proposed. Lozano *et al.* [14] and Lozano and Villa [15, 16] first introduced the concept of centralized resource allocation in an intra-organizational scenario. Since the development of CDEA models by Lozano *et al.* [14] and Lozano and Villa [15, 16], there have been a number of studies in the literature that deal with the problem of centralized resource allocation (e.g. [9, 10, 17–26]). These new DEA models consider the situation where there is a central decision-maker who supervises or 'owns' all of the operating units, and the total output and input are more important than the outputs and inputs for the individual units. CDEA is particularly relevant in situations where certain variables are controlled by a central authority, rather than by individual unit managers. With centralization, issues of overall system efficiency are resolved, rather than simply issues pertaining to individual levels.

Although sudden or generational changes can have huge effects on the operations of a firm, they can also cause organizational resistance and reduced performance

---

[2] First-order change implies changing individuals in a setting to adjust resources. In other words, by considering first-order change, changes to the resource itself, its perspectives and its direction designed to deliver better performance for a specific decision-making unit can be understood.

[3] Second-order-change means attending to systems and structures (i.e. an overall resources perspective) in problems relating to resource adjustment.

[9, 27, 28]. The use of several different strategies to evaluate the fluctuating objectives (i.e. to find a fit change policy) is more appropriate when resource reallocation is being considered. This chapter develops a systematic resource reallocation process to provide solutions that reduce organizational resistance. We will discuss three policies for resource reallocation. The first policy (referred to as the minor adjustment policy) states that the central decision-maker does not change the aggregated amounts of adjustable inputs. The second policy (referred to as the moderate adjustment policy) states that the aggregated amounts of some adjustable inputs can be reduced, but that others cannot be changed. The third policy (referred to as the major adjustment policy) states that the decision-maker will cut the aggregated amounts of adjustable inputs. In addition, this chapter provides some applications to illustrate how to allocate resources by applying CDEA models. The concept described in this chapter can be expanded to solve more complex problems.

This chapter is organized as follows. Following the introduction, the CDEA model for resource allocation is formulated. Then, some related applications in the transport industry are provided. Afterwards, the CDEA model is extended to deal with undesirable outputs. Finally, conclusions are drawn.

19.2   CENTRALIZED DEA IN RESOURCE ALLOCATION

Since the DMUs are under the supervision of a central decision-maker, a CDEA approach that takes into account an overall objective of the organization when allocating resources can be used. Following Lozano *et al.* [14] and Lozano and Villa [15, 16], a two-phase CDEA is constructed here. In the first phase, the organization tries to maximize the production of the aggregated output of all DMUs at a given level of any input. In the second phase, given the optimal solution for the aggregated-output expansion rate in the first phase, the minimized total slacks for the inputs are sought. In order to reduce organizational resistance, the minor adjustment policy, moderate adjustment policy and major adjustment policy are demonstrated in the second phase. The three resource adjustment policies are considered in the following scenarios. First, the minor adjustment policy assumes that the central decision-maker of the organization can transfer adjustable inputs among DMUs, but cannot change the aggregated amounts of adjustable inputs from their original aggregated level. Second, the moderate adjustment policy assumes that some of the adjustable inputs can be cut for each DMU and can be transferred among DMUs, while the others are allowed to be transferred among DMUs without changing their original aggregated level. The major adjustment policy assumes that all adjustable inputs can be reduced and transferred for maximum output efficiency. However, the individual amounts of the other non-adjustable inputs remain unchanged.

Before formulating the models, we shall introduce the notation to be used. Let n be the number of DMUs; m_a the number of adjustable inputs; m_{na} the number of non-adjustable inputs; s the number of outputs; j, k indices of a DMU; i_a an index of an adjustable input; i_{na} an index of a non-adjustable input; r an index of an output;

$x_{i_a j}$ the amount of input i_a consumed by DMU$_j$; $x_{i_{na} j}$ the amount of input i_{na} consumed by DMU$_j$; y_{rj} the amount of output r produced by DMU$_j$; θ_r the efficiency score of output r; $s_{i_a k}$ the slacks for the adjustable input i_1; and λ_{1k}, λ_{2k}, λ_{3k}, ..., λ_{jk} the vector of the intensity variable for projecting DMU$_k$.

19.2.1 Minor Adjustment

From the perspective of resource reallocation, the minor adjustment policy allows the central decision-maker to transfer adjustable inputs among DMUs at the original over-all level. As discussed previously, two phases are used in the minor adjustment policy. Phase I involves finding the output scalar values θ_r^{MI} ($r = 1, 2, ..., s$) maximized for the given levels of both aggregated adjustable inputs and individual adjustable inputs. Phase II seeks the maximized net slacks for the adjustable inputs, given the optimal solution for the aggregated-output expansion rate in the first phase. This model is built on the assumption of variable-returns-to-scale (VRS) characterizations with increasing returns to scale, constant returns to scale (CRS) and decreasing returns to scale.

We also have mentioned that the intensity variable has two subscripts in our model, similarly to conventional DEA. The major difference between CDEA and conventional DEA is that the former projects 'all' DMUs onto the frontier by using a DEA model, while the latter obtains the projection of each DMU by using the DEA model once. This means that CDEA generates all of the intensity-variable values for each DMU using a model, while conventional DEA uses the DEA model n times if $k = 1, ..., n$. The intensity variables for each DMU in CDEA are obtained by running the CDEA model one time only. In order to know the values of the intensity variables which each DMU$_k$ has, we must use each DMU$_k$ with its intensity variables λ_{jk} in the CDEA model. This implies that we will obtain all of the DMUs' intensity-variable values by running the CDEA model once. That is why the variable λ_{jk} has two subscripts in our model. The major difference between conventional DEA and CDEA models is that the former must run the DEA model n times, while the latter runs it once.

Furthermore, the rth output scalar value θ_r^{MI} can be obtained in the first-phase model for the minor adjustment policy by describing the ways in which the average sum of the expansion ratios of each output could be expanded. We evaluate the output-oriented efficiency of CDEA by solving the following linear program. This model yields a set of new output measures that render the output efficient.

19.2.1.1 Phase I Find the maximum average efficiency scores under the content of the minor adjustment policy:

$$\max \frac{1}{s}\sum_{r=1}^{s}\theta_r^{MI} \tag{19.1}$$

$$\text{s.t.}\sum_{k=1}^{n}\sum_{j=1}^{n}\lambda_{jk}x_{i_a j}=\sum_{k=1}^{n}x_{i_a k}, i_a=1,2,...,m_a \tag{19.2}$$

$$\sum_{j=1}^{n} \lambda_{jk} x_{i_a j} \leq x_{i_a k}, i_a = 1, 2, \ldots, m_a, k = 1, \ldots, n \tag{19.3}$$

$$\sum_{j=1}^{n} \lambda_{jk} x_{i_{na} j} \leq x_{i_{na} k}, i_{na} = 1, 2, \ldots, m_{na}, k = 1, \ldots, n \tag{19.4}$$

$$\sum_{k=1}^{n} \sum_{j=1}^{n} \lambda_{jk} y_{rj} \geq \theta_r^{MI} \sum_{k=1}^{n} y_{rk}, r = 1, 2, \ldots, s \tag{19.5}$$

$$\sum_{j=1}^{n} \lambda_{jk} y_{rj} \geq y_{rk}, r = 1, 2, \ldots, s, k = 1, \ldots, n \tag{19.6}$$

$$\sum_{j=1}^{n} \lambda_{jk} = 1, k = 1, \ldots, n \tag{19.7}$$

$$\lambda_{jk} \geq 0, \quad j = 1, \ldots, n, \quad k = 1, \ldots, n \tag{19.8}$$

Equation (19.1) seeks the optimum expansion of the aggregated output of all DMUs. The adjustable-input constraints, as shown in (19.2) and (19.3), allow the adjustable inputs to be transferred among DMUs. Furthermore, (19.2) limits the aggregated amount of adjustable inputs to the original level. Equation (19.3) ensures that the frontier number of inputs will be no larger than the observed level. Equation (19.4), which represents the reference point, is a linear combination of DMUs. Thus, the reference set may include DMUs that operate with a different amount of non-adjustable inputs from the assessed DMU. These non-adjustable input variables should appear as inequality constraints in a way similar to how Banker and Morey [29] proposed to handle non-discretionary inputs [30]. Equation (19.5) seeks to non-radially increase each output as much as possible, and ensures that each output remains in the feasible aggregated output set. The constraints in (19.5) ensure that these n projected points cannot lie outside the aggregated output set. The projected point of DMU_k is a linear combination of observed production points using λ_{jk}. Besides the aggregated output constraints, (19.6) imposes the restriction that the projected point for each DMU will be no less than the observed output quantities of DMU_k. Equation (19.6) guarantees that the above condition can be satisfied. In other words, the value of the expansion rate measures the maximum expansion of the total aggregated output required to bring it to the aggregated frontier of the output set for the input vector. Equation (19.7) shows that VRS is adopted in this model and any intensity variable used to project the DMU cannot be less than zero, as shown in (19.8).

19.2.1.2 Phase II   The optimal slack values of the adjustable inputs represent the quantities of inputs that can be transferred for each DMU at the given total levels of non-adjustable inputs for each DMU, as well as the maximum aggregated outputs that can be achieved. Hence, we find the slack variables of the adjustable inputs for the maximum aggregated outputs obtained in Phase I:

$$\max \sum_{i_a=1}^{m_a}\sum_{k=1}^{n}\frac{\left(s_{i_ak}^{\mathrm{MI}-}-s_{i_ak}^{\mathrm{MI}+}\right)}{x_{i_ak}} \tag{19.9}$$

$$\text{s.t.}\sum_{k=1}^{n}\sum_{j=1}^{n}\lambda_{jk}x_{i_aj}=\sum_{k=1}^{n}x_{i_ak},i_a=1,2,\ldots,m_a \tag{19.10}$$

$$\sum_{j=1}^{n}\lambda_{jk}x_{i_aj}=x_{i_ak}+s_{i_ak}^{\mathrm{MI}+}-s_{i_ak}^{\mathrm{MI}-},i_a=1,2,\ldots,m_a,k=1,\ldots,n \tag{19.11}$$

$$\sum_{j=1}^{n}\lambda_{jk}x_{i_{na}j}\le x_{i_{na}k},i_{na}=1,2,\ldots,m_{na},k=1,\ldots,n \tag{19.12}$$

$$\sum_{k=1}^{n}\sum_{j=1}^{n}\lambda_{jk}y_{rj}=\theta_r^{\mathrm{MI}*}\sum_{k=1}^{n}y_{rk},r=1,2,\ldots,s \tag{19.13}$$

$$\sum_{j=1}^{n}\lambda_{jk}y_{rj}\ge y_{rk},r=1,2,\ldots,s,k=1,\ldots,n \tag{19.14}$$

$$\sum_{j=1}^{n}\lambda_{jk}=1,k=1,\ldots,n \tag{19.15}$$

$$\lambda_{jk}\ge0,\ j=1,\ldots,n,\ \ k=1,\ldots,n \tag{19.16}$$

$$s_{i_ak}^{\mathrm{MI}-},s_{i_ak}^{\mathrm{MI}+}\ge0,\ \ i_a=1,\ldots,m_a,\ \ k=1,\ldots,n \tag{19.17}$$

Equations (19.10) and (19.11) are imposed to restrict the individual input sets to their original levels, and allow transfers into or out from other DMUs. Furthermore, (19.10) ensures that the aggregated amounts of adjustable inputs are equal to their original levels. That is, (19.10) implies that the total slack of the input transferred to all DMUs ($s_{i_ak}^{\mathrm{MI}+}$) should be equal to the total slack of the input transferred from all DMUs ($s_{i_ak}^{\mathrm{MI}-}$). Equation (19.13) can be seen as using the reallocation perspective to reach the ideal output under the CDEA perspective, and $\theta_r^{\mathrm{MI}*}$ is the optimum of Phase I. Furthermore, the constraints on the output variables should satisfy two conditions. First, the rth output expansion should be equal to $\theta_r^{\mathrm{MI}*}$. Second, the rth output level after aggregation cannot be less than the level of the rth output before planning. A restriction, (19.14), is added to guarantee that the above condition can be satisfied. In other words, the constraints ensure no worsening of the outputs appears in this model compared with the first model. In (19.13), the constraint $\sum_{k=1}^{n}\sum_{j=1}^{n}\lambda_{jk}y_{rj}=\theta_r^{\mathrm{MI}*}\sum_{k=1}^{n}y_{rk}$ is imposed instead of $\sum_{k=1}^{n}\sum_{j=1}^{n}\lambda_{jk}y_{rj}\ge\theta_r^{\mathrm{MI}*}\sum_{k=1}^{n}y_{rk}$, which implies that the total output obtained in Phase I is given at the maximum level.

Equations (19.12), (19.14) and (19.15) can be compared with (19.4), (19.6) and (19.7), respectively. Equations (19.16) and (19.17) show that the slacks and any vector for projecting the DMU cannot be less than zero.

The second-phase model provides information about the total reduction and the increases in adjustable inputs for each DMU, and the optimal slack variables $(s_{i_a k}^{MI+*}, s_{i_a k}^{MI-*})$ are obtained. The minor adjustment policy focuses on the net slack values of adjustable inputs equal to zero, as shown in (19.10). Therefore, the minor adjustment policy is concentrated on the adjustment of adjustable inputs. Thus, in the minor adjustment policy, for a specific DMU_k, the values of the adjustable inputs are given as

$$x_{i_a k}^{MI*} = x_{i_a k} + s_{i_a k}^{MI+*} - s_{i_a k}^{MI-*}, i_a = 1, 2, \ldots, m_a \tag{19.18}$$

19.2.2 Moderate Adjustment

The analysis of the moderate adjustment policy is concerned mainly with the reduction of some adjustable inputs, given that the other adjustable inputs cannot be changed, but the numbers of the other adjustable inputs for each DMU can be transferred. Since some of the adjustable inputs can be reduced and others cannot be changed, the adjustable inputs, $x_{i_a j}$ $(i_a = 1, 2, \ldots, m_a)$, can be divided further into two parts: the reducible inputs, $x_{i_a^r j}^r$ $(i_a^r = 1, 2, \ldots, m_a^r)$, and the non-reducible inputs, $x_{i_a^{nr} j}^{nr}$ $(i_a^{nr} = 1, 2, \ldots, m_a^{nr})$, where $m_a = m_a^r + m_a^{nr}$. There are also two phases involved when performing moderate-adjustment-policy analysis. Similarly to the minor adjustment policy in Phase I, the objective is to identify the maximum average expansion rate θ_r^{MO} $(r = 1, 2, \ldots, s)$ of the output quantities.

19.2.2.1 Phase I Find the maximum system efficiency score under the content of the moderate adjustment policy:

$$\max \frac{1}{s} \sum_{r=1}^{s} \theta_r^{MO} \tag{19.19}$$

$$\text{s.t.} \sum_{k=1}^{n} \sum_{j=1}^{n} \lambda_{jk} x_{i_a^{nr} j}^{nr} = \sum_{k=1}^{n} x_{i_a^{nr} k}^{nr}, i_a^{nr} = 1, 2, \ldots, m_a^{nr} \tag{19.20}$$

$$\sum_{j=1}^{n} \lambda_{jk} x_{i_a^{nr} j}^{nr} \le x_{i_a^{nr} k}^{nr}, i_a^{nr} = 1, 2, \ldots, m_a^{nr}, k = 1, \ldots, n \tag{19.21}$$

$$\sum_{k=1}^{n} \sum_{j=1}^{n} \lambda_{jk} x_{i_a^r j}^r \le \sum_{k=1}^{n} x_{i_a^r k}^r, i_a^r = 1, 2, \ldots, m_a^r \tag{19.22}$$

$$\sum_{j=1}^{n} \lambda_{jk} x_{i_a^r j}^r \le x_{i_a^r k}^r, i_a^r = 1, 2, \ldots, m_a^r, k = 1, \ldots, n \tag{19.23}$$

$$\sum_{j=1}^{n} \lambda_{jk} x_{i_{na}j} \le x_{i_{na}k}, i_{na} = 1, 2, \ldots, m_{na}, k = 1, \ldots, n \tag{19.24}$$

$$\sum_{k=1}^{n} \sum_{j=1}^{n} \lambda_{jk} y_{rj} \ge \theta_r^{MO} \sum_{k=1}^{n} y_{rk}, r = 1, 2, \ldots, s \tag{19.25}$$

$$\sum_{j=1}^{n} \lambda_{jk} y_{rj} \ge y_{rk}, r = 1, 2, \ldots, s, k = 1, \ldots, n \tag{19.26}$$

$$\sum_{j=1}^{n} \lambda_{jk} = 1, k = 1, \ldots, n \tag{19.27}$$

$$\lambda_{jk} \ge 0, \quad j = 1, \ldots, n, \quad k = 1, \ldots, n \tag{19.28}$$

Unlike the first-phase model for the minor adjustment policy, the constraint on the aggregated reducible inputs in this model is in an inequality form, because the aggregated amounts of reducible inputs can be changed in this model, indicating that the reducible inputs should be less than or equal to their original aggregated levels, as shown in (19.22). The non-reducible inputs are unchanged in the aggregated perspective; hence, the equality constraint is retained in (19.20). Equation (19.23) ensures that the frontier amounts of reducible inputs will be no larger than the observed levels. This model is similar to the first-phase model for the minor adjustment policy, and (19.19)–(19.21) and (19.24)–(19.28) can be compared to (19.1)–(19.8), respectively.

The rth output scalar value θ_r^{MO*} obtained in the third model is used further in the constraints on the rth output in the next model to calculate the required slack values. The objective in Phase II of the moderate adjustment policy is to determine the maximum number of reducible inputs that can be reduced, and the minimum number of non-reducible inputs that can be transferred, without changing the total number of non-reducible inputs. Thus, only the slack of the reducible inputs of all DMUs in the objective function is shown in (19.29).

19.2.2.2 Phase II Find the total slacks of both the reducible and the non-reducible inputs of all DMUs:

$$\max \sum_{i_a^r = 1}^{m_a^r} \sum_{k=1}^{n} \frac{\left(s_{i_a^t k}^{MO-} - s_{i_a^t k}^{MO+} \right)}{x_{i_a^t k}^r} \tag{19.29}$$

$$\text{s.t.} \sum_{k=1}^{n} \sum_{j=1}^{n} \lambda_{jk} x_{i_a^{nr}j}^{nr} = \sum_{k=1}^{n} x_{i_a^{nr}k}^{nr}, i_a^{nr} = 1, 2, \ldots, m_a^{nr} \tag{19.30}$$

$$\sum_{j=1}^{n} \lambda_{jk} x_{i_a^{nr}j}^{nr} = x_{i_a^{nr}k}^{nr} + s_{i_a^{nr}k}^{MO+} - s_{i_a^{nr}k}^{MO-}, i_a^{nr} = 1, 2, \ldots, m_a^{nr}, k = 1, \ldots, n \tag{19.31}$$

$$\sum_{k=1}^{n} \sum_{j=1}^{n} \lambda_{jk} x_{i_a^t j}^r = \sum_{k=1}^{n} \left(x_{i_a^t k}^r + s_{i_a^t k}^{MO+} - s_{i_a^t k}^{MO-} \right), i_a^r = 1, 2, \ldots, m_a^r \tag{19.32}$$

$$\sum_{j=1}^{n} \lambda_{jk} x_{i_a^r j}^{r} = x_{i_a^r k}^{r} + s_{i_a^r k}^{\mathrm{MO}+} - s_{i_a^r k}^{\mathrm{MO}-}, i_a^r = 1, 2, \ldots, m_a^r, k = 1, \ldots, n \tag{19.33}$$

$$\sum_{k=1}^{n} s_{i_a^r k}^{\mathrm{MO}+} \leq \sum_{k=1}^{n} s_{i_a^r k}^{\mathrm{MO}-}, i_a^r = 1, 2, \ldots, m_a^r \tag{19.34}$$

$$\sum_{j=1}^{n} \lambda_{jk} x_{i_{na} j} \leq x_{i_{na} k}, i_{na} = 1, 2, \ldots, m_{na}, k = 1, \ldots, n \tag{19.35}$$

$$\sum_{k=1}^{n} \sum_{j=1}^{n} \lambda_{jk} y_{rj} = \theta_r^{\mathrm{MO}*} \sum_{k=1}^{n} y_{rk}, r = 1, 2, \ldots, s \tag{19.36}$$

$$\sum_{j=1}^{n} \lambda_{jk} y_{rj} \geq y_{rk}, r = 1, 2, \ldots, s, k = 1, \ldots, n \tag{19.37}$$

$$\sum_{j=1}^{n} \lambda_{jk} = 1, k = 1, \ldots, n \tag{19.38}$$

$$\lambda_{jk} \geq 0, \quad j = 1, \ldots, n, \quad k = 1, \ldots, n \tag{19.39}$$

$$s_{i_a^r k}^{\mathrm{MO}-}, s_{i_a^{nr} k}^{\mathrm{MO}+}, s_{i_a^r k}^{\mathrm{MO}-}, s_{i_a^r k}^{\mathrm{MO}+} \geq 0, \quad i_a^{nr} = 1, \ldots, m_a^{nr}, i_a^r = 1, \ldots, m_a^r, k = 1, \ldots, n \tag{19.40}$$

For a specific DMU_k, the optimal slacks of the non-reducible inputs $(s_{i_a^{nr} k}^{\mathrm{MO}+})$ imply that some amounts of non-reducible inputs $(s_{i_a^{nr} k}^{\mathrm{MO}+})$ should be transferred from other DMUs to DMU_k. Otherwise, amounts of $s_{i_a^{nr} k}^{\mathrm{MO}-}$ should be transferred to other DMUs. The total slack of a specific non-reducible input transferred to all DMUs should be equal to the total slack of that non-reducible input transferred from all DMUs, as shown in (19.30). Equation (19.31) is added to guarantee that the non-reducible inputs for each DMU under the individual perspective equal the original value with a difference of $(s_{i_a^{nr} k}^{\mathrm{MO}+} - s_{i_a^{nr} k}^{\mathrm{MO}-})$. The constraints of (19.32) help determine if the current aggregated amounts of reducible inputs are appropriate from the centralized perspective.[4] Equation (19.33) is imposed to restrict the individual input sets. Equations (19.33) and (19.34) imply that the frontier amounts of reducible inputs will be no larger than the observed amounts of reducible inputs for DMU_k. Furthermore, (19.34) ensures that the total slacks of the reducible inputs transferred to all DMUs $(s_{i_a^r k}^{\mathrm{MO}+})$ should not be larger than the total slacks of the reducible inputs transferred from all DMUs $(s_{i_a^r k}^{\mathrm{MO}-})$. Reducible inputs can be cut from their current level, meaning that the aggregated amounts of reducible inputs should be less than or equal to their original levels

---

[4] Since the constraints in (19.32) are the sum of all the constraints in (19.33), they can be considered as redundant.

after resource reallocation. Equation (19.40) indicates that the slacks of the reducible and non-reducible inputs cannot be less than zero. Similarly, (19.29)–(19.30) and (19.35)–(19.39) can be compared to (19.9)–(19.17), respectively.

After re-evaluating Phase II, for a specific DMU_k, the values of the non-reducible inputs are given as

$$x_{i_a^{nr}k}^{MO*} = x_{i_a^{nr}k}^{nr} + s_{i_a^{nr}k}^{MO+*} - s_{i_a^{nr}k}^{MO-*}, i_a^{nr} = 1, 2, \ldots, m_a^{nr} \tag{19.41}$$

In other words, the values of $s_{i_a^{nr}k}^{MO+*}$ and $s_{i_a^{nr}k}^{MO-*}$ represent how many non-reducible inputs can be increased and decreased, respectively, with the slack variables of the reducible inputs as an analogy as follows:

$$x_{i_a^r k}^{MO*} = x_{i_a^r k}^{r} + s_{i_a^r k}^{MO+*} - s_{i_a^r k}^{MO-*}, i_a^r = 1, 2, \ldots, m_a^r \tag{19.42}$$

19.2.3 Major Adjustment

The analysis of the major adjustment policy is mainly concerned with the reduction of adjustable inputs, while increasing output levels. Two phases are also evaluated in the major-adjustment-policy analysis. Phase I involves finding the maximum average expansion rate θ_r^{MA} $(r = 1, 2, \ldots, s)$ of the output quantities.

19.2.3.1 Phase I Find the maximum efficiency scores under the content of the major adjustment policy:

$$\max \frac{1}{s} \sum_{r=1}^{s} \theta_r^{MA} \tag{19.43}$$

$$\text{s.t.} \sum_{k=1}^{n} \sum_{j=1}^{n} \lambda_{jk} x_{i_a j} \leq \sum_{k=1}^{n} x_{i_a k}, i_a = 1, 2, \ldots, m_a \tag{19.44}$$

$$\sum_{j=1}^{n} \lambda_{jk} x_{i_a j} \leq x_{i_a k}, i_a = 1, 2, \ldots, m_a, k = 1, \ldots, n \tag{19.45}$$

$$\sum_{j=1}^{n} \lambda_{jk} x_{i_{na} j} \leq x_{i_{na} k}, i_a = 1, 2, \ldots, m_a, k = 1, \ldots, n \tag{19.46}$$

$$\sum_{k=1}^{n} \sum_{j=1}^{n} \lambda_{jk} y_{rj} \geq \theta_r^{MA} \sum_{k=1}^{n} y_{rk}, r = 1, 2, \ldots, s \tag{19.47}$$

$$\sum_{j=1}^{n} \lambda_{jk} y_{rj} \geq y_{rk}, r = 1, 2, \ldots, s, k = 1, \ldots, n \tag{19.48}$$

$$\sum_{j=1}^{n} \lambda_{jk} = 1, k = 1, \ldots, n \tag{19.49}$$

$$\lambda_{jk} \geq 0, \quad j = 1, \ldots, n, \quad k = 1, \ldots, n \tag{19.50}$$

Equations (19.44) and (19.45) are similar to (19.22) and (19.23), implying that, under the major adjustment policy, all adjustable inputs can be cut. The constraints of (19.43) and (19.48)–(19.50) are comparable to the constraints of (19.1) and (19.4)–(19.8).

To find the maximum amounts of adjustable inputs that can be reduced and the minimum amounts of adjustable inputs that can be transferred, the objective function of Phase II for the major adjustment policy is treated as the maximum net slacks of all adjustable inputs. The optimal slack values of each DMU's adjustable inputs ($s_{i_a k}^{\mathrm{MA}+*}$ and $s_{i_a k}^{\mathrm{MA}-*}$) can be obtained by solving the next model, which represents the amounts of adjustable inputs that can be reduced or transferred for each DMU.

19.2.3.2 *Phase II* Find the slack values of all adjustable inputs:

$$\max \sum_{i_a = 1}^{m_a} \sum_{k=1}^{n} \frac{\left(s_{i_a k}^{\mathrm{MA}-} - s_{i_a k}^{\mathrm{MA}+} \right)}{x_{i_a k}} \tag{19.51}$$

$$\text{s.t.} \sum_{k=1}^{n} \sum_{j=1}^{n} \lambda_{jk} x_{i_a j} = \sum_{k=1}^{n} \left(x_{i_a k} + s_{i_a k}^{\mathrm{MA}+} - s_{i_a k}^{\mathrm{MA}-} \right), i_a = 1, 2, \ldots, m_a \tag{19.52}$$

$$\sum_{j=1}^{n} \lambda_{jk} x_{i_a j} = x_{i_a k} + s_{i_a k}^{\mathrm{MA}+} - s_{i_a k}^{\mathrm{MA}-}, i_a = 1, 2, \ldots, m_a, k = 1, \ldots, n \tag{19.53}$$

$$\sum_{k=1}^{n} s_{i_a k}^{\mathrm{MA}+} \leq \sum_{k=1}^{n} s_{i_a k}^{\mathrm{MA}-}, i_1 = 1, 2, \ldots, m_1 \tag{19.54}$$

$$\sum_{j=1}^{n} \lambda_{jk} x_{i_{na} j} \leq x_{i_{na} k}, i_{na} = 1, 2, \ldots, m_{na}, k = 1, \ldots, n \tag{19.55}$$

$$\sum_{k=1}^{n} \sum_{j=1}^{n} \lambda_{jk} y_{rj} = \theta_r^{\mathrm{MA}*} \sum_{k=1}^{n} y_{rk}, r = 1, 2, \ldots, s \tag{19.56}$$

$$\sum_{j=1}^{n} \lambda_{jk} y_{rj} \geq y_{rk}, r = 1, 2, \ldots, s, k = 1, \ldots, n \tag{19.57}$$

$$\sum_{j=1}^{n} \lambda_{jk} = 1, k = 1, \ldots, n \tag{19.58}$$

$$\lambda_{jk} \geq 0, \quad j = 1, \ldots, n, \quad k = 1, \ldots, n \tag{19.59}$$

$$s_{i_a k}^{\mathrm{MA}-}, s_{i_a k}^{\mathrm{MA}+} \geq 0, \quad i_a = 1, \ldots, m_a, \quad k = 1, \ldots, n \tag{19.60}$$

Equations (19.52)–(19.54) are also similar to (19.32)–(19.34), indicating that the amounts of adjustable inputs after resource allocation can be less than or equal to their original levels.[5] Finally, (19.51) and (19.55)–(19.60) can also be compared to (19.9) and (19.12)–(19.17), respectively.

Under the major adjustment policy, for a specific DMU_k, the values of adjustable inputs are given as

$$x_{i_a k}^{MA*} = x_{i_a k} + s_{i_a k}^{MA+*} - s_{i_a k}^{MA-*}, i_a = 1, 2, \ldots, m_a \qquad (19.61)$$

The values of $s_{i_a k}^{MA+*}$ and $s_{i_a k}^{MA-*}$ represent how many adjustable inputs can be increased ($s_{i_a k}^{MA+*}$) or decreased ($s_{i_a k}^{MA-*}$), respectively, for each DMU_k.

19.3 APPLICATIONS OF CENTRALIZED DEA IN RESOURCE ALLOCATION

This section presents two related applications in empirical studies of the transport industry. First, human resource rightsizing in airports will be illustrated. Second, resource allocation in container terminal operations will be explored.

19.3.1 Human Resource Rightsizing in Airports[6]

We present an example based on 18 Taiwanese airports controlled by the Taiwan Civil Aeronautics Administration (CAA) to investigate human resource rightsizing for regular and contracted employees. Regular employees of the Taiwan CAA are certified by national examinations; thus, their positions are protected by official employment laws. They cannot be dismissed from their jobs, except when they have violated laws or regulations. In contrast, contracted employees have employment periods of just one year. Following the findings of Hitt *et al.* [31], reduced contracted employee quotas seem to cause lower resistance than dismissing regular employees. Hence, three policies for manpower reallocation strategies will be discussed. The three strategies are:

1. *Long-term policy.* The CAA can reduce and transfer all contracted and regular manpower.
2. *Middle-term policy.* The CAA can cut the amount of contracted manpower in each airport and transfer contracted manpower among airports, while reduction of total regular manpower is disallowed.

---

[5] Since the constraints of (19.52) are the sum of all the constraints of (19.53), they can be considered as redundant.
[6] Adapted from Yu *et al.* [9].

3. *Short-term policy.* The CAA cannot change the amount of regular manpower in each airport from its original level and must maintain the total amount of contracted manpower unchanged. However, airports are allowed to transfer their contracted manpower to other airports without changing the aggregated amount of contracted manpower from the original level.

In addition, some assumptions and terms need to be clarified. First, owing to pre-emptive constraints (e.g. qualifications), contracted employees cannot replace regular employees, and vice versa; thus, there is no overlap in the assigned work of regular and contracted employees. Second, there is no cost for transferred and/or dismissed work of regular and contracted employees. Third, except for both regular and contracted employees, who must be represented by integer values, other input factors can be represented by non-integer values (i.e. this implies that the regular and contracted employees are non-separate resources). Fourth, only the output vector, as well as the numbers of contracted and regular employees, can be used in the aggregated view for analysis. This aggregated view refers to the utilization of the centralized perspective to sum up a specific resource for multiple airports and aggregate it into a centralized view. Taking this perspective, facility input variables might be treated as non-discretionary variables, with a special environmental variable that constrains manpower variables without any modifications to the three policies.

19.3.1.1 Input and Output Variables

The input and output variables of an airport that are adopted in this example are as follows:

1. *Adjustable inputs*: regular employees and contracted employees.
2. *Non-adjustable inputs*: runway areas, apron areas and terminal areas.
3. *Outputs*: flights, passengers and tons of cargo.

19.3.1.2 Numerical Results

The results of the long-, middle- and short-term policy analyses are summarized in Table 19.1. Under the long-term policy, the CAA has room to lay off 54 regular employees and 143 contracted employees without affecting the production of the maximum outputs. From a manpower perspective, only eight airports (Airports 1, 3, 6, 10–12, 14 and 16) show no changes in both inputs. Two airports (Airports 17 and 18) need to increase the numbers of both contracted employees and regular employees, while six of the 18 airports require a reduction of both regular and contracted employees (Airports 2, 4, 5, 7, 8 and 9). Two airports (Airports 13 and 15) need to increase regular employees, but decrease contracted employees.

Under the middle-term policy, the CAA could lay off 65 contracted employees for maximum output performance. These 65 contracted employees could be sourced from Airports 4, 5, 7, 8, 9, 13 and 15, which need to reduce their number of contracted employees by 68. Airports 16 and 17 need to increase their number of contracted employees by three; these could be transferred from Airports 4, 5, 7, 8, 9,

TABLE 19.1 Comparison of slack values for the three policies.[a]

Airport	Short-term policy		Middle-term policy		Long-term policy	
	Δx_r	Δx_c	Δx_r	Δx_c	Δx_r	Δx_c
1	0	0	0	0	0	0
2	0	0	0	0	-32	-70
3	0	0	0	0	0	0
4	0	0	1	-13	-10	-23
5	0	0	-3	-20	-3	-17
6	0	0	0	0	0	0
7	0	0	1	-3	-2	-4
8	0	0	-3	-8	-9	-10
9	0	0	-4	-22	-4	-22
10	0	0	0	0	0	0
11	0	0	0	0	0	0
12	0	0	0	0	0	0
13	0	0	0	-1	1	-1
14	0	0	0	0	0	0
15	0	2	1	-1	1	-1
16	0	0	2	2	0	0
17	0	-2	5	1	3	3
18	0	0	0	0	1	2
Total	0	0	0	-65	-54	-143

[a] Δx_r is the variation in the number of regular employees, and Δx_c is the variation in the number of contracted employees.

13 and 15. In addition, from the resource exchange perspective, for regular employees, Airports 5, 8 and 9 could transfer 10 employees to airports 4, 7, 15, 16 and 17.

Finally, under the short-term policy, the total numbers of both contracted and regular employees are fixed at their original levels, with only the number of contracted employees at some airports exhibiting change (Airports 15 and 17), indicating that any resistance that there may be to manpower adjustments will be caused largely by the short-term policy. The short-term analysis results indicate that Airport 17 can transfer two contracted employees to Airport 15. They also indicate the variation in the number of contracted employees for each airport; however, the results are indifferent to the specific airports involved in the transfer of employees.

Furthermore, the results of the three policies for the slacks are compared here. Table 19.1 also represents a fit change policy (i.e. a change from the short- and middle-term policies to the long-term policy), in addition to a rapid-change downsizing policy. The slacks of the long-term employee levels are generally larger than those of the middle- and short-term employee levels. If a rapid-change policy is conducted, a large number of contracted and regular staff will lose their jobs, possibly resulting in larger organizational resistance. In contrast, if a fit change policy is adopted, staff are reduced or adjusted smoothly and systematically (in a short-, middle- and long-term way), resulting in lower organizational resistance.

19.3.2   Resource Allocation in Container Terminal Operations[7]

To illustrate resource allocation by use of a CDEA model, an example based on five dedicated terminals supervised by a specific liner shipping company, which is one of the world's top 20 liner shipping companies, is used here. Two (named A and B) of the five dedicated terminals are in America, two (named D and E) are in Asia and one is in Europe (named C). Data related to the year 2011 were obtained from the five container terminals. Two strategies are considered.

1. *Minor adjustment policy*. The liner shipping company lets both labour and hauling equipment be transferred among terminals, and cannot change the aggregated amount of hauling equipment from its original aggregated level, but only allows the aggregated amount of labour to be reduced.
2. *Major adjustment policy*. Both labour and hauling equipment are transferable among terminals, and the aggregated amounts of labour and hauling equipment can be reduced.

However, the model for the minor adjustment policy is built on the assumption of VRS, while the model for the major adjustment policy is built on the assumption of CRS.

19.3.2.1   *Input and Output Variables*   The input and output variables for a container terminal that are used in this example are as follows:

1. *Adjustable inputs*: labour and hauling equipment.
2. *Non-adjustable inputs*: quay gantry cranes and marshalling yard.
3. *Output*: container throughput in twenty-foot equivalent units (TEU).

19.3.2.2   *Numerical Results*   The results of the analyses of the minor and major adjustment policies are shown in Table 19.2. Under the minor adjustment policy, the aggregated amount of the labour cost needs to be reduced by 46 984 689 USD without affecting the production of the maximum outputs. Terminal A requires a reduction in labour cost by 38 957 548 USD, and terminal B could decrease its labour cost by 8 027 141 USD. Although there is no need for the total number of pieces of hauling equipment to change, the hauling equipment needs to be reallocated among the terminals, as also shown in Table 19.2. The shipping company can maximize production in terms of the aggregated output for all terminals by transferring hauling equipment from terminal A to terminal B. Terminal A requires a reduction in hauling equipment by nine items, and terminal B should increase its hauling equipment by nine items, with terminals C, D and E needing no changes in labour and hauling equipment.

[7] Adapted from Chang *et al.* [10].

TABLE 19.2 **Comparison of slack values for the two policies.**[a]

Terminal	Minor adjustment policy		Major adjustment policy	
	Δx_l (USD)	Δx_h (items)	Δx_l (USD)	Δx_h (items)
A	−38 957 548	−9	−116 649 264	−28
B	−8 027 141	9	−4 928 290	5
C	0	0	0	0
D	0	0	0	0
E	0	0	0	0
Total	−46 984 689	0	−121 577 554	−23

[a] Δx_l is the variation in the number of labourers, and Δx_h is the variation in the number of items of hauling equipment.

Under the major adjustment policy, both labour and hauling equipment need to be reduced in order to maximize production in terms of the aggregated output of all terminals efficiently. Without affecting the production of the maximum outputs, the aggregated amount of the labour cost should be reduced by 121 577 554 USD, while the number of pieces of hauling equipment needs to be reduced by 23. There is a great need for terminal A to reduce its labour cost by 116 649 264 USD, and terminal B requires a reduction in its labour cost by 4 928 290 USD. The hauling equipment needs to be reallocated among the terminals: as shown in Table 19.2, terminal A requires a reduction in hauling equipment by 28 pieces, and terminal B is in need of five items.

In summary, the operations of terminals C, D and E are the most efficient. This means that the resource utilization of terminals C, D and E is at the optimum, based on both the minor and the major adjustment scenarios. The liner shipping company does not need to adjust any resources in them. Terminal A is the most inefficient terminal under the two scenarios. In order to improve the overall efficiency of the five terminals, the shipping company should reduce the resources in Terminal A. In terminal B, labour needs to be reduced, but the hauling equipment needs to be increased.

19.4 EXTENSION

In the above sections, we have assumed that all outputs are desirable. However, in many real situations, transport organizations will produce desirable and undesirable outputs simultaneously. For example, aircraft noise is a kind of pollution produced by an aircraft or its components, and has impacts on the communities surrounding an airport [32]. In order to deal with undesirable outputs, the original CDEA model needs to be modified to consider the trade-off between the utilization of desirable outputs and the control of undesirable outputs.

We use the Russell directional distance function (RDDF) to incorporate undesirable outputs into the CDEA model for resource allocation. Before formulating the new models, the outputs, y_{rj} $(r = 1, \ldots, s)$, must be divided further into desirable outputs, $y_{r_d j}$ $(r_d = 1, \ldots, s_d)$, and undesirable outputs, $y_{r_{nd} j}$ $(r_{nd} = 1, \ldots, s_{nd})$, where $s = s_d + s_{nd}$.

In addition, some more notation must be added. Let δ_{r_d} be the inflation of desirable output r_d, and $\varphi_{r_{nd}}$ be the deflation of undesirable output r_{nd}. The model for the minor adjustment policy can be modified as follows.

19.4.1 Phase I

$$\max \rho^{\text{MI}} = \frac{1}{2}\left(\frac{1}{s_d}\sum_{r_d=1}^{s_d}\delta_{r_d}^{\text{MI}} + \frac{1}{s_{nd}}\sum_{r_{nd}=1}^{s_{nd}}\varphi_{r_{nd}}^{\text{MI}}\right) \tag{19.62}$$

$$\text{s.t.}\sum_{k=1}^{n}\sum_{j=1}^{n}\lambda_{jk}x_{i_aj} = \sum_{k=1}^{n}x_{i_ak}, i_a = 1,2,\ldots,m_a \tag{19.63}$$

$$\sum_{j=1}^{n}\lambda_{jk}x_{i_aj} \le x_{i_ak}, i_a = 1,2,\ldots,m_a, k = 1,\ldots,n \tag{19.64}$$

$$\sum_{j=1}^{n}\lambda_{jk}x_{i_{na}j} \le x_{i_{na}k}, i_{na} = 1,2,\ldots,m_{na}, k = 1,\ldots,n \tag{19.65}$$

$$\sum_{k=1}^{n}\sum_{j=1}^{n}\lambda_{jk}y_{r_dj} \ge \left(1+\delta_{r_d}^{\text{MI}}\right)\sum_{k=1}^{n}y_{r_dk}, r_d = 1,2,\ldots,s_d \tag{19.66}$$

$$\sum_{j=1}^{n}\lambda_{jk}y_{r_dj} \ge y_{r_dk}, r_d = 1,2,\ldots,s_d, k = 1,\ldots,n \tag{19.67}$$

$$\sum_{k=1}^{n}\sum_{j=1}^{n}\lambda_{jk}y_{r_{nd}j} = \left(1-\varphi_{r_{nd}}^{\text{MI}}\right)\sum_{k=1}^{n}y_{r_{nd}k}, r_{nd} = 1,2,\ldots,s_{nd} \tag{19.68}$$

$$\sum_{j=1}^{n}\lambda_{jk}y_{r_{nd}j} \le y_{r_{nd}k}, r_{nd} = 1,2,\ldots,s_{nd}, k = 1,\ldots,n \tag{19.69}$$

$$\sum_{j=1}^{n}\lambda_{jk} = 1, k = 1,\ldots,n \tag{19.70}$$

$$\lambda_{jk} \ge 0, \quad j = 1,\ldots,n, \quad k = 1,\ldots,n \tag{19.71}$$

Equation (19.62) seeks the optimum expansion rate of the aggregated desirable outputs and contraction rate of the aggregated undesirable outputs. Equation (19.66) seeks to non-radially increase each desirable output as much as possible, whereas (19.68) seeks to non-radially decrease each undesirable output as much as possible. Equations (19.66) and (19.68) ensure that each desirable output and each undesirable output remain in the feasible aggregated desirable output set and aggregated undesirable output set, respectively. Equation (19.67) imposes the restriction that the projected point of each DMU must be no less than the observed desirable output quantities of DMU_k, whereas (19.69) imposes the restriction that the projected point for each DMU must be no more than the observed undesirable output quantities of DMU_k. Equations (19.63)–(19.65) and (19.70)–(19.71) can be compared with Equations (19.2)–(19.4) and (19.7)–(19.8), respectively.

19.4.2 Phase II

Find the slack variables of the adjustable inputs at $\delta_{r_d}^{\text{MI}*}$ and $\varphi_{r_{nd}}^{\text{MI}*}$ obtained in Phase I:

$$\max \sum_{i_a=1}^{m_a}\sum_{k=1}^{n}\frac{\left(s_{i_ak}^{\text{MI}-}-s_{i_ak}^{\text{MI}+}\right)}{x_{i_ak}} \tag{19.72}$$

$$\text{s.t.}\sum_{k=1}^{n}\sum_{j=1}^{n}\lambda_{jk}x_{i_aj}=\sum_{k=1}^{n}x_{i_ak},i_a=1,2,\ldots,m_a \tag{19.73}$$

$$\sum_{j=1}^{n}\lambda_{jk}x_{i_aj}=x_{i_ak}+s_{i_ak}^{\text{MI}+}-s_{i_ak}^{\text{MI}-},i_a=1,2,\ldots,m_a,k=1,\ldots,n \tag{19.74}$$

$$\sum_{j=1}^{n}\lambda_{jk}x_{i_{na}j}\le x_{i_{na}k},i_{na}=1,2,\ldots,m_{na},k=1,\ldots,n \tag{19.75}$$

$$\sum_{k=1}^{n}\sum_{j=1}^{n}\lambda_{jk}y_{r_dj}=\left(1+\delta_{r_d}^{\text{MI}*}\right)\sum_{k=1}^{n}y_{r_dk},r_d=1,2,\ldots,s_d \tag{19.76}$$

$$\sum_{j=1}^{n}\lambda_{jk}y_{r_dj}\ge y_{r_dk},r_d=1,2,\ldots,s_d,k=1,\ldots,n \tag{19.77}$$

$$\sum_{k=1}^{n}\sum_{j=1}^{n}\lambda_{jk}y_{r_{nd}j}=\left(1-\varphi_{r_{nd}}^{\text{MI}*}\right)\sum_{k=1}^{n}y_{r_{nd}k},r_{nd}=1,2,\ldots,s_{nd} \tag{19.78}$$

$$\sum_{j=1}^{n}\lambda_{jk}y_{r_{nd}j}\le y_{r_{nd}k},r_{nd}=1,2,\ldots,s_{nd},k=1,\ldots,n \tag{19.79}$$

$$\sum_{j=1}^{n}\lambda_{jk}=1,k=1,\ldots,n \tag{19.80}$$

$$\lambda_{jk}\ge 0,\ j=1,\ldots,n,\ \ k=1,\ldots,n \tag{19.81}$$

$$s_{i_ak}^{\text{MI}-},s_{i_ak}^{\text{MI}+}\ge 0,\ \ i_a=1,\ldots,m_a,\ \ k=1,\ldots,n \tag{19.82}$$

Equations (19.76) and (19.78) can be seen as using the reallocation perspective to reach the ideal desirable and undesirable outputs under the CDEA perspective.

Equations (19.77) and (19.79) are analogous to (19.67) and (19.69). In addition, (19.72)–(19.75) and (19.80)–(19.82) can be compared with (19.9)–(19.12) and (19.15)–(19.17), respectively. Finally, the optimal slack variables of the adjustable inputs can be obtained from the above model with undesirable outputs.

Similarly, by applying the objective function identified in (19.62) and the constraints identified in (19.20)–(19.24) and (19.66)–(19.71), the Phase I model for the moderate adjustment policy can be constructed, and the optimal slack values of the adjustable inputs can be obtained from Phase II, which is built by applying the objective function identified in (19.72) and the constraints identified in (19.30)–(19.35), (19.76)–(19.81) and (19.40).

Finally, the models for the major adjustment policy can be constructed. In Phase I, the objective function identified in (19.62) and the constraints identified in (19.44)–(19.46) and (19.66)–(19.71) are used to find the optimal expansion rate of desirable outputs and the optimal contract rate of undesirable outputs. In Phase II, the object function identified in (19.72) and the constraints identified in (19.52)–(19.55) and (19.76)–(19.82) are applied to obtain the optimal slack values of the adjustable inputs.

19.5   CONCLUSIONS

In this chapter, we have provided a systematic and centralized perspective on resource reallocation, and applied this perspective to construct two-phase CDEA models that illustrate various adjustment policies designed in order to lessen organizational resistance. In the resource reallocation process utilized by the authorities in an organization, the CDEA model for planning and reallocating resources is a valid tool for the investigation of both reduction and transformation of inputs among the units of the organization. Since the focus of the chapter was on providing the appropriate resource reallocation policies, we have developed a minor adjustment policy, moderate adjustment policy and major adjustment policy. These policies can provide a systematic resource reallocation process to reduce organizational resistance. They can also provide a step-by-step allocation path. In addition, we have chosen two related applications in transport organizations to investigate the applicability of resource reallocation by using CDEA models, and described the results of resource reallocation. Finally, we have modified the CDEA models to deal with undesirable outputs, because such undesirable outputs are jointly produced with desirable outputs in many transport organizations.

REFERENCES

[1] Kwak, N.K. and Lee, C. (1998) A multicriteria decision-making approach to university resource allocations and information infrastructure planning. *European Journal of Operational Research*, **110**, 234–242.

[2] Joglekar, N.R. and Ford, D.N. (2005) Product development resource allocation with foresight. *European Journal of Operational Research*, **160**, 72–78.

[3] Belfares, L., Klibi, W., Lo, N. and Guitouni, A. (2007) Multi-objectives Tabu Search based algorithm for progressive resource allocation. *European Journal of Operational Research*, **177**, 1779–1799.

[4] Hsu, L.H. (2003) Application of artificial intelligence to forecast the tourist arrivals to Taiwan. Portland International Conference on Management of Engineering and Technology, Portland, OR.

[5] Ogryczak, W., Wierzbicki, A. and Milewski, M. (2008) A multi-criteria approach to fair and efficient bandwidth allocation. *Omega*, **36**(3), 451–464.

[6] Thanassoulis, E. (1996) A data envelopment analysis approach to clustering operating units for resource allocation purpose. *Omega*, **24**, 463–476.

[7] Caballero, R., Galache, T., Gomez, T. and Molina, J. (2004) Budgetary allocations and efficiency in the human resources policy of university following multiple criteria. *Economics of Education Review*, **23**, 67–74.

[8] Casu, B. and Girardone, C. (2010) Integration and efficiency convergence in EU banking markets. *Omega*, **38**, 260–267.

[9] Yu, M.M., Chern, C.C. and Hsiao, B. (2013) Human resource rightsizing using centralized data envelopment analysis: Evidence from Taiwan's airports. *Omega*, **41**(1), 119–130.

[10] Chang, S.M., Wang, J.S., Yu, M.M. *et al.* (2015) An application of centralized data envelopment analysis in resource allocation in container terminal operations. *Maritime Policy & Management*, **42**(8), 776–788.

[11] Tzeng, G.H., Cheng, H.J. and Huang, T.D. (2007) Multi-objective optimal planning for designing relief delivery systems. *Transportation Research, Part E*, **43**, 673–686.

[12] Demirtas, E.A. and Üstün, Ö. (2008) An integrated multiobject decision making process for supplier selection and order allocation. *Omega*, **36**, 76–90.

[13] Morais, D.C. and Almeida, A.T. (2012) Group decision making on water resources based on analysis of individual rankings. *Omega*, **40**, 42–52.

[14] Lozano, S., Villa, G. and Adenso-Díaz, B. (2004) Centralized target setting for regional recycling operations using DEA. *Omega*, **32**, 101–110.

[15] Lozano, S. and Villa, G. (2004) Centralized resource allocation using data envelopment analysis. *Journal of Productivity Analysis*, **22**, 143–161.

[16] Lozano, S. and Villa, G. (2005) Centralized DEA models with the possibility of downsizing. *Journal of the Operational Research Society*, **56**, 357–364.

[17] Korhonen, P. and Syrjänen, M. (2004) Resource allocation based on efficiency analysis. *Management Science*, **50**(8), 1134–1144.

[18] Asmild, M., Paradi, J.C. and Pastor, J.T. (2009) Centralized resource allocation BCC models. *Omega*, **37**, 40–49.

[19] Lozano, S., Villa, G. and Brännlund, R. (2009) Centralised reallocation of emission permits using DEA. *European Journal of Operational Research*, **193**, 752–760.

[20] Hosseinzadeh Lotfi, F., Noora, A.A., Jahanshahloo, G.R. *et al.* (2010). Centralized resource allocation for enhanced Russell models. *Journal of Computational and Applied Mathematics*, **235**, 1–10.

[21] Lozano, S., Villa, G. and Canca, D. (2011) Application of centralized DEA approach to capital budgeting in Spanish ports. *Computers and Industrial Engineering*, **60**, 455–465.

[22] Hosseinzadeh Lotfi, F., Nematollahi, N., Behzadi, M.H. *et al.* (2012). Centralized resource allocation with stochastic data. *Journal of Computational and Applied Mathematics*, **236**, 1783–1788.

[23] Fang, L. (2013) A generalized DEA model for centralized resource allocation. *European Journal of Operational Research*, **228**, 405–412.

[24] Mar-Molinero, C., Prior, D., Segovia, M.M. and Portillo, F. (2014) On centralized resource utilization and its reallocation by using DEA. *Annals of Operations Research*, **221**(1), 273–283.

[25] Fang, L. (2015) Centralized resource allocation based on efficiency analysis for step-by-step improvement paths. *Omega*, **51**, 24–28.

[26] Fang, L. and Li, H.C. (2015) Cost efficiency in data envelopment analysis under the law of one price. *European Journal of Operational Research*, **240**, 488–492.

[27] Golembiewski, R., Billingsley, K. and Yeager, S. (1976) Measuring change and persistence in human affairs: Types of change generated by OD designs. *Journal of Applied Behavioral Science*, **12**(2), 133–157.

[28] Argris, C. and Schön, D. (eds) (1978) *Organizational Learning: A Theory of Action Perspective*, Addison-Wesley, Reading, MA.

[29] Banker, R.D. and Morey, R.C. (1986) Efficiency analysis for exogenously fixed inputs and outputs. *Operations Research*, **34**(4), 513–521.

[30] Syrjänen, M.J. (2004) Non-discretionary and discretionary factors and scale in data envelopment analysis. *European Journal of Operational Research*, **158**, 20–33.

[31] Hitt, M.A., Hoskisson, R.E. and Ireland, R.D. (1990) Acquisitive growth and commitment innovation in M-form firms. *Strategic Management Journal*, **11**, 29–47.

[32] Morrell, P. and Lu, C.H.Y. (2000) Aircraft noise social cost and charge mechanisms – a case study of Amsterdam Airport Schiphol. *Transportation Research Part D*, **5**, 305–320.

20

HOW TO DEAL WITH NON-CONVEX FRONTIERS IN DATA ENVELOPMENT ANALYSIS[1]

KAORU TONE

National Graduate Institute for Policy Studies, Tokyo, Japan

MIKI TSUTSUI

Central Research Institute of Electric Power Industry, Tokyo, Japan

20.1 INTRODUCTION

In data envelopment analysis (DEA), we are often puzzled by the large difference between the constant-returns-to-scale (CRS) score and the variable-returns-to-scale (VRS) score.[2] Several authors ([1–3], among others) have proposed solutions to this problem. In this chapter, we propose a different approach to solving this problem, and present our results. A further problem, which is closely related to the problem mentioned above, is the conventional assumption of a convex production possibility set. Several researchers have discussed non-convex production possibility issues

---

[1] Part of the material in this chapter is adapted from the *Journal of Optimization Theory and Applications*, Vol. 166, Tone K. and Tsutsui M., How to deal with non-convex frontiers in data envelopment analysis, (2014) 1002–1028, with permission from Springer.

[2] See Figure 20.9 for a comparison of CRS and VRS scores, where large differences are observed between the two scores.

Advances in DEA Theory and Applications: With Extensions to Forecasting Models,
First Edition. Edited by Kaoru Tone.

([4–7], among others). Among these publications, we refer two relevant articles. Dekker and Post [4] extended the standard assumption of a concave efficient frontier in DEA to a quasi-concavity model. However, this quasi-concavity assumption does not always work for identification of frontiers associated with real-world problems. Our approach is more general in dealing with the non-convexity issue. Kousmanen [5] utilized disjunctive programming for identification of a conditionally convex production set. He characterized various reference technologies by setting irrelevant intensity values to zero, while keeping the convexity condition in the chosen technology. His method is enumerative in the sense that the number of reference technologies is combinatorial. Our method differs from his in (a) introduction of clusters instead of enumeration, and (b) relaxation of the convexity condition on the intensity vector. As far as we know, no paper has discussed these subjects in the scale- and cluster-related context. In this chapter, we discuss the above two fundamental problems of DEA.

A further objective of this chapter is measurement of the scale elasticity of production. Most prior research into this subject has been based on the assumption of a convex production possibility set. We propose a new scheme for evaluation of the scale elasticity within a specific cluster containing each individual decision-making unit (DMU).

We refer to the seminal papers of Farrell [8] and Farrell and Fieldhouse [9] for a discussion of the case of economies and diseconomies of scale. Charnes *et al.* [10] extended Farrell's work on evaluation of a program and its managerial efficiency through experiments, where program follow-through (PFT) and non-follow-through (NFT) were treated as two separate clusters. Førsund *et al.* [11] revisited Farrell [8] and Farrell and Fieldhouse [9]. They pointed out that Farrell and Fieldhouse's grouping method creates efficient frontiers for each group. They also generalized this idea to multiple outputs and tried to represent frontier functions graphically, where EffiVision [12] was utilized. They discussed several economic concepts through this visualization.

Considering the scope of these previous studies, the novel features of this chapter are as follows:

1. We extend Farrell's approach to discriminate scale merits and scale demerits, by utilizing scale efficiency. We decompose the slacks of each DMU into scale-dependent and scale-independent parts.

2. We extend the clustering approach of Charnes *et al.* [10], by coupling that approach with scale merits and scale demerits. Thus, we can find non-convex frontiers.

3. Although non-convex frontiers can be identified by the free disposal hull (FDH) model [13], that model is a discrete model in the sense that the elements of the intensity vector are binary. In addition, scale effects are not involved. Our model permits continuous intensity vectors and can find non-convex frontiers by means of clusters.

This chapter is structured as follows. In Section 20.2, we describe the decomposition of the CRS slacks after introducing the basic notation, and define the scale-independent dataset. In Section 20.3, we introduce clusters and define the scale- and cluster-adjusted score (SAS). In Section 20.4, we explain our scheme using an artificial example. We develop the radial-model case in Section 20.5. In Section 20.6, we define the scale elasticity based on the scale-dependent dataset. A typical application, in this case concerning Japanese national university research output, follows in Section 20.7, and the last section concludes the chapter.

20.2 GLOBAL FORMULATION

In this section, we introduce the notation and basic tools, and discuss the decomposition of the slacks. Throughout Sections 20.2 to 20.4, we utilize the input-oriented slacks-based measure (SBM) [14], a non-radial model, for explanation of our model.

20.2.1 Notation and Basic Tools

Let the input and output data matrices be, respectively,

$$\mathbf{X} = \left(x_{ij}\right) \in \mathrm{IR}^{m \times n} \text{ and } \mathbf{Y} = \left(y_{rj}\right) \in \mathrm{IR}^{s \times n} \tag{20.1}$$

where m, s and n are the numbers of inputs, outputs and DMUs, respectively. We assume that the data are positive, that is, $\mathbf{X} > \mathbf{0}$ and $\mathbf{Y} > \mathbf{0}$.

Then, the production possibility sets for the CRS and VRS models are defined by

$$
\begin{aligned}
P_{\mathrm{CRS}} &= \{(\mathbf{x}, \mathbf{y}) \mid \mathbf{x} \geq \mathbf{X}\boldsymbol{\lambda}, \mathbf{y} \leq \mathbf{Y}\boldsymbol{\lambda}, \boldsymbol{\lambda} \geq \mathbf{0}\}, \\
P_{\mathrm{VRS}} &= \{(\mathbf{x}, \mathbf{y}) \mid \mathbf{x} \geq \mathbf{X}\boldsymbol{\lambda}, \mathbf{y} \leq \mathbf{Y}\boldsymbol{\lambda}, \mathbf{e}\boldsymbol{\lambda} = 1, \boldsymbol{\lambda} \geq \mathbf{0}\}
\end{aligned}
\tag{20.2}
$$

respectively, where $\mathbf{x}(>\mathbf{0}) \in \mathrm{IR}^m$, $\mathbf{y}(>\mathbf{0}) \in \mathrm{IR}^s$ and $\boldsymbol{\lambda}(\geq\mathbf{0}) \in \mathrm{IR}^n$ are the input, output and intensity vectors, respectively, and $\mathbf{e} \in \mathrm{IR}^n$ is the row vector with all elements equal to 1.

The input-oriented slacks-based measures for evaluation of the efficiency of each DMU $(\bar{\mathbf{x}}_k, \bar{\mathbf{y}}_k)$ $(k = 1, \dots, n)$, in the CRS and VRS models, are as follows:

$$[\mathrm{CRS}] \; \theta_k^{\mathrm{CRS}} = \min_{\boldsymbol{\lambda}, \mathbf{s}^-, \mathbf{s}^+} \left(1 - \frac{1}{m} \sum_{i=1}^{m} \frac{s_i^-}{x_{ik}}\right), \text{ s.t.}$$

$$\mathbf{X}\boldsymbol{\lambda} + \mathbf{s}^- = \mathbf{x}_k, \mathbf{Y}\boldsymbol{\lambda} - \mathbf{s}^+ = \mathbf{y}_k, \boldsymbol{\lambda} \geq \mathbf{0}, \mathbf{s}^- \geq \mathbf{0}, \mathbf{s}^+ \geq \mathbf{0} \tag{20.3}$$

$$[\mathrm{VRS}] \; \theta_k^{\mathrm{VRS}} = \min_{\boldsymbol{\lambda}, \mathbf{s}^-, \mathbf{s}^+} \left(1 - \frac{1}{m} \sum_{i=1}^{m} \frac{s_i^-}{x_{ik}}\right), \text{ s.t.}$$

$$\mathbf{X}\boldsymbol{\lambda} + \mathbf{s}^- = \mathbf{x}_k, \mathbf{Y}\boldsymbol{\lambda} - \mathbf{s}^+ = \mathbf{y}_k, \mathbf{e}\boldsymbol{\lambda} = 1, \boldsymbol{\lambda} \geq \mathbf{0}, \mathbf{s}^- \geq \mathbf{0}, \mathbf{s}^+ \geq \mathbf{0} \tag{20.4}$$

where λ is the intensity vector, and $\mathbf{s}^-$ and $\mathbf{s}^+$ are the input and output slacks, respectively. Although we present our model in the context of the input-oriented SBM model, we can also develop the model in the context of the output-oriented and non-oriented SBM models, as well as the radial models.

We define the scale efficiency (σ_k) of DMU$_k$ as

$$\sigma_k = \frac{\theta_k^{CRS}}{\theta_k^{VRS}} \tag{20.5}$$

We denote the optimal slacks of the CRS model by

$$\left(\mathbf{s}_k^{-*}, \mathbf{s}_k^{+*}\right) \tag{20.6}$$

Although we utilize the scale efficiency, CRS/VRS, as an index of the scale merits and scale demerits, we can make use of other indices that are appropriate for discriminating handicaps due to scale. However, the index must be normalized between 0 and 1, with a larger value indicating a better scale condition.

20.2.2   Uniqueness of Slacks

Although the CRS and VRS scores are determined uniquely, their slacks are not always unique in the SBM model. We can resolve this problem as follows:

1. *Priority.* We determine the priority (importance) of the input factors. For example, the most cost-influential input factor is identified as the first priority, because the reduction of its slack is most recommended. The second and others follow in this way.

2. *Multi-objective programming for the determination of slacks according to their priority.* We assume that the priority is $s_1^-, s_2^-, \ldots, s_m^-$ in that order. We maximize the slacks $s_1^-, s_2^-, \ldots, s_m^-$ using the multi-objective programming framework below:

$$\max_{\lambda, \mathbf{s}^-, \mathbf{s}^+} \left(s_1^-, s_2^-, \ldots, s_m^-\right)$$
$$\text{s.t. } \frac{1}{m}\sum_{i=1}^m \frac{s_i^-}{x_{ik}} = 1 - \theta_k^{CRS}, \mathbf{X}\lambda + \mathbf{s}^- = \mathbf{x}_k, \mathbf{Y}\lambda - \mathbf{s}^+ = \mathbf{y}_k, \lambda \geq 0, \mathbf{s}^- \geq 0, \mathbf{s}^+ \geq 0 \tag{20.7}$$

This notation indicates that we first maximize s_1^- subject to (20.7). Then, fixing s_1^- at the optimal value, we maximize s_2^- subject to (20.7). We repeat this process until s_{m-1}^- is reached.

For the VRS model (20.4), we can determine the slacks uniquely using the above procedure by including the convexity constraint $\mathbf{e}\lambda = 1$.

20.2.3 Decomposition of CRS Slacks

We decompose the CRS slacks into scale-independent and scale-dependent parts as follows:

$$
\begin{aligned}
\mathbf{s}_k^{-*} &= \sigma_k \mathbf{s}_k^{-*} + (1 - \sigma_k) \mathbf{s}_k^{-*} \\
\mathbf{s}_k^{+*} &= \sigma_k \mathbf{s}_k^{+*} + (1 - \sigma_k) \mathbf{s}_k^{+*}
\end{aligned}
\tag{20.8}
$$

If DMU_k satisfies $\sigma_k = 1$ (the so-called 'most productive scale size'), all its slacks are attributed to scale-independent slacks. However, if $\sigma_k < 1$, the slacks are decomposed into a scale-independent part and a scale-dependent part as follows:

$$
\begin{aligned}
&\text{Scale} - \text{independent slacks}: \left(\sigma_k \mathbf{s}_k^{-*}, \sigma_k \mathbf{s}_k^{+*} \right) \\
&\text{Scale} - \text{dependent slacks}: \left((1 - \sigma_k) \mathbf{s}_k^{-*}, (1 - \sigma_k) \mathbf{s}_k^{+*} \right)
\end{aligned}
\tag{20.9}
$$

20.2.4 Scale-Independent Dataset

We define the scale-independent data $(\bar{\mathbf{x}}_k, \bar{\mathbf{y}}_k)$ $(k = 1, \ldots, n)$ by subtracting and adding the scale-dependent slacks as follows:

$$
\begin{aligned}
&\text{Scale} - \text{independent input}: \bar{\mathbf{x}}_k = \mathbf{x}_k - (1 - \sigma_k) \mathbf{s}_k^{-*} \\
&\text{Scale} - \text{independent output}: \bar{\mathbf{y}}_k = \mathbf{y}_k + (1 - \sigma_k) \mathbf{s}_k^{+*}
\end{aligned}
\tag{20.10}
$$

This process is illustrated in Figure 20.1. The scale-independent dataset $(\bar{\mathbf{X}}, \bar{\mathbf{Y}})$ is defined by

$$
(\bar{\mathbf{X}}, \bar{\mathbf{Y}}) = \left\{ (\bar{\mathbf{x}}_j, \bar{\mathbf{y}}_j) \mid j = 1, \ldots, n \right\}
\tag{20.11}
$$

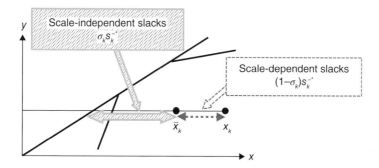

Figure 20.1 Scale-independent input.

We define the production possibility sets $P(\mathbf{X}, \mathbf{Y})$ and $P(\bar{\mathbf{X}}, \bar{\mathbf{Y}})$ for $(\mathbf{x}_j, \mathbf{y}_j)$ and $(\bar{\mathbf{x}}_j, \bar{\mathbf{y}}_j)$ $(j = 1, \ldots, n)$, respectively, as follows:

$$
\begin{aligned}
P(\mathbf{X}, \mathbf{Y}) &= \left\{ (\mathbf{x}, \mathbf{y}) \middle| \mathbf{x} \ge \sum\nolimits_{j=1}^{n} \mathbf{x}_j \lambda_j, \mathbf{0} \le \mathbf{y} \le \sum\nolimits_{j=1}^{n} \mathbf{y}_j \lambda_j, \lambda \ge \mathbf{0} \right\} \\
P(\bar{\mathbf{X}}, \bar{\mathbf{Y}}) &= \left\{ (\bar{\mathbf{x}}, \bar{\mathbf{y}}) \middle| \bar{\mathbf{x}} \ge \sum\nolimits_{j=1}^{n} \bar{\mathbf{x}}_j \lambda_j, \mathbf{0} \le \bar{\mathbf{y}} \le \sum\nolimits_{j=1}^{n} \bar{\mathbf{y}}_j \lambda_j, \lambda \ge \mathbf{0} \right\}
\end{aligned}
\tag{20.12}
$$

Lemma 20.1 $P(\mathbf{X}, \mathbf{Y}) = P(\bar{\mathbf{X}}, \bar{\mathbf{Y}}).$

Proof   We define the scale-independent DMU, $(\bar{\mathbf{x}}_j, \bar{\mathbf{y}}_j)$ $(j = 1, \ldots, n)$, by

$$
\begin{aligned}
\bar{\mathbf{x}}_j &= \mathbf{x}_j - \left(1 - \sigma_j \right) \mathbf{s}_j^{-*} \\
\bar{\mathbf{y}}_j &= \mathbf{y}_j + \left(1 - \sigma_j \right) \mathbf{s}_j^{+*}
\end{aligned}
\tag{20.13}
$$

If $\sigma_j = 1$, then we have $\bar{\mathbf{x}}_j = \mathbf{x}_j$ and $\bar{\mathbf{y}}_j = \mathbf{y}_j$. If $\sigma_j < 1z$, then

$$
\begin{aligned}
\bar{\mathbf{x}}_j &= \mathbf{x}_j - \left(1 - \sigma_j \right) \mathbf{s}_j^{-*} \ge \mathbf{x}_j - \mathbf{s}_j^{-*} \\
\bar{\mathbf{y}}_j &= \mathbf{y}_j + \left(1 - \sigma_j \right) \mathbf{s}_j^{+*} \le \mathbf{y}_j + \mathbf{s}_j^{+*}
\end{aligned}
\tag{20.14}
$$

where $\left(\mathbf{x}_j - \mathbf{s}_j^{-*}, \mathbf{y}_j + \mathbf{s}_j^{+*} \right)$ is the projection of $(\mathbf{x}_j, \mathbf{y}_j)$ onto the frontiers of $P(\mathbf{X}, \mathbf{Y})$. Thus $(\bar{\mathbf{x}}_j, \bar{\mathbf{y}}_j)$ $(j = 1, \ldots, n)$ belongs to $P(\mathbf{X}, \mathbf{Y})$. Hence, efficient frontiers are common to $P(\mathbf{X}, \mathbf{Y})$ and $P(\bar{\mathbf{X}}, \bar{\mathbf{Y}})$.   ∎

20.3   IN-CLUSTER ISSUE: SCALE- AND CLUSTER-ADJUSTED DEA SCORE

In this section, we introduce the clusters of DMUs and define the SAS.

20.3.1   Clusters

We classify DMUs into several clusters depending on their characteristics. The clusters can be determined by using a clustering method (in the field of statistics) appropriate to the problem concerned, or supplied exogenously (see Section 20.7 for an example).

Farrell and Fieldhouse [9] used the *grouping method* for their study of farm survey data for England and Wales for the period 1952–1953. They divided all observations

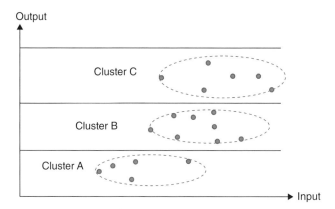

Figure 20.2 Clustering by output size.

(208) into 10 groups (clusters) according to output (gross sales). The method certainly depends on adequate datasets with sufficient observations in each size group. See Figure 20.2 for an illustration, where three clusters are shown, depending on the output size (a variation of Figure 20.1 in Førsund *et al.* [11]).

Environmental factors can be utilized for classification in addition to input/output factors. Charnes *et al.* [10] divided DMUs into two groups by PFT and NFT properties. Several authors have discussed environmental factors, for example Avkiran [15], Paradi *et al.* [16] and Cook [17], among others.

However, if the above clustering methods are unavailable, clusters can be determined *a posteriori* depending on the degree of scale efficiency. An example of the latter case, as a supplementary tool, is described in Appendix 20.A.

Since both the clusters and the scale efficiencies critically affect the results of the proposed scheme, we need to handle these matters deliberately, referring to the above literature, and we need to try many clustering cases to obtain a reasonable conclusion.

We denote the cluster containing DMU_j by $Cluster(j)$, where $j = 1, \ldots, n$.

20.3.2   Solving the CRS Model in the Same Cluster

We solve the CRS model for each DMU $(\bar{\mathbf{x}}_k, \bar{\mathbf{y}}_k)$ $(k = 1, \ldots, n)$, referring to the $(\bar{\mathbf{X}}, \bar{\mathbf{Y}})$ in the same cluster (k). The solution is formulated as follows:

$$
\begin{aligned}
\min_{\boldsymbol{\mu}, \mathbf{s}^{cl-}, \mathbf{s}^{cl+}} \quad & \left(1 - \frac{1}{m}\sum_{i=1}^{m} \frac{s_i^{cl-}}{\bar{x}_{ik}}\right) \\
\text{s.t.} \quad & \bar{\mathbf{X}}\boldsymbol{\mu} + \mathbf{s}^{cl-} = \bar{\mathbf{x}}_k, \bar{\mathbf{Y}}\boldsymbol{\mu} - \mathbf{s}^{cl+} = \bar{\mathbf{y}}_k, \\
& \mu_j = 0 \ (\forall j : \mathrm{Cluster}(j) \neq \mathrm{Cluster}(k)), \\
& \boldsymbol{\mu} \geq \mathbf{0}, \mathbf{s}^{cl-} \geq \mathbf{0}, \mathbf{s}^{cl+} \geq \mathbf{0}
\end{aligned}
\tag{20.15}
$$

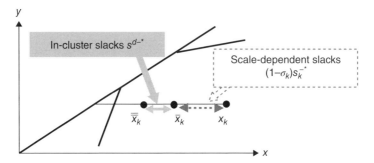

Figure 20.3   Scale- and cluster-adjusted input.

We denote the optimal in-cluster slacks by $\left(s_k^{cl-*}, s_k^{cl+*}\right)$.[3] By summing the scale-dependent slacks and the in-cluster slacks, we define the total number of slacks as

$$
\begin{aligned}
\text{Total input slacks}: \bar{\mathbf{s}}_k^- &= (1-\sigma_k)\mathbf{s}_k^{-*} + \mathbf{s}_k^{cl-*} \\
\text{Total output slacks}: \bar{\mathbf{s}}_k^+ &= (1-\sigma_k)\mathbf{s}_k^{+*} + \mathbf{s}_k^{cl+*}
\end{aligned}
\tag{20.16}
$$

The scale-and cluster-adjusted data $\left(\bar{\bar{\mathbf{x}}}_k, \bar{\bar{\mathbf{y}}}_k\right)$ (projection) is defined by

$$
\begin{aligned}
\bar{\bar{\mathbf{x}}}_k &= \mathbf{x}_k - \bar{\mathbf{s}}_k^- = \mathbf{x}_k - (1-\sigma_k)\mathbf{s}_k^{-*} - \mathbf{s}_k^{cl-*} \\
\bar{\bar{\mathbf{y}}}_k &= \mathbf{y}_k + \bar{\mathbf{s}}_k^+ = \mathbf{y}_k + (1-\sigma_k)\mathbf{s}_k^{+*} + \mathbf{s}_k^{cl+*}
\end{aligned}
\tag{20.17}
$$

Figure 20.3 illustrates the scale- and cluster-adjusted input.

At this point, we have removed the scale demerits and in-cluster slacks from the dataset. Thus, we have obtained the scale-free and in-cluster slacks-free (projected) dataset:

$$
\left(\bar{\bar{\mathbf{X}}}, \bar{\bar{\mathbf{Y}}}\right) = \left\{\left(\bar{\bar{\mathbf{x}}}_j, \bar{\bar{\mathbf{y}}}_j\right) \mid j = 1, \ldots, n\right\}
\tag{20.18}
$$

20.3.3   Scale- and Cluster-Adjusted Score

In the input-oriented case, the SAS is defined by

$$
\theta_k^{SAS} = 1 - \frac{1}{m}\sum_{i=1}^{m} \frac{\bar{s}_{ik}^-}{x_{ik}} = 1 - \frac{1}{m}\sum_{i=1}^{m} \frac{s_{ik}^{cl-*} + (1-\sigma_k) s_{ik}^{-*}}{x_{ik}}
\tag{20.19}
$$

---

[3] We apply the same priority rule for slacks as defined in Section 20.2.2.

The reason we utilize the above scheme is as follows. First, we wish to eliminate scale demerits from the CRS slacks. For this purpose, we decompose the CRS slacks into scale-dependent and scale-independent parts, by recognizing that the scale demerits are represented by $1 - \sigma_k$. If $\sigma_k = 1$, the DMU has no scale demerits, and its slacks are attributed to itself. If $\sigma_k = 0.25$, 75% of the slacks are attributed to the DMU's scale demerits. After removing the scale-dependent slacks, we evaluate the DMU within its cluster and determine the in-cluster slacks. If the DMU is efficient within its cluster, its in-cluster slacks are zero, while, if it is inefficient, the DMU has in-cluster slacks with respect to the efficient DMUs. Finally, we sum the in-cluster and scale-dependent slacks to obtain the total amount of slacks. Using the total slacks value determined, we define the SAS.

Proposition 20.1 The SAS is not less than the CRS score:

$$\theta_k^{SAS} \geq \theta_k^{CRS} \ (\forall k) \tag{20.20}$$

Proposition 20.2 If all DMUs belong to the same cluster, it holds that $\theta_k^{SAS} = \theta_k^{CRS} \ (\forall k)$.

This implies that no non-convex frontiers exist when all DMUs belong to the same cluster.

Proposition 20.3 If $\theta_k^{CRS} = 1$, then it holds that $\theta_k^{SAS} = \theta_k^{CRS}$, but not vice versa.

Proposition 20.4 The SAS decreases with increasing input and decreasing output, as long as both DMUs remain in the same cluster.

Proposition 20.5 The projected DMU $(\bar{\bar{\mathbf{x}}}_k, \bar{\bar{\mathbf{y}}}_k)$ is efficient under the SAS model among the DMUs in the cluster it belongs to. It is also CRS and VRS efficient among the DMUs in its cluster.

Proofs of these propositions are given in Appendix 20.B.

20.3.4 Summary of the SAS Computation

We can summarize the SAS computation as follows:

Step 1. Input data (**X**, **Y**, Cluster). The clusters can be supplied exogenously by some clustering method, including the use of experts' knowledge, or determined internally depending on the degree of scale efficiency.

Step 2. For $k = 1, \ldots, n$, solve (20.3) and (20.4) to obtain the CRS and VRS scores, θ_k^{CRS} and θ_k^{VRS}, respectively. Define the scale efficiency $\sigma_k = \theta_k^{CRS} / \theta_k^{VRS}$.

Step 3. Using the optimal slacks $(\mathbf{s}_k^{-*}, \mathbf{s}_k^{+*})$ for the CRS model, define the scale-dependent slacks as $((1 - \sigma_k)\mathbf{s}_k^{-*}, (1 - \sigma_k)\mathbf{s}_k^{+*})$.

Step 4. Define the scale-independent dataset $(\bar{\mathbf{X}},\bar{\mathbf{Y}})=\{(\bar{\mathbf{x}}_k,\bar{\mathbf{y}}_k)|k=1,\ldots,n\}$ by

$$
\begin{aligned}
\text{Scale}-\text{independent input:}\ \bar{\mathbf{x}}_k &= \mathbf{x}_k-(1-\sigma_k)\mathbf{s}_k^{-*}\\
\text{Scale}-\text{independent output:}\ \bar{\mathbf{y}}_k &= \mathbf{y}_k+(1-\sigma_k)\mathbf{s}_k^{+*}
\end{aligned}
\tag{20.21}
$$

Step 5. Solve the CRS model (20.15) for each $(\bar{\mathbf{x}}_k,\bar{\mathbf{y}}_k)$, referring to the $(\bar{\mathbf{X}},\bar{\mathbf{Y}})$ in the same cluster (k), and obtain the optimal in-cluster slacks $\left(\mathbf{s}_k^{cl-*},\mathbf{s}_k^{cl+*}\right)$.

Step 6. Define the SAS by

$$
\theta_k^{SAS}=1-\frac{1}{m}\sum_{i=1}^{m}\frac{s_{ik}^{cl-*}+(1-\sigma_k)\,s_{ik}^{-*}}{x_{ik}}
\tag{20.22}
$$

Step 7. Obtain the scale- and cluster-adjusted input and output (projection) $(\bar{\bar{\mathbf{x}}}_k,\bar{\bar{\mathbf{y}}}_k)$ by use of

$$
\begin{aligned}
\bar{\bar{\mathbf{x}}}_k &= \mathbf{x}_k-(1-\sigma_k)\mathbf{s}_k^{-*}-\mathbf{s}_k^{cl-*}\\
\bar{\bar{\mathbf{y}}}_k &= \mathbf{y}_k+(1-\sigma_k)\mathbf{s}_k^{+*}+\mathbf{s}_k^{cl+*}
\end{aligned}
\tag{20.23}
$$

Step 8. Define the scale- and cluster-adjusted dataset by $\left(\bar{\bar{\mathbf{X}}},\bar{\bar{\mathbf{Y}}}\right)=\{(\bar{\bar{\mathbf{x}}}_j,\bar{\bar{\mathbf{y}}}_j)$ $|j=1,\ldots,n\}$.

20.3.5   Global Characterization of SAS-Projected DMUs

From Proposition the SAS-projected $(\bar{\bar{\mathbf{x}}}_k,\bar{\bar{\mathbf{y}}}_k)$ is positioned on the convex frontier within its containing cluster. We can determine whether or not it is located on the global convex frontier by solving the following linear program:

$$
\begin{aligned}
\underline{u}_0(\bar{u}_0)=\min_{\mathbf{v},\mathbf{u},u_0}\left(\max_{\mathbf{v},\mathbf{u},u_0}\right)u_0\\
\text{s.t.}\ \mathbf{v}\bar{\bar{\mathbf{x}}}_k=1,\mathbf{u}\bar{\bar{\mathbf{y}}}_k-u_0=1,\\
-\mathbf{v}\bar{\bar{\mathbf{x}}}_j+\mathbf{u}\bar{\bar{\mathbf{y}}}_j-u_0\le0\ (\forall j),\\
\mathbf{v}\ge0,\mathbf{u}\ge0,\ u_0:\text{free in sign}
\end{aligned}
\tag{20.24}
$$

1. If this program is infeasible, then there is no supporting hyperplane of $\left(\bar{\bar{\mathbf{X}}},\bar{\bar{\mathbf{Y}}}\right)$ at $(\bar{\bar{\mathbf{x}}}_k,\bar{\bar{\mathbf{y}}}_k)$ which is located on the globally *non-convex* frontiers.
2. If this program is feasible, let an optimal solution be $\left(u_0^*,\mathbf{v}^*,\mathbf{u}^*\right)$, with $u_0^*=\underline{u}_0$ (or $\bar{u}_0$). Then, the hyperplane $-\mathbf{v}^*\mathbf{x}+\mathbf{u}^*\mathbf{y}-u_0^*=0$ is a supporting hyperplane

of $\left(\bar{\bar{\mathbf{X}}},\bar{\bar{\mathbf{Y}}}\right)$ at $(\bar{\bar{\mathbf{x}}}_k,\bar{\bar{\mathbf{y}}}_k)$. Hence, $(\bar{\bar{\mathbf{x}}}_k,\bar{\bar{\mathbf{y}}}_k)$ is located on the *convex* frontiers of the SAS-projected DMUs. Furthermore, we can characterize some of the convex frontiers specifically.

3. If there is a vector $\mathbf{v} \geq \mathbf{0}$ such that $\mathbf{v}\bar{\bar{\mathbf{x}}}_k = 1$ and $\mathbf{v}\bar{\bar{\mathbf{x}}}_j \geq 1$ $(\forall j)$, then $(\bar{\bar{\mathbf{x}}}_k,\bar{\bar{\mathbf{y}}}_k)$ is located on the boundary of $\left(\bar{\bar{\mathbf{X}}},\bar{\bar{\mathbf{Y}}}\right)$ and has a supporting hyperplane which is *vertical* with respect to the input axes.

4. If there is a vector $\mathbf{u} \geq \mathbf{0}$ such that $\mathbf{u}\bar{\bar{\mathbf{y}}}_k = 1$ and $\mathbf{u}\bar{\bar{\mathbf{y}}}_j \leq 1$ $(\forall j)$, then $(\bar{\bar{\mathbf{x}}}_k,\bar{\bar{\mathbf{y}}}_k)$ is located on the boundary of $\left(\bar{\bar{\mathbf{X}}},\bar{\bar{\mathbf{Y}}}\right)$ and has a supporting hyperplane which is *horizontal* with respect to the input axes.

5. If $(\bar{\bar{\mathbf{x}}}_k,\bar{\bar{\mathbf{y}}}_k)$ satisfies the above two conditions 3 and 4, then $(\bar{\bar{\mathbf{x}}}_k,\bar{\bar{\mathbf{y}}}_k)$ is located on the corner of $\left(\bar{\bar{\mathbf{X}}},\bar{\bar{\mathbf{Y}}}\right)$ and has supporting hyperplanes which are *vertical* or *horizontal* with respect to the input axes.

We note that the difference between the VRS score and the SAS score is unrelated to the global characteristics of the SAS-projected DMU.

20.4 AN ILLUSTRATIVE EXAMPLE

In this section, we present an artificial example with a single input and a single output. Table 20.1 shows 10 DMUs with input x and output y, while Figure 20.4 gives a graphical interpretation of the same data. These DMUs display a typical S-shaped curve.

Initially, we solved the input-oriented CRS and VRS models, and obtained the scale efficiency and CRS slacks, which were then decomposed into scale-independent

TABLE 20.1 Example.

DMU	Input	Output	Cluster
A	2	1	a
B	3	1.2	a
C	4	2	c
D	4.5	3	d
E	5	5	e
F	6	5.8	e
G	7	6.3	g
H	8	6.7	h
I	9	6.9	i
J	10	7	j

Figure 20.4   Data plot of DMUs defined in Table 20.1.

TABLE 20.2   CRS, VRS, scale efficiency and slacks for the DMUs in Table 20.1.

DMU	CRS-I	VRS-I	Scale efficiency	CRS slacks	Scale-independent slacks	Scale-dependent slacks
A	0.5	1	0.5	1	0.5	0.5
B	0.4	0.717	0.558	1.8	1.0047	0.7953
C	0.5	0.688	0.727	2	1.4545	0.5455
D	0.667	0.778	0.857	1.5	1.2857	0.2143
E	1	1	1	0	0	0
F	0.967	1	0.967	0.2	0.1933	0.0067
G	0.9	1	0.9	0.7	0.63	0.07
H	0.838	1	0.838	1.3	1.0888	0.2113
I	0.767	1	0.767	2.1	1.61	0.49
J	0.7	1	0.7	3	2.1	0.9

and scale-dependent parts. Table 20.2 shows the solutions and the decomposed parts. Since the output, y, has no slacks in this example, they are not included in the table.

In the second phase, we deleted the scale-dependent slacks from the data and obtained the dataset $(\bar{\mathbf{X}}, \bar{\mathbf{Y}})$. We solved the CRS model within each cluster and determined the in-cluster slacks. By summing the scale-dependent slacks and the in-cluster slacks, we obtained the total slacks. Table 20.3 shows the data obtained.

Finally, we computed the adjusted score, θ^{SAS}, and the projected input and output. These values are given in Table 20.4, while Figure 20.5 displays the results graphically. In the table, 'Frontier' indicates the global characteristics of the SAS-projected DMUs, identified using (20.24).

TABLE 20.3 Calculated $(\bar{X}, \bar{Y})$, in-cluster slacks and total slacks.

DMU	Cluster	Scale-independent input	Scale-independent output	Scale-dependent slacks	In-cluster slacks	Total slacks
A	a	1.5	1	0.5	0	0.5
B	a	2.2047	1.2	0.7953	0.4047	1.2
C	c	3.4545	2	0.5455	0	0.5455
D	d	4.2857	3	0.2143	0	0.2143
E	e	5	5	0	0	0
F	e	5.9933	5.8	0.0067	0.1933	0.2
G	g	6.93	6.3	0.07	0	0.07
H	h	7.7888	6.7	0.2113	0	0.2113
I	i	8.51	6.9	0.49	0	0.49
J	j	9.1	7	0.9	0	0.9

TABLE 20.4 Adjusted score (SAS) and projected input and output.

DMU	SAS-I	Projection input	Projection output	Frontier	Cluster
A	0.75	1.5	1	Vertical	a
B	0.6	1.8	1.2	Non-convex	a
C	0.8636	3.4545	2	Non-convex	c
D	0.9524	4.2857	3	Non-convex	d
E	1	5	5	Convex	e
F	0.9667	5.8	5.8	Convex	e
G	0.99	6.93	6.3	Non-convex	g
H	0.9736	7.7888	6.7	Convex	h
I	0.9456	8.51	6.9	Convex	i
J	0.91	9.1	7	Horizontal	j

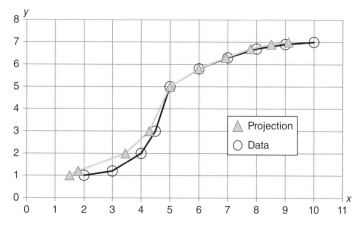

Figure 20.5 Projection and data.

TABLE 20.5   Comparison of the CRS scores, VRS scores and SAS.

DMU	CRS-I	VRS-I	SAS-I
A	0.5	1	0.75
B	0.4	0.7167	0.6
C	0.5	0.6875	0.8636
D	0.6667	0.7778	0.9524
E	1	1	1
F	0.9667	1	0.9667
G	0.9	1	0.99
H	0.8375	1	0.9736
I	0.7667	1	0.9456
J	0.7	1	0.91

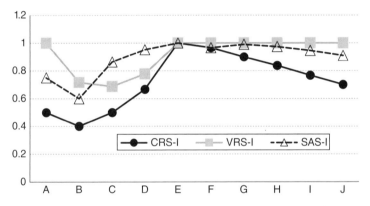

Figure 20.6   Comparison of the CRS scores, VRS scores and SAS.

In Table 20.5 and Figure 20.6, we compare the input-oriented CRS score, VRS score and SAS. The SASs of DMUs C and D have larger values than the scores given by the VRS model. This reflects the non-convex characteristics of the dataset. The projected frontiers are a mixture of non-convex and convex parts.

20.5   THE RADIAL-MODEL CASE

In this section, we develop our model to the case of radial models. We utilize the input-oriented radial measures of the CCR [18] and BCC [19] models for the efficiency evaluation of each DMU $(\bar{\mathbf{x}}_k, \bar{\mathbf{y}}_k)$ $(k = 1, \ldots, n)$ as follows:

$$[\text{CCR}]\ \theta_k^{\text{CCR}} = \min_{\lambda} \theta, \quad \text{s.t. } \mathbf{X}\lambda \leq \theta \mathbf{x}_k, \mathbf{Y}\lambda \geq \mathbf{y}_k, \ \lambda \geq \mathbf{0}, \ \theta : \text{free} \qquad (20.25)$$

$$[\text{BCC}]\ \theta_k^{\text{BCC}} = \min_{\lambda} \theta, \quad \text{s.t. } \mathbf{X}\lambda \leq \theta \mathbf{x}_k, \mathbf{Y}\lambda \geq \mathbf{y}_k,\ \mathbf{e}\lambda = 1,\ \lambda \geq \mathbf{0},\ \theta : \text{free} \qquad (20.26)$$

where $\lambda \in \text{IR}^n$ is the intensity vector.

Although we present our model in the case of the input-oriented radial model, we can develop the model for the output-oriented radial model as well.

We define the scale efficiency (σ_k) of DMU$_k$ by

$$\sigma_k = \frac{\theta_k^{\text{CCR}}}{\theta_k^{\text{BCC}}} \qquad (20.27)$$

20.5.1 Decomposition of CCR Slacks

We decompose the CRS score into scale-independent and scale-dependent parts as follows.

The radial input slack can be defined as

$$\mathbf{s}_k^- = \left(1 - \theta_k^{\text{CCR}}\right)\mathbf{x}_k \in R^m \qquad (20.28)$$

We decompose this radial input slack into scale-dependent and scale-independent slacks as follows:

$$\mathbf{s}_k^- = \left(1 - \sigma_k\right)\mathbf{s}_k^- + \sigma_k \mathbf{s}_k^- \qquad (20.29)$$

$$\begin{aligned}\text{Scale-dependent input slack}: \mathbf{s}_k^{\text{ScaleDep}-} &= \left(1 - \sigma_k\right)\mathbf{s}_k^- = \left(1 - \sigma_k\right)\left(1 - \theta_k^{\text{CCR}}\right)\mathbf{x}_k\\ \text{Scale-independent input slack}: \mathbf{s}_o^{\text{ScaleIndep}-} &= \sigma_k \mathbf{s}_k^- = \sigma_k\left(1 - \theta_k^{\text{CCR}}\right)\mathbf{x}_k\end{aligned} \qquad (20.30)$$

20.5.2 Scale-Adjusted Input and Output

We define the scale-adjusted input $\bar{\mathbf{x}}_k$ and output $\bar{\mathbf{y}}_k$ by

$$\begin{aligned}\bar{\mathbf{x}}_k &= \mathbf{x}_k - \mathbf{s}_k^{\text{ScaleDep}-} = \left(\sigma_k + \theta_k^{\text{CCR}} - \sigma_k \theta_k^{\text{CCR}}\right)\mathbf{x}_k\\ \bar{\mathbf{y}}_k &= \mathbf{y}_k\end{aligned} \qquad (20.31)$$

We define the scale-adjusted score by

$$\theta_k^{\text{scale}} = \sigma_k + \theta_k^{\text{CCR}} - \sigma_k \theta_k^{\text{CCR}} \qquad (20.32)$$

We have the following propositions.

Proposition 20.6 $\qquad\qquad 1 \geq \theta_k^{\text{scale}} \geq \max\left(\theta_k^{\text{CCR}}, \sigma_k\right)$ $\qquad\qquad$ (20.33)

Proposition 20.7 $\qquad\qquad \theta_k^{\text{scale}} = 1$ if and only if $\sigma_k = 1$ $\qquad\qquad$ (20.34)

Proofs of these propositions are given in Appendix 20.B.

20.5.3 Solving the CCR Model in the Same Cluster

We introduce the clusters of DMUs in the same manner as mentioned in the non-radial (SBM) case. We solve the input-oriented CCR model for each DMU $(\bar{\mathbf{x}}_k, \bar{\mathbf{y}}_k)$ $(k = 1, \ldots, n)$, referring to the $(\bar{\mathbf{X}}, \bar{\mathbf{Y}})$ in the same cluster (k), which can be formulated as follows:

$$
\begin{aligned}
&\theta_k^{cl*} := \min_{\boldsymbol{\mu}} \theta_k^{cl} \\
&\text{s.t. } \bar{\mathbf{X}}\boldsymbol{\mu} - \theta_k^{cl}\bar{\mathbf{x}}_k \leq \mathbf{0}, \\
&\bar{\mathbf{Y}}\boldsymbol{\mu} \geq \bar{\mathbf{y}}_k, \\
&\mu_j = 0 \ (\forall j : \text{Cluster}(j) \neq \text{Cluster}(k)) \\
&\boldsymbol{\mu} \geq \mathbf{0}, \ \theta_k^{cl} : \text{free}
\end{aligned}
\tag{20.35}
$$

The scale-cluster-adjusted data (projection) $(\bar{\bar{\mathbf{x}}}_k, \bar{\bar{\mathbf{y}}}_k)$ is defined by

$$
\text{Scale} - \text{cluster} - \text{adjusted input (Projected input)}: \bar{\bar{\mathbf{x}}}_k = \theta_k^{cl*}\bar{\mathbf{x}}_k = \theta_k^{cl*}\theta_k^{\text{scale}}\mathbf{x}_k,
$$
$$
\text{Projected output}: \bar{\bar{\mathbf{y}}}_k = \mathbf{y}_k
\tag{20.36}
$$

At this point, we have deleted the scale demerits and in-cluster slacks from the dataset. Thus, we have obtained a scale-free and in-cluster slacks-free (projected) dataset $\left(\bar{\bar{\mathbf{X}}}, \bar{\bar{\mathbf{Y}}}\right)$.

20.5.4 Scale- and Cluster-Adjusted Score

In the input-oriented case, the SAS is defined by

$$
\text{SAS}: \theta_k^{\text{SAS}} = \theta_k^{cl*}\theta_k^{\text{scale}}
\tag{20.37}
$$

In this case, the SAS is the product of the in-cluster efficiency and the scale-adjusted score. This differs from the definition of the SAS in the non-radial case (20.19), where the SAS is defined by using the sum of the scale-dependent slacks and in-cluster slacks.

Similarly to Propositions 20.1 to 20.4, we have the following propositions.

Proposition 20.8 The SAS is not less than the CCR score:

$$\theta_k^{SAS} \geq \theta_k^{CCR} \qquad (20.38)$$

Proposition 20.9 If $\theta_k^{CCR} = 1$, then $\theta_k^{SAS} = \theta_k^{CCR}$, but not vice versa.

Proposition 20.10 The SAS decreases with an increase in the input and with a decrease in the output as long as both DMUs remain in the same cluster.

Proposition 20.11 The SAS-projected DMU $(\bar{\bar{\mathbf{x}}}_k, \bar{\bar{\mathbf{y}}}_k)$ is radially efficient under the SAS model among the DMUs in the cluster that it belongs to. It is also CCR and BCC efficient among the DMUs in its cluster.

20.6 SCALE-DEPENDENT DATASET AND SCALE ELASTICITY

Thus far, we have discussed the efficiency score considerations of our proposed scheme. In this section, we consider scale elasticity. Many papers have discussed this subject in the guise of the globally-convex-frontier assumption [5–7,20–24]. However, in the case of non-convex frontiers, we believe there is a need for further research. Based on the decomposition of the CRS slacks mentioned in Section 20.2, we have developed a new scale elasticity that can cope with non-convex frontiers.

20.6.1 Scale-Dependent Dataset

We subtract or add scale-independent slacks from or to the dataset, and thus define the scale-dependent dataset $(\hat{\mathbf{x}}_k, \hat{\mathbf{y}}_k)$:

$$\text{Scale-dependent input: } \hat{\mathbf{x}}_k = \mathbf{x}_k - \sigma_k \mathbf{s}_k^{-*}$$
$$\text{Scale-dependent output: } \hat{\mathbf{y}}_k = \mathbf{y}_k + \sigma_k \mathbf{s}_k^{+*} \qquad (20.39)$$

Figure 20.7 illustrates an example of this scheme.

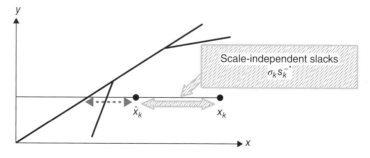

Figure 20.7 Scale-dependent input.

We define the scale-dependent set $(\hat{\mathbf{X}}, \hat{\mathbf{Y}}) = \{(\hat{\mathbf{x}}_k, \hat{\mathbf{y}}_k | k = 1, \ldots, n)\}$. We first project $(\hat{\mathbf{x}}_k, \hat{\mathbf{y}}_k)$ onto the VRS frontier of $(\hat{\mathbf{X}}, \hat{\mathbf{Y}})$ in the same cluster. Thus, we denote the projection $\left(\hat{\mathbf{x}}_k^{\text{Proj}}, \hat{\mathbf{y}}_k^{\text{Proj}}\right)$ by

$$(\hat{\mathbf{x}}_k, \hat{\mathbf{y}}_k) \rightarrow \left(\hat{\mathbf{x}}_k^{\text{Proj}}, \hat{\mathbf{y}}_k^{\text{Proj}}\right) \tag{20.40}$$

20.6.2   Scale Elasticity

The scale elasticity, or 'degree of scale economies', is defined as the ratio of marginal product to average product. In the single-input/output case, if an output y is produced by an input, x, we define the scale elasticity as

$$\varepsilon = \frac{dy}{dx} \bigg/ \frac{y}{x} \tag{20.41}$$

In the case of multiple-input–output environments, the scale elasticity is determined by solving linear programs related to the supporting hyperplane at the respective efficient point [25, pp. 147–149].

The projection set $(\hat{\mathbf{X}}^{\text{Proj}}, \hat{\mathbf{Y}}^{\text{Proj}})$ defined above has at least convex frontiers within each cluster; we can find a supporting hyperplane at $\left(\hat{\mathbf{x}}_k^{\text{Proj}}, \hat{\mathbf{y}}_k^{\text{Proj}}\right)$ that supports all projected DMUs in the cluster and has the minimum deviation t from the cluster. This scheme can be formulated as follows:

$$
\begin{aligned}
\min_{\mathbf{v}, \mathbf{u}, u_0} \ & t \\
\text{s.t.} \ & \mathbf{v}\hat{\mathbf{x}}_k^{\text{Proj}} = 1, \mathbf{u}\hat{\mathbf{y}}_k^{\text{Proj}} - u_0 = 1, \\
& -\mathbf{v}\hat{\mathbf{x}}_j^{\text{Proj}} + \mathbf{u}\hat{\mathbf{y}}_j^{\text{Proj}} - u_0 + w_j = 0 \ (\forall j : \text{Cluster}(j) = \text{Cluster}(k)), \\
& -w_j + t \geq 0 \ (\forall j : \text{Cluster}(j) = \text{Cluster}(o)), \\
& \mathbf{v} \geq \mathbf{0}, \mathbf{u} \geq \mathbf{0}, \ w_j \geq 0 (\forall j), \ t \geq 0, u_0 : \text{free in sign}
\end{aligned}
\tag{20.42}
$$

Let the optimal u_0 be u_0^*. We define the scale elasticity of DMU $(\mathbf{x}_k, \mathbf{y}_k)$ by Scale elasticity:

$$\varepsilon_k = \frac{1}{1 - u_0^*} \tag{20.43}$$

If u_o^* is not uniquely determined, we check its minimum and maximum while keeping t at the optimum value.

The reasons behind applying the scheme outlined above are as follows:

1. Conventional methods assume a global convex production possibility set to identify the returns-to scale (RTS) characteristics of each DMU. However, as we observed, the dataset does not always exhibit convexity. Moreover, the RTS property is a local property, but is not global, as indicated by (20.41). Hence, we need to investigate this issue within the individual cluster containing the DMU, after deleting the scale-independent slacks.

2. Conventional methods usually find multiple optimum values of u_0^*, and there may be a large difference between the minimum and maximum values. The scale elasticity ε_k defined above remains between the minimum and maximum, but has a much smaller range of allowed values.

20.7 APPLICATION TO A DATASET CONCERNING JAPANESE NATIONAL UNIVERSITIES

In this section, we apply our scheme to a dataset comprising information about research output from the faculties of medicine of 37 Japanese national universities.

20.7.1 Data

Table 20.6 shows the dataset concerning the research output of Japanese national universities with a faculty of medicine in 2008 (Report by the Council for Science and Technology Policy, Japanese Government, 2009). We chose two inputs, namely subsidy and number of faculty members, and three outputs, namely number of publications, number of JSPS (Japan Society for Promotion of Sciences) grants and number of funded research projects. Since there are large differences in size among the 37 universities, we classified them into four clusters, A, B, C and D, determined by the sum of the number of JSPS grants and the number of funded research projects. Cluster A was defined as the set of universities with a sum larger than 2000, cluster B between 2000 and 1000, cluster C between 1000 and 500, and cluster D less than 500. The average values for each cluster were 3225 for A, 1204 for B, 653 for C and 348 for D. Determination of the effect of size was one of the objectives of this application.

Figure 20.8 shows the 37 universities, considering the numbers of faculty (input) and of publications (output). Globally non-convex characteristics are observed. A large difference is observed between the big seven universities (cluster A) and the other universities (clusters B, C and D). We can observe similar tendencies when considering other inputs and outputs.

TABLE 20.6   Dataset for Japanese national universities.

DMU	Input		Output			
University	Subsidy	Faculty	Publication	JSPS fund	No. of funded res.	Cluster
A1	96,174	4,549	6,359	2,896	2280	A
A2	60,868	3,562	4,776	2,304	1504	A
A3	50,717	2,619	3,786	1,952	1382	A
A4	50,615	2,877	4,009	1,941	1357	A
A5	42,398	2,207	2,605	1,396	1186	A
A6	41,014	2,086	2,560	1,310	922	A
A7	35,985	1,792	2,443	1,351	796	A
B1	48,106	1,667	1,549	911	507	B
B2	28,896	1,814	1,362	811	543	B
B3	22,898	1,567	1,089	751	401	B
B4	18,245	1,303	1,143	606	453	B
B5	18,255	1,505	1,264	606	430	B
C1	19,200	1,129	803	537	314	C
C2	17,565	1,010	722	446	302	C
C3	20,467	1,224	706	428	317	C
C4	16,124	1,151	582	309	418	C
C5	14,515	867	643	351	321	C
C6	17,154	1,084	685	378	284	C
C7	13,196	898	481	325	329	C
C8	12,357	830	446	242	357	C
C9	14,850	799	628	266	319	C
C10	13,138	855	576	353	228	C
C11	16,884	1,121	531	311	265	C
C12	14,589	970	562	277	274	C
C13	14,436	976	550	311	229	C
D1	10,631	629	293	199	231	D
D2	11,319	795	465	190	233	D
D3	10,202	657	300	170	240	D
D4	10,953	668	311	184	191	D
D5	13,017	859	382	201	159	D
D6	11,355	775	339	191	156	D
D7	11,522	779	391	162	171	D
D8	10,637	785	287	174	142	D
D9	8,936	656	267	157	153	D
D10	11,054	692	343	158	134	D
D11	10,888	749	323	157	132	D
D12	10,686	645	254	152	135	D

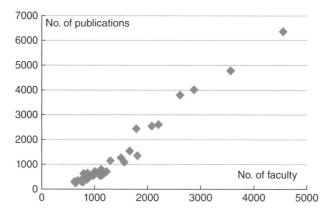

Figure 20.8 Plot of the number of faculty (input) versus number of publications (output).

20.7.2 Adjusted Score (SAS)

Table 20.7 compares the CRS scores, VRS scores and SAS for this dataset by means of the non-radial model, and Figure 20.9 displays the results graphically. In the table, 'Frontier' indicates the global characterization of the SAS-projected DMUs obtained from (20.26), with abbreviations H = horizontal, C = convex, N = non-convex and V = vertical.

The SASs of B1, B2 and B3 are much larger than the VRS score, demonstrating the non-convex structure of the dataset. The universities in cluster A are judged to be almost efficient when considering the adjusted scores. Table 20.8 summarizes the averages of the CRS scores, VRS scores and SAS for each cluster. For cluster A universities, the differences between the three scores are small, and these universities have the highest scores for each model. For cluster B universities, the average SAS is larger than the average VRS score. This indicates the existence of non-convex frontiers around cluster B universities. For cluster C universities, the discrepancy between the CRS and VRS scores becomes large, and the average SAS is between these values, but closer to the VRS score. For cluster D universities, the discrepancy between the scores is the largest, indicating the smallest scale efficiency. The SASs are positioned almost equally between the CRS and VRS scores. The average SAS decreases monotonically from cluster A to cluster D.

20.7.3 Scale Elasticity

Table 20.9 shows the scale elasticity ε calculated using (20.43).

TABLE 20.7   Comparison of the CRS scores, VRS scores, SAS and frontier.

DMU	CRS-I	VRS-I	SAS-I	Front
A1	0.925	1	0.925	H
A2	0.976	1	0.976	C
A3	1	1	1	C
A4	1	1	1	C
A5	1	1	1	C
A6	0.842	0.904	0.842	C
A7	1	1	1	C
B1	0.613	0.678	0.963	N
B2	0.665	0.764	0.858	N
B3	0.748	0.876	0.963	N
B4	0.779	1	0.951	N
B5	0.74	1	0.932	C
C1	0.682	0.9	0.923	N
C2	0.626	0.892	0.889	N
C3	0.527	0.734	0.729	N
C4	0.801	0.856	0.904	N
C5	0.74	0.971	0.938	N
C6	0.548	0.815	0.769	N
C7	0.787	0.999	0.955	N
C8	1	1	1	C
C9	0.755	1	0.94	N
C10	0.626	1	0.86	N
C11	0.501	0.726	0.651	N
C12	0.599	0.854	0.764	N
C13	0.511	0.843	0.719	N
D1	0.73	1	0.927	N
D2	0.641	0.986	0.874	N
D3	0.76	1	0.943	N
D4	0.578	0.951	0.803	N
D5	0.394	0.814	0.643	N
D6	0.435	0.88	0.69	N
D7	0.471	0.916	0.701	N
D8	0.409	0.865	0.648	N
D9	0.523	1	0.773	C
D10	0.403	0.952	0.656	N
D11	0.385	0.899	0.616	C
D12	0.421	0.95	0.638	V

We observe that, for cluster A universities, the scale elasticity is almost unity, with a maximum value of 1.0669 and a minimum value of 0.961. This cluster exhibits constant returns to scale. The universities in clusters B, C and D have an elasticity value higher than unity, and the average elasticity increases from cluster B to cluster D. These universities have increasing-returns-to-scale characteristics.

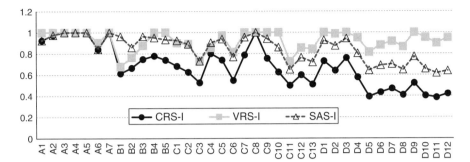

Figure 20.9 Comparison of the CRS scores, VRS scores and SAS.

TABLE 20.8 Average scores.

Cluster	CRS-I	VRS-I	SAS-I
A	0.9632	0.9862	0.9632
B	0.7087	0.8635	0.9334
C	0.6693	0.8916	0.8492
D	0.5124	0.9344	0.7426

TABLE 20.9 Scale elasticity values.

DMU	Scale elasticity	DMU	Scale elasticity	DMU	Scale elasticity	DMU	Scale elasticity
A1	0.961	B1	1.1522	C1	1.137	D1	1.6564
A2	0.9954	B2	1.0915	C2	1.422	D2	1.0532
A3	1.0267	B3	1.1965	C3	1.296	D3	1.7399
A4	1.0299	B4	1.3262	C4	1.152	D4	3.1328
A5	1.0525	B5	1.2003	C5	1.416	D5	1.9453
A6	1.051			C6	1.33	D6	2.034
A7	1.0669			C7	1.197	D7	1.9234
				C8	1.139	D8	3.5783
				C9	1.311	D9	2.1912
				C10	1.56	D10	2.0527
				C11	2.043	D11	2.1179
				C12	2.02	D12	2.1913
				C13	1.56		
Ave.	1.0262	Ave.	1.1933	Ave.	1.429	Ave.	2.1347
Max	1.0669	Max	1.3262	Max	2.043	Max	3.5783
Min	0.961	Min	1.0915	Min	1.137	Min	1.0532
StDev	0.0369	StDev	0.0863	StDev	0.303	StDev	0.6563

20.8 CONCLUSIONS

Most DEA models assume convex efficient frontiers for evaluation of DEA scores. However, in real-world situations, there exist non-convex frontiers which cannot be identified by the traditional models. Non-convex frontiers result from many factors, for example region, environment, ownership, size of enterprise, category of business and age. If we categorize DMUs into several classes by these factors and evaluate efficiencies within each class, the scores gained from such classification are local and we cannot obtain a global (overall) measurement of their performance.

We have developed a scale- and cluster-adjusted DEA model assuming scale efficiency and clustering of DMUs. The scale- and cluster-adjusted score reflects the inefficiency of the DMUs after removing the inefficiencies caused by scale demerits and accounting for in-cluster inefficiency. This model can identify non-convex (S-shaped) frontiers reasonably well. We have also proposed a new scheme for the evaluation of scale elasticity. We have applied this model to a dataset comprising the research input and output of Japanese universities.

The major implications of this study are as follows:

1. By using this model, we become free from the big differences typically observed between CRS and VRS scores. Many practitioners are puzzled as to which one is to be applied to their problem. Our approach will be of help with this problem when several clusters exist. Hence, the use of DEA will become more convenient and simple.
2. We do not require any statistical tests of the range of the intensity vector λ.
3. The model can cope with non-convex frontiers, for example S-shaped curves. In such cases, it is observed that even the VRS scores could become too stringent to be applied in the case of some DMUs.

Although we have presented the scheme in input-oriented form, we can extend it to output-oriented and non-radial non-oriented (both-oriented) models, as well as to directional distance models.

The main purpose of this paper was to introduce a DEA model that can cope with non-convex frontiers by recognizing the impact of scale efficiency (scale merits and scale demerits) and clusters. In our model, clustering plays a fundamental role, which is as important as the selection of DMUs, input/output items and DEA models.

Future research subjects include studies of alternative scale efficiency measures, rather than using the CRS/VRS ratios and clustering methods. Applications to negative data, cost, revenue and profit models are also potential subjects for future research. We believe that this study introduces a new roadmap for DEA research.[4]

---

[4] Software for non-convex DEA models is included in DEA-Solver Pro V13 (http://www.saitech-inc.com). See also Appendix A.

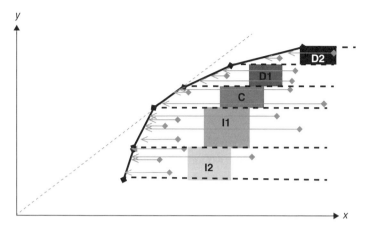

Figure 20.10 Clustering by the degree of scale efficiency.

APPENDIX 20.A CLUSTERING USING RETURNS TO SCALE AND SCALE EFFICIENCY

We already know the RTS characteristics of each DMU, that is, increasing returns to scale (IRS), CRS or decreasing returns to scale (DRS), as they are obtained from the VRS solution by projecting VRS-inefficient DMUs onto the VRS-efficient frontiers. We first classify the CRS DMUs as cluster C. Then, we classify the IRS DMUs depending on the degree of scale efficiency, σ. For example, IRS DMUs with $1 > \sigma \geq 0.8$ may be classified as I1, IRS DMUs with $0.8 > \sigma \geq 0.6$ as I2 and so on. DRS DMUs with $1 > \sigma \geq 0.8$ may be classified as D1 and so on, as above. We determine the number of clusters and their bandwidth by considering the number of DMUs in the clusters. Each cluster is expected to have at least as many DMUs as a few times the sum of the input and output factors. Figure 20.10 illustrates this point. This figure corresponds to the input-oriented case, where DMUs with highly different input scales may be classified into the same cluster. If such a classification is inappropriate, we may try the output-oriented, non-oriented or directional distance model to determine the clusters.

APPENDIX 20.B PROOFS OF PROPOSITIONS

Proposition 20.1 $\qquad \theta_k^{SAS} \geq \theta_k^{CRS} \quad (k = 1, \ldots, n)$

Proof The CRS scores for $(\mathbf{x}_k, \mathbf{y}_k)$ and $(\bar{\mathbf{x}}_k, \bar{\mathbf{y}}_k)$, respectively, are defined by

[CRS]

$$\theta_k^{CRS} = \min_{\lambda, \mathbf{s}^-, \mathbf{s}^+} \left(1 - \frac{1}{m} \sum_{i=1}^{m} \frac{s_i^-}{x_{ik}} \right) \text{s.t. } \mathbf{X}\lambda + \mathbf{s}^- = \mathbf{x}_k, \mathbf{Y}\lambda - \mathbf{s}^+ = \mathbf{y}_k, \lambda \geq \mathbf{0}, \mathbf{s}^- \geq \mathbf{0}, \mathbf{s}^+ \geq \mathbf{0}$$

$$(20.B1)$$

and

$$[\text{SAS}] \quad \theta_k^{\text{SAS}} = \min_{\boldsymbol{\mu}, \mathbf{s}^{cl-}, \mathbf{s}^{cl+}} \left(1 - \frac{1}{m} \sum_{i=1}^{m} \frac{s_i^{cl-} + (1-\sigma_k)\, s_i^{-*}}{\bar{x}_{ik}} \right)$$

$$\text{s.t.} \ \bar{\mathbf{X}}\boldsymbol{\mu} + \mathbf{s}^{cl-} = \bar{\mathbf{x}}_k, \ \bar{\mathbf{Y}}\boldsymbol{\mu} - \mathbf{s}^{cl+} = \bar{\mathbf{y}}_k, \ \mu_j = 0 (\forall j : \text{Cluster}(j) \neq \text{Cluster}(k)),$$

$$\boldsymbol{\mu} \geq \mathbf{0}, \ \mathbf{s}^{cl-} \geq \mathbf{0}, \ \mathbf{s}^{cl+} \geq \mathbf{0} \tag{20.B2}$$

We prove this proposition for two individual cases.

(Case 1) All DMUs belong to the same cluster.
In this case (20.B2) becomes

$$[\text{SAS}] \quad \theta_k^{\text{SAS}} = \min_{\boldsymbol{\lambda}, \mathbf{t}^-, \mathbf{t}^+} \left(1 - \frac{1}{m} \sum_{i=1}^{m} \frac{t_i^- + (1-\sigma_k)\, s_k^{-*}}{\bar{x}_{ik}} \right) \tag{20.B3}$$

$$\text{s.t.} \quad \bar{\mathbf{X}}\boldsymbol{\lambda} + \mathbf{t}^- = \bar{\mathbf{x}}_k, \ \bar{\mathbf{Y}}\boldsymbol{\lambda} - \mathbf{t}^+ = \bar{\mathbf{y}}_k, \boldsymbol{\lambda} \geq \mathbf{0}, \mathbf{t}^- \geq \mathbf{0}, \mathbf{t}^+ \geq \mathbf{0}$$

Let $(\boldsymbol{\lambda}^*, \mathbf{t}^{-*}, \mathbf{t}^{+*})$ be an optimal solution of (20.B2). Since $P(\mathbf{X}, \mathbf{Y}) = P(\bar{\mathbf{X}}, \bar{\mathbf{Y}})$ by Lemma , and both sets have the same efficient DMUs which span $(\bar{\mathbf{x}}_j, \bar{\mathbf{y}}_j)$, we have

$$\mathbf{X}\boldsymbol{\lambda}^* + \mathbf{t}^{-*} = \bar{\mathbf{x}}_k = \mathbf{x}_k - (1-\sigma_k)\mathbf{s}_k^{-*}$$
$$\mathbf{Y}\boldsymbol{\lambda}^* - \mathbf{t}^{+*} = \bar{\mathbf{y}}_k = \mathbf{y}_k + (1-\sigma_k)\mathbf{s}_k^{+*} \tag{20.B4}$$

Hence, we have

$$\mathbf{X}\boldsymbol{\lambda}^* + \mathbf{t}^{-*} + (1-\sigma_k)\mathbf{s}_k^{-*} = \mathbf{x}_k$$
$$\mathbf{Y}\boldsymbol{\lambda}^* - \mathbf{t}^{+*} - (1-\sigma_k)\mathbf{s}_k^{+*} = \mathbf{y}_k \tag{20.B5}$$

This indicates that $(\boldsymbol{\lambda}^*, \mathbf{t}^{-*} + (1-\sigma_k)\mathbf{s}_k^{-*}, \mathbf{t}^{+*} + (1-\sigma_k)\mathbf{s}_k^{+*})$ is feasible for (B1), and hence its objective function value is not less than the optimal value, θ_k^{CRS}:

$$\theta_k^{\text{SAS}} = 1 - \frac{1}{m} \sum_{i=1}^{m} \frac{t_i^{-*} + (1-\sigma_k)\, s_{ik}^{-*}}{x_{ik}} \geq \theta_k^{\text{CRS}} \tag{20.B6}$$

Conversely, $\mathbf{t}^{-*} = \sigma_k \mathbf{s}_k^{-*}$ and $\mathbf{t}^{+*} = \sigma_k \mathbf{s}_k^{+*}$ are feasible for the SAS, and hence it holds that $\theta_k^{\text{SAS}} = \theta_k^{\text{CRS}} \ (k = 1, ..., n)$.

(Case 2) Multiple clusters exist.
In this case, we have constraints additional to (20.B3) to define a restriction on clusters, as follows:

$$[\text{SAS}] \quad \theta_k^{\text{SAS}} = \min\left(1 - \frac{1}{m}\sum_{i=1}^{m}\frac{t_i^- + (1-\sigma_k)\,s_i^{-*}}{\bar{x}_{ik}}\right)$$

$$\text{s.t.} \quad \bar{\mathbf{X}}\boldsymbol{\lambda} + \mathbf{t}^- = \bar{\mathbf{x}}_k\,, \bar{\mathbf{Y}}\boldsymbol{\lambda} - \mathbf{t}^+ = \bar{\mathbf{y}}_k, \lambda_j = 0\;(\forall j : \text{Cluster}(j) \ne \text{Cluster}(k)),$$

$$\boldsymbol{\lambda} \ge \mathbf{0}, \mathbf{t}^- \ge \mathbf{0}, \mathbf{t}^+ \ge \mathbf{0}$$

$$(20.\text{B7})$$

Since adding constraints results in an increase of the objective value, it holds that

$$\theta_k^{\text{SAS}} \ge \theta_k^{\text{CRS}} \qquad (20.\text{B8})$$

∎

Proposition 20.3 If $\theta_o^{\text{CRS}} = 1$, then it holds that $\theta_o^{\text{SAS}} = \theta_o^{\text{CRS}}$, but not vice versa.

Proof If $\theta_k^{\text{CRS}} = 1$, then we have $\mathbf{s}_k^{-*} = \mathbf{0}$ and $\mathbf{s}_k^{+*} = \mathbf{0}$. Hence, we have total slacks $= 0$ and $\theta_k^{\text{SAS}} = 1$. The converse is not always true, as demonstrated in the example below, where all DMUs belong to an independent cluster.

DMU	(I) x	(O) y	Cluster
A	2	2	a
B	4	2	b
C	6	2	c

DMU	CRS-I	SAS-I	Cluster
A	1	1	a
B	0.5	1	b
C	0.3333	1	c

∎

Proposition 20.4 The SAS decreases with increasing input and decreasing output as long as both DMUs remain in the same cluster.

Proof Let $(\mathbf{x}_p, \mathbf{y}_p)$ and $(\mathbf{x}_q, \mathbf{y}_q)$, with $\mathbf{x}_p \le \mathbf{x}_q$ and $\mathbf{y}_p \ge \mathbf{y}_q$, be the original and varied DMUs, respectively, in the same cluster. Let $\mathbf{x}_q = \mathbf{x}_p + \boldsymbol{\delta}_p^-\;\left(\boldsymbol{\delta}_p^- \ge \mathbf{0}\right)$ and $\mathbf{y}_q = \mathbf{y}_p - \boldsymbol{\delta}_p^+\;\left(\boldsymbol{\delta}_p^+ \ge \mathbf{0}\right)$, and let the optimal solution for $(\mathbf{x}_p, \mathbf{y}_p)$ be $\left(\theta_p^{\text{SAS}}, \boldsymbol{\lambda}_p^*, \mathbf{s}_p^{-*}, \mathbf{s}_p^{+*}\right)$. We have $\mathbf{X}\boldsymbol{\lambda}_p^* + \mathbf{s}_p^{-*} = \mathbf{x}_p = \mathbf{x}_q - \boldsymbol{\delta}_p^-,\ \mathbf{Y}\boldsymbol{\lambda}_p^* - \mathbf{s}_p^{+*} = \mathbf{y}_p = \mathbf{y}_q + \boldsymbol{\delta}_p^+$. Hence $\left(\mathbf{s}_p^{-*} + \boldsymbol{\delta}_p^-, \mathbf{s}_p^{+*} + \boldsymbol{\delta}_p^+\right)$ is a feasible slack for $(\mathbf{x}_q, \mathbf{y}_q)$. We have

$$\theta_p^{\text{SAS}} = 1 - \frac{1}{m}\sum_{i=1}^{m}\frac{s_{ip}^{-*}}{x_{ip}} = 1 - \frac{1}{m}\sum_{i=1}^{m}\frac{s_{ip}^{-*}}{x_{iq} - \delta_p^-} \ge 1 - \frac{1}{m}\sum_{i=1}^{m}\frac{s_{ip}^{-*} + \delta_p^-}{x_{iq}} \ge \theta_q^{\text{SAS}}$$

$$(20.\text{B9})$$

∎

Proposition 20.5 The projected DMU $(\bar{\bar{\mathbf{x}}}_k, \bar{\bar{\mathbf{y}}}_k)$ is efficient under the SAS model among the DMUs in its containing cluster. It is CRS and VRS efficient among the DMUs in its cluster.

Proof From the definition of $(\bar{\bar{\mathbf{x}}}_k, \bar{\bar{\mathbf{y}}}_k)$, it is SAS efficient. Thus, it is CRS and VRS efficient in its cluster. ∎

Proposition 20.6

$$1 \geq \theta_k^{\text{scale}} \geq \max\left(\theta_k^{\text{CCR}}, \sigma_k\right)$$

Proof $\sigma_k + \theta_k^{\text{CCR}} - \sigma_k \theta_k^{\text{CCR}} = \sigma_k\left(1 - \theta_k^{\text{CCR}}\right) + \theta_k^{\text{CCR}} = \theta_k^{\text{CCR}}\left(1 - \sigma_k\right) + \sigma_k \geq \max\left\{\sigma_k, \theta_k^{\text{CCR}}\right\}$

$$(20.\text{B}10)$$

This term is increasing in σ_k and is equal to 1 when $\sigma_k = 1$. ∎

Proposition 20.7 $\theta_k^{\text{scale}} = 1$ if and only if $\sigma_k = 1$ $(20.\text{B}11)$

Proof If $\sigma_k = 1$, it holds that $\theta_k^{\text{scale}} = \sigma_k + \theta_k^{\text{CCR}} - \sigma_k \theta_k^{\text{CCR}} = 1$. Conversely, if $\theta_k^{\text{scale}} = \sigma_k + \theta_k^{\text{CCR}} - \sigma_k \theta_k^{\text{CCR}} = 1$, we have $\sigma_k\left(1 - \theta_k^{\text{CCR}}\right) = 1 - \theta_k^{\text{CCR}}$. Hence, if $\theta_k^{\text{CCR}} < 1$, then it holds that $\sigma_k = 1$. If $\theta_k^{\text{CCR}} = 1$, then we have $\theta_k^{\text{BCC}} = 1$ and $\sigma_k = 1$. ∎

REFERENCES

[1] Avkiran, N.K. (2001) Investigating technical and scale efficiencies of Australian universities through data envelopment analysis. *Socio-Economic Planning Science*, **35**, 57–80.

[2] Avkiran, N.K., Tone, K., Tsutsui, M. (2008) Bridging radial and non-radial measures of efficiency in DEA. *Annals of Operations Research*, **164**, 127–138.

[3] Bogetoft, P., Otto, L. (2010) *Benchmarking with DEA, SFA, and R*. Springer.

[4] Dekker, D., Post, T. (2001) A quasi-concave DEA model with an application for bank branch performance evaluation. *European Journal of Operational Research*, **132**, 296–311.

[5] Kousmanen, T. (2001) DEA with efficiency classification preserving conditional convexity. *European Journal of Operational Research*, **132**, 326–342.

[6] Podinovski, V.V. (2004) Local and global returns to scale in performance measurement. *Journal of the Operational Research Society*, **55**, 170–178.

[7] Olesen, O.B., Petersen, N.C. (2013) Imposing the regular ultra Passum law in DEA models. *Omega: The International Journal of Management Science*, **41**, 16–27.

[8] Farrell, M.J. (1957) The measurement of productive efficiency. *Journal of the Royal Statistical Society, Series A (General)*, **120**(3), 253–281.

[9] Farrell, M.J., Fieldhouse, M. (1962) Estimating efficient production functions under increasing returns to scale. *Journal of the Royal Statistical Society, Series A (General)*, **125**(2), 252–267.

[10] Charnes, A., Cooper, W.W., Rhodes, E. (1981) Evaluating program and managerial efficiency: An application of data envelopment analysis to Program Follow Through. *Management Science*, **27**, 668–678.

[11] Førsund, F.R., Kittelsen, S.A.C., Krivonozhko, V.E. (2009) Farrell revisited – Visualizing properties of DEA production frontiers. *Journal of the Operational Research Society*, **60**, 1535–1545.

[12] Krivonozhko, V.E., Utkin, O.B., Volodin, A.V., Sablin, I.A., Patrin, M. (2004) Constructions of economic functions and calculation of marginal rates in DEA using parametric optimization methods. *Journal of the Operational Research Society*, **55**, 1049–1058.

[13] Deprins, D., Simar, L., Tulkens, H. (1984) Measuring labor efficiency in post office, in *The Performance of Public Enterprises: Concepts and Measurement* (eds P. Marchand, P. Pestieau and H. Tulkens), North-Holland, pp. 243–267.

[14] Tone, K. (2001) A slacks-based measure of efficiency in data envelopment analysis. *European Journal of Operational Research*, **130**, 498–509.

[15] Avkiran, N.K. (2011) Applications of data envelopment analysis in the service sector, in *Handbook on Data Envelopment Analysis* (eds W.W. Cooper, L.M. Seiford and J. Zhu), Springer, Chapter 15.

[16] Paradi, J.C., Yang, Z., Zhu, H. (2011) Assessing bank and bank branch performance – Modeling considerations and approaches, in *Handbook on Data Envelopment Analysis* (eds W.W. Cooper, L.M. Seiford and J. Zhu), Springer, Chapter 13.

[17] Cook, W.D. (2011) Qualitative data in DEA, in *Handbook on Data Envelopment Analysis* (eds W.W. Cooper, L.M. Seiford and J. Zhu), Springer, Chapter 6.

[18] Charnes, A., Cooper, W.W., Rhodes, E. (1978) Measuring the efficiency of decision making units. *European Journal of Operational Research*, **2**, 429–444.

[19] Banker, R.D., Charnes, A. Cooper, W.W. (1984) Some models for estimating technical and scale inefficiencies in data envelopment analysis. *Management Science*, **30**, 1078–1092.

[20] Banker, R.D., Thrall, R.M. (1992) Estimation of returns to scale using data envelopment analysis. *European Journal of Operational Research*, **62**, 74–84.

[21] Banker, R.D., Cooper, W.W., Seiford, L.M., Thrall, R.M., Zhu, J. (2004) Returns to scale in different DEA models. *European Journal of Operational Research*, **154**, 345–362.

[22] Färe, R., Primond, D. (1995) *Multi-Output Production and Duality: Theory and Application*. Kluwer Academic.

[23] Førsund, F.R., Hjalmarsson, L. (2004) Are all scales optimal in DEA? Theory and empirical evidence. *Journal of Productivity Analysis*, **21**, 25–48.

[24] Førsund, F.R., Hjalmarsson, L. (2004) Calculating scale elasticity in DEA models. *Journal of the Operational Research Society*, **55**, 1012–1038.

[25] Cooper, W.W., Seiford, L.M., Tone, K. (2007) *Data Envelopment Analysis: A Comprehensive Text with Models, Applications, References and DEA-Solver Software*. Springer.

21

USING DEA TO ANALYZE THE EFFICIENCY OF WELFARE OFFICES AND INFLUENCING FACTORS: THE CASE OF JAPAN'S MUNICIPAL PUBLIC ASSISTANCE PROGRAMS

MASAYOSHI HAYASHI

Graduate School of Economics, University of Tokyo, Tokyo, Japan

21.1 INTRODUCTION

Like most OECD countries, Japan is experiencing substantial changes in its socio-economic structure due to the growing number of low-wage workers and the rapid pace of population aging. While this necessitates a series of reforms in public programs targeted at such disadvantaged households, public funds for such programs are limited, restricting the scope of possible reforms. Given this lack of resources, it is therefore important to achieve higher efficiency in welfare program implementation. We are thus naturally interested in examining the performance of welfare organizations and exploring factors that may affect the efficient implementation of their programs.

Data envelopment analysis (DEA) is one of the standard tools used for examining efficiency. While a number of DEA studies have examined public sector activities,

Advances in DEA Theory and Applications: With Extensions to Forecasting Models,
First Edition. Edited by Kaoru Tone.
© 2017 John Wiley & Sons Ltd. Published 2017 by John Wiley & Sons Ltd.

those addressing social assistance are limited [1–5]. This may partly be due to the traditional reluctance to conduct economic evaluation as a part of social policy [2]. In addition, DEA studies of social assistance have addressed varied concerns. While some are interested in the efficiency of welfare offices providing social assistance to a given number of welfare recipients [1, 2], others focus on the efficiency of social expenditure meant to reduce poverty [3, 4]. The efficiency of social assistance spending across different socio-economic environments has also been studied [5].

In this study, we examine the efficiency of welfare offices, not the efficiency of welfare spending. This focus means that the studies of Martin [1] and Ayala *et al.* [2] are of direct relevance to our analysis. Martin [1] may be among the first to have applied DEA to welfare programs, using data from social assistance offices in Oregon. In the same vein, Ayala *et al.* [2] evaluated the efficiency of 41 social services agencies in Madrid, Spain. We aim to improve on these studies by investigating the efficiency of social welfare offices in the Japanese system of local public administration. The Japanese case indeed merits analysis. First, no studies have utilized DEA to investigate efficiency issues in the Japanese social assistance program. Second, reasonably good data are available for the Japanese case, pertaining to caseloads by category for multiple output variables and public employment by type for input variables, both of them at the municipal level. We have also taken advantage of this availability to conduct a second-stage regression (2SR) analysis of the efficiency score for social welfare offices, to examine the factors influencing the efficiency of municipal programs.

This chapter is organized as follows. In Section 21.2, we describe the Japanese system of social assistance and elaborate on the activities of Japanese social welfare offices to set up an input–output model that yields DEA efficiency scores. We then obtain relevant efficiency scores and conduct an analysis. Section 21.3 then explores the effects of external factors on the efficiency scores. For this purpose, we utilize a 2SR analysis. In so doing, we elaborate on the issues concerning 2SR and employ several estimation methods proposed in the literature to obtain a set of estimates and compare the results. In Section 21.4, we extend the analysis in Section 21.3 to perform quantile regressions. We do so in the anticipation that external factors will exert different impacts on the efficiency, depending on the level of the latter. Section 21.5 concludes this study.

21.2 INSTITUTIONAL BACKGROUND, DEA, AND EFFICIENCY SCORES

A DEA study starts by specifying decision-making units (DMUs) and variables for inputs and outputs. Therefore, this section specifies the DMUs and the variables for inputs and outputs for the current analysis. Finding the relevant inputs and outputs for welfare office activities requires an examination of the nature of the actual system. We therefore elaborate on our DEA model, while discussing the institutional mechanism of the Japanese system of social assistance.

21.2.1 DMUs

Public assistance (PA), *Seikatsu Hogo* in Japanese, acts as the last social safety net covering those excluded from the upper layers of social programs and is implemented by local governments in the country. The Public Assistance Law (PAL) allows the Ministry of Health, Labor and Welfare (MHLW) to mandate local governments to implement PA programs. There are two levels of local government in Japan: prefectures and municipalities (cities, towns, villages, and Tokyo Metropolitan Special Wards (TMSWs)). The Social Welfare Law (SWL) requires cities (including TMSWs) and prefectures to set up social welfare offices, through which they implement social programs, including PA programs. The SWL does not require towns and villages to do so. In towns and villages that do not have their own welfare offices, prefectural welfare offices cover their population. We thus used 658 cities in 2010 as our DMUs, excluding villages and towns that have their own social welfare offices. Note that we could not utilize all those cities that have their own welfare offices, since the necessary data required for our analysis was lacking for some of them. Note also that since a small number of cities have more than one social welfare office, the number of DMUs is not necessarily identical to the number of individual welfare offices. Nonetheless, we do not consider this a major problem, as municipalities, not individual welfare offices, make decisions about human resource allocation concerning welfare offices.

21.2.2 Outputs and Inputs

A function of social welfare offices is to provide PA for those who are in need of it. The PA is intended to guarantee the minimum cost of living for Japanese citizens. Through the PAL, the central government sets uniform procedures for localities to follow when they provide PA benefits. That is, local governments do *not* set the eligibility standard or the benefit levels for their PA programs. The PA benefits are equal to the minimum cost of living in excess of what an individual earns with his/her best effort. The MHLW determines the minimum costs of living, allowing for differences in cost due to regional price differences, the formula for which applies uniformly across the nation. To receive benefits, applicants are supposed to exhaust their available resources. The PA program therefore requires local welfare offices to conduct a careful examination, or "means test," of the financial situation of the applicants.

It is then natural to employ welfare caseloads (the number of recipients obtaining assistance) as our choice of the output variable. Indeed, Martin [1] and Ayala *et al.* [2] made analogous choices. However, they also used other variables. In addition to caseloads, Martin [1] used the number of job placements, successful exits (the number of recipients who had been off assistance during the last 18 months at least), and child support benefits. Meanwhile, Ayala *et al.* [2] additionally used the median length of time taken to process applications. Our omission of these additional outputs may be justifiable in the Japanese institutional context. First, Japanese welfare offices do not implement active labor market programs that are comparable to those in other OECD countries. Second, child support benefits are irrelevant in the Japanese case, since another municipal branch is responsible for them. Third, the processing time may be reflected in the caseload size, since a shorter processing time leads to a larger caseload size in a given time period. As we mentioned above, the activities of Japanese welfare offices center on means testing, delivery of benefits, and monitoring of the recipients.

While we focus on welfare caseloads as the output in the current study, a single type of caseload may not suffice. Since the needs of the PA recipients vary depending on their characteristics, the services required for different categories of PA recipients must also be different. Fortunately, we had access to disaggregated caseload data for five categories of recipient households: those made up only of (i) the elderly (those consisting only of people aged 65 years and above), and those headed by (ii) single mothers, (iii) the handicapped, (iv) the sick and injured, and (v) others. The categorization is lexicographic, starting from (i) and proceeding to (v). In FY2010, elderly households constituted the largest proportion (43%), followed by households headed by the injured and sick (23%), and the disabled (11%). The remaining consisted of households headed by single mothers (8%) and others (16%). These categories of recipients apparently require different types of casework. We therefore used their caseloads as different multiple (five) outputs in our analysis.

A natural choice for the inputs is the size of the employment at welfare offices, since "labor" is a straightforward input in this production process. Indeed, Ayala *et al.* [2] and analogous DEA studies on public employment offices [6–8] used the sizes of office staff by type as prime inputs. Our data allow us to differentiate the staff into caseworkers and administrative staff. It is also natural to consider "capital"-type production inputs. For example, Martin [1] considered the number of offices, while Althin *et al.* [8] employed office space. However, the data for office space are not available in the current case. We thus have to content ourselves with the use of a single type of input, i.e., labor. This may not be a serious problem, though, since labor is more relevant than capital for the analysis of welfare programs, given the labor-intensive character of social services [2].

21.2.3 Efficiency Scores

Given the nature of the PA system in Japan, the size of the need is largely exogenous for municipalities. Although it might be possible for welfare offices to implement programs to reduce need, the chances of this are slim. Thus, the concept of efficiency in this analysis concerns how efficiently welfare offices manage a given level of caseloads without reducing the services for the recipients. We attempt this using the input-oriented efficiency score. Put more formally, the efficiency score E is the maximal contraction of all inputs x ((i) caseworkers and (ii) other staff) which still allows us to produce a given combination of outputs y (caseloads for (i) the elderly, (ii) single mothers, (iii) the handicapped, (iv) the sick and injured, and (v) others): $E \equiv \min\{E > 0 \mid (Ex, y) \in T\}$, where T is a technology set. The following measurement then utilizes the Charnes–Cooper–Rhodes (CCR) model [9, 10] to obtain the score E based on the Farrell index of input efficiency.

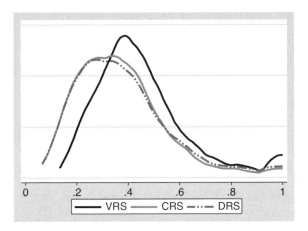

Figure 21.1   Distributions of efficiency scores.

A linear programming exercise yields three types of efficiency score (E_{VRS}, E_{CRS}, E_{DRS}), which are based on the concepts of variable returns to scale (VRS), constant returns to scale (CRS), and decreasing returns to scale (DRS), respectively. Figure 21.1 shows their kernel distributions. There are noticeable differences between the scores based on variable returns to scales (E_{VRS}) and those based on the other two types of returns (E_{CRS}, E_{DRS}). As expected, E_{VRS} tends to yield higher efficiency scores. The distribution of E_{VRS} has an average of 0.459 with a standard deviation (s.d.) of 0.191, and ranges from 0.133 to unity. Meanwhile, the distributions of E_{CRS} and E_{DRS} have an average of 0.376 and 0.387 with an s.d. of 0.183 and 0.196, and range from 0.064 and 0.064, respectively, to unity. We performed bootstrapped tests for returns-to-scale as indicated by Simar and Wilson [11, 12], using (i) the ratio of means, (ii) the mean of ratios, and (iii) the mean of ratios of DEA scores less unity (the number of replications was 3000). The tests rejected both of the null hypotheses of constant returns and decreasing returns to scale at the standard levels of statistical significance. In the next analysis, we therefore proceeded with E_{VRS}, the score based on variable returns to scale. There were indeed large differences in the efficiency score between municipalities. Only 29 municipalities (4.4%) were located on the frontier with a full score of unity, while the other 629 municipalities (95.6%) were off the frontier. As Figure 21.1 indicates, the majority of municipalities had smaller efficiency scores, suggesting sizable room for efficiency improvement.

21.3   EXTERNAL EFFECTS ON EFFICIENCY

21.3.1   Adjustments for Environmental/External Factors

In the previous section, we noted the large variations in the efficiency scores among municipalities. We might then rank individual DMUs according to the efficiency scores so that the less efficient DMUs could use those scores as a benchmark to improve their efficiency. However, doing so may not be appropriate, since Japanese social welfare offices apparently operate in nonhomogeneous environments, which are likely to exert different impacts on the performance of the DMUs. In other words, the "inefficiency" may originate from factors outside the control of the DMUs, thus limiting the scope for efficiency improvement that Figure 21.1 might otherwise indicate. The ranking of the DMUs would then require adjustments of the efficiency scores to allow for different external (environmental) factors.[1]

While the evaluation of individual DMUs is the primary objective of DEA, it may also be meaningful to examine if and how external factors affect the efficiency. Indeed, it is important to examine the factors that affect PA implementation at social welfare offices and to find the directions and degrees of their impacts on efficiency. For example, if the central government can change such external factors, this could improve the overall efficiency of municipal PA programs. For this reason, we discuss only two-stage models in the following sections, and leave the adjustment of efficiency scores to future studies.

21.3.2   The Second-Stage Regression Model

We specified our 2SR model as a typical linear-in-parameters form:

$$E_i = z_i \cdot \beta + u_i \tag{21.1}$$

where E_i is an efficiency measure obtained in the first stage, z_i is a vector of external factors, β is a vector of coefficients, u_i is an error term, and i indexes the DMUs. For the efficiency measure (i.e., dependent variable), we could alternatively use the efficiency score (E_{VRS}) and its reciprocal (the distance function $1/E_{VRS}$). For the factors in z (i.e., explanatory variables), we considered the municipal population (on a log scale), the surface area (on a log scale), the fiscal capacity index (FCI), the obligatory expenses ratio (OER), the local allocation tax (LAT) received (as a binary variable), and the caseload growth rate. The variables are defined later in this section. We selected these six factors for the following reasons.

First, as the literature on local public finance shows, there exist economies of scale in local expenditure up to a certain level of population [19]. Larger localities tend to provide more categories of services than smaller ones [20], yielding economies of *scope* in more populous municipalities. Using the savings from these two scale economies, localities with a higher population could invest more resources so that welfare offices may become more efficient. This line of reasoning suggests that a larger population would lead to higher efficiencies.

Second, the spatial size of a given locality should also matter. Since caseworkers visit their PA recipients within a given period, the more widespread the locations of

---

[1] There are three approaches to such adjustments [13]. First, one-stage models regard external factors as additional inputs in the standard DEA model but obtain the efficiency scores with special restrictions on them [14, 15]. Second, two-stage models obtain the efficiency scores without external factors in the first stage, and regress the scores on a set of external factors in the second stage [16]. They then use the second-stage estimates to adjust the efficiency scores. Third, adjusted-values models utilize the estimates for the effects of external factors on slacks to adjust the values of discretionary variables and obtain DEA efficiency scores with these slack-adjusted values [17, 18].

the recipients, the more the time spent on their cases. Since more time spent on a case implies that the caseworker will handle a smaller number of cases, spacious jurisdictions imply less efficiency in PA services.

Third, fiscal capacity is also a concern. The Japanese government estimates an index for the "fiscal capacity" of localities with data obtained from its system of central grants, namely, the LAT. The amount of LAT a locality receives is the nonnegative difference between its standard fiscal demand (SFD) and standard fiscal revenue (SFR). The SFD estimates the level of local expenditure required to maintain the standard level of public services, while the SFR estimates standardized local revenue. The system then defines the FCI as the three-year average of the ratio of the SFR to the SFD. Another index of fiscal capacity is the OER, which shows the percentage of expenses that a locality cannot easily adjust, including personnel expenses, local debt service payments, and other expenses that the central government requires them to incur. A larger fiscal capacity implies more fiscal abundance, from which localities could spare more resources to provide welfare offices with more caseworkers and administrative staff for a given level of caseloads. This may or may not imply that a larger fiscal capacity would lead to lower efficiency.

However, a drawback of using the FCI as an index of fiscal capacity is its negative correlation with the LAT grant a locality receives. The LAT may adversely affect the efficiency of local spending [21], which then suggests that a large value of the FCI may be associated with a smaller value of the efficiency score. To control the effect of receiving transfers, the regression presented below allows for the receipt of LAT grants. In addition to its claimed adverse effects, the LAT compensates for the local burden of PA expenditure. While the central government disburses 75% of local PA expenditure by means of matching grants to localities, the SFD allows for the rest of the cost. In other words, while LAT recipients enjoy an increase in PA benefits covered by central grants, the nonrecipients have to meet 25% of that increase out of their own pocket. This would imply that receiving LAT grants adversely affects the efficiency score.

Lastly, the speed of caseload changes affects the efficiency. Roughly speaking, efficiency is the ratio of output to input. Since the inputs are the numbers of caseworkers and other staff members, the efficiency tends to increase if the adjustments of the inputs are slow relative to the changes in the outputs. The analysis presented below allows for this aspect by including the rate of increase in PA caseloads from FY2008 to FY2009 as a measured input at the beginning of FY2010.

21.3.3 Econometric Issues

It is important when estimating 2SR models to recognize that the DEA scores obtained in the first stage are *estimates* [22]. We can frame the issue as a typical case of measurement errors in the dependent variables (e.g., [23, pp. 76–77]). The 2SR typically assumes the following data generation process (DGP):

$$E_i^* = z_i \cdot \beta + \varepsilon_i \qquad (21.2)$$

where E_i^* is the *true* value of the efficiency score. Since E_i^* is not observable, the estimated score E_i is a surrogate for E_i^*. Defining the measurement error in the dependent variable as $s_i \equiv E_i - E_i^*$, we can express (21.2) as (21.1) with $u_i \equiv s_i + \varepsilon_i$. If E_i is consistent, s_i approaches zero (i.e., E_i approaches E_i^*) as the sample size approaches infinity. Since E_i is indeed consistent [24, 25], the existence of s_i does not affect the *asymptotic* distribution of estimators for β. In other words, treating DEA scores as *estimates* does not pose an issue if we have a suitably large sample.

On the other hand, we are not sure how large a suitable sample would be, since the convergence of E_i to E_i^* becomes slower as the number of inputs and outputs in the DEA model increases [25]. This "curse of dimensionality" might make the asymptotic approximation fail even with a relatively large sample. In addition, the very calculation of DEA scores for individual DMUs creates correlations among them [26].[2] The curse of dimensionality and the correlation among the scores make u_i nonspherical through s_i in a finite sample, even if ε_i is spherical (i.e., independent and identically distributed). Furthermore, it is very likely that ε_i is also nonspherical since we typically use a sample of cross-sectional data. We thus adjust the covariance matrix of the β estimates to arrive at a valid inference. As the pattern of the nonspherical error u is unknown, our choices include utilizing a heteroskedastic consistent covariance matrix estimator [27] or bootstrapping the covariance matrix [22].

Another econometric issue concerns the method used to estimate (21.1). The most straightforward is the ordinary least squares (OLS) estimator (e.g. [16]), which does not explicitly allow for the fact that the dependent variable is bounded: $E_{\mathrm{VRS}} \in (0, 1]$ or $1/E_{\mathrm{VRS}} \in [1, \infty)$. To allow for these bounds, Bjurek *et al.* [28] estimated the 2SR model with censoring at unity (Tobit estimator). While a number of studies use this method, the Tobit estimator also has its shortcomings when applied to the 2SR model [27]. Therefore, Simar and Wilson [22] modeled it as a linear model with truncation, whereas Hoff [29] and Ramalho *et al.* [30] departed from the linear specification to utilize the fractional response (FR) model of Papke and Wooldrige [31].

21.3.4 Estimation Results

To estimate (21.1), we employed all of the four models mentioned above: the (i) OLS, (ii) Tobit, (iii) truncation, and (iv) FR models.[3] Obviously, the FR model is not applicable to cases that employ $1/E_{\mathrm{VRS}}$ as the dependent variable. For the covariance matrices of these estimators, we bootstrapped the standard errors with 3000 replications to allow for their possible inconsistency and finite sample bias. In addition, we utilized the double bootstrap procedure (Algorithm 2) of Simar and Wilson [22], which essentially replaces the original E_{VRS} with a bias-corrected bootstrapped E_{VRS} when bootstrapping the truncated regression of (21.1). The numbers of replications for the bias-corrected E_{VRS} and 2SR were 200 and 3000, respectively. Note that, since this method is not computationally applicable to an efficiency score with bounds (0, 1] (see [32–34]), we applied it only to the case with the distance function.

Table 21.1 lists the results. The first four columns are for cases using the efficiency score, and the last four are for cases using its reciprocal (the distance function). These

---

[2] Perturbations of DMUs lying on the estimated frontier change the scores of some other DMUs.
[3] We used the logistic distribution for the cumulative distribution that shapes the fractional response.

TABLE 21.1 Estimation results.[a]

Dependent variable	Efficiency score (E_{VRS})				Distance function ($1/E_{VRS}$)			
Model	E-OLS	E-Tobit	E-FR	E-TC	D-OLS	D-Tobit	D-TC	D-TC/SW
ln(population)	.075***	.078***	.075***	.061***	−.269***	−.294***	−.368***	−.362***
	(.014)	(.014)	(.014)	(.011)	(.062)	(.067)	(.104)	(.113)
ln(surface area)	−.027***	−.028***	−.027***	−.026***	.143***	.149***	.229***	.246***
	(.009)	(.009)	(.009)	(.007)	(.043)	(.046)	(.071)	(.083)
FCI	−.264***	−.270***	−.263***	−.246***	1.267***	1.320***	1.929***	2.070***
	(.056)	(.057)	(.056)	(.043)	(.294)	(.301)	(.448)	(.501)
OER	.143	.149	.143	.109	−1.168	−1.222	−1.751	−1.967
	(.166)	(.173)	(.170)	(.145)	(.947)	(.979)	(1.431)	(1.659)
LAT receipt	−.140***	−.145***	−.140***	−.108***	.561**	.602**	.750*	1.000**
	(.049)	(.051)	(.049)	(.064)	(.274)	(.285)	(.436)	(.506)
Caseload growth	−.021	−.016	−.022	−.064	.046	.082	−.082	.098
	(.095)	(.099)	(.096)	(.088)	(.472)	(.491)	(.691)	(.794)
Constant	−.075	−.099	—	−.039	4.538***	4.729***	4.852***	4.904***
	(.171)	(.181)		(.158)	(.895)	(.949)	(1.438)	(1.609)

[a] The sample sizes were 658 for E-OLS, E-Tobit, E-FR, D-OLS, and D-Tobit. The truncation regressions (E-TRC and D-TRC) excluded DMUs with E_{VRS} (1/E_{VRS}) = 1, trimming the sample size down to 629. ***, **, and *indicate $p \leq .01$, $.01 < p \leq .05$, and $.05 < p \leq .10$, respectively. Bootstrapped standard errors are in parentheses (3000 replications). D-TC/SW utilized the r-DEA package of Simm and Besstremyannaya [34] to obtain the bootstrapped bias-corrected efficiency scores as suggested by Simar and Wilson [22] with 200 replications, and then used these scores for the bootstrapped truncated regression with 3000 replications. E-FR lists the marginal effects evaluated at the sample averages.

results are robust in the sense that the statistical significance does not change between different estimation methods for a given dependent variable (E_{VRS} or $1/E_{VRS}$). In all cases, the population, surface area, FCI, and LAT received are all statistically significant, while the OER and caseload changes are not. The results show that a larger population, a smaller surface area, a lower fiscal capacity, and the nonreceipt of LAT grants tend to increase the efficiency as expected, whereas the other two variables do not affect the efficiency. In particular, the insignificance of caseload changes implies that the inputs (caseworkers and other staff members) adjust smoothly to the changes in the outputs (PA caseloads).

Furthermore, the marginal effects of these external factors are very similar among the E-OLS, E-Tobit, and E-FR models and between the D-OLS and D-Tobit models (where "E" and "D" denote efficiency and distance, respectively). On the other hand, the differences in values between those from the truncated regressions and those from other models are rather conspicuous, albeit not so large, with statistically significant coefficients. These differences are likely to be due to different samples rather than the different DGP or estimation method (truncated or not), as the E-TRC and D-TRC models exclude 29 DMUs that have full efficiency scores of unity. In addition, while the D-TRC/SW model does not exclude them, it uses bootstrapped bias-corrected scores whose values not only are different from the standard ones but also differ from unity.

McDonald [27] argued that the idea of efficiency scores as estimates of "true" scores, as suggested by Simar and Wilson [22], "would lead to considerable complexity and perhaps only minor changes in inference." Qualitatively, his argument seems to apply to our cases. All the estimation methods, including OLS, show that a larger population, a smaller surface area, a lower FCI, and nonreceipt of LAT grants would increase the efficiency, whereas the others would not affect the efficiency. Quantitatively, however, the results are somewhat different. In particular, the effects of population, FCI, and LAT receipt differ in the truncation regressions, although the effect of surface area does not change much.

21.4   QUANTILE REGRESSION ANALYSIS

21.4.1   Different Responses along the Quantiles of Efficiency

All the preceding models, except the FR model, assumed that the marginal effects of the external factors on efficiency were constant on average. Such effects may plausibly differ among DMUs with different levels of efficiency, however. This section therefore presents a 2SR using quantile regression (QR) to address the possible different responses of the efficiency scores to the external factors.

With QR, we can estimate the responses of the efficiency score to changes in the external factors across the conditional quantiles of the former. In QR analysis, we first define a conditional quantile function of E as $Q_\tau(E_i \mid z_i) \equiv F^{-1}(\tau \mid z_i)$, where $F(\tau \mid z_i)$ is the cumulative distribution function of E at quantile τ, conditioned on a given set of external factors z_i. We then specify a linear regression model as $E = z \cdot \beta_\tau + u$, where β_τ is a vector of coefficients that vary across quantiles, and u is an error term. The QR estimator of β_τ is a sample analogue of $b_\tau \equiv \text{argmin}_b\ \text{E}\{\rho_\tau(E = z \cdot b)\}$, where

ρ_τ is a check function defined as $\rho_\tau \equiv 1[E - z \cdot b > 0] \cdot \tau \cdot |E - z \cdot b| + 1[E - z \cdot b \leq 0] \cdot (1 - \tau) \cdot | E - z \cdot b|$. This asymmetric weighting scheme results in a minimand that selects conditional quantiles.

We used the super-efficiency score as the dependent variable in the QR analysis. We obtained the super-efficiency score of a given DMU, K, by gauging it against another efficiency frontier calculated with a group of DMUs that *excluded* K [10]. In our case, the super-efficiency score equals the standard efficiency score for DMUs that are *off* the frontier, and takes a value of more than unity for DMUs that are *on* the frontier. Since the scores could take a value exceeding unity, it was convenient for us to use them as the dependent variable of our QR model. Note, however, that we might not be able to calculate super-efficiency scores for all DMUs. As we could not obtain a score for one DMU, the sample size for the second-stage QR was 657.

21.4.2 Results

Table 21.2 lists the results of the QR analysis at the 0.15, 0.25, 0.50, 0.75, and 0.85 quantiles. As a benchmark, we also list the OLS estimates obtained with the super-efficiency scores. The results of the OLS model show that the directional impacts of the external factors are qualitatively the same as those for the E-OLS model in Table 21.1. However, their magnitudes (in absolute value) are larger, reflecting the changes in the efficiency scores from unity for the DMUs on the frontier. The coefficient estimates for the five quantiles are indeed different from those obtained from the OLS models. In addition, the coefficient values vary across the quantiles.

To better understand the changes in coefficients across quantiles, Figure 21.2 plots the coefficient estimates at 18 quantiles (0.05, 0.10, …, 0.90, and 0.95) along with

TABLE 21.2 Estimation results.[a]

	OLS	Quantile				
		.15	.25	.50	.75	.85
ln(population)	.098[***]	.022	.047[***]	.055[***]	.083[***]	.122[***]
	(.021)	(.017)	(.018)	(.016)	(.017)	(.022)
ln(surface area)	−.041[***]	−.017[**]	−.030[***]	−.028[***]	−.025	−.046[***]
	(.013)	(.009)	(.010)	(.010)	(.016)	(.016)
FCI	−.361[***]	−.157[***]	−.268[***]	−.194[***]	−.285[***]	−.477[***]
	(.090)	(.052)	(.066)	(.061)	(.082)	(.100)
OER	−.242	.349[*]	.313[*]	.026	.342	.053
	(.251)	(.172)	(.176)	(.198)	(.264)	(.375)
LAT receipt	−.161[***]	−.065	−.097[**]	−.099	−.256[***]	−.274[***]
	(.055)	(.042)	(.043)	(.075)	(.097)	(.095)
Caseload growth	−.037	−.039	−.083	−.076	−.064	.265
	(.104)	(.155)	(.082)	(.089)	(.160)	(.185)
Constant	−.252	.010	−.040	.164	−.144	−.014
	(.295)	(.230)	(.237)	(.230)	(.246)	(.321)

[a] The sample size was 658. Standard errors were bootstrapped with 3000 replications. ***, **, and * indicate $p \leq .01$, $.01 < p \leq .05$, and $.05 < p \leq .10$, respectively.

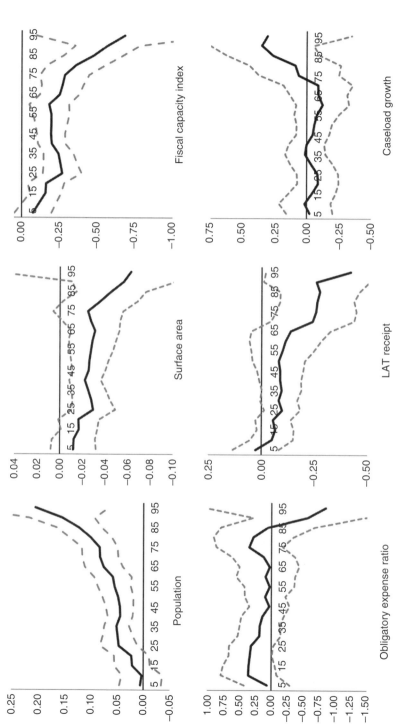

Figure 21.2 QR coefficient estimates and 95% confidence intervals. The solid lines in the panels connect the QR coefficient estimates at 18 quantiles (0.05, 0.10, ..., 0.90, and 0.95). The dotted lines show 95% confidence intervals of the QR estimates.

their 95% confidence intervals. In general, the effects tend to increase in absolute value toward the upper quantiles if they have a large statistical significance (population, surface area, FCI, and LAT), while they seem to change relatively little if the effects are not statistically significant (OER and caseload growth). These findings suggest that the impacts on efficiency are larger for those municipalities located closer to the edge of the frontier.

21.5 CONCLUDING REMARKS

In this study, we obtained the Farrell scores for the input-oriented efficiency of municipal PA programs in Japan and explored the effects of a set of external factors on the efficiency scores. We showed that the efficiency varies across municipalities, implying a large potential for efficiency improvement. Nonetheless, such disparities may be due to variations in external factors that municipalities cannot control. Employing a 2SR analysis with a variety of estimators, we then examined how a set of external factors would affect the efficiency. Our results indicated that surface area, fiscal capacity, and receipt of LAT grants decreased the efficiency, while population improved it. Furthermore, a QR analysis with super-efficiency scores showed that the marginal effects of the external factors on the efficiency would become larger in absolute value for the upper quantiles of the efficiency scores. This may then imply that when we compare efficiency scores among DMUs, we should adjust the scores, taking account of their effects across the quantiles of the efficiency scores. While there is a large body of literature on the adjustment of efficiency scores for variations in external factors (e.g., [13]), to the best of our knowledge, no study explicitly allows for such differentiated effects across quantiles. Our next task should then be to elaborate further on the QR approach in a 2SR analysis and possibly construct adjusted efficiency scores that allow for such quantile effects.

ACKNOWLEDGEMENTS

The author is grateful to Kaoru Tone and the participants of "Workshop 2015: Advances in DEA Theory and Applications with Extensions to Forecasting Models" held at the National Graduate Institute for Policy Studies on December 1 and 2, 2015. In addition, he would like to thank Katsumi Shimotsu for his helpful comments. Parts of this study were financially supported by a Grant-in-Aid for Scientific Research (A-15070000033 and B-15H03359) from the Japan Society for the Promotion of Science.

REFERENCES

[1] Martin, L.L. (2002) Comparing the performance of multiple human service providers using Data Envelopment Analysis. *Administration in Social Work*, **26**(4), 45–60.

[2] Ayala, L., Pedraja, F., and Salinas-Jiménez, J. (2008) Performance measurement of local welfare programmes: Evidence from Madrid's regional government. *Environment and Planning C*, **26**(5), 906–923.

[3] Enache, C. (2012) The efficiency of expenditure-related redistributive policies in the European countries. *Timisoara Journal of Economics*, **5**(18), 380–394.

[4] Habibov, N. and Fan, L. (2010) Comparing and contrasting poverty reduction performance of social welfare programs across jurisdictions in Canada using Data Envelopment Analysis (DEA): An exploratory study of the era of devolution. *Evaluation and Program Planning*, **33**(4), 45–60.

[5] Broersma, L., Edes, A.J.E., and van Dijk, J. (2013) Have Dutch municipalities become more efficient in managing the costs of social assistance dependency? *Journal of Regional Science*, **53**(2), 274–291.

[6] Sheldon, G.M. (2003) The efficiency of public employment services: A nonparametric matching function analysis for Switzerland. *Journal of Productivity Analysis*, **20**(1), 49–70.

[7] Vassiliev, A., Luzzi, G.F., Flückiger, Y., and Ramirez, J.V. (2006) Unemployment and employment offices' efficiency: What can be done? *Socio-Economic Planning Sciences*, **40**(3), 169–186.

[8] Althin, R., Behrenz, L., Färe, R., Grosskopf, S., and Mellander, E. (2010) Swedish employment offices: A new model for evaluating effectiveness. *European Journal of Operational Research*, **207**(3), 1535–1544.

[9] Charnes, A., Cooper, W.W., and Rhodes, E. (1978) Measuring the efficiency of decision making units. *European Journal of Operational Research*, **2**(6), 429–444.

[10] Bogetoft, P. and Otto, L. (2011) *Benchmarking with DEA, SFA, and R*. Springer, New York.

[11] Simar, L. and Wilson, P.W. (2002) Non-parametric tests of returns to scale. *European Journal of Operational Research*, **139**(1), 115–132.

[12] Simar, L. and Wilson, P.W. (2011) Inference by the *m* out of *n* bootstrap in nonparametric frontier models. *Journal of Productivity Analysis*, **36** (1), 33–53.

[13] Cordero, J.M., Pedraja, F., and Santín, D. (2009) Alternative approaches to include exogenous variables in DEA measures: A comparison using Monte Carlo. *Computers & Operations Research*, **36**(10), 2699–2706.

[14] Banker, R.D. and Morey, R.C. (1986) Efficiency analysis for exogenously fixed inputs and outputs. *Operational Research*, **34**(4), 513–521.

[15] Ruggiero, J. (1996) On the measurement of technical efficiency in the public sector. *European Journal of Operational Research*, **90**(3), 553–565.

[16] Ray, S.C. (1991) Resource use efficiency in public schools: A study of Connecticut data. *Management Science*, **37**(12), 1620–1628.

[17] Muñiz, M. (2002) Separating managerial inefficiency and external conditions in data. *European Journal of Operational Research*, **143**(3), 625–643.

[18] Fried, H., Schmidt, S., and Yaisawarng S. (1999) Incorporating the operating environment into a nonparametric measure of technical efficiency. *Journal of Productivity Analysis*, **12**(3), 249–267.

[19] Duncombe, W.D. and Yinger, J. (1993) An analysis of returns to scale in public production with an application to fire protection. *Journal of Public Economics*, **52**(1), 49–72.

[20] Oates, W.E. (1988) On the measurement of congestion in the provision of local public goods. *Journal of Urban Economics*, **24**(1), 85–94.

[21] Otsuka, A., Goto, M., and Sueyoshi, T. (2014) Cost-efficiency of Japanese local governments: Effects of decentralization and regional integration. *Regional Studies, Regional Science*, **1**(1), 207–220.

[22] Simar, L. and Wilson, P.W. (2007) Estimation and inference in two-stage, semi-parametric models of production process. *Journal of Econometrics*, **136**(1), 31–64.

[23] Wooldridge, J.M. (2010) *Econometric Analysis of Cross Section and Panel Data*, 2nd edn, MIT Press, Cambridge, MA.

[24] Banker, R.D. (1993) Maximum likelihood, consistency and data envelopment analysis: A statistical foundation. *Management Science*, **39**(10), 1265–1273.

[25] Kneip, A., Park, B.U., and Simar, L. (1998) A note on the convergence of nonparametric DEA estimators for production efficiency scores. *Econometric Theory*, **14**(6), 783–793.

[26] Xue, M. and Harker, P.T. (1999) Overcoming the inherent dependency of DEA efficiency scores: A bootstrap approach. Working Paper Series 99-17, Financial Institutions Center, Wharton School, University of Pennsylvania.

[27] McDonald, J. (2009) Using least squares and tobit in second stage DEA efficiency analyses. *European Journal of Operational Research*, **197**(2), 792–798.

[28] Bjurek, H., Kjulin, U., and Gustafsson, B. (1992) Efficiency, productivity and determinants of inefficiency at public day care centers in Sweden. *Scandinavian Journal of Economics*, **94**(2), 173–187.

[29] Hoff, A. (2007) Second stage DEA: Comparison of approaches for modelling the DEA score. *European Journal of Operational Research*, **181**(1), 425–435.

[30] Ramalho, E.A., Ramalho, J.J.S., and Henriques, P.D. (2010) Fractional regression models for second stage DEA efficiency analysis. *Journal of Productivity Analysis*, **34**(3), 239–255.

[31] Papke, L.E. and Wooldridge, J.M. (1996) Econometric methods for fractional response variables with an application to 401(K) plan participation rate. *Journal of Applied Econometrics*, **11**(6), 619–632.

[32] Simar, L. and Wilson, P.W. (2008) Statistical inference in nonparametric frontier models: Recent developments and perspectives, in *The Measurement of Productive Efficiency and Productivity Growth* (eds Fried, H.O., Lovell, C.A.K., and Schmidt, S.S.), Oxford University Press, New York, pp. 421–521.

[33] Besstremyannaya, G. and Simm, J. (2015) Robust non-parametric estimation of cost efficiency with an application to banking industry. CEFIR/NES Working Paper Series, No. 217, Centre for Economic and Financial Research at New Economic School.

[34] Simm, J., and Besstremyannaya, G. (2015) *rDEA: Robust Data Envelopment Analysis (DEA) for R*, https://cran.r-project.org/web/packages/rDEA/index.html (accessed 8 April 2016).

22

DEA AS A KAIZEN TOOL: SBM VARIATIONS REVISITED

KAORU TONE

National Graduate Institute for Policy Studies, Tokyo, Japan

22.1 INTRODUCTION

The original slacks-based measure (SBM) model evaluates the efficiency of decision-making units (DMUs) by referring to the furthest frontier point within some range. This results in the worst score for a DMU, and the projection may go to a remote point on the efficient frontiers, which may be inappropriate as a reference. Tone [1] developed four variants of the SBM model where the main concern is to search for the nearest point on the efficient frontiers of the production possibility set.

We depict the relationship between the ordinary SBM (SBM-Min), CCR and SBM-Max models in Figure 22.1. The projections of the inefficient DMU P are Q, R and S by SBM-Min, CCR and SBM-Max, respectively. Mathematically, finding S is an NP-hard problem, because it is a problem of maximization of a convex function over a convex region. However, the projected point S indicates that we can attain an efficient status with less input reduction and less output expansion than in the ordinary SBM (i.e. SBM-Min) models. Thus, the projection in the SBM-Max model represents a practical 'Kaizen' (improvement) by data envelopment analysis (DEA).

Referring to these variations, several authors have published new models. Among them, I introduce two important papers.

Fukuyama *et al.* [2] developed a least-distance efficiency measure with strong/weak monotonicity of the ratio form measure under several norms, including the 1-norm, 2-norm and ∞-norm. This model utilizes mixed-integer linear programming

Advances in DEA Theory and Applications: With Extensions to Forecasting Models,
First Edition. Edited by Kaoru Tone.

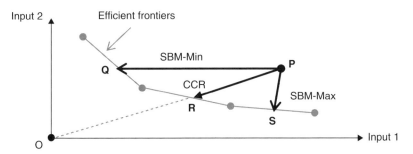

Figure 22.1   Comparison of SBM, CCR and SBM-Max models.

(MILP) to identify efficiency frontiers, and hence a computational difficulty arises for large-scale problems.

Hadi-Vencheh *et al.* [3] developed a new SBM model to find the nearest point on the efficient frontiers. They utilized a multiplier form model to find all supporting hyperplanes. This model also utilizes software which uses fractional coefficients (high-precision arithmetic) to avoid data loss. Hence, the computational time increases for large-scale problems.

In order to apply DEA models to real-world problems, we need to solve many instances with different input/output factors and sets of DMUs before attaining the final scheme of evaluation. For this purpose, an acceptable computation time and easily accessible software are desirable. The motivation and purpose of this chapter is to obtain the 'nearly' closest points on the efficient frontiers with foreseeable computational loads using only popular linear programming codes.

The rest of this chapter is organized as follows. Section 22.2 introduces the ordinary SBM-Min model briefly. Section 22.3 presents the new SBM-Max model. Observations on this new model are provided in Section 22.4. Two numerical examples are presented in Section 22.5. Section 22.6 concludes the chapter. Although we present the model in non-oriented mode, we can treat input- and output-oriented models as well. As to returns-to-scale characteristics, we present the constant-returns-to-scale (CRS) case. However, we can deal with a variable-returns-to-scale (VRS) model as well.

22.2   THE SBM-MIN MODEL

The SBM model was introduced in [4,5]. It has three variations, input-, output- and non-oriented. The non-oriented model is both input- and output-oriented.

Let the set of DMUs be $J = \{1,2,\ldots,n\}$, and let each DMU make use of m inputs to produce s outputs. We denote the vectors of inputs and outputs for DMU_j by $\mathbf{x}_j = \left(x_{1j}, x_{2j}, \ldots, x_{mj}\right)^T$ and $\mathbf{y}_j = \left(y_{1j}, y_{2j}, \ldots, y_{sj}\right)^T$, respectively. We define the input and output matrices $\mathbf{X}$ and $\mathbf{Y}$ by

$$\mathbf{X} = (\mathbf{x}_1, \mathbf{x}_2, \ldots, \mathbf{x}_n) \in R^{m \times n} \text{ and } \mathbf{Y} = (\mathbf{y}_1, \mathbf{y}_2, \ldots, \mathbf{y}_n) \in R^{s \times n} \tag{22.1}$$

We assume that all data are positive, i.e. $\mathbf{X} > 0$ and $\mathbf{Y} > 0$.

22.2.1 Production Possibility Set

The production possibility set is defined using a non-negative combination of the DMUs in the set J as

$$P = \left\{ (\mathbf{x}, \mathbf{y}) \,\middle|\, \mathbf{x} \ge \sum\nolimits_{j=1}^{n} \lambda_j \mathbf{x}_j, \ \mathbf{0} \le \mathbf{y} \le \sum\nolimits_{j=1}^{n} \lambda_j \mathbf{y}_j, \ \lambda \ge \mathbf{0} \right\} \tag{22.2}$$

$\lambda = (\lambda_1, \lambda_2, \ldots, \lambda_n)^T$ is called the intensity vector.

The inequalities in (22.2) can be transformed into equalities by introducing slacks as follows:

$$\begin{aligned}
\mathbf{x} &= \sum\nolimits_{j=1}^{n} \lambda_j \mathbf{x}_j + \mathbf{s}^- \\
\mathbf{y} &= \sum\nolimits_{j=1}^{n} \lambda_j \mathbf{y}_j - \mathbf{s}^+ \\
\mathbf{s}^- &\ge \mathbf{0}, \ \mathbf{s}^+ \ge \mathbf{0}
\end{aligned} \tag{22.3}$$

where $\mathbf{s}^- = (s_1^-, s_2^-, \ldots, s_m^-)^T \in R^m$ and $\mathbf{s}^+ = (s_1^+, s_2^+, \ldots, s_s^+)^T \in R^s$ are called the input and output slacks, respectively.

22.2.2 Non-Oriented SBM

The non-oriented or both-oriented SBM efficiency $\rho_o^{\min}$ is defined by

$$[\text{SBM-Min}] \ \rho_o^{\min} = \min_{\lambda, \mathbf{s}^-, \mathbf{s}^+} \frac{1 - \dfrac{1}{m} \sum_{i=1}^{m} \dfrac{s_i^-}{x_{io}}}{1 + \dfrac{1}{s} \sum_{r=1}^{s} \dfrac{s_r^+}{y_{ro}}}$$

subject to $\tag{22.4}$

$$\begin{aligned}
x_{io} &= \sum\nolimits_{j=1}^{n} x_{ij} \lambda_j + s_i^- \quad (i = 1, \ldots, m) \\
y_{ro} &= \sum\nolimits_{j=1}^{n} y_{rj} \lambda_j - s_r^+ \quad (r = 1, \ldots, s) \\
\lambda_j &\ge 0 \ (\forall j), \ s_i^- \ge 0 \ (\forall i), \ s_r^+ \ge 0 \ (\forall r)
\end{aligned}$$

Definition 22.1 (SBM-efficient)

DMU$_o = (\mathbf{x}_o, \mathbf{y}_o)$ is called SBM-efficient if $\rho_o^{\min} = 1$ holds.

This means that $\mathbf{s}^- = \mathbf{0}$ and $\mathbf{s}^{+*} = \mathbf{0}$, i.e. all input and output slacks are zero.

[SBM-Min] can be transformed into a linear program using the Charnes–Cooper transformation as follows:

$$[\text{SBM-Min-LP}] \quad \tau^* = \min_{t,\Lambda,S^-,S^+} t - \frac{1}{m}\sum_{i=1}^{m}\frac{S_i^-}{x_{io}}$$

subject to

$$1 = t + \frac{1}{s}\sum_{r=1}^{s}\frac{S_r^+}{y_{ro}}$$

$$tx_{io} = \sum_{j=1}^{n}x_{ij}\Lambda_j + S_i^- \quad (i=1,\ldots,m) \tag{22.5}$$

$$ty_{ro} = \sum_{j=1}^{n}y_{rj}\Lambda_j - S_r^+ \quad (r=1,\ldots,s)$$

$$\Lambda_j \geq 0 \; (\forall j), \; S_i^- \geq 0 \; (\forall i), \; S_r^+ \geq 0 \; (\forall r), \; t > 0$$

Let an optimal solution be $(\tau^*, t^*, \Lambda^*, S^{-*}, S^{+*})$. Then, we have an optimal solution of [SBM-Min] as defined by

$$\rho_o^{\min} = \tau^*, \; \lambda^* = \Lambda^*/t^*, \; s^{-*} = S^{-*}/t^*, \; s^{+*} = S^{+*}/t^* \tag{22.6}$$

22.3   THE SBM-MAX MODEL

In this section, we introduce the new non-oriented SBM-Max model.

Step 1. Solve SBM-Min
First, we solve the ordinary SBM (SBM-Min) model as represented by the program
(22.4) for DMU $(\mathbf{x}_o, \mathbf{y}_o)$ $(o = 1,\ldots,n)$. Let an optimal solution be $(\lambda^*, s^{-*}, s^{+*})$.

Step 2. Define Efficient DMUs
We define the set R^{eff} of all efficient DMUs as

$$R^{\text{eff}} = \left\{ j \middle| \rho_j^{\min} = 1, j = 1,\ldots,n \right\} \tag{22.7}$$

We denote these efficient DMUs by $\left(\mathbf{x}_1^{\text{eff}}, \mathbf{y}_1^{\text{eff}}\right), \left(\mathbf{x}_2^{\text{eff}}, \mathbf{y}_2^{\text{eff}}\right), \ldots, \left(\mathbf{x}_{Neff}^{\text{eff}}, \mathbf{y}_{Neff}^{\text{eff}}\right)$, where
Neff is the number of efficient DMUs.

Step 3. Local Reference Set
For an inefficient DMU $(\mathbf{x}_o, \mathbf{y}_o)$, we define the local reference set R_o^{local}, i.e. the set
of efficient DMUs for DMU $(\mathbf{x}_o, \mathbf{y}_o)$, by (22.8):

$$R_o^{\text{local}} = \left\{ j \middle| \lambda_j^* > 0, j = 1,\ldots,n \right\} \tag{22.8}$$

Step 4. Pseudo-Max Score

For each inefficient DMU, i.e. $\rho_o^{\min} < 1$, we solve the following program:

$$[\text{Pseudo-1}] \quad \max \frac{1 - \dfrac{1}{m}\sum_{i=1}^{m}\dfrac{s_{io}^-}{x_{io}}}{1 + \dfrac{1}{s}\sum_{r=1}^{s}\dfrac{s_r^+}{y_{ro}}}$$

subject to

$$\mathbf{x}_o = \sum_{j \in R_o^{\text{local}}} \mathbf{x}_j \lambda_j + \mathbf{s}^-$$

$$\mathbf{y}_o = \sum_{j \in R_o^{\text{local}}} \mathbf{y}_j \lambda_j - \mathbf{s}^+$$

$$\mathbf{s}^-, \mathbf{s}^+, \boldsymbol{\lambda} \geq \mathbf{0}$$

(22.9)

Let the optimal slacks be $(\mathbf{s}^{-*}, \mathbf{s}^{+*})$. We solve the following program with variables $(\boldsymbol{\lambda}, \mathbf{s}^-, \mathbf{s}^+)$:

$$[\text{Pseudo-2}] \quad \min \frac{1 - \dfrac{1}{m}\sum_{i=1}^{m}\dfrac{s_{io}^-}{x_{io} - s_i^{-*}}}{1 + \dfrac{1}{s}\sum_{r=1}^{s}\dfrac{s_r^+}{y_{ro} + s_r^{+*}}}$$

subject to

$$\mathbf{x}_o - \mathbf{s}^{-*} = \sum_{j \in R^{\text{eff}}} \mathbf{x}_j^{\text{eff}} \lambda_j + \mathbf{s}^-$$

$$\mathbf{y}_o + \mathbf{s}^{+*} = \sum_{j \in R^{\text{eff}}} \mathbf{y}_j^{\text{eff}} \lambda_j - \mathbf{s}^+$$

$$\mathbf{s}^-, \mathbf{s}^+, \boldsymbol{\lambda} \geq \mathbf{0}$$

(22.10)

Let the optimal slacks be $(\mathbf{s}^{-**}, \mathbf{s}^{+**})$. We define the pseudo-max score $\rho_o^{\text{pseudo max}}$ by

$$[\text{Pseudo-Max}] \quad \rho_o^{\text{pseudo max}} = \frac{1 - \dfrac{1}{m}\sum_{i=1}^{m}\dfrac{s_{io}^{-*} + s_{io}^{-**}}{x_{io}}}{1 + \dfrac{1}{s}\sum_{r=1}^{s}\dfrac{s_r^{+*} + s_r^{+**}}{y_{ro}}}$$

(22.11)

Step 5. Distance and SBM-Max Score

For each inefficient DMU $(\mathbf{x}_o, \mathbf{y}_o)$, i.e. $\rho_o^{\min} < 1$, we calculate the distance between $(\mathbf{x}_o, \mathbf{y}_o)$ and $(\mathbf{x}_h^{\text{eff}}, \mathbf{y}_h^{\text{eff}})$ $(h = 1, \ldots, Neff)$ as

$$[\text{Distance}] \quad d_h = \sum_{i=1}^{m}\frac{\left| x_{ih}^{\text{eff}} - x_{io} \right|}{x_{io}} + \sum_{i=1}^{s}\frac{\left| y_{ih}^{\text{eff}} - y_{io} \right|}{y_{io}}$$

(22.12)

This distance is units-invariant.

Step 5.1. Reorder the Distances

We renumber the efficient DMUs in ascending order of d_h, so that

$$d_1 \leq d_2 \leq \ldots \leq d_{Neff} \tag{22.13}$$

We define the set R_h by

$$R_h = \{1, \ldots, h\} \quad (h = 1, \ldots, Neff) \tag{22.14}$$

Step 5.2. Find Slacks and Max-Score for the Set R_h

We evaluate the efficiency score of the inefficient DMU $(\mathbf{x}_o, \mathbf{y}_o)$, referring to the set R_h, by solving the following program:

$$
[\text{Max-1}] \quad \max_{\lambda, s^-, s^+} \frac{1 - \dfrac{1}{m}\sum_{i=1}^{m} \dfrac{s_{io}^-}{x_{io}}}{1 + \dfrac{1}{s}\sum_{r=1}^{s} \dfrac{s_r^+}{y_{ro}}}
$$

subject to

$$\mathbf{x}_o = \sum_{j \in R_h} \mathbf{x}_j^{\text{eff}} \lambda_j + \mathbf{s}^-$$

$$\mathbf{y}_o = \sum_{j \in R_h} \mathbf{y}_j^{\text{eff}} \lambda_j - \mathbf{s}^+$$

$$\mathbf{s}^-, \mathbf{s}^+, \lambda \geq \mathbf{0}$$

(22.15)

a. If this program is infeasible, we define $\rho_{oh}^* = 0$. Otherwise, let the optimal slacks be $(\mathbf{s}^{-*}, \mathbf{s}^{+*})$.

b. If the optimal objective value is 1, i.e. $\mathbf{s}^{-*} = \mathbf{0}$ and $\mathbf{s}^{+*} = \mathbf{0}$, we define $\rho_{oh}^* = 0$. This indicates that DMU $(\mathbf{x}_o, \mathbf{y}_o)$ can be expressed as a non-negative combination of DMUs in R_h and hence, in view of $\rho_o^{\min} < 1$, it is inside the production possibility set.

c. If the optimal objective value is less than 1, we again solve the following program with the variables $(\lambda, \mathbf{s}^-, \mathbf{s}^+)$:

$$
[\text{Max-2}] \quad \min_{\lambda, s^-, s^+} \frac{1 - \dfrac{1}{m}\sum_{i=1}^{m} \dfrac{s_{io}^-}{x_{io} - s_i^{-*}}}{1 + \dfrac{1}{s}\sum_{r=1}^{s} \dfrac{s_r^+}{y_{ro} + s_r^{+*}}}
$$

subject to

$$\mathbf{x}_o - \mathbf{s}^{-*} = \sum_{j \in R^{\text{eff}}} \mathbf{x}_j^{\text{eff}} \lambda_j + \mathbf{s}^-$$

$$\mathbf{y}_o + \mathbf{s}^{+*} = \sum_{j \in R^{\text{eff}}} \mathbf{y}_j^{\text{eff}} \lambda_j - \mathbf{s}^+$$

$$\mathbf{s}^-, \mathbf{s}^+, \lambda \geq \mathbf{0}$$

(22.16)

Let the optimal slacks be $(\mathbf{s}^{-**}, \mathbf{s}^{+**})$. We define ρ_{oh}^{*} by

$$
[\rho_{oh}^{*}] \quad \rho_{oh}^{*} = \frac{1 - \dfrac{1}{m} \sum_{i=1}^{m} \dfrac{s_{io}^{-*} + s_{io}^{-**}}{x_{io}}}{1 + \dfrac{1}{s} \sum_{r=1}^{s} \dfrac{s_{r}^{+*} + s_{r}^{+**}}{y_{ro}}} \qquad (22.17)
$$

We assign a value of ρ_{oh}^{*} to the max-score referring to the set R_h.

Step 5.3. SBM-Max and Projection

Finally, we define the max-score $\rho_{o}^{\max}$ of the inefficient DMU $(\mathbf{x}_o, \mathbf{y}_o)$ by

$$
[\text{SBM} - \text{Max}] \quad \rho_{o}^{\max} = \max\left\{\rho_{o}^{\text{pseudo max}}, \rho_{o1}^{*}, \ldots, \rho_{oNeff}^{*}\right\} \qquad (22.18)
$$

We also keep the slacks $(\mathbf{s}^{-**}, \mathbf{s}^{+**})$ corresponding to the maximum $\rho_{o}^{\max}$. The projection of DMU $(\mathbf{x}_o, \mathbf{y}_o)$ onto the efficient frontiers is given by

$$
[\text{Projection}] \quad \mathbf{x}_{o}^{*} = \mathbf{x}_o - \mathbf{s}^{-*} - \mathbf{s}^{-**}, \mathbf{y}_{o}^{*} = \mathbf{y}_o + \mathbf{s}^{+*} + \mathbf{s}^{+**} \qquad (22.19)
$$

The projected point $\left(\mathbf{x}_{o}^{*}, \mathbf{y}_{o}^{*}\right)$ is efficient with respect to the efficient DMU set R^{eff}. However, it does not always satisfy the Pareto–Koopmans efficiency condition.

22.4 OBSERVATIONS

In this section, we discuss several characteristics of the algorithm presented in Section 22.3.

22.4.1 Distance and Choice of the Set R_h

The set R_h plays a central role in choosing reference DMUs for inefficient DMUs. Because our main concern is the projection to the nearest point on the efficient frontiers, we evaluate the distance between the DMU $(\mathbf{x}_o, \mathbf{y}_o)$ and the efficient DMUs by use of (22.12), and choose the shortest-distance DMU as the first candidate DMU. Then, we expand the reference set in ascending order of distance. Thus, we can expect a point close to an efficient point on the frontiers with high probability. If a tie occurs in the distances, we can choose any one at random.

22.4.2 The Role of Programs (22.10) and (22.16)

For example, Program (22.16) is necessary to project the point $(\mathbf{x}_o - \mathbf{s}^{-*}, \mathbf{y}_o + \mathbf{s}^{+*})$ onto the efficient frontiers. Thus, $\left(\mathbf{x}_{o}^{*} = \mathbf{x}_o - \mathbf{s}^{-*} - \mathbf{s}^{-**}, \mathbf{y}_{o}^{*} = \mathbf{y}_o + \mathbf{s}^{+*} + \mathbf{s}^{+**}\right)$ is the projected point on the efficient frontiers, and it is expected to be close to DMU $(\mathbf{x}_o, \mathbf{y}_o)$ by the selection rule for R_h.

22.4.3 Computational Amount

The computations needed for this algorithm for an inefficient DMU are as follows.

Let t_1 and t_2 be the CPU times for solving a linear programming (LP) problem with $(m + s)$ rows and n columns, and $(m + s)$ rows and *Neff* columns, respectively. Since the solution time for an LP problem is proportional to the number of columns, we can estimate roughly that $t_1 = (n/Neff)t_2$.

1. Program (22.4) or (22.5) needs $n * t_1$ CPU time.
2. Programs (22.9) and (22.10) need at most $2 * (n - Neff) * t_2$ CPU time.
3. Programs (22.15) and (22.16) need at most $1.5 * (n - Neff) * Neff * t_2$ CPU time, because the index of the member of R_h in (22.15) varies from 1 to *Neff*.

However, if Step 5.2(b) occurs for some set R_h, we can skip the computations for the succeeding programs (22.15) and (22.16) for $h + 1, \ldots, Neff$.

Overall, the total time for the LP computation is at most

$$T = n*t_1 + (n - Neff)*t_2 + 1.5*(n - Neff)*Neff*t_2$$
$$= [n + (2 + 1.5*Neff)*(n - Neff)*(Neff/n)]*t_1 \tag{22.20}$$

Thus, the computational amount is of polynomial order and we do not need other software, for example MILP or fractional arithmetic.

22.4.4 Consistency with the Super-Efficiency SBM Measure

The SBM-Max model aims at getting to the nearest point on the efficient frontiers. This concept is in line with the super-efficiency SBM model [6], which solves the following program for an efficient DMU $(\mathbf{x}_o, \mathbf{y}_o)$ to measure the minimum ratio-scale distance from the efficient frontier excluding the DMU $(\mathbf{x}_o, \mathbf{y}_o)$:

$$[\text{Super-SBM}] \quad \delta^* = \min_{\lambda, s^-, s^+} \frac{1 + \dfrac{1}{m} \sum_{i=1}^{m} \dfrac{s_i^-}{x_{io}}}{1 - \dfrac{1}{s} \sum_{r=1}^{s} \dfrac{s_r^+}{y_{ro}}}$$

subject to

$$\mathbf{x}_o = \sum_{j=1, j \neq o}^{n} \mathbf{x}_j \lambda_j - \mathbf{s}^-$$

$$\mathbf{y}_o = \sum_{j=1, j \neq o}^{n} \mathbf{y}_j \lambda_j + \mathbf{s}^+$$

$$\lambda \geq \mathbf{0}, \mathbf{s}^- \geq \mathbf{0}, \mathbf{s}^+ \geq \mathbf{0}$$

$$(22.21)$$

We can solve the super-efficiency SBM model by applying the LP code just once, because this problem belongs to the class of convex programming, i.e. minimization of a convex function over a convex region. However, the SBM-Max problem cannot be solved in this manner, because it is a maximization of a convex function over a convex region.

22.4.5 Addition of Weights to Input and Output Slacks

We can assign weights (w^- and w^+) to the input and output slacks in the objective function of the above SBM models corresponding to the relative importance of items as follows:

$$[\text{Weighted-SBM}] \; \rho^* = \min_{\lambda, s^-, s^+} \frac{1 - \frac{1}{m} \sum_{i=1}^{m} \frac{w_i^- s_i^-}{x_{io}}}{1 + \frac{1}{s} \sum_{r=1}^{s} \frac{w_r^+ s_r^+}{y_{ro}}} \tag{22.22}$$

with $\sum_{i=1}^{m} w_i^- = m$ and $\sum_{r=1}^{m} w_r^+ = s$. The weights should reflect the intentions of the decision-makers. We can define input- and output-oriented weighted-SBM models by omitting the denominator and numerator, respectively, of the objective function in (22.22).

22.5 NUMERICAL EXAMPLES

In this section, we show two numerical examples: the first one is illustrative and the other deals with real data. All computations were executed using a PC with an Intel Core i7-3770 CPU operating at 3.40 GHz with 16 GB RAM and Microsoft Excel VBA (Visual Basic for Applications). An LP software package (using the revised simplex method) was coded by the author. We checked the results of the first example using LINGO (LINDO Systems Inc.) and obtained the same figures.

22.5.1 An Illustrative Example

We considered the same data as in [3]. Table 22.1 displays the data, with two inputs (Doctor and Nurse) and two outputs (Outpatient and Inpatient).

22.5.1.1 Solution of SBM-Min Model   First, we solved the SBM-Min model and obtained the results shown in Table 22.2.

We find four efficient DMUs, i.e. $R^{\text{eff}} = \{A, B, D, L\}$.

22.5.1.2 The Case of Inefficient DMU I   We present the case of inefficient DMU *I*, step by step.

1. Steps 1, 2 and 3: $\rho_I^{\min} = 0.9016$, $R_I^{\text{local}} = \{A, L\}$, $R^{\text{eff}} = \{A, B, D, L\}$.
2. Step 4: $\rho_I^{\text{pseudo max}} = 0.9016$.

TABLE 22.1   Illustrative example.

DMU	(I) Doctor	(I) Nurse	(O) Outpatient	(O) Inpatient
A	20	151	100	90
B	19	131	150	50
C	25	160	160	55
D	27	168	180	72
E	22	158	94	66
F	55	255	230	90
G	33	235	220	88
H	31	206	152	80
I	30	244	190	100
J	50	268	250	100
K	53	306	260	147
L	38	273	250	133

TABLE 22.2   Results of SBM-Min model.

DMU	Score	Rank	Reference	(Lambda)		
A	1	1	A	1		
B	1	1	B	1		
C	0.8265	8	B	0.449	L	0.371
D	1	1	D	1		
E	0.7277	11	B	0.667	L	0.246
F	0.6857	12	A	0.092	L	0.883
G	0.8765	6	B	0.16	L	0.784
H	0.7713	9	L	0.755		
I	0.9016	5	A	0.233	L	0.667
J	0.7653	10	B	0.152	L	0.909
K	0.8619	7	B	0.15	L	1.049
L	1	1	L	1		

3. Step 5.1: the distances from efficient DMUs are $d_A = 1.28816$, $d_B = 1.54030$, $d_D = 0.74411$ and $d_L = 1.03131$. Thus, we have $R_1 = \{D\}, R_2 = \{D,L\}, R_3 = \{D,L,A\}$ and $R_4 = \{D,L,A,B\}$.

4. Step 5.2: we solve (22.16) and (22.17), and find $\rho_{I1}^* = 0.859885, \rho_{I2}^* = 0.910900$, $\rho_{I3}^* = 0.921168$ and $\rho_{I4}^* = 0.920198$.

5. Step 5.3: from (22.18), we find $\rho_I^{\max} = \max\left\{\rho_I^{\text{pseudo max}}, \rho_{I1}^*, \rho_{I2}^*, \rho_{I3}^*, \rho_{I4}^*\right\} = 0.921168 = \rho_{I3}^*$ with the reference set $R_3 = \{A,D,L\}$. Its projection is $\left(x_{1I}^* = 30, x_{2I}^* = 205.53, y_{1I}^* = 190, y_{2I}^* = 100\right)$ with the slacks $\left(s_1^- = 0, s_2^- = 38.47, s_1^+ = 0, s_2^+ = 0\right)$. The SBM-Min model has $\rho_I^{\min} = 0.9016$ with slacks $\left(s_1^- = 0, s_2^- = 26.767, s_1^+ = 0, s_2^+ = 9.667\right)$. This indicates that the SBM-Min

model requires a reduction of Nurse by 26.767 and an increase of Inpatient by 9.667 to attain an efficient status, whereas the SBM-Max model requires a reduction of Nurse by 38.47 to attain that status.

22.5.1.3 Comparison of SBM-Max, Pseudo-Max and SBM-Min Scores

Table 22.3 compares the results for the SBM-Max, SBM-Pseudo and SBM-Min scores. The inefficient DMUs increase in efficiency from SBM-Min to SBM-Max.

Table 22.4 shows $\rho_o^{\text{pseudo max}}, \rho_{o1}^*, \ldots, \rho_{o4}^*$ for the inefficient DMUs. The shaded portions indicate the maximum values. The SBM-Max scores were found at several stages for R_h.

TABLE 22.3 Comparisons.

DMU	SBM-Max	Rank	Pseudo	Rank	SBM-Min	Rank
A	1	1	1	1	1	1
B	1	1	1	1	1	1
C	0.87507	8	0.855	8	0.8265	8
D	1	1	1	1	1	1
E	0.7682	11	0.7391	11	0.7277	11
F	0.72648	12	0.6868	12	0.6857	12
G	0.93688	5	0.9052	5	0.8765	6
H	0.80918	10	0.7714	10	0.7714	9
I	0.92117	6	0.9016	6	0.9016	5
J	0.81032	9	0.7898	9	0.7653	10
K	0.88894	7	0.8622	7	0.8619	7
L	1	1	1	1	1	1
Average	0.8947		0.8759		0.8681	
Max	1		1		1	
Min	0.7265		0.6868		0.6857	
St. Dev.	0.0982		0.1114		0.115	

TABLE 22.4 ρ values.

DMU	$\rho_o^{\text{pseudo max}}$	ρ_{o1}^*	ρ_{o2}^*	ρ_{o3}^*	ρ_{o4}^*
C	0.854953846	0.875070028	0.875070028	0.875070028	0.875070028
E	0.7391066	0.768203137	0.768203137	0.768203137	0.768203137
F	0.686814742	0.686814742	0.726479403	0.726479403	0.726479403
G	0.905158931	0.936879433	0.936879433	0.936879433	0.936879433
H	0.771353638	0.809180119	0.809180119	0.809180119	0.809180119
I	0.901628474	−10	0.910900045	0.921167545	0.920188082
J	0.789823609	0.757308083	0.810323383	0.810323383	0.810323383
K	0.862207404	0.862207404	0.866150331	0.888935642	0.888935642

22.5.1.4 Comparison of Average Differences between SBM-Max and SBM-Min Figure 22.2 shows the average of the percentage deviations, |Data − Projection| * 100/Data. Notice that large differences exist in SBM-Min, while small differences exist in SBM-Max. Table 22.5 reports the data and projections along with the deviations (%) in the case of SBM-Max.

22.5.2   Japanese Municipal Hospitals

The data were collected from the Annual Databook of Local Public Enterprise published by the Ministry of Internal Affairs and Communications of the Japanese Government, 2005.

22.5.2.1   Data

- Number of DMUs: 707 hospitals ($n = 707$).
- Number of inputs: 5. (1) Number of beds (Bed), (2) expenses for outsourcing (Outsource), (3) number of doctors (Doctor), (4) number of nurses (Nurse) and (5) expenses for other medical materials (Material) ($m = 5$).
- Number of outputs: 4. (1) Revenue from operations per day (Operation), (2) revenue from first consultation per day (1st time), (3) revenue from return to clinic per day (Follow-up) and (4) revenue from hospitalization per day (Hotel) ($s = 4$).

Table 22.6 shows statistics of the dataset.

22.5.2.2   SBM Scores   The SBM-Min model found that 66 hospitals among the 707 were efficient ($Neff = 66$). Table 22.7 compares the three scores. We found large differences between the SBM-Max and SBM-Min models.

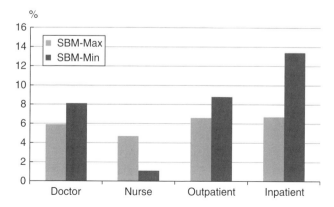

Figure 22.2   Average deviations (%).

TABLE 22.5 Data and projections by use of SBM-Max.

DMU	Score	Rank	Doctor			Nurse			Outpatient			Inpatient		
			Data	Proj.	Diff. (%)	Data	Proj.	Diff. (%)	Data	Proj.	Diff. (%)	Data	Proj.	Diff. (%)
A	1	1	20	20	0	151	151	0	100	100	0	90	90	0
B	1	1	19	19	0	131	131	0	150	150	0	50	50	0
C	0.8751	8	25	25	0	160	155.6	−2.78	160	166.7	4.17	55	66.7	21.21
D	1	1	27	27	0	168	168	0	180	180	0	72	72	0
E	0.7682	11	22	20.9	−4.88	158	158	0	94	104.6	11.32	66	94.2	42.69
F	0.7265	12	55	34.5	−37.27	255	214.7	−15.82	230	230	0	90	92	2.22
G	0.9369	5	33	33	0	235	205.3	−12.62	220	220	0	88	88	0
H	0.8092	10	31	30	−3.23	206	186.7	−9.39	152	200	31.58	80	80	0
I	0.9212	6	30	30	0	244	205.5	−15.77	190	190	0	100	100	0
J	0.8103	9	50	43.1	−13.86	268	268	0	250	287.1	14.86	100	115	14.86
K	0.888	7	53	46.5	−12.26	306	306	0	260	289.1	11.2	147	147	0
L	1	1	38	38	0	273	273	0	250	250	0	133	133	0

TABLE 22.6 Statistics of dataset ($n = 707$).

	Bed	Outsource	Doctor	Nurse	Material	Operation	1st time	Follow-up	Hotel
Max	1063	2 231 247	215.562	955.464	3 395 791	1.7E+07	1 432 079	3 359 160	1.8E+07
Min	25	7767	0.98	11	9197	8979	2706	13 636	109 650
Average	255.924	312 686	33.2783	175.709	491 909	2 128 506	211 538	415 872	3 263 845
SD	191.764	334 184	34.1647	149.678	598 474	2 533 050	212 439	327 749	3 020 809

TABLE 22.7   Comparison of the three scores.

	SBM-Max	Pseudo	SBM-Min
Average	0.7835	0.6997	0.4515
Max	1	1	1
Min	0.1889	0.0394	0.0118
St. Dev.	0.1339	0.211	0.229

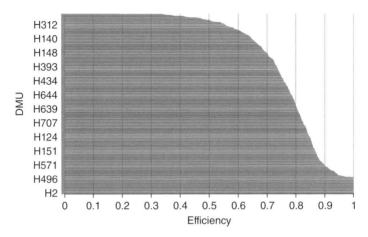

Figure 22.3   Distribution of SBM-Max scores.

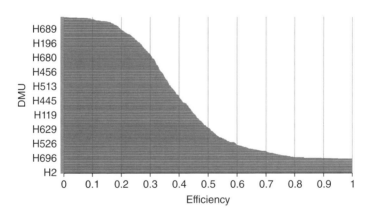

Figure 22.4   Distribution of SBM-Min scores.

Figures 22.3 (SBM-Max) and 22.4 (SBM-Min) show the respective scores of the 707 hospitals in ascending order, where we can observe big differences. Table 22.8 shows the average of the percentage deviations, $|$ Data − Projection $|$ * 100/Data. It can

TABLE 22.8   Average deviation (%).

	SBM-Max	SBM-Min
Bed	13.1644	3.4014
Outsource	20.0291	24.8771
Doctor	12.0707	9.9367
Nurse	9.6586	5.7022
Material	10.7101	8.144
Operation	12.7155	48.2925
1st time	9.4884	407.243
Follow-up	13.3537	192.406
Hotel	17.4181	1.1889

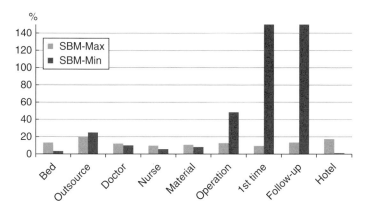

Figure 22.5   Average deviations (%) of inputs and outputs (cut at 150%).

be observed that large differences exist in SBM-Min, while there are only small differences in SBM-Max.

Figure 22.5 illustrates the deviations graphically. Large differences are found in SBM-Min, while balanced deviations are found in SBM-Max.

22.5.2.3   Computational Time   The computation time increases as the number of efficient DMUs (*Neff*) increases, because the number of facets increases accordingly and we need to solve an additional *Neff* linear programs. In this example, we had:

1. CPU time for SBM-Min and SBM-Pseudo = 12 seconds.
2. CPU time for SBM-Max = 179 seconds.

SBM-Max needs about 15 times as much computation time as SBM-Min and SBM-Pseudo. This number is reasonable and consistent with the formula (22.20).

22.6 CONCLUSIONS

In this chapter, we have developed the SBM-Max model, which attempts to find nearly the closest reference point on the efficient frontiers so that slacks are minimized while the scores are maximized. Sacrificing rigorous solutions, the proposed model utilizes a standard LP code and finds approximate solutions in an allowable (polynomial) time.

Many applications of SBM-Min models have been developed over the world. According to the Google Citation Index, 1648 articles cited Tone [4] on March 14 2016. Also, many DEA models have been developed based on this model. Above all, network SBM (NSBM) [7], dynamic SBM (DSBM) [8], dynamic network SBM (DNSBM) [9] and Malmquist SBM [10] are representative. Revisions of these models based on the SBM-Max model are imperative future research subjects.[1]

REFERENCES

[1] Tone, K. (2010) Variations on the theme of slacks-based measure of efficiency in DEA. *European Journal of Operational Research*, **200**, 901–907.

[2] Fukuyama, H., Masaki, H., Sekitani, K., Shi, J. (2014) Distance optimization approach to ratio-form efficiency measures in data envelopment analysis. *Journal of Productivity Analysis*, **42**, 175–186.

[3] Hadi-Vencheh, A., Jablonsky, J., Esmaeilzadeh, A. (2015) The slack-based measure model based on supporting hyperplanes of production possibility set. *Expert Systems with Applications*, DOI: http://dx.doi.org/10.1016/j.eswa.2015.03.032.

[4] Tone, K. (2001) A slacks-based measure of efficiency in data envelopment analysis. *European Journal of Operational Research*, **130**, 498–509.

[5] Pastor, J.T., Ruiz, J.L., Sirvent, I. (1999) An enhanced DEA Russell graph efficiency measure. *European Journal of Operational Research*, **115**, 596–607.

[6] Tone, K. (2002) A slacks-based measure of super-efficiency in data envelopment analysis. *European Journal of Operational Research*, **143**, 32–41.

[7] Tone, K., Tsutsui, M. (2009) Network DEA: A slacks-based measure approach. *European Journal of Operational Research*, **197**, 243–252.

[8] Tone, K., Tsutsui, M. (2010) Dynamic DEA: A slacks-based measure approach. *Omega*, **38**, 145–156.

[9] Tone, K., Tsutsui, M. (2014) Dynamic DEA with network structure: A slacks-based measure approach. *Omega*, **42**, 124–131.

[10] Cooper, W.W., Seiford, L.M., Tone, K. (2007) *Data Envelopment Analysis: A Comprehensive Text with Models, Applications, References and DEA-Solver Software*, 2nd edn, Springer.

---

[1] Software for SBM-Max models is included in DEA-Solver Pro V13 (http://www.saitech-inc.com). See also Appendix A.

PART III

DEA FOR FORECASTING AND DECISION-MAKING (PAST–PRESENT–FUTURE SCENARIO)

23

CORPORATE FAILURE ANALYSIS USING SBM

JOSEPH C. PARADI

Centre for Management of Technology and Entrepreneurship, University of Toronto, Toronto, ON, Canada

XIAOPENG YANG

Centre for Management of Technology and Entrepreneurship, University of Toronto, Toronto, ON, Canada

KAORU TONE

National Graduate Institute for Policy Studies, Tokyo, Japan

23.1 INTRODUCTION

Corporate failure analysis is one of the crucial factors affecting company management and individual investors, as it tells the current corporate health status of a company and often forecasts the firm's future developmental trend. Therefore, the motivation of corporate failure analysis is the need to predict the financial stress that a company faces, by employing mathematical modelling.

In the realm of corporate failure research, a wide spectrum of studies has been published; one of the most used types is based on a set of financial ratios. Most of the ratios used [1] are obtained from the financial statements of the company. The typical method for bankruptcy prediction, the Altman Z score [2], was proposed by Edward Altman. He combined several important financial ratios by using multiple discriminant analysis, and generated a score which is the weighted sum of these ratios. The operational status of a company is classified into three classes by the value of this score, namely, troubled, healthy and a middle status, the 'grey area'. This method

Advances in DEA Theory and Applications: With Extensions to Forecasting Models,
First Edition. Edited by Kaoru Tone.

became popular in evaluating the potential corporate financial stress of various companies; specifically, it is effective for manufacturing companies as most of the ratios rely on the size of the firm's assets [3,4]. However, besides manufacturing firms, non-manufacturing entities represent a large number of firms. Most of these non-manufacturing firms are service-oriented and affect our daily life, but do not actually rely on asset size [5]. To make his methodology suitable for non-manufacturing firms, Altman proposed a second model, which was named the Altman Z'' model [6]. By testing the effectiveness of the Z'' method on non-manufacturing firms, Altman selected optimal parameters to make the model satisfy both manufacturing and non-manufacturing industries. Nevertheless, this model is still substantially based on asset size, notwithstanding the fact that the majority of firms today are mainly focused on services, their most important asset is their people and they do not have a large real asset base [5]. It follows that an investigation of the Altman Z'' model for the non-manufacturing sector was deemed necessary.

To solve the problem of predicting corporate failure for non-manufacturing firms, still following the ratios proposed by Altman, we introduce a new method using data envelopment analysis (DEA). Since the first DEA model introduced by Charnes *et al.* [7] based on Farrell's research [8], DEA has developed into a popular non-parametric approach to productivity, efficiency and effectiveness evaluation in many industries, covering finance, logistics, management and so on [9–12]. The main framework of DEA is a fractional linear programming technique that maximizes the productivity and efficiency of the firm under evaluation, which is referred to in DEA as a decision-making unit (DMU), and restricting other DMUs to certain limits. The flexible structure of DEA has three benefits in predicting non-manufacturing corporate failure. Firstly, the selection of inputs and outputs of the DMU is flexible, which allows us to select related ratios from Altman's method, as it was accepted in his research that these ratios were the most effective in corporate failure prediction. Therefore, the ratios that we selected from Altman's model were the most influential ones that cover different aspects of the operating status of firms. Secondly, the selection of inputs and outputs in DEA allows us to choose preferable attributes, so that we can eliminate the 'asset' factor from the ratios and make the DEA method suitable for non-manufacturing industry. Finally, DEA is non-parametric, which is quite different from Altman's methods, which are parametric. In Altman's methods, in order to evaluate the status of a firm, a big enough training dataset is necessary to obtain appropriate parameters (he used data on thousands of firms). By introducing DEA, such complicated procedures can be avoided. Considering these merits of DEA, we provide a new approach to corporate failure analysis using the slacks-based measure (SBM) model of DEA.

23.2 LITERATURE REVIEW

Understanding the previous literature on corporate failure analysis is essential to our current research, and helps to clarify the differences between our research and the published studies and to highlight our contributions. In this section, we introduce the most influential methods and applications in corporate failure analysis, such as Beaver's univariate model, Altman's multivariate model and some other studies.

23.2.1   Beaver's Univariate Model

One of the first attempts to predict corporate failure was carried out by William Beaver in 1967 [1], in which Beaver defined failure as 'the inability of a firm to pay its financial obligations as they mature' and a financial ratio as 'a quotient of two numbers, where both numbers consist of financial statement items'. Beaver also proposed 'predictive ability', which is essentially the usefulness of a data item for identifying an event before it occurs. Beaver collected data from Moody's industrial manual between 1954 and 1964, inclusive. Each failed firm from Moody's was compared with a healthy firm in the same industry of a comparable asset size. At the time, there were statistics-based reasons to believe that the larger of two firms would have less probability of failure even if they had identical financial ratios. Therefore, he believed that firms of different asset sizes could not be accurately compared [13]. Beaver compiled 30 ratios and showed the following 14 to be the most effective:

- cash flow/total debt;
- current assets/current liabilities;
- net income/total assets;
- quick assets/current liabilities;
- total debt/total assets;
- cash/current liabilities;
- current assets/total assets;
- current assets/sales;
- quick assets/total assets;
- quick assets/sales;
- working capital/total assets;
- working capital/sales;
- cash/total assets;
- cash/sales.

By comparing the above ratios, Beaver selected 'cash flow/total debt' as the best predictor, and 'total debt/total assets' as the second best. He concluded that 'the most crucial factor was the net liquid asset flow supplied to the reservoir while the size of the reservoir was the least important factor'.

Beaver also visited the concept of the likelihood ratio (LR), which is the ratio of these two values:

$$P(NF) = \text{percentage for non-failed firms}$$

$$P(F) = \text{percentage for failed firms}$$

$$LR = \frac{P(F)}{P(NF)} \tag{23.1}$$

A likelihood ratio could be found for every interval for each of the financial ratios in each year before bankruptcy. However, Beaver was inconclusive in his analysis of these ratios. He stated that in the year before failure, the likelihood ratio mirrored the financial ratio; however, in the years before that, the results varied greatly. He also stated that though his work was univariate, it would be valuable to consider a multivariate approach [1]. This is where Altman stepped in.

23.2.2 Altman's Multivariate Model

The univariate approach proposed by Beaver selects only the most crucial factor in corporate failure analysis, which does not correspond to the real case in many applications. In 1968, Edward Altman attempted the first multivariate approach to bankruptcy prediction. The analysis technique that he adopted was multiple discriminant analysis (MDA).

In Altman's time, MDA was not as popular as regression analysis and was used mainly in biological and behavioural sciences [2]. MDA is a statistical technique used to classify an observation into one of several 'a priori' groupings dependent upon the observation's individual characteristics. It is usually used to classify a variable into a qualitative group, for example male or female, or bankrupt or non-bankrupt. The process used for MDA is first to establish groups, which may be more than two in size, and then collect data for objects within each of those groups. Then a linear combination is created from the data collected that best discriminates between the groups. This is done by assigning coefficients to each data item. In the case of bankruptcy, a coefficient is assigned to each financial ratio chosen and the output of the linear combination is a number that can classify a firm as 'bankrupt' or 'non-bankrupt'. MDA allows the entire profile of variables to be analysed simultaneously rather than individually [6].

To develop the model, Altman took a sample of 66 corporations with 33 firms in the bankrupt group and 33 in the non-bankrupt group. All bankrupt firms were manufacturers that filed a bankruptcy petition under Chapter 11 of the National Bankruptcy Act between 1946 and 1965. The non-bankrupt firms were selected by a paired sample method (similar to that of Beaver). A list of 22 potential ratios was compiled, which were split into five standard ratio categories: liquidity, profitability, leverage, solvency and activity ratios. From the list of 22, five ratios were selected to be able to do the best overall job in collectively predicting bankruptcy. These were selected based on (i) the statistical significance of various potential functions, while determining the relative contribution of each individual variable; (ii) the intercorrelation between the variables; (iii) the predictive accuracy of various profiles; and (iv) judgement based on the analysis [2].

Altman's multivariate model is as follows:

$$Z = 1.2T_1 + 1.4T_2 + 3.3T_3 + 0.6T_4 + 0.999T_5 \qquad (23.2)$$

where

$$T_1 = \frac{\text{Working capital}}{\text{Total assets}}, \; T_2 = \frac{\text{Retained earnings}}{\text{Total assets}}, \; T_3 = \frac{\text{Earnings before income and taxes}}{\text{Total assets}}$$

$$T_4 = \frac{\text{Market value of equity}}{\text{Total liabilities}}, T_5 = \frac{\text{Sales}}{\text{Total assets}}$$

with cut-off zones

$$Z > 2.99 : \text{Safe}$$

$$1.81 < Z < 2.99 : \text{Grey area}$$

$$Z < 1.81 : \text{Distress zone}$$

Altman found a classification accuracy of 83.5% for his model and showed that his model could predict bankruptcy up to three years before the bankruptcy date. Altman also stated in his research that companies could be categorized into three zones by selected cut-off points, that is, safe ($Z > 2.6$), grey ($1.1 < Z < 2.6$) and distress ($Z < 1.1$).

23.2.3 Subsequent Models

Later, in 1972, Edward Deakin revisited Beaver's analysis [14]. He used the 14 ratios that Beaver found to be most effective and attempted to use a discriminant analysis similar to Altman's Z score method. Deakin also attempted to look at data up to five years before the date of bankruptcy. In his analysis, he found that the significance of each ratio changed across the five years. He also found that he was only able to get significant prediction results for up to three years before the date of bankruptcy.

In 1980, James Ohlson attempted an alternative method of bankruptcy prediction using a probabilistic approach [15]. He looked at data between 1970 and 1976. Essentially, in this method he looked at a vector of financial ratios, determined a vector of parameters for those ratios and looked at the probability of bankruptcy for those ratios and parameters. He then attempted to find a cut-off probability point between zero and one for bankruptcy and non-bankruptcy. The ratios that Ohlson employed were:

- Size = log(total assets/GNP price-level index);
- TLTA = total liabilities divided by total assets;
- WCTA = working capital divided by total assets;
- CLCA = current liabilities divided by current assets;
- ONENEG = one if total liabilities exceed total assets, zero otherwise;
- NITA = net income divided by total assets;
- FUTL = funds provided by operations divided by total liabilities;
- INTWO = one if net income was negative for the last two years, zero otherwise;
- CHIN = $(\text{NI}_t - \text{NI}_{t-1})/(|\text{NI}_t| + |\text{NI}_{t-1}|).$, where NI_t is the net income for the most recent period. The denominator acts as a level indicator. This variable is thus intended to measure change in net income.

Ohlson, however, did not find promising results with this model as compared with Altman's model, and thus this model is not commonly used today.

In 1984, Zmijewski explored the potential methodological drawbacks of the previous bankruptcy prediction techniques [16]. His main issue was that previous studies that had used non-random samples, that is, bankrupt and non-bankrupt groups had been predelineated before modelling. Zmijewski attempted to use random sampling and incorporated a probit model to test for bankruptcy. The firms chosen for this study were from the American and New York Stock Exchanges with SIC codes of less than 6000 and were obtained between 1972 and 1978. What Zmijewski did was to create a variable B, where if $B > 0$ then the company was at risk of bankruptcy. His model is below:

$$B^* = a_0 + a_1 \text{ROA} + a_2 \text{FINL} + a_3 \text{LIQ} + u \qquad (23.3)$$

$$P(B^* > 0) = P(-u < a_0 + a_1 \text{ROA} + a_2 \text{FINL} + a_3 \text{LIQ})$$

where ROA = net income to total assets (return on assets), FINL = total debt to total assets (financial leverage), LIQ = current assets to current liabilities (liquidity) and u = normally distributed error term.

However, Zmijewski found that his results were qualitatively similar to those that used non-random sampling and that there was no apparent improvement in the overall classification rates [16].

In the 1990s, there were many critiques of bankruptcy prediction. In 1993, Su-Jane Hsieh criticized methods for determining the cut-off point for bankruptcy [17]. Some issues that were pointed out were the fact that the cut-off point was determined by trial and error, not by statistics, and that the cut-off point was determined without considering the relative losses for Type I and Type II errors. Hsieh derived a modified Bayesian decision model to estimate an optimal cut-off point for bankruptcy prediction models [17]. A function was added in this model to account for the error costs of Type I and Type II errors, and attempted to minimize these costs and not simply the probability of the error. However, although Hsieh came up with this method for determining the cut-off point, it has never actually been applied to previous bankruptcy models to determine its effectiveness versus the common trial and error approach.

In 2001, John Grice and Michael Dugan noted another drawback, that models may not be as effective outside the time period in which the model was created [18]. That same year, Tyler Shumway attempted to create a bankruptcy prediction method using a hazard model to account for changes over time [19]. He collected data for over 31 years and used the same ratios that Altman had used in his Z score model. Shumway's model, although it showed results better than Altman's in the first year before bankruptcy, had a significant decline in accuracy before the second year before bankruptcy. And, of course, predicting bankruptcy only one year ahead is not very useful, as by then the financial stress is relatively easily observed even without any models.

It can be seen that many bankruptcy models have used Altman's model as a benchmark for bankruptcy prediction. In 2001 another study was done by John Grice, along with Robert Ingram, to look at the generalizability of the Altman Z score model [20]. Grice looked at data between 1988 and 1991 and again showed that Altman's model was not as accurate during that time as it was at the time that it was developed. It was also shown that Altman's model was significantly more effective in predicting bankruptcy for a sample of specifically manufacturing firms than for a general dataset of companies.

In 2004, a study was carried out by Sudheer Chava and Robert Jarrow to look at industry effects in bankruptcy prediction [21]. Data were collected from 1962 to 1999 and firms were taken from the AMEX, NYSE and NASDAQ listings. This study looked at both yearly and monthly intervals and showed that monthly intervals had the potential for being better predictors of failure if the data could be collected. A hazard model was run on the variables from Altman's model [2], Zmijewski's model [16] and Shumway's model [19] and showed that industry groupings had a significant effect on the slope and intercept coefficients in these models.

In 2004, Stephen Hillegeist, Elizabeth Keating, Donald Cram and Kyle Lundstedt attempted to use an options pricing model to look at the probability of bankruptcy [4]. However, again this model looked only at manufacturing firms to compare the results with Altman's Z score [2] and Ohlson's model [15], and it was suggested by the authors that the coefficients should be updated to provide industry adjustments.

From our literature review, it can be seen that the type of industry is a factor in bankruptcy prediction. Based on Altman's Z score method, a large number of related studies were developed by employing different ratios [14–21], of which the majority still focused on manufacturing companies. Consequently, Altman proposed his lesser-known Z'' score method to address this deficiency and deal specifically with non-manufacturing industry. The method is shown below:

$$Z'' = 6.56T''_1 + 3.26T''_2 + 6.72T''_3 + 1.05T''_4 \tag{23.4}$$

where

$$T''_1 = \frac{\text{Working capital}}{\text{Total assets}}, T''_2 = \frac{\text{Retained earnings}}{\text{Total assets}}$$

$$T''_3 = \frac{\text{EBIT}}{\text{Total assets}}, T''_4 = \frac{\text{Book value of equity}}{\text{Total liabilities}}$$

with cut-off zones

$$Z > 2.6 : \text{Safe}$$

$$1.1 < Z < 2.6 : \text{Grey area}$$

$$Z < 1.1 : \text{Distress zone}$$

Altman revised the coefficients and items in the former Z score model to form a Z'' score model. Similarly to the Z score method, firms are classified into three areas. Even though the Z'' score model can be called an attempt to examine alternative industries compared with the Z score model, it still shows a major influence of the firms' asset size. Given this, a non-parametric method, that is, DEA, which is flexible in attribute selection, was considered in the present research.

Recently, DEA has been welcomed as a method for corporate failure prediction in some comparisons with various traditional methods [22–25]. Cielen *et al.* compared a linear programming model, a decision tree method and DEA from the methodological viewpoint for corporate failure prediction, and concluded that there were no large accuracy discrepancies between linear programming models and DEA, but both of those methods outperformed the decision tree method [26]. On the other hand, Sueyoshi *et al.* applied DEA-DA (DEA with discriminant analysis) to bankruptcy assessment and compared it with the DEA method, and found that DEA-DA was more appropriate for datasets over time [27]. Furthermore, a novel DEA method that integrated it with rough set theory (RST) and support vector machines (SVM) was used to increase the accuracy of prediction of corporate failure [28]. These studies utilized different methods and compared them with DEA, emphasizing the predominance of DEA in corporate failure prediction. However, as aforementioned, none of these studies focuses on the prediction of failure of non-manufacturing firms, which have a small asset size compared with other industries, and deserve more attention.

23.3 METHODOLOGY

Inside DEA, there are many technical details affecting the selection of models, such as returns to scale, and radial or non-radial models. The first DEA model was CRS, which is a constant-returns-to-scale DEA model [7]. From CRS, many other DEA models have been developed, and most of these are radial models. However, the radial DEA models, such as the CRS and the variable-returns-to-scale (VRS) [29] models, are limited by the fact that they do not account for mix inefficiencies. In this case, the company under examination is not limited to 'proportional attributes change', but is evaluated by the general deviation from the best firms. It follows that the SBM model [30], which accounts for mix inefficiencies, is more suitable for the current study.

23.3.1 Slacks-Based Measure

Before we present the utilization of SBM in corporate failure analysis, we briefly introduce the SBM model in this section. Assume there are n DMUs in the current system, and each of them has m inputs and s outputs. Therefore, the output vectors and input vectors for these DMUs can be expressed as an $(m \times n)$ matrix X and an $(s \times n)$ matrix Y, respectively. We use DMU_o to denote the DMU currently under evaluation. Then the efficiency score of DMU_o can be expressed by the following model:

$$\rho = \min_{\lambda, s^-, s^+} \frac{1 - \dfrac{1}{m} \displaystyle\sum_{i=1}^{m} \dfrac{s_i^-}{x_{io}}}{1 + \dfrac{1}{s} \displaystyle\sum_{r=1}^{s} \dfrac{s_r^+}{y_{ro}}}$$

$$\text{s.t.} \quad x_o - s^- = X\lambda$$

$$y_o + s^+ = Y\lambda$$

$$\lambda \geq 0, \ s^- \geq 0, \ s^+ \geq 0$$

(23.5)

In the above model, $x_o = (x_{1o}, x_{2o}, \ldots, x_{mo})^\mathrm{T}$ and $y_o = (y_{1o}, y_{2o}, \ldots, y_{so})^\mathrm{T}$ are the input and output vectors of DMU_o. Slack vectors are defined by $s^- \in R^m$ and $s^+ \in R^s$, which can be explained as input excesses and output shortfalls referring to the efficient frontier. The production possibility set P is defined as follows:

$$P = \{(x,y) \mid x \geq X\lambda, y \leq Y\lambda, \lambda \geq 0\} \tag{23.6}$$

The combination $(X\lambda, Y\lambda)$ defined by the production possibility set is formed by a non-negative vector λ, and it always outperforms (x_o, y_o). Tone [30] concluded that the above SBM model satisfied the following four properties: (P1) *Units invariance*: the optimal value of the objective function is independent of the units in which the inputs and outputs are measured. (P2) *Monotonicity*: the efficiency of a DMU decreases monotonically with an increase in any slack for either the input or the output. (P3) *Reference set dependence*: the efficiency of a DMU should be measured only by referring to its corresponding reference set. (P4) *Charnes–Cooper transformation*: the original non-linear SBM model in (23.5) can be transformed into a linear one using the Charnes–Cooper transformation.

The upper limit of the objective function in (23.5) is 1, and this can be interpreted as meaning that the ratio of the mean input and output mix inefficiencies has an upper limit of 1. If the optimal solution for an inefficient DMU_o of (23.5) is denoted as $(\rho^*, \lambda^*, s^{-*}, s^{+*})$, DMU_o can be improved to be efficient by reducing its input excesses and augmenting its output shortfalls as follows:

$$\begin{aligned} \hat{x}_o &= x_o - s^{-*} \\ \hat{y}_o &= y_o + s^{+*} \end{aligned} \tag{23.7}$$

Here, $(\hat{x}_o, \hat{y}_o)$ is usually considered to be an improving target, and it is defined by projecting DMU_o to a given point on the efficiency frontier. The reference set of DMU_o is constituted by all the positive elements in the vector λ^*. In cases where it is only necessary to investigate the slacks in the inputs, the input-oriented SBM model is usually utilized. The input-oriented SBM model is actually the numerator of the SBM model, with corresponding modifications to the constraints that can be expressed as follows:

$$\rho = \min_{\lambda, s^-, s^+} 1 - \frac{1}{m} \sum_{i=1}^{m} \frac{s_i^-}{x_{io}}$$

$$\text{s.t.} \quad x_o - s^- = X\lambda \tag{23.8}$$

$$y_o \leq Y\lambda$$

$$\lambda \geq 0, \; s^- \geq 0$$

By mathematical manipulations similar to those mentioned before, we can obtain the output-oriented SBM model, but we will not discuss this further here. There are also many other variations of the SBM model concerning returns to scale, super-efficiency, Russell measure and so on. For a detailed introduction to these subjects, see [31].

23.3.2 Model Development

Since we are using SBM scores instead of Altman's Z'' scores to measure the health status of a company, the first step is to design the structure of the DMUs. Complying with the ratios used in Altman's method, we considered splitting these ratios and extracting useful numerators and denominators from them as independent inputs and outputs. This means that all of the numerators of the ratios were considered to be outputs and the denominators were defined as inputs in the SBM model. The ratios were split rather than being input directly, as it has been shown that ratios used as inputs or outputs in DEA models can affect the results.

Originally, earnings before interest and taxes (EBIT) was used in Altman's model, and was calculated as revenue minus expenses, excluding tax and interest; it was used to characterize a company's profitability. Owing to data availability, EBIT was substituted for operating income, which is used interchangeably across much of the accounting and investing world. Operating income is calculated as gross income less operating expenses, depreciation and amortization. This excludes taxes and interest expenses, just as with EBIT. The only difference between EBIT and operating income is that operating income is considered an official measure under Generally Accepted Accounting Principles (GAAP), whereas EBIT is not. But this does not affect the use of operating income instead of EBIT.

Moreover, as one of the main purposes of our research, we need to see how accurately bankruptcy can be predicted regardless of asset size. So, additionally, the attribute 'total liabilities' was removed and 'working capital' was split into 'current assets' and 'current liabilities'. To test the relevance of human capital, which is important to smaller non-manufacturing firms in our model, the number of employees and the number of shareholders were added to the model. The number of employees was added to introduce a measure of human capital (the most important 'asset' in a non-manufacturing firm) as a contributor to the efficiency of a company. The number of shareholders was added because, for many smaller non-manufacturing firms, the shareholders have decision-making powers and invest both time and money that contribute to the success of a firm. In this sense, the number of shareholders can also be seen as a reflection of the financial well-being of a company as viewed by the public.

Another problem we met with was that many bankrupt companies had negative values for retained earnings (RE), operating income (OI) and book value of equity (BE), to which the SBM model was not applicable. Thus each output was split into positive and negative parts. For example, RE was split into RE^+ and RE^-, where RE^+ was defined as an output in its usual sense, but RE^- was defined as an input. This method says essentially that RE^+ is an output and therefore should be made as large as possible to improve the company's operating efficiency. However, RE^- is viewed as an input which should be minimized. The inputs and outputs of the model after revision are shown in Table 23.1.

Generally, the calculation results obtained from DEA models are affected by the relationship between the number of DMUs and the dimensions of the DMUs, and this topic has taken a variety of forms in the DEA literature [32–35]. Although we did

TABLE 23.1 Classification of inputs and outputs.

Inputs	Outputs
Current liabilities (CL)	Current assets (CA)
Negative retained earnings (RE$^-$)	Positive retained earnings (RE$^+$)
Negative operating income (OI$^-$)	Positive operating income (OI$^+$)
Negative book value of equity (BVE$^-$)	Positive book value of equity (BVE$^+$)
Number of employees (EM)	Number of shareholders (SH)

attempt to use the normal SBM model, that is, without orientation, to calculate the scores, the number of DMUs applicable to our study was between 23 and 42, which is quite limited, considering the above 10 attributes. The numbers of either bankrupt or non-bankrupt DMUs in each year were changed owing to the lack of available financial data. We give a detailed description of the data in Section 23.4. As a result, many DMUs obtained an efficiency score of '1', which was relatively undiscriminating for judging bankruptcy. Given this, we adopted a rough rule of thumb for guidance in deciding the number of DMUs and their dimensions as follows [31]:

$$n \geq \max\{m \times s,\ 3(m+s)\} \tag{23.9}$$

where n, m and s are the numbers of DMUs, inputs and outputs, respectively.

From the above equation, it can be observed that the number of DMUs in our case should be at least 30; however, most of the time the scale of the DMUs was smaller than 30. Therefore, we used the input-oriented SBM model as shown in problem (23.8) in the actual calculations to comply with the constraints in (23.9). Undoubtedly, the output-oriented SBM model should also be feasible and give satisfactory results. Furthermore, various studies have concentrated on generating new datasets to overcome the problem of insufficient DMUs, for which we will not offer a detailed discussion here [32, 36, 37].

23.4 APPLICATION TO BANKRUPTCY PREDICTION

DEA is capable of calculating efficiency scores using various DEA models by providing appropriate classifications of inputs and outputs, and attribute values of DMUs. An efficiency score ranging from 0 to 1 can then be assigned to each DMU, in this case a non-manufacturing company, to describe its overall health status. For a group of companies, we need first to clarify the thresholds used to categorize what kind of companies are considered to be healthy, what kind of companies are considered to be bankrupt, and what between these two categories is the intermediate state. This means that there is a necessity to select two values in the interval [0, 1] as cut-off points to categorize the companies under analysis into three zones, that is, safe, distress and grey, similarly to Altman's models. To obtain the cut-off points, the data sample collected was divided into two groups. The first group was a training set, used to define appropriate cut-off values. Then we applied the input-oriented SBM model to the second group, and compared the results obtained by SBM and the results provided by Altman's method to validate our methodology.

23.4.1 Data Acquisition

In this research, the data that we utilized were collected from two sources. Part of the data was retrieved from the Mergent Online database [38], and the second data source was a professional company which focused mainly on bankrupt companies filing in North America dating back to the 1980s, selected by SIC (Standard Industrial Classification) codes. The list of companies was narrowed down to those classified as non-manufacturing or service-based firms. Those companies must also have filed for bankruptcy between the years 2000 and 2006. The reason for these dates was that more recent filings could be more easily obtained, and more easily compared with current companies. Bankruptcy filings from 2007 to the present were not selected, owing to the economic recession that was taking place, and hence we decided that the data could not reflect the real situation in that period. The companies considered to be bankrupt during that period could be more so for external reasons, which was not the main purpose of the current research.

For each bankrupt company, financial data were collected for up to five years before the date of bankruptcy being filed, as it had been shown that there was potential to predict bankruptcy up to five years in advance [1,39]. Some companies did not have a full five years of data, and thus only had the number of years before bankruptcy collected. Whenever it was possible to identify them, companies that had filed for bankruptcy but did not fail were excluded from the study. Many of these companies filed for bankruptcy for reasons other than complete insolvency: some liquidations were due to legal issues, and others filed because they were suffering financial distress, in an attempt to reorganize and restructure their corporate strategy and alleviate the debt. Data from the full balance sheets, income statements, cash flow statements and retained earnings were collected. From the balance sheet, current assets, total assets, current liabilities, total liabilities, retained earnings and shareholders' equity values were extracted. From the income statement, the operating profit was calculated using the formula Net Sales – Cost of goods – Expenses. The number of employees and number of shareholders were also collected.

Once the data had been collected for the bankrupt companies, healthy companies were then found. A healthy company was chosen for every bankrupt company based on SIC number and on its years of health. Healthy companies had to be in existence for at least five years after the bankruptcy of their bankrupt counterpart. Healthy companies also had not to have filed for bankruptcy during the time that they were being compared with their bankrupt counterpart. The same financial data were collected for the healthy company as for the bankrupt counterpart for the same years. For example, if a bankrupt company filed for bankruptcy in 2002, financial data were collected for 1997–2001. The healthy company would have to have been in existence and not to have filed for bankruptcy between the years from 1996 to 2006. In some cases, a

TABLE 23.2 Numbers of companies in Group 1.

Year before bankruptcy	Number of bankrupt companies	Number of non-bankrupt companies
1	40	29
2	34	28
3	31	26
4	32	24
5	26	23

TABLE 23.3 Numbers of companies in Group 2.

Year before bankruptcy	Number of bankrupt companies	Number of non-bankrupt companies
1	42	35
2	38	34
3	39	34
4	32	30
5	26	27

suitable healthy match could not be found, and thus the number of bankrupt companies exceeded the number of non-bankrupt ones. There might be a more scientific data selection method than the one that we were using, as in most cases the number of bankrupt companies and the number of non-bankrupt companies are not equal: in the real world, healthy companies are much more numerous than bankrupt ones.

The numbers of bankrupt and non-bankrupt companies used in the first group to determine cut-off points are shown in Table 23.2. The numbers of bankrupt and non-bankrupt companies in the second group are listed in Table 23.3.

23.4.2 Analysis of Results

The companies in group 1 were evaluated by an input-oriented SBM model for five years, and the results are shown in Table 23.4. Once each company had been assigned an efficiency score, a measure of bankruptcy status had to be determined. For each year, every possible cut-off point was tested at increments of 0.05 from 0 to 1 to determine the bankrupt and non-bankrupt classification accuracy at those potential cut-off points. Figure 23.1 shows the accuracy percentages versus the crossover points for the first year. For example, for a cut-off point of zero, no bankrupt companies would be classified as bankrupt and all non-bankrupt companies would be classified as non-bankrupt. With an increasing cut-off value, the accuracy of identifying non-bankrupt companies increases, but the accuracy of finding bankrupt companies decreases. The only point which we should choose to maintain the highest accuracy for both bankrupt and non-bankrupt companies is the crossover point of the two curves. Here that point is 0.55, where the bankrupt and non-bankrupt accuracies are 67.50% and 68.97% separately.

TABLE 23.4 SBM scores of companies in Group 1.[a]

DMU	Year 1	Year 3	Year 3	Year 4	Year 5	DMU	Year 1	Year 2	Year 3	Year 4	Year 5
1	0.3332	1.0000	0.5301			39	1.0000	1.0000	1.0000	1.0000	1.0000
2	0.7337	0.8203	0.9105	0.7400		40	1.0000	1.0000			0.8786
3	0.1195	0.3279	0.7355	0.6680	1.0000	41					1.0000
4	0.2617	0.3405	0.3021	1.0000	0.5443	42	0.0969	0.5254	0.4908	0.6073	0.5270
5	0.4825	1.0000	1.0000	1.0000	0.5391	43	0.2961	0.3298	0.4771	0.5084	0.0330
6	0.0236	0.2667	1.0000	0.3258	0.8414	44	1.0000	0.2365	0.2225	0.0429	0.5512
7	0.2536	0.6710	0.5042	0.7558		45	0.5290	0.8027	0.8824	0.7540	0.4682
8	0.6622	0.1387				46	0.4031			0.6297	
9	1.0000	1.0000	1.0000	1.0000	1.0000	47	0.5218	0.4664	0.6330	0.6744	
10						48	1.0000		0.3492		
11	0.4366	0.8556	0.7703	0.8345	0.8099	49	0.6908	0.6041	0.6496	0.8010	0.8090
12	0.8311	0.8645	0.9449	0.8502	0.9320	50	0.3193	0.4351	0.3290		
13	0.8183	0.8977	1.0000	1.0000	1.0000	51	0.1367	1.0000	1.0000	1.0000	1.0000
14	0.8698	0.7154	0.6873	0.7951	1.0000	52	1.0000	1.0000	1.0000	1.0000	1.0000
15	1.0000	1.0000	1.0000	0.5749	0.6314	53	1.0000	1.0000	1.0000	1.0000	1.0000
16						54	0.5078	0.5366	0.2890	0.2829	0.5013
17	0.4274	1.0000	1.0000	1.0000	1.0000	55	0.2760	0.3823	0.4903	0.8728	0.6392
18	0.7348	0.9226	0.8880	0.7959	0.8237	56	0.4215	1.0000	1.0000	0.7239	0.8620
19	0.9407	0.8759	0.8587	0.6425	0.6166	57	1.0000	1.0000			
20	0.0413	0.2572	0.3821			58	0.0462	0.5322	0.5602	0.6218	0.2517
21	0.2785	0.5957				59	0.2532	0.6502	0.8040	0.8601	1.0000
22	0.3011	0.7659	0.6567	0.9349	0.8341	60	0.7830			0.7981	0.6919
23						61	0.2890	0.3216	0.1119	0.0850	0.1131
24	0.5562	0.7208	0.6755	0.4497	0.6608	62	0.5458	0.6431	0.5632	0.5382	0.6441
25	0.5054	0.6033	1.0000	1.0000	1.0000	63	0.2915	0.5893	1.0000	1.0000	
26	1.0000	0.6232	0.5007			64	1.0000				
27	0.2838	0.3743	0.3619	0.2937	0.3991	65	0.6373	0.7706	0.0762		
28	0.2521	0.3761	0.4744	0.4888	1.0000	66	0.5512	0.5520		0.0797	

29	0.4845	0.7973	0.9323	0.3988	
30	0.0797	0.3908	0.2136	1.0000	0.4411
31	0.5730	0.7260	1.0000	0.6105	
32	0.2830			0.4717	
33	0.1054	0.5417	0.4941	0.7215	0.5147
34	0.2527				0.7296
35					
36	1.0000	1.0000	1.0000	1.0000	1.0000
37	0.6267	0.6967	1.0000	1.0000	1.0000
38	1.0000	1.0000	1.0000	1.0000	

67	1.0000	1.0000	1.0000	1.0000	1.0000
68					
69	0.2781	0.3720	0.8169	0.6701	0.5106
70	0.6131	0.4972	0.4968	0.3019	0.8132
71	0.8047	0.7483	1.0000	0.7641	1.0000
72	0.3157	0.5317	0.7623	0.7681	0.1575
73	1.0000	0.3065	0.3086	0.0871	
74	1.0000	1.0000	1.0000	1.0000	1.0000
75	1.0000	1.0000	1.0000	1.0000	1.0000

[a] Some companies may not have efficiency scores owing to bankruptcy or a lack of available data for that year.

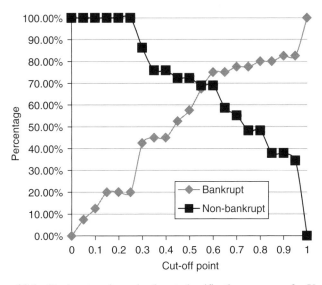

Figure 23.1   Bankrupt and non-bankrupt classification accuracy for Year 1.

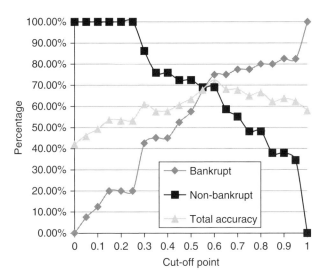

Figure 23.2   Selection of bottom and top cut-off points for Year 1.

To categorize all the companies into three zones, that is, safe, grey and distress, we need to choose two cut-off points. If we plot the curve of total accuracy, where both bankrupt and non-bankrupt companies are correctly categorized, as in Figure 23.2, we can find two points where we achieve a relatively high total accuracy around the point 0.55. One of those points is at 0.5, located to the left, with 63.77% overall accuracy.

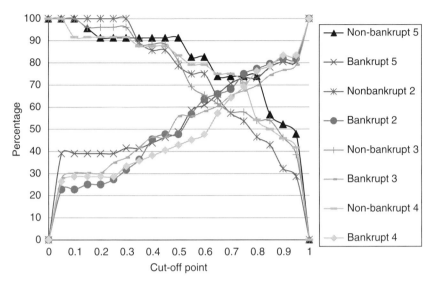

Figure 23.3   Cut-off points from Year 2 to Year 5 before bankruptcy.

TABLE 23.5   Cut-off points for SBM model.

Interval	Classification
$\theta \geq 0.80$	Safe area
$0.65 < \theta < 0.80$	Grey area
$\theta \leq 0.65$	Distress area

Here, the bankrupt companies have a classification accuracy of 57.50% and the non-bankrupt companies have a classification accuracy of 72.41%. This point was thus considered to be the bottom cut-off point to discriminate between 'distress' and 'grey' zones. In the same way, we fixed the top cut-off point at 0.6, where the total accuracy reaches another high value. At this point, the classification accuracy for bankrupt companies is 75.00%, and for non-bankrupt companies the classification accuracy is 68.97%. Therefore, this point was regarded as the boundary that separated 'grey' and 'safe' zones.

However, this is only the process for selecting cut-off points for one year before bankruptcy. In the same way, we can plot the bankrupt and non-bankrupt percentage curves for the other four years before bankruptcy, as shown in Figure 23.3. As we were more concerned about the classification accuracy for bankrupt companies than for non-bankrupt ones, we have shifted these points up. The finalized cut-off points, obtained by comparing the values over the five years, are indicated in Table 23.5.

We then calculated the SBM efficiency scores for all companies in group 2, as shown in Table 23.6. Based on the cut-off points that we obtained from group 1,

TABLE 23.6 SBM scores of companies in Group 2.

DMU	Year 1	Year 2	Year 3	Year 4	Year 5	DMU	Year 1	Year 2	Year 3	Year 4	Year 5
1	0.4671	0.4759	0.8892	0.6681		45	0.4436	0.3974	0.3599	0.3616	0.3761
2	1.0000	1.0000	1.0000	0.2977	0.2322	46		1.0000	0.5314	0.3608	1.0000
3	0.1937	0.1681	0.4395	1.0000	1.0000	47	0.0804	0.4013	0.5014	0.3332	0.5777
4	1.0000	1.0000	1.0000	1.0000	1.0000	48	0.8082	0.8554	1.0000	1.0000	1.0000
5	1.0000	1.0000	1.0000			49	1.0000				
6	0.8034	0.7558	0.7553	0.7346	0.7413	50	0.1158		1.0000		
7					1.0000	51	0.6579	1.0000	0.6973	1.0000	1.0000
8	0.3124	0.1162	0.2337	0.0780	0.3352	52	1.0000	1.0000	1.0000	1.0000	
9	0.7817	1.0000	1.0000	1.0000	1.0000	53	0.1583	0.1210			
10	1.0000	1.0000	1.0000	1.0000	1.0000	54					
11	1.0000	0.7980	0.8862	0.7704	1.0000	55	0.5178	0.8522	1.0000	0.6867	0.8104
12	0.8331	0.8436	0.8684	0.7623	0.8106	56	1.0000	1.0000	1.0000	1.0000	1.0000
13					1.0000	57	0.2558	0.2681	0.2842	0.3289	0.3169
14	0.7181	0.7218	0.7402	0.7745	0.7522	58					
15	1.0000	1.0000	1.0000	1.0000	1.0000	59	0.3390	1.0000	1.0000	1.0000	1.0000
16	1.0000			1.0000		60	1.0000	1.0000	1.0000	1.0000	1.0000
17	0.5071					61	0.6462	0.6888	0.7641		
18	1.0000	1.0000	1.0000	1.0000	1.0000	62	0.4321	0.4194	0.2214		0.0854
19	0.7484	0.7180	1.0000	1.0000	1.0000	63	0.6317	0.5374	0.3726	0.4164	0.8570
20	1.0000	1.0000	1.0000	1.0000		64	0.4923	1.0000	0.5715	1.0000	
21	0.7149	0.7908	0.8061	0.8125	0.7356	65	0.2869	0.0583			
22	0.5457	0.7382	0.7422	0.7398	0.7827	66	0.3773	1.0000	1.0000		
23	0.5489	0.5966	0.7836	0.7915	0.8103	67		1.0000			
24	0.9612	1.0000	1.0000	1.0000		68	0.4057	1.0000	1.0000	1.0000	0.3515
25	0.4143	0.8312	0.7234			69	0.5086	0.2944	0.2849	0.5566	
26		0.7259	0.6940	0.8135	0.9357	70	1.0000	1.0000	1.0000	1.0000	1.0000
27	0.2709	0.3241	1.0000	0.3093		71	1.0000	1.0000	1.0000	1.0000	1.0000
28	1.0000	1.0000	1.0000	0.7705		72	0.7755	0.7534	0.8137		

29	1.0000	1.0000	1.0000	1.0000	1.0000					
30	0.2523	0.4529	0.6730	0.7045	0.7190					
31	1.0000	1.0000	0.2037	0.2280	0.2174					
32	0.9831	1.0000	1.0000							
33	0.0642	0.2547	0.4764	0.2980	0.4990					
34	0.0849	0.7486	1.0000							
35	0.7423	0.8097	0.8222	0.8360	0.8322					
36	0.6434	0.6370			0.5614					
37	1.0000	1.0000	1.0000	1.0000						
38	0.3789	0.5783	0.6040	0.6271	0.8892					
39	0.7512	0.7386	0.7556	0.7820	0.9853					
40	0.2567		0.6749	0.6628	0.7110					
41	1.0000	1.0000	1.0000	1.0000	1.0000					
42	0.3825	0.6303	0.3268	0.2887	1.0000					
43	0.4176	0.5269	0.0453	0.0398	0.2596					
44	0.2386	0.0130								

73	0.6239	0.8440	0.7942	0.7603	0.8238
74	1.0000	1.0000	0.9698	1.0000	0.7284
75					
76			0.0678	0.2259	
77	1.0000		1.0000		
78	0.3308	0.1287	0.7783	0.7480	
79	0.1245	0.0174	0.1528	0.0210	0.2408
80	0.0178	0.6094	0.2678	0.7109	0.7527
81	0.0365	0.4840	0.5679	0.5794	0.4947
82	0.2689	0.3145	0.4683	0.5283	0.5244
83	0.0090	0.7060	0.6571	0.5638	
84	0.5370		0.4407	0.6724	0.2925
85					
86	0.9606	0.7091	1.0000	1.0000	
87	0.2930	0.6173	0.5420		

the classification accuracy for group 2 was estimated as shown in Table 23.7. Classification accuracy results for group 2 could also be obtained by Altman's Z'' model, and these are shown in Table 23.8. By comparing the results in Tables 23.7 and 23.8, we find that some of the values of the classification accuracy obtained from the SBM model are lower than those obtained from Altman's model. However, most of the values obtained by SBM show better performance than Altman's model. If we investigate the overall classification accuracy, including both bankrupt and non-bankrupt companies, and plot the results as shown in Figure 23.4, it is apparent that the SBM model performs much better than Altman's model. Moreover, the longer the time before bankruptcy, the higher the accuracy that SBM can provide.

23.5 CONCLUSIONS

This research has surveyed the related literature on bankruptcy prediction, stretching from Beaver's univariate model to Altman's Z'' model, then proposed an approach of utilizing a non-parametric method, that is, the SBM model in DEA, to predict corporate failure. To deal with negative factors in this study, we split such factors into positive and negative parts, which could be a viable option when needed in DEA analyses. Based on this methodological revision to SBM, we also validated our method with two groups of bankrupt and non-bankrupt firms. The second group was examined using cut-off points obtained from the first group.

   The overall accuracy of the SBM model was obviously higher than that of the Altman Z'' model, which showed that the total assets or liabilities of a company were actually not necessary for predicting bankruptcy, and that SBM could be a more appropriate method for corporate failure prediction for non-asset-heavy firms. The results are significant for companies such as non-manufacturing or retail companies which do not investments in large hard assets and are not suitable for using Altman's Z'' model. The overall classification results showed that the Altman Z'' model had good prediction accuracy in the years close before bankruptcy, but still lower than the SBM model developed here, which, in fact, shows a dramatically higher accuracy than Altman's Z'' model further from bankruptcy, unveiling a company's health status in advance, which should be more important for the company management (where they could change the course of the firm before it is too late) or for investors or lenders (where they could force a change in management, or simply withdraw their investment while there is time).

   This research has many useful conclusions but, as usual, there are suggestions for further work, including the following. (i) Employing alternative DEA models or constraint conditions, particularly using the assurance region model, which put more restrictions on the variable weights and may lead to more accurate results. (ii) The prediction accuracy may be affected by different approaches to selecting inputs and outputs, and therefore different or related financial factors may bring lead to prediction accuracy. (iii) Owing to the lack of available data, the number of DMUs used in this study was insufficient for a more comprehensive assessment of the model. With

TABLE 23.7   Classification accuracy (%) for group 2 determined by cut-off points.

Year	1	2	3	4	5
Bankrupt accuracy	78.6	57.9	46.2	53.1	38.5
Non-bankrupt accuracy	62.9	61.8	73.5	66.7	70.4
Total accuracy	71.4	59.7	58.9	59.7	54.7
Bankrupt accuracy including grey area	85.7	68.4	69.2	78.1	57.7
Non-bankrupt accuracy including grey area	77.1	88.2	88.2	93.3	81.5
Total accuracy including grey area	81.8	77.8	78.1	85.5	69.8
Total bankrupt	53.3	36.1	30.1	30.7	28.3
Total non-bankrupt	36.4	45.8	50.7	43.6	56.6
Total within grey area	10.4	18.1	19.2	25.8	15.1

TABLE 23.8   Results of Altman Z'' model for group 2.

Year	1	2	3	4	5
Bankrupt accuracy	77.8	59.1	50.0	41.5	35.1
Non-bankrupt accuracy	47.5	52.5	55.0	52.5	63.9
Total accuracy	63.5	55.9	52.4	46.9	49.3
Bankrupt accuracy including grey area	88.9	86.4	70.5	70.7	83.8
Non-bankrupt accuracy including grey area	60.0	72.5	75.0	75.0	88.9
Total accuracy including grey area	72.9	69.1	59.5	60.5	67.1
Total bankrupt	61.2	45.2	39.3	34.6	30.1
Total non-bankrupt	29.4	34.5	44.1	46.9	52.1
Total within grey area	11.8	23.8	20.2	25.9	36.9

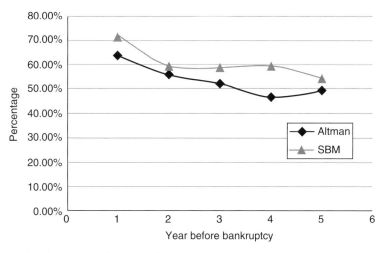

Figure 23.4   Comparison of total classification accuracy between Altman's method and SBM.

a larger number of DMUs, the cut-off points for bankruptcy prediction will become more accurate. (iv) Innovative approaches to determine the cut-off points could be explored. The trial and error approach is simple and intuitive; however, a different and more statistically sound method could be developed. Decision trees were considered but not employed; however, this could also be considered in future research.

Both the previous univariate models and Altman's Z and Z'' models focus mostly on the asset size of firms, and use parametric methods, that is, a weighted sum of asset-based items, which is more likely to result in an empirical selection process for cut-off points, but is not based on the reality of corporate structures. It follows that the DEA technique, a non-parametric method, could solve the problem, resulting in a rather practical approach to predicting corporate failure, especially for non-manufacturing firms. In closing, we hope that this research will be insightful and informative to future researchers.

REFERENCES

[1] Beaver, W.H. (1967) Financial ratios as predictor of failure. *Journal of Accounting Research*, **5**, 71–111.

[2] Altman, E. (1968) Financial ratios, discriminant analysis and the prediction of corporate bankruptcy. *Journal of Finance*, **23**(4), 589–609.

[3] Grice, J.S. and Ingram, R.W. (2001) Tests of the generalizability of Altman's bankruptcy prediction model. *Journal of Business Research*, **54**(1), 53–61.

[4] Hillegeist, S.A., Keating, E.K., Cram, D.P. and Lundstedt, K.G. (2004) Assessing the probability of bankruptcy. *Review of Accounting Studies*, **9**(1), 5–34.

[5] Hitchcock, D. and Willard, M. (2015) *The Business Guide to Sustainability: Practical Strategies and Tools for Organizations,* 3rd edn, Routledge, New York.

[6] Altman, E. (2002) *Bankruptcy, Credit Risk and High Yield Junk Bonds,* Blackwell, Malden, MA.

[7] Charnes, A., Cooper, W.W. and Rhodes, E. (1978) Measuring the efficiency of decision making units. *European Journal of Operational Research*, **2**(6), 429–444.

[8] Farrell, M.J. (1957) The measurement of productive efficiency. *Journal of the Royal Statistical Society, Series A*, **120**(3), 253–290.

[9] Emrouznejad, A., Parker, B.R. and Tavares, G. (2008) Evaluation of research in efficiency and productivity: A survey and analysis of the first 30 years of scholarly literature in DEA. *Socio-Economic Planning Sciences*, **42**(3), 151–157.

[10] Paradi, J.C. and Zhu, H. (2013) A survey on bank branch efficiency and performance research with data envelopment analysis. *Omega*, **41**(1), 61–79.

[11] Liu, J.S., Lu, L.Y., Lu, W.M. and Lin, B.J. (2013) Data envelopment analysis 1978–2010: A citation-based literature survey. *Omega*, **41**(1), 3–15.

[12] Yang, X. and Morita, H. (2013) Efficiency improvement from multiple perspectives: An application to Japanese banking industry. *Omega*, **41**(1), 501–509.

[13] Alexander, S.S. (1949) The effect of size of manufacturing corporation on the distribution of the rate of return. *Review of Economics and Statistics*, **31**, 229–235.

[14] Deakin, E. (1972) A discriminant analysis of predictors of business failure. *Journal of Accounting Research*, **10**(1), 167–179.

[15] Ohlson, J.A. (1980) Financial ratios and the probabilistic prediction of bankruptcy. *Journal of Accounting Research*, **18**(1), 109–131.

[16] Zmijewski, M.E. (1984) Methodological issues related to the estimation of financial distress prediction models. *Journal of Accounting Research*, **22**, 59–82.

[17] Hsieh, S.J. (1993) A note on the optimal cutoff point in bankruptcy prediction models. *Journal of Business Finance and Accounting*, **30**(3), 457–464.

[18] Grice, J.S. and Dugan, M. (2001) The limitations of bankruptcy prediction models: Some cautions for the researcher. *Review of Quantitative Finance and Accounting*, **17**(2), 151–166.

[19] Shumway, T. (2001) Forecasting bankruptcy more accurately: A simple hazard model. *Journal of Business*, **74**(1), 101–124.

[20] Grice, J.S. and Ingram, R.W. (2001) Tests of the generalizability of Altman's bankruptcy prediction model. *Journal of Business Research*, **54**(1), 53– 61.

[21] Chava, S. and Jarrow, R. (2004) Bankruptcy prediction with industry effects. *Review of Finance*, **8**(4), 537–569.

[22] Premachandra, I., Chen, Y. and Watson, J. (2011) DEA as a tool for predicting corporate failure and success: A case of bankruptcy assessment. *Omega*, **39**(6), 620–626.

[23] Li, Z., Crook, J. and Andreeva, G. (2014) Chinese companies distress prediction: An application of data envelopment analysis. *Journal of the Operational Research Society*, **65**, 466–479.

[24] Shetty, U., Pakkala, T. and Mallikarjunappa, T. (2012) A modified directional distance formulation of DEA to assess bankruptcy: An application to IT/ITES companies in India. *Expert Systems with Applications*, **39**(2), 1988–1997.

[25] Xu, X. and Wang, Y. (2009) Financial failure prediction using efficiency as a predictor. *Expert Systems with Applications*, **36**(1), 366–373.

[26] Cielen, A., Peeters, L. and Vanhoof, K. (2004) Bankruptcy prediction using a data envelopment analysis. *European Journal of Operational Research*, **154**(2), 526–532.

[27] Sueyoshi, T. and Goto, M. (2009) Methodological comparison between DEA (data envelopment analysis) and DEA–DA (discriminant analysis) from the perspective of bankruptcy assessment. *European Journal of Operational Research*, **199**(2), 561–575.

[28] Yeh, C.C., Chi, D.J. and Hsu, M.F. (2010) A hybrid approach of DEA, rough set and support vector machines for business failure prediction. *Expert Systems with Applications*, **37**(2), 1535–1541.

[29] Banker, R.D., Charnes, A. and Cooper, W.W. (1984) Some models for estimating technical and scale inefficiencies in data envelopment analysis. *Management Science*, **30**(9), 1078–1092.

[30] Tone, K. (2001) A slacks-based measure of efficiency in data envelopment analysis. *European Journal of Operational Research*, **130**(3), 498–509.

[31] Cooper, W.W., Seiford, L.M. and Tone, K. (2007) *Data Envelopment Analysis: A Comprehensive Text with Models, Applications, References and DEA-Solver Software*, 2nd edn, Springer, New York.

[32] Staat, M. (2001) The effect of sample size on the mean efficiency in DEA: Comment. *Journal of Productivity Analysis*, **15**(2), 129–137.

[33] Zhang, Y. and Bartels, R. (1998) The effect of sample size on the mean efficiency in DEA with an application to electricity distribution in Australia, Sweden and New Zealand. *Journal of Productivity Analysis*, **9**(3), 187–204.

[34] Smith, P. (1997) Model misspecification in Data Envelopment Analysis. *Annals of Operation Research*, **73**(0), 233–252.

[35] Banker, R.D., Chang, H. and Cooper, W.W. (1996) Simulation studies of efficiency, returns to scale and misspecification with nonlinear functions in DEA. *Annals of Operations Research*, **66**(4), 233–253.

[36] Panagiotis, Z. (2012) *Dealing with Small Samples and Dimensionality Issues in Data Envelopment Analysis,* http://mpra.ub.uni-muenchen.de/39226/(accessed 16 February 2016).

[37] Sergio, P. and Daniel, S. (2009) How to generate regularly behaved production data? A Monte Carlo experimentation on DEA scale efficiency measurement. *European Journal of Operational Research*, **199**(1), 303–310.

[38] Mergent, I. (2011) Mergent Online, http://www.mergentonline.com/(accessed 16 February 2016).

[39] Merwin, C.L. (1942) *Financing Small Corporations in Five Manufacturing Industries, 1926–36,* National Bureau of Economic Research, New York.

24

RANKING OF BANKRUPTCY PREDICTION MODELS UNDER MULTIPLE CRITERIA[1]

JAMAL OUENNICHE
Business School, University of Edinburgh, Edinburgh, UK

MOHAMMAD M. MOUSAVI
Business School, University of Edinburgh, Edinburgh, UK

BING XU
School of Social Sciences, Heriot-Watt University, Edinburgh, UK

KAORU TONE
National Graduate Institute for Policy Studies, Tokyo, Japan

24.1 INTRODUCTION

Corporate failure often occurs when a firm experiences serious losses and/or becomes insolvent with liabilities that are disproportionate to its assets. Corporate failure may result from one or a combination of internal and external factors, for example, managerial errors due to insufficient or inappropriate industry experience, risk-seeking managers, lack of commitment and motivation to lead the company efficiently, refusal

[1] Part of the material in this chapter is adapted from Mousavi M.M., Ouenniche J. and Xu B. (2015) Performance evaluation of bankruptcy prediction models: an orientation-free super-efficiency DEA-based framework, *International Review of Financial Analysis*, **42**, 64–74, with permission from Elsevier.

Advances in DEA Theory and Applications: With Extensions to Forecasting Models,
First Edition. Edited by Kaoru Tone.
© 2017 John Wiley & Sons Ltd. Published 2017 by John Wiley & Sons Ltd.

or failure to adjust managerial and operational structures of the firm to new realities, inefficient or inappropriate corporate policies, economic climate, changes in legislation, or industry decline – see, for example, van Gestel *et al.* [1].

Bankruptcy induces substantial costs to the business community such as court costs, lawyer costs, lost sales, lost profits, higher cost of credit, inability to issue new securities and lost investment opportunities (e.g. [2–4]) – for a detailed review of the costs of bankruptcy, we refer the reader to Branch [5]. Therefore, the design of reliable models to predict bankruptcy is crucial for auditing business risks and assisting managers to prevent the occurrence of failure, and assisting stakeholders to assess and select firms to collaborate with or invest in (e.g. [6, 7]).

Given the importance of bankruptcy prediction, there is a considerable amount of literature focusing on both financial and non-financial information, and proposing new bankruptcy prediction models to classify firms as healthy or non-healthy (e.g. [7–9]). With the increasing number of quantitative models available, one of the challenging issues faced by both academics and professionals is how to evaluate these competing models and select the best one(s).

Our survey of the literature on bankruptcy prediction reveals that although some studies tend to use several performance criteria and, for each criterion, one or several measures to evaluate the performance of competing prediction models, the assessment exercise is generally restricted to the ranking of models by a single measure of a single criterion at a time. For example, Theodossiou [10] compared the performance of linear probability models, logit models and probit models using an equally weighted average of Type I and Type II errors as a measure of the correctness of categorical prediction, the Brier score (BS) as a measure of the quality of the estimates of the probability of default, and the pseudo-R^2 statistic as a measure of information content, and found that logit models outperformed both linear probability models and probit models on all measures; however, with respect to the pseudo-R^2 statistic and an equally weighted average of Type I and Type II errors, probit models outperformed linear probability models, but linear probability models outperformed probit models on the BS. Bandyo-padhyay [11] compared the performance of several multivariate discriminant analysis (MDA) models using Type I and Type II errors, and compared the performance of several logit models using the overall correct classification (OCC), the receiver operating characteristic (ROC) measure, the pseudo-R^2 statistic and the log-likelihood (LL) statistic, and found that the rankings of models differed with respect to different measures. Tinoco and Wilson [12] compared the performance of several logit models with different categories of explanatory variables using the ROC, Gini index, and the Kolmogorov–Smirnov (KS) statistic as measures of discriminatory power and the Hosmer–Lemeshow statistic as a measure of calibration accuracy, and found that the rankings of models differed with respect to different criteria and their measures.

In sum, a performance evaluation exercise under multiple criteria remains unidimensional in nature, on one hand, and the 'big picture' is not taken into account, in that only a single criterion or a very restricted number of criteria are used, on the other hand. The drawback of the commonly used approach to the evaluation of the relative performance of competing bankruptcy prediction models is that the rankings corresponding to different criteria or measures are often different, which results in a situation where one cannot make an informed decision as to which model performs best when all criteria and their measures are taken into consideration. This methodological issue has been pointed out by Xu and Ouenniche [13–15] and Ouenniche et al. [16, 17], who proposed several multicriteria frameworks based on DEA and multicriteria decision analysis (MCDA) for assessing the performance of prediction models for crude oil prices and their volatility. In the bankruptcy prediction area, DEA has been used either to classify firms into healthy and non-healthy categories (e.g. [18–21]) or to compute aggregate efficiency scores to be used within statistical or stochastic modelling and prediction frameworks (e.g. [22–25]). Unlike these uses of DEA in bankruptcy research, in this chapter we report on the use of DEA in the performance evaluation of competing bankruptcy prediction models as suggested by Mousavi et al. [26], along with some elements of answers to two research questions related to the design of bankruptcy prediction models: (i) do some modelling frameworks perform better than others by design? and (ii) to what extent do the choice and/or the design of explanatory variables and their nature affect the performance of modelling frameworks?

The remainder of this chapter is organized as follows. In Section 24.2, we survey and classify the literature on bankruptcy prediction models. In Section 24.3, we present the proposed multicriteria methodology, namely, an orientation-free super-efficiency SBM framework to evaluate the relative performance of competing prediction models of bankruptcy. In Section 24.4, we present and discuss our empirical findings. Finally, Section 24.5 concludes the chapter.

24.2 AN OVERVIEW OF BANKRUPTCY PREDICTION MODELS

Bankruptcy prediction models can be divided into four main categories according to the type of information they are fed with, namely, accounting-information-based models, market-information-based models, accounting- and market-information-based models, and accounting-, market- and macroeconomic-information-based models. These models can also be classified into several categories according to the underlying type of modelling framework, namely, discriminant analysis models, single-period probability models, multiperiod probability models and stochastic models. In this chapter, we focus on assessing the relative performance of accounting-based models, market-based models and hybrids. A generic framework for implementing these bankruptcy prediction models can be summarized in the following two-phase procedure:

- *Phase 1*. Use a quantitative modelling framework to estimate the probability of default.
- *Phase 2*. Classify firms into two or more risk classes (e.g. risky versus non-risky or bankrupt versus non-bankrupt) using one or several cut-off points or thresholds depending on whether one classifies firms into two classes or more than two classes.

In the following, we provide a brief description of such bankruptcy prediction models along with a discussion of their main similarities and differences.

24.2.1 Discriminant Analysis Models

Discriminant analysis (DA), first proposed by Fisher [27], is a collection of classification methods which aim at partitioning observations into two or more subsets or groups so as to maximize within-group similarity and minimize between-group similarity, where 'similarity' is measured by some sort of distance between observations (e.g. the Mahalanobis distance). Univariate DA was first applied to bankruptcy prediction by Beaver [28], and MDA was first applied to bankruptcy prediction by Altman [29]. A generic MDA model can be summarized as follows:

$$z = f\left(\sum_{j=1}^{P} \beta_j x_j\right) \tag{24.1}$$

where z is commonly referred to as a score or z-score, the x_j are explanatory variables, the β_j represent the coefficients of the explanatory variables in the model and f denotes a mapping of $\beta' x$ on the set of real numbers $\mathfrak{R}$, often referred to as a classifier, which can be either linear or non-linear. Note that in comparing MDA models with other subcategories of statistical models, one typically needs to estimate the probability of default (PD), which is used as an input to many performance measures. In this chapter, we follow Hillegeist et al. [30] in using a logit transformation:

$$PD = \frac{e^z}{1 + e^z} \tag{24.2}$$

Note that, under the normality assumption, the MDA and logit approaches are closely related [31]. For a two-group classification problem, the classifier f is often a simple function that maps all observations or cases with discriminant or z-score values above a certain threshold or cut-off point to the first group and all other cases to the second group, where the cut-off point – often referred to as the cutting score or the critical z-score, is the average of the centroids of the groups, if the group sizes are equal, or a weighted average of them, if the group sizes are unequal, where the centroid of a group refers to the vector of group means of the explanatory variables. In the literature on bankruptcy prediction, MDA models differ mainly with respect to the choice of the explanatory variables and the form of the classifier (see Table 24.1), and are part of most comparative analysis exercises; our comparative analysis is no exception.

24.2.2 Probability Models

As compared with discriminant analysis, regression models for categorical variables (e.g. logit and probit) – also known as probability models – allow one to overcome some of the limitations of discriminant analysis. For example, within a regression

TABLE 24.1 Original statistical models for bankruptcy prediction.

Model	Variables
	Discriminant analysis
Altman [29]	Working capital/total assets
	Earnings before interest and taxes/total assets
	Market value of equity/total debt
	Sales/total assets
Altman [37]	Working capital/total assets
	Retained earnings/total assets
	Earnings before interest and taxes/total assets
	Book value equity/total liabilities
	Sales/total assets
Lis (1972), cited in Taffler [38]	Working capital/total assets
	Earnings before interest and taxes/total assets
	Market value of equity/total liabilities
	Net wealth/total assets
Taffler [38]	Profit before tax/current liabilities
	Current liabilities/total assets
	Current assets/total liabilities
	Number of credit intervals
	Probability models
Theodossiou [10], linear probability model	Working capital to total assets
	Net income to total assets
	Long-term debt to total assets
	Total debt to total assets
	Retained earnings to total assets
Ohlson [39], logit model	Total liabilities to total assets
	Working capital to total assets
	Current liabilities to current assets
	OENEG = 0 if total liabilities exceed total assets and 1 otherwise
	Net income to total assets
	Funds from operations (operating income minus depreciation) to total liabilities
	INTWO = 1 if net income has been negative for the last 2 years and 0 otherwise
	CHIN $= (NI_t - NI_{t-1})/(|NI_t| + |NI_{t-1}|)$, where NI_t denotes the net income for the last period – this variable is a proxy for the relative change in net income
	Size = log (total assets/GNP price-level index)
Zmijewski [40], probit model	Net income/total assets
	Total liabilities/total assets
	Current assets/current liabilities
Bemmann [41], logit model	Total liabilities/total assets

(*continued overleaf*)

TABLE 24.1   (*continued*)

Model	Variables
	Survival analysis
Shumway [32]	Net income/total liabilities
	Total liabilities/total assets
	RealSize = log(number of outstanding shares multiplied by year-end share price divided by total market value)
	LagExRet = cumulative annual return in year $t-1$ minus the value-weighted FTSE index return in year $t-1$
	LagSigma = standard deviation of residuals derived from regressing monthly stock return on market return in year $t-1$
	Stochastic models
Hillegeist *et al.* [30] and Bharath and Shumway [36], BSM-based models	Market value of equity
	Market value of assets
	Continuously compounded expected return on assets
	Continuous dividend rate expressed in terms of market value of assets
	Face value of debt maturing at time t
	Asset volatility
	Time to debt maturity, considered as 1 year

framework for discrete response variables, the normality and homoscedasticity assumptions are relaxed, on one hand, and a knowledge of prior probabilities of belonging to each group and of misclassification costs is not required, on the other hand. The generic model for binary variables can be stated as follows:

$$\text{PD} = \text{Prob}(y = 1) = F(\beta, x) \tag{24.3}$$

where y denotes the categorical response variable, x denotes the vector of explanatory variables, β denotes the vector of coefficients of x in the model and F is a function – commonly referred to as the link function – that maps any real number, for example a score $\beta^t x$, onto a probability. The choice of F determines the type of probability model. For example, the normal probability model, known as probit, assumes that the link function is the cumulative standard normal distribution, Φ; that is, $F(\beta, x) = \Phi^{-1}(\beta^t x)$. The logistic probability model, known as logit, assumes that the link function is the cumulative logistic distribution function, Λ; that is, $F(\beta, x) = \Lambda^{-1}(\beta^t x)$ or, equivalently,

$$\text{PD} = \Lambda(\beta^t x) = \frac{e^{\beta^t x}}{1 + e^{\beta^t x}} \tag{24.4}$$

Finally, the linear probability model assumes that the link function is linear; that is, $F(\beta, x) = \beta^t x$ or, equivalently,

$$PD = \beta^t x \qquad (24.5)$$

In the literature on bankruptcy prediction, the logit is the most popular probability model, and logit models differ only with respect to the choice of the explanatory variables (see Table 24.1), and are part of most comparative analysis exercises.

24.2.3 Survival Analysis Models

Discriminant analysis models and probability models (e.g. the linear probability model, logit and probit) are cross-sectional models and as such fail to take account of differences in firms' performance or risk profile over time; in sum, the PD provided by these static models is time-independent. In order to overcome this issue, one can use a dynamic methodology such as survival analysis. Survival analysis is concerned with the analysis of the time to events. In this chapter, we limit ourselves to a single event of interest, namely, bankruptcy or failure. Two functions are of special interest in survival analysis, namely, the survival function and the hazard function. The survival function, $S(t)$, is a function of time and represents the probability that the time of failure is later than some specified time t; that is, $S(t) = P(T > t)$, where T is a random variable describing the time of failure for an observation or firm. In sum, the survival function provides survival probabilities, or the probabilities of survival past specified times. On the other hand, the hazard function, $H(t)$, is also a function of time and represents the failure or hazard rate at time t conditional on survival until t or later; that is,

$$H(t) = \lim_{\Delta t \to 0} \frac{P(t \le T \le \Delta t | T \ge t)}{\Delta t} = -\frac{S'(t)}{S(t)} \qquad (24.6)$$

where $S'(t)$ denotes the derivative of the survival function S with respect to time and Δt denotes a change in t. As far as the application of survival analysis to bankruptcy prediction is concerned, the aim is to model the relationship between survival time and a set of explanatory variables. The most commonly used hazard model for bankruptcy modelling and prediction is the discrete-time hazard model proposed by Shumway [32], where the survival and hazard functions are defined as follows:

$$S(t,x;\theta) = 1 - \sum_{j < t} f(j,x;\theta) \quad \text{and} \quad H(t,x;\theta) = \frac{f(t,x;\theta)}{S(t,x;\theta)} \qquad (24.7)$$

Here, $f(t, x; \theta)$ denotes the probability mass function of the discrete random variable 'failure time' t, defined as the time when a firm leaves the sample; x is a vector of explanatory variables used to predict bankruptcy; and θ is the vector of parameters of the mass function f. Shumway [32] estimated this discrete-time hazard model using an estimation procedure similar to the one used for estimating the parameters of a multiperiod logit model – this choice was motivated by a proposition whereby he proved that a multiperiod logit model is equivalent to a discrete-time hazard model with a hazard function chosen as the cumulative distribution function of $f(t, x; \theta)$. He compared the performance of the discrete-time hazard model with MDA models, logit models and probit models based on the OCC and proved its superiority for his dataset. Following the lead of Hillegeist *et al.* [30], the probability of default at the time period *t* was estimated as follows:

$$PD = \frac{e^{H_0(t) + \beta' x_t}}{1 + e^{H_0(t) + \beta' x_t}} \quad (24.8)$$

where $H_0(t)$ denotes the unconditional hazard function – commonly referred to as the baseline hazard.

24.2.4 Stochastic Models

Most bankruptcy prediction models make use of accounting ratios as explanatory variables, which leads to a number of issues or criticisms. For example, accounting statements only present a firm's historical performance and may not be informative in predicting the future; the 'true' asset values may be very different from the book values; and accounting numbers can be manipulated by management (e.g. [7,33]). In order to overcome these drawbacks, one can make use of market-based explanatory variables. The rationale behind the use of market-based explanatory variables is that, in an efficient market, stock prices will reflect both the information contained in the accounting statements and the information contained in the future expected cash flows. Furthermore, market variables are unlikely to be influenced by firm's accounting policies. In this subsection, a category of such models is presented, namely, Black–Scholes–Merton (BSM)-based bankruptcy prediction models. Before presenting such bankruptcy prediction models, few comments are worthy of consideration. First, in practice, stochastic processes are often used to model the behaviour of stock prices, and a specific type of stochastic process, namely, the Itô process, has proven to be a valid modelling framework for derivatives, where an Itô process refers to a generalized Wiener process with both the drift and the variance rate being dependent on the underlying stock price and on time. Second, the basic BSM model is concerned with modelling the price of an option as a function of the underlying stock price and of time using an Itô process modelling framework. Third, in the Itô process modelling framework, the natural logarithms of stock prices are normally distributed. Last, but not least, the BSM model can be linked to the probability of a firm filing for bankruptcy; to be more specific, based on the observation by Merton [34] that holding the equity of a firm can be viewed as taking a long position in a call option, the PD can be viewed as the probability that the call option will expire worthless; that is, the value of the firm's assets (V_A) is less than the face value of its liabilities at the end of the holding period. Based on the above observations, McDonald [35] derived the following expression for the probability of default or bankruptcy, $P(V_A < D)$:

$$PD = \Phi\left(-\frac{\ln(V_A/D) + (\mu - \delta - 0.5\sigma^2) \times T}{\sigma\sqrt{T}} \right) \quad (24.9)$$

where $\Phi(.)$ denotes the cumulative distribution function of the standard normal distribution, V_A is the value of the firm's assets, μ is the firm's expected return, σ^2 is the volatility of the firm's assets, δ is the dividend rate and is typically proxied by the ratio of dividends to the sum of total liabilities and the market value of equity, D is the firm's debt and is proxied by its liabilities, and T denotes both the time to expiry of the option and the debt maturity time, and is assumed to be one year. In order to operationalize this BSM-based model for bankruptcy prediction, one needs to estimate V_A, μ and σ, as these parameters are not directly observable. Hillegeist et al. [30] first estimated V_A and σ by solving the following system of equations:

$$\begin{cases} V_E = V_A e^{-\delta T}\Phi(d_1) - De^{-rT}\Phi(d_2) + (1 - e^{-\delta T})\Phi(d_1)V_A \\ \sigma_E = \dfrac{V_A e^{-\delta T}\Phi(d_1)\sigma}{V_E} \end{cases} \qquad (24.10)$$

where the first equation is referred to as the call option equation, the second equation is referred to as the optimal hedge equation, V_E denotes the market value of common equity at the time of estimation, σ_E denotes the annualized standard deviation of daily stock returns over the 12 months prior to estimation, r denotes the risk-free interest rate, and d_1 and d_2 are computed as follows:

$$d_1 = \frac{\ln(V_A/D) + (r - \delta - 0.5\sigma^2) \times T}{\sigma\sqrt{T}}, \quad d_2 = d_1 - \sigma\sqrt{T} \qquad (24.11)$$

Then, μ is estimated as follows and is restricted to lie between r and 100%:

$$\mu = \frac{V_{A,t} + \text{Dividends} - V_{A,t-1}}{V_{A,t-1}} \qquad (24.12)$$

where $V_{A,t}$ denotes the current value of the firm's assets and $V_{A,t-1}$ denotes the previous year's value of the firm's assets. Alternatively, Bharath and Shumway [36] estimated V_A and σ as follows:

$$V_A = V_E + D, \quad \sigma = \frac{V_E}{V_A}\sigma_E + \frac{D}{V_A}\sigma_D \qquad (24.13)$$

where $\sigma_D = 0.05 + 0.25\sigma_E$. The firm's expected return, μ, is proxied by either the risk-free rate r or the previous year's stock return, restricted to lie between r and 100%.

In the next section, we shall describe the DEA framework proposed for assessing the relative performance of bankruptcy prediction models based on these modelling frameworks.

24.3 A SLACKS-BASED SUPER-EFFICIENCY FRAMEWORK FOR ASSESSING BANKRUPTCY PREDICTION MODELS

In this section, we discuss how one might adapt a DEA framework to assess the relative performance of competing bankruptcy prediction models. DEA is a generic framework and, as such, its implementation for our relative performance evaluation exercise requires three decisions to be made: (i) what are the units to be assessed, or decision-making units (DMUs)? (ii) what are the inputs and the outputs? and (iii) what is the appropriate DEA formulation to solve? Answers to these questions are provided in the next three subsections.

24.3.1 What Are the Units To Be Assessed, or DMUs?

In this section, we have chosen to assess the relative performance of the most popular accounting-based bankruptcy prediction models, market-based bankruptcy prediction models and hybrid models. The accounting-based bankruptcy prediction models considered in our comparative analysis include the MDA models proposed by Altman [29,37], Lis (1972, cited in [38]) and Taffler [38]; the logit model proposed by Ohlson [39]; the probit model proposed by Zmijewski [40]; the linear probability model proposed by Theodossiou [10]; and the MDA models proposed by Altman [29,37] and Lis (1972, cited in [38]), reproduced or implemented in a logit framework. The market-based bankruptcy prediction models considered in our comparative analysis include the BSM-based models proposed by Bharath and Shumway [36] and Hillegeist *et al.* [30]. The hybrid models considered in our comparative analysis include the survival analysis model proposed by Shumway [32] and estimated as a multiperiod logit model. We refer to the above-mentioned models as the 'original' models; see Table 24.1 for details.

   We also included in our comparative analysis three additional categories of models that we refer to as original models refitted, reworked models in a logit framework and new models. As the name suggests, the original models refitted include the above-mentioned models refitted with our sample data (i.e. those of Altman [29], Lis, Altman [37], Ohlson [39], Taffler [38], Zmijewski [40] and Shumway [32]). The reworked models in a logit framework refer to the original non-logit models implemented or replicated in a logit framework with the same original explanatory variables (i.e. those of Altman [29, 37], Lis, Taffler [38] and Zmijewski [40], and the total liabilities/total assets (TLTA) model of Bemmann [41]). Finally, the category of new models consists of MDA, logit, probit, linear probability and survival analysis models, where the explanatory variables were chosen from a list of variables using stepwise procedures. The list of variables consisted of accounting-based ratios and market-based variables chosen by repeated use of factor analysis on an initial list of 74 accounting ratios and three market-based variables, where the factors were selected so that both the absolute values of their loadings were greater than 0.5 and their communalities were greater than 0.8, and the stopping criterion was either no improvement in the total explained variance or that no more variables were excluded. The factor analysis was run using principal component analysis with VARIMAX as a factor extraction method. The list of variables consists of the variables that make up the factors. The new models are summarized in Table 24.2.

In sum, a total of 30 models were assessed in our comparative analysis. Note that all chosen models were tested out-of-sample and the training period ranged from 1989 to 2001, including 1571 failure and 5615 non-failure firm–year observations.

TABLE 24.2   New models for bankruptcy prediction.

Model	Variables
MDA model	Working capital to total assets
	Net income to capital
	Net income to current assets
	Net income to equity
	Total debt to equity
	EBIT to total assets
	Total liabilities to total assets
	Inventory to working capital
	Inventory to sales
	Quick assets to sales
	Current liabilities to inventory
	Total liabilities to working capital
	Net worth to total assets
	Real size
	LagExRet
	Sigma
Linear probability model	Working capital to total assets
	Total liabilities to total assets
	Inventory to working capital
	Real size
	LagExRet
	Sigma
Logit model	Working capital to total assets
	Real size
	LagExRet
	Sigma
Probit model	Working capital to total assets
	Real size
	LagExRet
	Sigma
Survival analysis model	Working capital to total assets
	Net income to equity
	Net worth to total assets
	Current liabilities to total assets
	Real size
	LagExRet
	Long-term debt to total assets

24.3.2 What Are the Inputs and the Outputs?

The inputs and outputs are the performance measures of the relevant criteria for assessing bankruptcy prediction models. In our analysis, we focused on the discriminatory power, the calibration accuracy or quality of estimates of the probabilities of default, the information content, and the correctness of categorical-prediction criteria and their measures. The discriminatory-power criterion refers to the ability of a model to discriminate between the good cases and the bad ones, where a case refers to a firm. The calibration accuracy criterion refers to the quality of estimation of the probability of default. The information content criterion refers to the extent to which the output of a model (e.g. score or PD) carries enough information for bankruptcy prediction. The correctness of the categorical-prediction criterion refers to the ability of a model to produce forecasts that are consistent with the actuals in that the forecasts reveal firms as healthy or non-healthy when the actuals are healthy or non-healthy, respectively. In addition, the inputs and outputs are chosen according to the principle of the less for inputs and the more for outputs, the better; therefore, the inputs and outputs refer to the performance metrics to be minimized and maximized, respectively – see Table 24.3 for a description of the performance metrics. Note that, unless an application of DEA involves undesirable outputs, the principle of the less (or the more) the better is commonly used across the literature on DEA applications to select inputs (or outputs, respectively) according to the economic theory of production.

In our comparative analysis of models, we used the KS statistic, the area under the ROC (AUC) (also known as the c-statistic), the Gini index and the information value (IV) to measure the discriminatory-power criterion; we used the BS to measure the calibration accuracy criterion; we used the LL statistic and pseudo-coefficient of determination (pseudo-R^2) to measure the information content criterion; and we used Type I errors, Type II errors, the misclassification rate (MR), the sensitivity (Sen), the specificity (Spe) and the OCC to measure the correctness of the categorical-prediction criterion – see Table 24.3 for descriptions of these measures.

24.3.3 What Is the Appropriate DEA Formulation To Solve?

Although basic DEA models could be used to classify competing bankruptcy prediction models into efficient and non-efficient ones and rank them according to their scores, one cannot differentiate between efficient ones as they all receive a score of 1. In many application areas, decision-makers are interested in obtaining a complete ranking in order to refine the evaluation of DMUs, and our application is no exception. Here, we propose an orientation-free slacks-based super-efficiency DEA framework for assessing the relative performance of competing bankruptcy prediction models. We have deliberately chosen an orientation-free analysis over an input-oriented or output-oriented analysis because, in our application of evaluating the performance of bankruptcy prediction models, input-oriented and output-oriented analyses are not relevant. In addition, any type of oriented analysis would be inappropriate for the following reasons. First, under the variable-returns-to-scale (VRS) assumption, which is the case for our data on bankruptcy prediction models, input-oriented

TABLE 24.3 Performance criteria and their measures for assessing bankruptcy prediction models.

Measure of criterion	Formula and definition
	Discriminatory power
Information value	The IV measures the relative distance between an empirical probability distribution and a theoretical one. The measure provided here is based on a discrete approximation to the density functions of the good and bad cases:

$$IV = \sum_{i=1}^{I} (g_i/n_G - b_i/n_B)\ln\left(\frac{g_i/n_G}{b_i/n_B}\right)$$

where g_i and b_i denote the number of good and bad cases, respectively, in band I, $\sum_{i=1}^{I} g_i = n_G$, and $\sum_{i=1}^{I} b_i = n_B$.

| Kolmogorov–Smirnov statistic | The KS statistic measures the distance between the empirical cumulative distribution functions of the samples of good and bad cases: |

$$KS = \max_s (F(s|B) - F(s|G))$$

where $F(.|G)$ and $F(.|B)$ denote the empirical cumulative distribution functions of the samples of good and bad cases, respectively.

| Receiver operating characteristic | The ROC curve is a plot of the hit rate against the false alarm rate for all cut-off points. A good prediction model would generate an ROC curve very far away from the diagonal, which suggests that the larger the area under the ROC curve, often referred to as AUC or AUROC, and measured by a statistic called the concordance or c-statistic, the better the prediction model performance: |

$$C\_Statistic = \frac{U}{N_G \cdot N_B}$$

where U denotes the Mann–Whitney U statistic, N_G denotes the number of good cases in the sample and N_B denotes the number of bad cases in the sample.

| Gini coefficient | The Gini coefficient, G, is a measure of the area between the ROC curve and the diagonal, also referred to as Somer's D: |

$$G = 2 \cdot C\_Statistic - 1$$

| | *Information content* |
| Log-likelihood statistic | The LL statistic is a measure of goodness of fit and is computed as the natural logarithm of the maximum value of the likelihood function of a model, where the likelihood function is a function of the parameters of the model which are determined so that the model is in maximum 'agreement' with the data. In our empirical investigation, we computed LL values as suggested by Hillegeist *et al.* [30]. The formulas are model-dependent and are not presented for reasons of space. |

(continued overleaf)

TABLE 24.3   (*continued*)

Measure of criterion	Formula and definition
Pseudo- R^2	The pseudo- R^2 is a measure of the strength of association between the output of a logistic regression model and the set of explanatory variables, and its value lies between 0 and 1, with higher values indicating a better likelihood of the logit model with intercept and predictors, i.e. better 'agreement' of the selected model with the observed data:

$$R^2 = 1 - \frac{\text{LL(Logit model with intercept and predictors)}}{\text{LL(Logit model with intercept only)}}$$

where LL denotes the log-likelihood of a model.

Correctness of categorical predictions

Sensitivity	Given a specific cut-off score s_c, the sensitivity S_e, also referred to as the hit rate, is defined as the fraction of the bad cases that would have scores below the cut-off and would therefore be rightly rejected, i.e. the proportion of bad cases that are predicted as bad:

$$S_e = \frac{N_{B|B}}{N_{B|B} + N_{B|G}}$$

where $N_{B|B}$ denotes the number of bad cases predicted as bad, and $N_{B|G}$ denotes the number of bad cases predicted as good.

Specificity	Given a specific cut-off score s_c, the specificity S_p is defined as the fraction of the good cases that have scores above the cut-off and would therefore be rightly accepted, i.e. the proportion of good cases that are predicted as good:

$$S_p = \frac{N_{G|G}}{N_{G|G} + N_{G|B}}$$

where $N_{G|G}$ denotes the number of good cases predicted as good, and $N_{G|B}$ denotes the number of good cases predicted as bad.

Type I error	The Type I error is the proportion of bad cases misclassified as good cases:

$$\text{Type I error} = 1 - \text{Specificity}$$

Type II error	The Type II error is the proportion of good cases misclassified as bad cases, also referred to as the false alarm rate:

$$\text{Type II error} = 1 - \text{Sensitivity}$$

efficiency scores can be different from output-oriented efficiency scores, which may lead to different rankings. Second, radial super-efficiency DEA models may be infeasible for some efficient DMUs; therefore, ties would persist in the rankings. Third, radial super-efficiency DEA models ignore potential slacks in inputs and outputs and thus may overestimate the efficiency score by ignoring mix efficiency. The proposed framework is a three-stage process and can be summarized as follows:

Stage 1: returns-to-scale (RTS) analysis. Perform RTS analysis to find out whether to solve a DEA model under constant-returns-to-scale (CRS) conditions, VRS conditions, non-increasing returns-to-scale (NIRS) conditions or non-decreasing returns-to-scale (NDRS) conditions – see Banker *et al.* [42] for details.

Stage 2: classification of DMUs. For each DMU_k $(k = 1,\ldots,n)$, solve the following slacks-based measure (SBM) model [43]:

$$\text{Min}: \rho_k = \left(1 - \frac{1}{m}\sum_{i=1}^{m} \frac{s_{i,k}^-}{x_{i,k}}\right) \Big/ \left(1 + \frac{1}{s}\sum_{i=1}^{s} \frac{s_{r,k}^+}{y_{r,k}}\right)$$

$$\text{s.t.}: \sum_{j=1}^{n} \lambda_j x_{i,j} + s_{i,k}^- = x_{i,k}; \forall i \qquad (24.14)$$

$$\sum_{j=1}^{n} \lambda_j y_{r,j} - s_{r,k}^+ = y_{r,k}; \forall r$$

$$\lambda_j \geq 0; \forall j; s_{i,k}^- \geq 0; \forall i; s_{r,k}^+ \geq 0; \forall r$$

where n denotes the number of DMUs, m is the number of inputs, s is the number of outputs, $x_{i,j}$ is the amount of input i used by DMU_j, $y_{r,j}$ is the amount of output r produced by DMU_j, λ_j is the weight assigned to DMU_j in constructing its ideal benchmark, and $s_{i,k}^-$ and $s_{r,k}^+$ are slack variables associated with the first and the second sets of constraints, respectively. If the optimal objective function value ρ_k^* is equal to 1, then DMU_k is classified as efficient; otherwise, it is classified as inefficient. Note that the model (24.14) is solved as it is if stage 1 reveals that CRS conditions hold; otherwise, one has to impose one of the following additional constraints depending on whether VRS, NIRS or NDRS conditions, respectively, prevail:

$$\sum_{j=1}^{n} \lambda_j = 1; \quad \sum_{j=1}^{n} \lambda_j \geq 1; \quad \sum_{j=1}^{n} \lambda_j \leq 1 \qquad (24.15)$$

Stage 3: break efficiency ties. For each efficient DMU_k, solve the following slacks-based super-efficiency DEA model, first proposed by Tone [44]:

$$\text{Min}: \delta_k = \left(\frac{1}{m}\sum_{i=1}^{m} \frac{(x_{i,k} + t_{i,k}^-)}{x_{i,k}}\right) \Big/ \left(\frac{1}{s}\sum_{i=1}^{s} \frac{(y_{r,k} - t_{r,k}^+)}{y_{r,k}}\right)$$

$$\text{s.t.}: \sum_{j=1; j \neq k}^{n} \lambda_j x_{i,j} \leq x_{i,k} + t_{i,k}^-; \forall i \qquad (24.16)$$

$$\sum_{j=1; j \neq k}^{n} \lambda_j y_{r,j} \geq y_{r,k} - t_{r,k}^+; \forall i$$

$$\lambda_j \geq 0; \forall j \neq k; t_{i,k}^- \geq 0; \forall i; t_{r,k}^+ \geq 0; \forall r$$

where $t_{i,k}^-$ and $t_{r,k}^+$ denote the amount by which input i and output r of the efficient DMU$_k$ should be increased and decreased, respectively, to reach the frontier constructed by the remaining DMUs. Note that the model (24.16) is solved as it is if stage 1 reveals that CRS conditions hold; otherwise, one has to impose an additional constraint from amongst those in (24.16) as outlined in stage 2. The super-efficiency scores δ_k^* are then used to rank order the efficient DMUs.

At this stage, it is worth mentioning that unlike radial super-efficiency DEA models (e.g. [45]), slacks-based super-efficiency models are always feasible [44,46]. Note that the slacks-based super-efficiency models of Tone [44] and Du *et al.* [46] are identical with respect to their constraints in that one can be obtained from the other using a simple transformation of variables. However, for applications where positive input and output data is a requirement, Du *et al.* [46] provided a variant of the model solved in stage 3 to accommodate this situation. In the next section, we shall use the methodology described above to rank-order competing bankruptcy prediction models and discuss the empirical results obtained, using UK data for the period 1989–2006.

24.4   EMPIRICAL RESULTS FROM SUPER-EFFICIENCY DEA

In our empirical investigation, we first generated monocriterion rankings of the 30 models under evaluation (see Figure 24.1 – this is a typical output from existing studies) to highlight the problems with using a monocriterion methodology to rank-order competing bankruptcy prediction models; that is, the models are ranked in ascending order of the relevant measure of each of the criteria under consideration if the measure is to be minimized, or in descending order if the measure is to be maximized. Indeed, monocriterion or single-criterion rankings tend to have many ties (e.g. the monocriterion rankings corresponding to Type I errors, sensitivity and information value). In addition, one can clearly see that the monocriterion rankings can be different from one performance criterion to another – see, for example, Theodossiou [10], Bandyopadhyay [11], and Tinoco and Wilson [12].

For our dataset, most monocriterion rankings are different; in fact, the monocriterion rankings based on T1 and Sen differ from those based on T2, MR, OCC and Spe, which differ from those based on the area under the ROC curve and the Gini index, and the latter also differ from those based on the KS statistic, IV, BS, LL statistic and pseudo-R^2. Notice that the monocriterion ranking based on IV does not discriminate between the eight worst-ranked models, because the probabilities of default produced by these models are all very close to zero and thus belong to the same band in the discrete approximation to the density functions of the good and bad cases.

For our dataset, the monocriterion rankings suggest that, for all performance measures except IV and BS, the new models outperform all of the original models, the original models refitted and the reworked models, with the exception of the logit model of Shumway [32]. Therefore, the selection of explanatory variables using factor analysis along with stepwise procedures seems to enhance the performance of models regardless of their underlying modelling framework. In addition, the use of a mixture

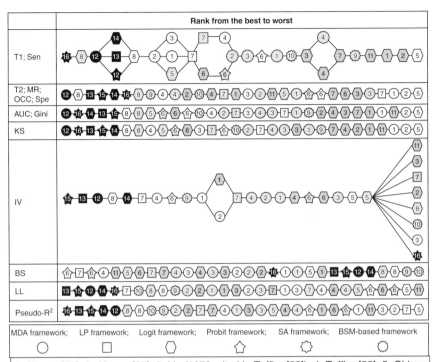

Figure 24.1 Monocriterion rankings of bankruptcy prediction models.

of accounting-based and market-based information improves bankruptcy prediction. Furthermore, it seems that these new models are doing a better job of classifying firms than of producing their probabilities of default.

Also, for most performance measures, notice that in general refitting models seems to improve their ranks, which suggests that the nature of the information in the training sample under consideration and the period of study do, as expected, tend to affect the performance of bankruptcy models. Recall that most of the original models were fitted to US data; therefore, when refitted to UK data, they tend to do better at predicting bankruptcy for UK firms.

On the other hand, for most performance measures, reworking the original MDA, probit and linear probability models with the same explanatory variables in a logit framework seems to improve the ranks – with the exception of the MDA models of Lis (1972) cited in Taffler [38] and of Taffler [38], which were originally fitted to UK data, which suggest that this improvement in the rankings could be due to a change in the training sample, in the modelling framework or in both. Also, for most performance measures, when all logit-framework-based models are compared, the multiperiod logit model of Shumway [32] seems to outperform the others, which suggests as expected that its dynamic nature improves bankruptcy prediction.

Finally, using only market-based data does not seem to provide good enough information to classify a firm as risky or not; in fact, BSM-based models do not make the top five. However, Hillegeist et al.'s model [30] seems always to outperform Bharath and Shumway's model [36].

At this stage, we would like to remind the reader that monocriterion rankings should not be discarded, as they convey valuable information; however, from both practical and methodological perspectives, one cannot make an informed decision as to which model performs best under multiple criteria. In order to address this issue, one needs a single ranking that takes account of multiple criteria, which we provide using the proposed DEA framework.

The multicriteria rankings of the above-mentioned 30 models are provided in Figure 24.2 for different combinations of measures of the four criteria under consideration, where the models are ranked in descending order of the corresponding SBM super-efficiency DEA scores. The empirical results reveal that the multicriteria rankings differ from the monocriterion ones. In addition, the multicriteria rankings have no ties, which suggests that the choice of the SBM super-efficiency DEA framework is an effective one in that it helps to get rid of ties between bankruptcy prediction models. Furthermore, we have considered several measures of the performance criteria under consideration to find out about the robustness of the multicriteria rankings with respect to the choice of measures.

For our dataset, and regardless of the combination of performance metrics used, the multicriteria rankings suggest that some of the new models are always amongst the top-ranked ones. In addition, the selection of explanatory variables using factor analysis along with stepwise procedures seems always to improve MDA and survival-analysis-based bankruptcy prediction. Also, with the exception of combinations of metrics including T1 and BS or BS and Sen simultaneously, the selection of explanatory variables using factor analysis along with stepwise procedures seems always to improve the performance of linear probability models in predicting bankruptcy. However, the new way of selecting explanatory variables does not seem to advantage the logit modelling framework or the probit modelling framework – although, for the logit framework, the new models do better than the original ones. In addition, in general, the use of a mixture of accounting-based and market-based information improves bankruptcy prediction in most modelling frameworks.

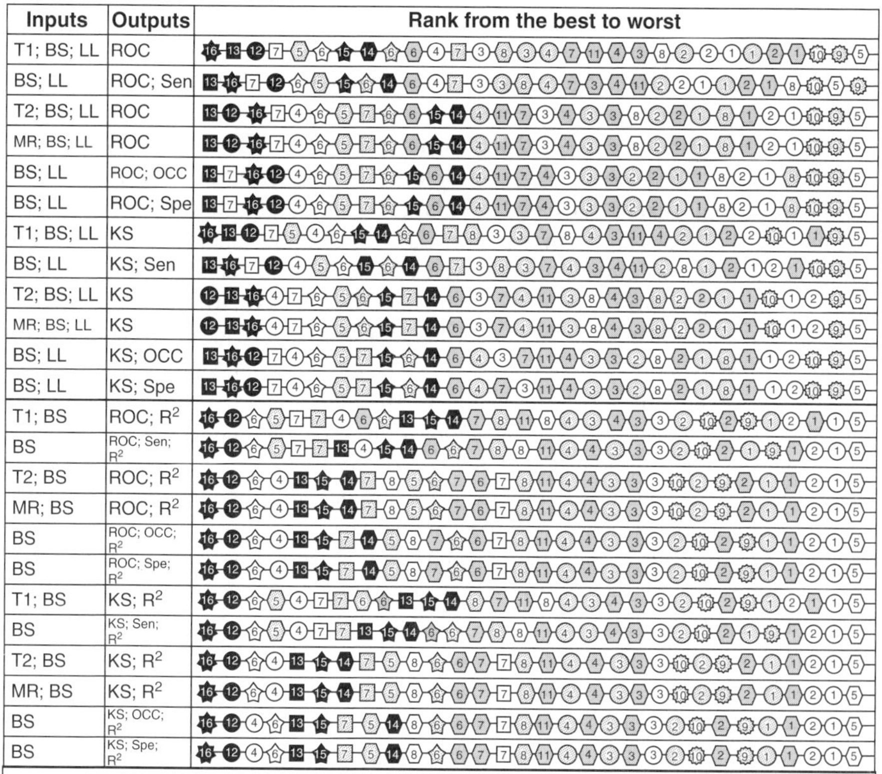

The notation of models is the same with Figure 24.1. This figure presents the multicriteria rankings of 30 competing bankruptcy models, where models are ranked from best to worst using SBM super-efficiency DEA scores computed using several combinations of measures of different criteria; T1, T2, MR, Sen, Spe and OCC are used as measures of correctness of categorical prediction; AUC, the Gini coefficient, the KS statistic and IV are used as measures of discriminatory power; BS is used as a measure of calibration accuracy; and LL and pseudo-R^2 are used as measures of information content. Different shapes represent different modelling frameworks, namely, MDA, linear probability (LP), logit analysis (LA), probit analysis (PA), survival analysis (SA) and BSM-based. White, dotted white, grey and black shapes represent the original models, the original models refitted, the reworked models with the same explanatory variables and the new models, respectively.

Figure 24.2 Multicriteria rankings of bankruptcy prediction models.

Also, for most combinations of performance measures, notice that, with the exception of the MDA models of Altman [29, 37] and the logit model of Ohlson [39], refitting models does not seem to improve their ranks – these conclusions are different from the ones derived from the analysis of the monocriterion rankings. Therefore, in a multicriteria setting, refitting models is not necessarily a means for improvement.

On the other hand, regardless of the combination of performance metrics, reworking the original MDA models with the same explanatory variables in a logit framework seems to improve their ranks – with the exception of the MDA model of Taffler [38].

As to reworking the original linear probability models with the same explanatory variables in a logit framework, it seems that for most combinations of performance metrics the ranks have improved. Notice, however, that reworking the original probit model did not lead to any improvement in the multicriteria rankings. Therefore, in a multicriteria setting, reworking the models could be a means for improvement of some modelling frameworks such as MDA models. Also, regardless of the combination of performance metrics, when all logit-framework-based models are compared, the multiperiod logit model of Shumway [32] does not seem to perform as well as in the unidimensional case. The refitted logit model of Ohlson [39], however, seems to be superior to the remaining logit models, followed by the reworked probit model of Zmijewski [40].

Finally, using only market-based data does not seem to provide good enough information to classify a firm as risky or not; in fact, BSM-based models do not make the top five. However, Hillegeist et al.'s model [30] seems always to outperform Bharath and Shumway's model [36]. This is amongst the very few findings of the monocriterion analysis that still hold in the multicriteria case.

To sum up, the conclusions derived from the analysis of the monocriterion rankings are not always consistent with their multicriteria counterparts. Therefore, multicriteria rankings help to better apprehend the relative performance of bankruptcy prediction models. The multicriteria rankings of the best and the worst models do not seem to be too sensitive to changes in most combinations of performance metrics. However, overall, the multicriteria rankings of the models under consideration tend to be sensitive to some extent to the choice of performance measures, which suggests that in practice one would have to carefully select these measures to reflect the application context and the purpose of the use of bankruptcy prediction models; in other words, the choice of performance metrics should be 'fit for purpose'.

Last, but not least, our findings suggest the following answers to our research questions. First, the survival analysis model tends be superior, followed by linear probability and multivariate discriminant analysis models; therefore, some modelling frameworks perform better than others by design, as survival analysis models are dynamic and have the modelling ability to take on board both accounting-based and market-based information. Second, the numerical results seem to suggest that the choice and/or design of explanatory variables and their nature affect the performance of different modelling frameworks to varying extents. To be more specific, most modelling frameworks improved in performance when a mixture of accounting-based and market-based information was taken into account, where survival analysis, linear probability and multivariate discriminant analysis models benefited the most from the new way of selecting explanatory variables.

24.5   CONCLUSION

Prediction of corporate failure is one of the major activities in auditing firms' risks and uncertainties. The design of reliable models to predict bankruptcy is crucial for many decision-making processes. Although a large number of models have been designed to predict bankruptcy, the relative performance evaluation of competing prediction models remains an exercise that is unidimensional in nature, which results in conflicting rankings of models from one performance criterion to another. In this research, we have proposed an orientation-free super-efficiency data envelopment analysis model to overcome this methodological issue; in sum, the proposed framework delivers a single ranking based on multiple performance criteria. In addition, we performed an exhaustive comparative analysis of the six most popular bankruptcy-modelling frameworks, resulting in 30 prediction models for UK firms, including our own models, organized into four categories, namely, original models, original models refitted, reworked models in a logit framework with the same original explanatory variables and new models. We used four criteria which are commonly used in the literature, namely, discriminatory power, calibration accuracy, information content and correctness of categorical prediction. We considered several measures for each criterion to find out about the robustness of multidimensional rankings with respect to different combinations of measures. Furthermore, we addressed two important research questions: namely, do some modelling frameworks perform better than others by design? and to what extent do the choice and/or the design of explanatory variables and their nature affect the performance of modelling frameworks?

Our main findings may be summarized as follows. First, the proposed multidimensional framework provides a valuable tool to apprehend the true nature of the relative performance of bankruptcy prediction models. Second, the multidimensional rankings of the best and the worst models do not seem to be too sensitive to changes in most combinations of performance metrics. Third, the numerical results seem to suggest that the survival analysis model tends be superior, followed by the linear probability and multivariate discriminant analysis models; therefore, some modelling frameworks perform better than others by design, as survival analysis models are dynamic and have the modelling ability to take on board both accounting-based and market-based information. Fourth, the numerical results seem to suggest that the choice and/or the design of explanatory variables and their nature affect the performance of different modelling frameworks to varying extents. To be more specific, most modelling frameworks improved in performance when a mixture of account-based and market-based information was taken into account, where survival analysis, linear probability and multivariate discriminant analysis models benefited the most from the new way of selecting explanatory variables.

REFERENCES

[1] van Gestel, T., Baesens, B., van Dijcke, P., Garcia, J., Suykens, J.A.K. and Vanthienen, J. (2006) A process model to develop an internal rating system: Sovereign credit ratings. *Decision Support Systems*, **42**(2), 1131–1151.

[2] Bris, A., Welsh, I. and Zhu, N. (2006) Chapter 7 liquidation versus chapter 11 reorganization. *Journal of Finance*, **61**(3), 1253–1303.

[3] Davydenko, S.A., Strebulaev, I.A. and Zhao, X. (2012) A market-based study of the cost of default. *Review of Financial Studies*, **25**(10), 2955–2999.

[4] Elkamhi, R., Ericsson, J. and Parsons, C.A. (2012) The cost and timing of financial distress. *Journal of Financial Economics*, **105**(1), 62–81.

[5] Branch, B. (2002) The costs of bankruptcy: A review. *International Review of Financial Analysis*, **11**(1), 39–57.

[6] Ahn, B.S., Cho, S.S. and Kim, C.Y. (2000) The integrated methodology of rough set theory and artificial neural network for business failure prediction. *Expert Systems with Applications*, **18**(2), 65–74.

[7] Balcaen, S. and Ooghe, H. (2006) 35 years of studies on business failure: An overview of the classic statistical methodologies and their related problems. *British Accounting Review*, **38**(1), 63–93.

[8] Aziz, M.A. and Dar, H.A. (2006) Predicting corporate bankruptcy: Where we stand? *Corporate Governance*, **6**(1), 18–33.

[9] Ravi Kumar, P. and Ravi, V. (2007) Bankruptcy prediction in banks and firms via statistical and intelligent techniques – a review. *European Journal of Operational Research*, **180**(1), 1–28.

[10] Theodossiou, P. (1991) Alternative models for assessing the financial condition of business in Greece. *Journal of Business Finance & Accounting*, **18**(5), 697–720.

[11] Bandyopadhyay, A. (2006) Predicting probability of default of Indian corporate bonds: Logistic and Z-score model approaches. *Journal of Risk Finance*, **7**(3), 255–272.

[12] Tinoco, M.H. and Wilson, N. (2013) Financial distress and bankruptcy prediction amongst listed companies using accounting, market and macroeconomic variables. *International Review of Financial Analysis*, **30**, 394–419.

[13] Xu, B. and Ouenniche, J. (2011) A multidimensional framework for performance evaluation of forecasting models: Context-dependent DEA. *Applied Financial Economics*, **21**, 1873–1890.

[14] Xu, B. and Ouenniche, J. (2012) A data envelopment analysis-based framework for the relative performance evaluation of competing crude oil prices' volatility forecasting models. *Energy Economics*, **34**, 576–583.

[15] Xu, B. and Ouenniche, J. (2012) Performance evaluation of competing forecasting models: A multidimensional framework based on MCDA. *Expert Systems with Applications*, **39**, 8312–8324.

[16] Ouenniche, J., Xu, B. and Tone, K. (2014) Forecasting models evaluation using a slacks-based context-dependent DEA framework. *Journal of Applied Business Research*, **30**(5), 1477–1484.

[17] Ouenniche, J., Xu, B. and Tone, K. (2014) Relative performance evaluation of competing crude oil prices' volatility forecasting models: A slacks-based super efficiency DEA model. *American Journal of Operations Research*, **4**(4), 235–245.

[18] Shetty, U., Pakkala, T.P.M. and Mallikarjunappa, T. (2012) A modified directional distance formulation of DEA to assess bankruptcy: An application to IT/ITES companies in India. *Expert Systems with Applications*, **39**(2), 1988–1997.

[19] Premachandra, I.M., Bhabra, G.S. and Sueyoshi, T. (2009) DEA as a tool for bankruptcy assessment: A comparative study with logistic regression technique. *European Journal of Operational Research*, **193**(2), 412–424.

[20] Premachandra, I.M., Chen, Y. and Watson, J. (2011) DEA as a tool for predicting corporate failure and success: A case of bankruptcy assessment. *Omega*, **39**(6), 620–626.

[21] Paradi, J.C., Asmild, M. and Simak, P.C. (2004) Using DEA and worst practice DEA in credit risk evaluation. *Journal of Productivity Analysis*, **21**(2), 153–165.

[22] Psillaki, M., Tsolas, I.E. and Margarits, D. (2010) Evaluation of credit risk based on firm performance. *European Journal of Operational Research*, **201**(3), 873–881.

[23] Yeh, C.C., Chi, D.J. and Hsu, M.F. (2010) A hybrid approach of DEA, rough set and support vector machines for business failure prediction. *Expert Systems with Applications*, **37**(2), 1535–1541.

[24] Xu, X. and Wang, Y. (2009) Financial failure prediction using efficiency as a predictor, *Expert Systems with Applications*, **36**(1), 366–373.

[25] Li, Z., Crook, J. and Andreeva, G. (2013) Chinese companies distress prediction: An application of data envelopment analysis. *Journal of the Operational Research Society*, **65**(3), 466–479.

[26] Mousavi, S.M.M., Ouenniche, J. and Xu, B. (2015) Performance evaluation of bankruptcy prediction models: An orientation-free DEA-based framework. *International Review of Financial Analysis*, **42**, 64–75.

[27] Fisher, R.A. (1938) The statistical utilization of multiple measurements. *Annals of Human Genetics*, **8**(4), 376–386.

[28] Beaver, W.H. (1966) Financial ratios as predictors of failure. *Journal of Accounting Research*, **4**, 71–111.

[29] Altman, E. (1968) Financial ratios, discriminant analysis and the prediction of corporate bankruptcy. *Journal of Finance*, **23**(4), 589–609.

[30] Hillegeist, S.A., Keating, E.K., Cram, D.P. and Lundstedt, K.G. (2004). Assessing the probability of bankruptcy. *Review of Accounting Studies*, **9**(1), 5–34.

[31] McFadden, D. (1976) A comment on discriminant analysis 'versus' logit analysis. *Annals of Economic and Social Measurement*, **5**(4), 511–523.

[32] Shumway, T. (2001) Forecasting bankruptcy more accurately: A simple hazard model. *Journal of Business*, **74**(1), 101–124.

[33] Agarwal, V. and Taffler, R. (2008) Comparing the performance of market-based and accounting-based bankruptcy prediction models. *Journal of Banking & Finance*, **32**(8), 1541–1551.

[34] Merton, R.C. (1974) On the pricing of corporate debt: The risk structure of interest rates. *Journal of Finance*, **29**(2), 449–470.

[35] McDonald, R. (2002) *Derivative Markets, Addison-Wesley, Boston, MA.*

[36] Bharath, S.T. and Shumway, T. (2008) Forecasting default with the Merton distance to default model. *Review of Financial Studies*, **21**(3), 1339–1369.

[37] Altman, E. (1983) *Corporate Financial Distress: A Complete Guide to Predicting, Avoiding and Dealing with Bankruptcy, John Wiley & Sons, Inc., New York.*

[38] Taffler, R.J. (1984) Empirical models for the monitoring of UK corporations. *Journal of Banking & Finance*, **8**(2), 199–227.

[39] Ohlson, J.A. (1980) Financial ratios and the probabilistic prediction of bankruptcy. *Journal of Accounting Research*, **18**(1),109–131.

[40] Zmijewski, M.E. (1984) Methodological issues related to the estimation of financial distress prediction models. *Journal of Accounting Research*, **22**, 59–82.

[41] Bemmann, M. (2005) Improving the comparability of insolvency predictions. Dresden Economics Discussion Paper Series, No. 08/2005.

[42] Banker, R.D., Cooper, W.W., Seiford, L.M., Thrall, R.M. and Zhu, J. (2004) Returns to scale in different DEA models. *European Journal of Operational Research*, **154**(2), 345–362.

[43] Tone, K. (2001) A slacks-based measure of efficiency in data envelopment analysis. *European Journal of Operational Research*, **130**(3), 498–509.

[44] Tone, K. (2002) A slacks-based measure of super-efficiency in data envelopment analysis. *European Journal of Operational Research*, **143**(1), 32–41.

[45] Andersen, P. and Petersen, N.C. (1993) A procedure for ranking efficient units in data envelopment analysis. *Management Science*, **39**(10), 1261–1264.

[46] Du, J., Liang, L. and Zhu, J. (2010) A slacks-based measure of super-efficiency in data envelopment analysis: A comment. *European Journal of Operational Research*, **204**(3), 694–697.

25

DEA IN PERFORMANCE EVALUATION OF CRUDE OIL PREDICTION MODELS[1]

JAMAL OUENNICHE

Business School, University of Edinburgh, Edinburgh, UK

BING XU

School of Social Sciences, Heriot-Watt University, Edinburgh, UK

KAORU TONE

National Graduate Institute for Policy Studies, Tokyo, Japan

25.1 INTRODUCTION

Predicting the future has always fascinated human beings. Over time, mankind has developed a variety of formal and informal frameworks for devising predictions. The formal frameworks can be divided into three broad categories, namely, qualitative prediction frameworks, quantitative prediction frameworks and hybrid prediction frameworks. Prediction frameworks are as good as their performance turns out to be. However, in real-life settings, one cannot afford to implement forecasts and wait to 'observe' or measure how good they were. In fact, in most real-life applications, prediction systems are designed and tested before being implemented. In this chapter, we focus on the assessment of the performance of quantitative prediction systems.

---

[1] Part of the material in this chapter is adapted from Ouenniche J., Xu B. and Tone K. (2014) Relative performance evaluation of competing crude oil prices' volatility forecasting models: a slacks-based superefficiency DEA model, **4**(4), 235–245, with permission from Scientific Research.

In general, the quality of such prediction systems is assessed within a simulation environment based on what is known as out-of-sample testing. In sum, the performance of prediction systems – whether based on a single model or multiple models – is assessed by simulating their potential behaviour in the 'future' as represented by the test set or test period. To be more specific, the sample of relevant historical data available is typically divided into an initialization set and a holdout set. The initialization set is used to estimate the parameters of the prediction system by fitting a model or models to the initialization dataset, whereas the holdout set (the simulated future) is used to test the performance of the system. Typically, the fitted system is used to produce forecasts of the observations in the holdout set, and the differences between the actual values and the forecasts, commonly referred to as errors, are fed into a measurement system to assess the quality of the forecasts. The measurement system is designed around performance criteria. The conventional performance criterion used in many application fields is accuracy. However, nowadays the assessment of prediction systems makes use of a variety of additional criteria – see Table 25.1 for a summary of performance criteria. The reader is referred to Xu [1] and Xu and Ouenniche [2] for detailed discussions of the measures used.

When designing a new prediction system or re-engineering an existing one, one typically considers an initial pool of prediction models, assesses their relative performance and then chooses the model or subset of models to use. Unlike the design of quantitative prediction models, which has attracted the attention of a large number of academics and professionals for some time, the performance evaluation of competing prediction models has not received as much attention. Although most published research involves using several performance criteria and measures to compare prediction models, the performance evaluation exercise remains monocriterion-based in that prediction models are ranked by a single measure of a single criterion at a time, which often results in different rankings for different criteria and measures. Consequently, despite the exercise being multicriteria-based, a monocriterion-based framework is what has so far been used in assessing the relative performance of models. The importance of addressing this methodological issue lies in the practice-driven needs to devise a single multicriteria ranking to guide decision making, on one hand, and to automate some aspects of the design of prediction systems, on the other hand.

In order to illustrate the problem with the current monocriterion approach, we shall use the literature on predicting crude oil price volatilities. For example, Day and Lewis [3] used historical volatility models, generalized autoregressive conditional heteroscedasticity (GARCH) models and implied volatility models to forecast the volatility of crude oil daily prices. They used both goodness-of-fit and biasedness criteria and several metrics (i.e. the mean error (ME), mean absolute error (MAE), root mean squared error (RMSE) and coefficient of determination (R^2)) to evaluate their competing forecasting models, but their out-of-sample results were mixed with respect to which model outperformed the rest, as a result of the unidimensional nature of their rankings. Sadorsky [4] forecasted the volatility of daily futures prices of WTI crude oil. He used several time series volatility models, namely, random walk (RW), simple moving

TABLE 25.1 Classification and definitions of criteria and subcriteria used in forecasting.

Criterion or subcriterion	Definition
Reliability	Multidimensional criterion consisting of five subcriteria, namely, theoretical relevance, validity, accuracy, informational efficiency and degree of uncertainty of the output of a model or forecast.
Theoretical relevance	Degree of suitability of a model for a given dataset and a given forecasting exercise; that is, its ability, from a design perspective, to take into account all features of the dataset under consideration (e.g. patterns, turning points, structural change) as well as its suitability for a specific forecasting horizon – the relevance of the forecasting horizon lies in the fact that, by design, some forecasting models are more suitable for a specific time horizon than others.
Validity	Refers to whether the assumptions underlying a model hold or not. From a methodological perspective, invalid models should be discarded from further consideration; however, in practice, things are not either black or white. For example, within a linear regression analysis framework, one needs to test whether or not the residuals are normally distributed, using several normality tests which may lead to different conclusions; in this case, one may conclude that the normal distribution is a reasonable approximation and consider the model as valid.
Accuracy	Refers to the ability of a model to reproduce the past. Accuracy is a multidimensional construct that has three main facets, namely, the goodness-of-fit dimension, the biasedness dimension and the correct-sign dimension.
Goodness-of-fit	Refers to how close the forecasts are to the actual values.
Biasedness	Refers to whether a model tends to systematically overestimate or underestimate the forecasts.
Correct sign	Refers to the ability of a model to forecast the correct sign; that is, to produce forecasts that are consistent with the actuals in that the forecasts reveal an increase or decrease in value when the actuals increase or decrease, respectively, in value – this criterion is particularly important in investment environments.
Informational efficiency	Refers to the ability of a model to capture all elements of information in the data.
Degree of uncertainty of forecast	Refers to the likelihood that the forecast and the actual values will be close to each other.
Cost	The cost of a forecasting model refers to the extent to which it is relatively cheap to acquire and use, and can be used to discriminate between competing models. It includes several

(*continued overleaf*)

TABLE 25.1 (*continued*)

Criterion or subcriterion	Definition
	categories of costs, for example costs of development/ purchase and maintenance of forecasting software; costs of data purchase/collection, storage and pre-processing; costs of training analysts to use a model or method effectively; costs related to the time required to obtain a forecast; costs of repeated application of the method. These cost elements vary in importance and magnitude depending on whether the forecasting method is quantitative or qualitative.
Benefits	Refers to the (expected) benefits that would result from the use of a model in generating forecasts such as cost savings and improved decisions.
Complexity	Refers to the complexity of a model or method – also referred to in the literature as ease of use or ease of implementation of a model/method. In this chapter, the complexity of a forecasting model or method refers to the extent to which it is easy to understand by users/managers and to interpret its results or, equivalently, the level of conceptual and technical knowledge/expertise required for effective use of the model/method.
Universality	Refers to the extent to which a model is widely used in practice or the familiarity of the audience with it. Note, however, that for quantitative models, this factor may depend largely on the availability of a model or method in popular software packages.
Ability to incorporate managerial judgement	Refers to the ability of a model or method to incorporate managerial judgement; that is, integration of subjective information to produce a forecast.

average (SMA), single exponential smoothing (SES), autoregressive (AR), linear regression and stochastic volatility (SV) models, and evaluated their performance using various measures (i.e. mean squared error (MSE), MAE, mean percentage error (MPE) and mean absolute percentage error (MAPE)) and statistical tests (i.e. the Diebold–Mariano (DM) test, a modified DM test and regression tests for biasedness). The overall results were again mixed as a result of the unidimensional nature of the rankings. Sadorsky [5] forecasted the volatility of several petroleum futures returns (i.e. WTI crude oil, heating oil No. 2, unleaded gasoline and natural gas) using a large number of models (i.e. RW, SES, linear regression, AR, vector autoregressive (VAR), GARCH and state space models). Although he evaluated these competing forecasting models using several performance measures (i.e. MSE, MAE and Theil U coefficients) and statistical tests (i.e. the DM test and correct-sign tests), the overall results were again inconsistent because of the unidimensional nature of the rankings. Agnolucci [6] used different types of GARCH models and implied volatility models to forecast the volatility of daily WTI futures prices, but empirical results revealed that their performance was inconsistent with respect to different measures (e.g. MAE and MSE) and statistical tests (e.g. regression-based test for biasedness), again because of the unidimensional nature of the rankings. Marzo and Zagaglia [7] used several GARCH models to forecast the volatility of daily futures prices of crude oil traded on NYMEX. They did not find a constantly superior model based on several different performance measures (i.e. MAE, MSE, heteroscedasticity-adjusted MSE and success ratio) and statistical tests (e.g. direction accuracy test and DM test), again because of the unidimensional nature of their rankings.

To the best of our knowledge, the only papers that have both raised concerns about the above-mentioned methodological issue and addressed it are the ones by Xu [1], Xu and Ouenniche [2, 8, 9] and Ouenniche *et al.* [10, 11]. This chapter is an account of some of our contributions based on data envelopment analysis (DEA). DEA is a mathematical-programming-based multicriteria framework for the relative performance evaluation of a set of entities commonly referred to as decision-making units (DMUs). DEA is a generic framework and, as such, its implementation for a specific relative performance evaluation or benchmarking exercise requires a number of decisions to be made, namely, the choice of the units to be assessed, the choice of the relevant inputs and outputs to be used, and the choice of the appropriate DEA model. In order to present and discuss how one might adapt this framework to measure and evaluate the relative performance of competing prediction models, we survey and classify the literature on performance criteria and discuss how continuous and discrete metrics can be designed to measure these criteria, on one hand, and we use crude oil prices to demonstrate the use of DEA in evaluating and selecting competing forecasting models, on the other hand.

The remainder of this chapter is organized as follows. In Section 25.2, we provide an overview of crude oil prices and their volatilities. In Section 25.3, we present the DEA-based frameworks proposed to date for assessing the relative performance of prediction models of crude oil price volatility. Finally, Section 25.4 concludes the chapter.

25.2   AN OVERVIEW OF CRUDE OIL PRICES AND THEIR VOLATILITIES

Oil is an important source of energy that drives modern economies, and large swings in its price can have a substantial adverse impact on both oil importers and exporters. Therefore, a proactive knowledge of future movements of oil prices and their volatility can lead to better decisions in various areas such as macroeconomic policy making, risk management, options pricing and portfolio management. In fact, with respect to macroeconomics and policy making, swings or volatility in crude oil prices tends to negatively affect the economy in many ways [12–14]; for example, high oil prices tend to increase production costs or decrease production output, which affect oil-importing economies' output as measured by GDP. In addition, a large oil price volatility tends to raise uncertainty, which affects consumers' consumption and investment behaviour and often results in reduced or postponed purchases of goods and investment in equipment [15–17]. Furthermore, central banks take explicit account of the volatility of commodities in establishing their monetary policies; therefore, reliable forecasts of oil price volatility are crucial for macroeconomic policy makers in setting policies to stabilize the economy. As to investment risk management, the standard approach in the financial industry to modelling risk in the framework of a parametric approach is to use value-at-risk (VaR) as a proxy to measure the risk of financial instruments (e.g. stocks; bonds; commodities, including crude oil; and futures and options), which requires a reliable estimate of volatility. With respect to options pricing and portfolio management, forecasting oil price volatility is a critical activity for investors faced with a massive growth in the trading of crude oil and its underlying derivative securities. For instance, investors or portfolio managers need to forecast the expected volatility over the lifetime of a futures or option contract to assist them in designing hedging strategies, on one hand, and to adjust their investment portfolios if the crude oil market becomes very unstable, on the other hand.

The price of crude oil is highly dependent on its grade, as measured by its specific gravity, sulphur content and geographical location. There are two main types of crude oil that are commonly used as benchmarks or references with respect to price, namely, West Texas Intermediate (WTI) and Brent Crude Oil. Regardless of the quality of crude oil and where it was extracted from, nowadays crude oil prices tend to move together [18].

Daily spot prices of crude oil seem to have been greatly influenced by exogenous events (Figure 25.1). First, global macroeconomic conditions are believed to have a significant impact on oil demand and subsequently on its prices (e.g. [12]); for example, the 1997 Asian crisis was followed by a price fall in 1997–1998; an unexpected increase in demand for crude oil from some emerging markets (e.g. China and India) was followed by a price surge in 2002–2008; and the relatively recent credit crunch was followed by a price fall in late 2008–2009. Second, political instabilities in both OPEC regions (e.g. Iraq and Venezuela) and non-OPEC regions (e.g. Bolivia, Turkey and Russia) led to reduced production, which increased precautionary demand as well as prices [19]. Third, changes in quotas or production polices were followed by a price fall in 1997 and a price rise in 2005. Fourth, environmental events such as Hurricane Katrina in August 2005 led to a reduction of approximately 20% in the Gulf of Mexico's oil and gas production, which was followed by a price increase [20].

As crude oil prices are level non-stationary, in the literature there is a tendency to study their level-stationary equivalent, namely, returns (Figure 25.2). We have computed daily WTI crude oil returns as follows: $R_t = \ln(P_t/P_{t-1}) \times 100$, where P_t denotes the WTI crude oil price on day t – obviously these returns do not contain unit roots, which was confirmed by the augmented Dickey–Fuller test, the Phillips–Peron test and the Kwiatkowski–Phillips–Schmidt–Shin test.

The volatility of crude oil returns can be measured in several ways. In fact, one can measure volatility over the time unit under consideration (e.g. day, week or month) by any dispersion measure, such as variance or standard deviation, mean absolute

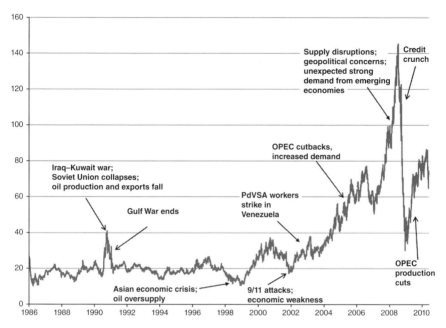

Figure 25.1   Daily WTI crude oil prices from January 1986 to May 2010 and major events.

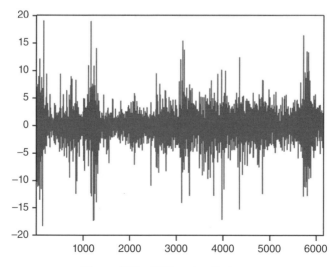

Figure 25.2   WTI crude oil returns.

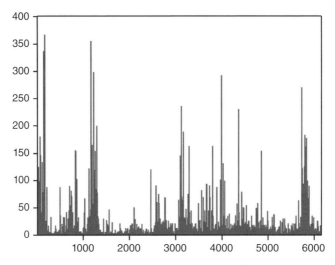

Figure 25.3   WTI crude oil volatility.

deviation, or range of returns – as long as the relevant data are available (e.g. intra-day returns for daily volatility). However, such measures are affected by outliers; therefore, their use is only appropriate when the distribution of returns is symmetric or normal. In addition, when volatility is computed using high-frequency data (e.g. intra-day returns), the volatility obtained can end up being very noisy as a result of market microstructure effects such as non-synchronous trading, discrete price observations, intraday periodic volatility patterns and bid–ask bounce. Since crude oil daily returns are not normally distributed – as confirmed by the Jarque–Bera test of normality – we opted for an alternative approach to modelling daily volatility that consisted of using daily squared returns R_t^2 as a proxy (e.g. [4–6, 21]). One should, however, be aware that squared daily returns provide a noisy proxy, although they remain an unbiased estimator [22]. Daily squared returns are depicted in Figure 25.3. The same tests as were performed on returns were performed on squared returns and the results revealed that such volatility proxy series are also level-stationary and autocorrelated.

In the next section, we present DEA-based performance analytics of prediction models for crude oil price volatility.

25.3   ASSESSMENT OF PREDICTION MODELS OF CRUDE OIL PRICE VOLATILITY

In this section, we use prediction models of crude oil price volatility to illustrate another use of a DEA-based multicriteria performance evaluation framework for prediction models of continuous variables. Xu and Ouenniche [8] proposed an oriented super-efficiency-based DEA methodology for assessing the relative performance of competing prediction models of crude oil price volatility. Later, Ouenniche *et al.* [10] proposed an orientation-free context-dependent DEA framework, namely, a slacks-based CDEA framework, and used it to rank prediction models of crude oil price volatility. Finally, Ouenniche *et al.* [11] proposed an orientation-free super-efficiency DEA framework, namely, a slacks-based super-efficiency DEA framework, to rank prediction models of crude oil price volatility. The material in this section is based on Ouenniche *et al.* [11]. In the rest of this section, we first present the models that were used in this study to predict crude oil price volatility (Section 25.3.1). Then, we specify suitable performance criteria and the measures of them that were used to evaluate the chosen prediction models of crude oil price volatility (Section 25.3.2). In Section 25.3.3, we present the slacks-based super-efficiency DEA framework used in our empirical analysis. Finally, we discuss the empirical results from our slacks-based super-efficiency DEA framework in Section 25.3.4.

25.3.1   Forecasting Models of Crude Oil Volatility – DMUs

As far as the literature on forecasting oil price volatility is concerned, quantitative prediction models can be divided into three main categories, namely, time series volatility models [4–7, 21, 23], implied volatility models [3, 6] and hybrid models [24, 25]. Time series volatility models can be decomposed further into three subcategories, namely, historical volatility models, GARCH models and SV models. Historical volatility models are averaging methods that use volatility estimates (e.g. standard deviation of past returns over a fixed interval) as input and assume that conditional variances are level-stationary – these models can be divided further into two subcategories depending on whether they use a pre-specified weighting scheme (e.g. RW, historical mean (HM), SMA and SES) or not (e.g. AR and autoregressive moving average (ARMA)). GARCH models consist of two equations, where one models the conditional mean and the other models the conditional variance; they use returns as input, and assume that conditional variances are level-stationary. These models can be divided further into two subcategories depending on the nature of their memory, namely, short-memory models (e.g. GARCH, GARCH-in-mean (GARCH-M), power ARCH (PARCH), exponential GARCH (EGARCH) and threshold GARCH (TGARCH)), which assume that the autocorrelation function (ACF) of the conditional variance decays exponentially, and long-memory models (e.g. component GARCH (CGARCH)), which assume that the ACF of the conditional variance decays slowly. GARCH models have been widely used in the literature owing to their ability to capture some peculiar features of financial data such as volatility clustering or pooling, leverage effects, and leptokurtosis, which are typical of crude oil prices – see for example Agnolucci [6] and Kang *et al.* [21]. SV models can be viewed as variants of GARCH models where the conditional variance equation has an additional error term – see Ghysels *et al.* [26] for a detailed discussion of SV models and their relation to GARCH models. On the other hand, implied volatility models are forward-looking models in that they use information about market-traded options in combination with an options pricing model (e.g. the Black–Scholes model) to derive volatility. Finally, hybrid volatility models are combinations of different models (e.g. the regime-switching GARCH used by Fong and See [24]); the design of these models has been motivated by the highly volatile nature of crude oil prices. For a general discussion of volatility models, the reader is referred to Poon and Granger [27].

In sum, in our survey of the literature on crude oil price volatility forecasting, time series models tend to be the popular ones. We found 14 time series models that turned out to be valid for our dataset and were included in our performance evaluation exercise, namely, RW, HM, SMA with averaging periods α of 20 and 60 (SMA(20) and SMA(60)), SES, AR with orders 1 and 5 (AR(1) and AR(5)), ARMA(1, 1), GARCH (1, 1), GARCH-M(1, 1), EGARCH(1, 1), TGARCH(1, 1), PARCH(1, 1) and CGARCH(1, 1). See Table 25.2 for a general description of these models.

25.3.2 Performance Criteria and Their Measures: Inputs and Outputs

Our review of the literature on forecasting the volatility of crude oil prices revealed that three performance criteria have typically been used, namely, goodness-of-fit, biasedness and correct sign – see Xu and Ouenniche [8]. Note that, depending on the application context, the features of the data and the decision-makers' preferences as to how to penalize large, small, positive and negative errors, different metrics can be used. In this chapter, goodness-of-fit is measured by one of the following metrics: MSE, MSVolScE, MAVolScE, MMEU and MMEO; biasedness is measured by one of the following metrics: ME or mean volatility scaled error (MVolScE); and the correct sign is measured by the percentage of correct direction change predictions (PCDCP) – see Table 25.3 for a description of the metrics used in our performance evaluation exercise to measure these criteria.

25.3.3 Slacks-Based Super-Efficiency Analysis

In this subsection, we present an extension of the work by Xu and Ouenniche [8], which overcomes the following issues. First, under the variable-returns-to-scale (VRS) assumption, input-oriented efficiency scores can be different from output-oriented efficiency scores, which may lead to different rankings. Second, radial super-efficiency DEA models may be infeasible for some efficient DMUs; therefore, ties would persist in the rankings. Third, radial super-efficiency DEA only takes account of technical efficiency. Finally, in many applications such as ours, the choice of an orientation in DEA is rather superfluous. In sum, we propose an orientation-free super-efficiency DEA framework, namely, a slacks-based super-efficiency DEA framework, for assessing the relative performance of competing volatility forecasting models. The proposed framework is a three-stage process and can be summarized as follows:

Stage 1: returns-to-scale (RTS) analysis. Perform RTS analysis to find out whether to solve a DEA model under constant-returns-to-scale (CRS) conditions, VRS conditions, non-increased returns-to-scale (NIRS) conditions or non-decreased returns-to-scale (NDRS) conditions – see Banker *et al.* [28] for details.

TABLE 25.2   Prediction models of crude oil price volatility.

Forecasting model	Formulation and comments
	$$\hat{\sigma}_t = \sigma_{t-1}; \forall t > 1$$
Random walk (RW)	RW can be viewed as an averaging method with a pre-specified weighting scheme designed so that all weight is put on the most recent historical observation. By design, RW is suitable for forecasting a time series with no trend and no seasonality.
	$$\hat{\sigma}_t = \frac{1}{t-1}\sum\nolimits_{i=1}^{t-1}\sigma_i; \forall t > 1$$
Historical mean (HM)	HM is an averaging method with a pre-specified weighting scheme designed so that all available historical observations are equally weighted. By design, HM is suitable for forecasting a time series with no trend and no seasonality.
	$$\hat{\sigma}_t = \frac{1}{\alpha}\sum\nolimits_{i=t-\alpha}^{t-1}\sigma_i; \forall t > \alpha$$
Simple moving average (SMA)	SMA is an averaging method with a pre-specified weighting scheme designed so that only the most recent α historical observations are used to forecast the next period and those α observations are equally weighted. By design, SMA is suitable for forecasting a time series with no trend and no seasonality.
	$$\hat{\sigma}_t = \lambda\sigma_{t-1} + (1-\lambda)\hat{\sigma}_{t-1}; \forall t > 1, \lambda \in (0,1)$$
Simple exponential smoothing (SES)	SES can be viewed as an averaging method with a pre-specified weighting scheme designed so that the weights decrease exponentially as the observations get older. By design, SES is suitable for forecasting a time series with no trend and no seasonality.
	$$\hat{\sigma}_t = \mu + \sum\nolimits_{i=1}^{p}\phi_i\sigma_{t-i}$$
Autoregressive model of order p (AR(p))	AR expresses a forecast as a linear function of previous values of the time series through the use of the response variable lagged by one or more time periods, say p, as explanatory variables. These models assume that the response variable is stationary and generate forecasts by relying heavily on autocorrelation patterns in the time series, but no particular pattern is assumed. The parameters of the model are usually estimated using a non-linear least squares method.
	$$\hat{\sigma}^{\text{'}}_t = \mu + \sum\nolimits_{i=1}^{p}\phi_i\sigma_{t-i} + \sum\nolimits_{i=1}^{q}\omega_i\sigma\varepsilon_{t-i}$$
Autoregressive moving average model of orders p and q (ARMA(p, q))	ARMA expresses a forecast as a linear function of previous values of the time series and previous errors or residuals, where the response variable lagged by one or more time periods, say p, and the errors lagged by one or more time

(continued overleaf)

TABLE 25.2   (*continued*)

Forecasting model	Formulation and comments
	periods, say q, are used as explanatory variables. These models also assume that the response variable is stationary and generate forecasts by relying heavily on autocorrelation patterns in the time series and its forecasting errors, but no particular pattern is assumed. The parameters of the model are usually estimated using the method of maximum likelihood.

$$r_t = \beta_0 + \sum\nolimits_{i=1}^{k} \beta_i X_{i,t} + \varepsilon_t, \;\; \varepsilon_t \approx N(0, \sigma_t)$$
$$\hat{\sigma}_t = \alpha_0 + \sum\nolimits_{i=1}^{p} \alpha_i \varepsilon_{t-i}^2 + \sum\nolimits_{i=1}^{q} \lambda_i \sigma_{t-i}$$

| Generalized autoregressive conditional heteroscedasticity of orders p and q (GARCH (p, q)) | The GARCH model consists of two equations, commonly referred to as the mean equation and the variance equation, where the mean equation regresses the response variable, for example returns, on a set of explanatory variables (which may include lagged values of the response variable) and an error term that is assumed to be normally distributed with mean zero and to be heteroscedastic, and the variance equation regresses the variance of the error term of the mean equation on a set of p lagged squared errors (often referred to as news about volatility) and a set of q lagged variances. Notice that the variance equation is a function of the magnitudes of lagged residuals and not their signs, which enforces a symmetric response of the volatility to positive and negative shocks. |

$$r_t = \beta_0 + \sum\nolimits_{i=1}^{k} \beta_i X_{i,t} + \gamma \sigma_t + \varepsilon_t, \;\; \varepsilon_t \approx N(0, \sigma_t)$$
$$\hat{\sigma}_t = \alpha_0 + \sum\nolimits_{i=1}^{p} \alpha_i \varepsilon_{t-i}^2 + \sum\nolimits_{i=1}^{q} \lambda_i \sigma_{t-i}$$

| GARCH-M(p, q) | GARCH-in-mean is an extension of GARCH that allows the mean of a time series to depend on its conditional variance, and thus models risk–return trade-offs. In general, it is expected that investors should receive a higher return by taking additional risk, which would be the case if the estimate of γ, say $\hat{\gamma}$, was statistically significant and positive. |

$$r_t = \beta_0 + \sum\nolimits_{i=1}^{k} \beta_i X_{i,t} + \varepsilon_t, \;\; \varepsilon_t \approx N(0, \sigma_t)$$
$$\log(\hat{\sigma}_t) = \alpha_0 + \sum\nolimits_{i=1}^{p} \alpha_i \left| \frac{\varepsilon_{t-i}}{\sigma_{t-i}} \right| + \sum\nolimits_{i=1}^{q} \lambda_i \; \log(\sigma_{t-i}) + \sum\nolimits_{i=1}^{r} \gamma_i \frac{\varepsilon_{t-i}}{\sigma_{t-i}}$$

| EGARCH(p, q) | Exponential GARCH is an extension of GARCH that is designed to take account of volatility asymmetry, commonly referred to as the leverage effect; that is, negative shocks increase volatility more than positive shocks of equal magnitude. In fact, the exponential leverage effect is captured by the log of the conditional variance, which guarantees that |

TABLE 25.2 (*continued*)

Forecasting model	Formulation and comments
	the forecasts are non-negative; therefore, there is no need to impose an estimation constraint to avoid negative variance. Notice that the conditional variance depends on both the size and the sign of the standardized errors. A statistically significant γ_i such that $\gamma_i < 0$ indicates the presence of a leverage effect.
	$$r_t = \beta_0 + \sum_{i=1}^{k} \beta_i X_{i,t} + \varepsilon_t, \quad \varepsilon_t \approx N(0, \sigma_t)$$ $$\hat{\sigma}_t = \alpha_0 + \sum_{i=1}^{p} \alpha_i \varepsilon_{t-i}^2 + \sum_{i=1}^{q} \lambda_i \sigma_{t-i} + \sum_{i=1}^{p} \gamma_i \varepsilon_{t-i}^2 I_{t-i}$$
TGARCH(p, q)	Threshold GARCH is an extension of GARCH that is designed to take account of the leverage effect through the additional term in the variance equation, where $I_{t-i} = 1$ if $\varepsilon_{t-i} < 0$ and 0 otherwise. A statistically significant γ_i such that $\gamma_i > 0$ indicates the presence of a leverage effect.
	$$r_t = \beta_0 + \sum_{i=1}^{k} \beta_i X_{i,t} + \varepsilon_t, \quad \varepsilon_t \approx N(0, \sigma_t)$$ $$\sigma_t^s = \alpha_0 + \sum_{i=1}^{p} \alpha_i \left(|\varepsilon_{t-i}| - \gamma_i \varepsilon_{t-i} \right)^s + \sum_{i=1}^{q} \lambda_i \sigma_{t-i}^s$$ $$s > 0 \text{ and } |\gamma_i| \leq 1$$
APARCH(p, q)	Asymmetric power ARCH is an extension of GARCH that is designed to take account of volatility asymmetry. In addition, the power s is estimated instead of imposed; consequently, APARCH nests several GARCH models such as GARCH and TGARCH. A statistically significant γ_i such that $\gamma_i > 0$ indicates the presence of a leverage effect.
	$$r_t = \beta_0 + \sum_{i=1}^{k} \beta_i X_{i,t} + \varepsilon_t, \quad \varepsilon_t \approx N(0, \sigma_t)$$ $$\hat{\sigma}_t = m_t + u_t$$ $$m_t = \alpha_0 + \rho m_{t-1} + \phi \left(\varepsilon_{t-1}^2 - \sigma_{t-1} \right)$$ $$u_t = \sum_{i=1}^{p} \alpha_i \left(\varepsilon_{t-i}^2 - m_{t-i} \right) + \sum_{i=1}^{q} \lambda_i \left(\sigma_{t-i} - m_{t-i} \right)$$
CGARCH(p, q)	Component GARCH is an extension of GARCH that, as opposed to GARCH, EGARCH and TGARCH, has the ability to capture long-memory volatility. In fact, CGARCH models volatility as the sum of a permanent process m_t (e.g. a time-varying trend) and a transitory mean-reverting process u_t. In addition, it allows for mean reversion to a varying level m_t that evolves slowly in an autoregressive manner and is driven by the volatility prediction error.

TABLE 25.3   Performance measures of prediction models of crude oil price volatility.[a]

Performance measure and its formulation

Mean error (ME) assumes that the cost of errors is symmetrical; that is, positive and negative errors of the same magnitude are equally weighted:

$$ME = \frac{1}{T}\sum_{i=1}^{T} e_t$$

Mean squared error (MSE) penalizes large errors (e.g. $e_t > 1$, $e_t < -1$) more than small ones (e.g. $-1 \leq e_t \leq 1$). Therefore, decision-makers may use this measure if several small errors are preferable to a few large ones:

$$MSE = \frac{1}{T}\sum_{i=1}^{T} e_t^2$$

Mean absolute error (MAE) assumes that errors of the same magnitude are assigned the same weight regardless of their signs. Notice that MAE is less sensitive to large errors than MSE:

$$MAE = \frac{1}{T}\sum_{i=1}^{T} |e_t|$$

Mean mixed error underestimation penalized (MMEU) is an asymmetric measure that allows one to express his or her preferences by penalizing underpredictions, where positive large errors are penalized more heavily:

$$MMEU = \frac{1}{T}\sum_{i=1}^{T} \begin{cases} |e_t|, & if\ e_t < 0 \\ \sqrt{e_t}, & if\ 0 \leq e_t \leq 1 \\ e_t^2, & if\ e_t > 1 \end{cases}$$

Mean mixed error overestimation penalized (MMEO) is an asymmetric measure that allows one to express his or her preferences by penalizing overpredictions, where negative large errors are penalized more heavily:

$$MMEO = \frac{1}{T}\sum_{i=1}^{T} \begin{cases} |e_t|, & if\ e_t > 0 \\ \sqrt{e_t}, & if\ -1 \leq e_t \leq 0 \\ e_t^2, & if\ e_t < -1 \end{cases}$$

Mean volatility-adjusted or scaled errors (MVolScE) is an alternative measure to ME, where errors are adjusted for volatility as measured by the variance of observations over the whole horizon (i.e. $t_1 = 1$ and $t_2 = T$) or part of it, depending on the type of implementation. This measure of biasedness proves useful when the series volatility is important enough to distort the picture conveyed by ME, which is the case with most financial time series:

$$MVolScE_{t1,t2} = \frac{1}{T}\sum_{i=1}^{T} \frac{e_t}{S_{t1,t2}^2}; \quad S_{t1,t2}^2 = \frac{1}{t_2-t_1}\sum_{k=t_1}^{t_2} \left(Y_k - \bar{Y}_{t1,t2}\right)^2;$$

$$\bar{Y}_{t1,t2} = \frac{1}{t_2-t_1+1}\sum_{k=t_1}^{t_2} Y_k; t_2 > t_2 \geq 1$$

Mean squared volatility-scaled errors (MSVolScE) is an alternative measure to MSE, where squared errors are adjusted for volatility, and proves useful when the series volatility is important enough to distort the picture conveyed by MSE:

$$MVolScE_{t1,t2} = \frac{1}{T}\sum_{i=1}^{T} \frac{e_t^2}{S_{t1,t2}^2}$$

TABLE 25.3   (*continued*)

Performance measure and its formulation

Mean absolute volatility-scaled errors (MAVolScE) is an alternative measure to MAE, where
absolute errors are adjusted for volatility, and proves useful when the series volatility is
important enough to distort the picture conveyed by MAE:

$$\text{MVolScE}_{t1,t2} = \frac{1}{T}\sum_{i=1}^{T}\left|\frac{e_t}{S_{t1,t2}^2}\right|$$

Percentage of correct direction change predictions (PCDCP) computes the proportion of correct
direction change predictions by a forecasting model, where n denotes the number of observations
and z_t is a binary variable set equal to 1 if $(\sigma_t - \sigma_{t-1})\cdot(\hat{\sigma}_t - \hat{\sigma}_{t-1}) > 0$ and 0 otherwise.

$$\text{PCDCP} = \sum_{i=1}^{T}\frac{z_t}{n}$$

[a] Here, $e_t = \sigma_t - \hat{\sigma}_t$, σ_t is the original series and $\hat{\sigma}_t$ is the forecasted series.

Stage 2: classification of DMUs. For each DMU_k $(k = 1,\dots,n)$, solve the following
slacks-based measure (SBM) model [29]:

$$\text{Min}: \rho_k = \left(1 - \frac{1}{m}\sum_{i=1}^{m}\frac{s_{i,k}^-}{x_{i,k}}\right) \bigg/ \left(1 + \frac{1}{s}\sum_{i=1}^{s}\frac{s_{r,k}^+}{y_{r,k}}\right)$$

$$\text{s.t.}: \sum_{j=1}^{n}\lambda_j x_{i,j} + s_{i,k}^- = x_{i,k}; \forall i \tag{25.1}$$

$$\sum_{j=1}^{n}\lambda_j y_{r,j} - s_{r,k}^+ = y_{r,k}; \forall r$$

$$\lambda_j \geq 0; \forall j; s_{i,k}^- \geq 0; \forall i; s_{r,k}^+ \geq 0; \forall r$$

where n denotes the number of DMUs, m is the number of inputs, s is the number of
outputs, $x_{i,j}$ is the amount of input i used by DMU_j, $y_{r,j}$ is the amount of output r produced by DMU_j, λ_j is the weight assigned to DMU_j in constructing its ideal benchmark,
and $s_{i,k}^-$ and $s_{r,k}^+$ are slack variables associated with the first and the second sets of
constraints, respectively. If the optimal objective function value ρ_k^* is equal to 1,
then DMU_k is classified as efficient; otherwise, it is classified as inefficient. Note
that the model (25.1) above is solved as it is if stage 1 reveals that CRS conditions
hold; otherwise, one has to impose one of the following additional constraints
depending on whether VRS, NIRS or NDRS conditions prevail, respectively:

$$\sum_{j=1}^{n}\lambda_j = 1; \quad \sum_{j=1}^{n}\lambda_j \geq 1; \quad \sum_{j=1}^{n}\lambda_j \leq 1 \tag{25.2}$$

Stage 3: break efficiency ties. For each efficient DMU_k, solve the following slacks-
based super-efficiency DEA model, first proposed by Tone [30]:

$$\text{Min}: \delta_k = \left(\frac{1}{m}\sum_{i=1}^{m}\frac{\left(x_{i,k} + t_{i,k}^-\right)}{x_{i,k}}\right) \bigg/ \left(\frac{1}{s}\sum_{i=1}^{s}\frac{\left(y_{r,k} - t_{r,k}^+\right)}{y_{r,k}}\right)$$

$$\text{s.t.} : \sum_{j=1;j\neq k}^{n} \lambda_j x_{i,j} \leq x_{i,k} + t_{i,k}^{-}; \forall i \tag{25.3}$$

$$\sum_{j=1;j\neq k}^{n} \lambda_j y_{r,j} \geq y_{r,k} - t_{r,k}^{+}; \forall i$$

$$\lambda_j \geq 0; \forall j \neq k; t_{i,k}^{-} \geq 0; \forall i; t_{r,k}^{+} \geq 0; \forall r$$

where $t_{i,k}^{-}$ and $t_{r,k}^{+}$ denote the amount by which input i and output r of the efficient DMU_k should be increased and decreased, respectively, to reach the frontier constructed by the remaining DMUs. Note that the model (25.2) above is solved as it is if stage 1 reveals that CRS conditions hold; otherwise, one has to impose an additional constraint from amongst (25.2) as outlined in stage 2. The super-efficiency scores δ_k^* are then used to rank-order the efficient DMUs.

At this stage, it is worth mentioning that unlike radial super-efficiency DEA models (e.g. [31]), slacks-based super-efficiency models are always feasible [30, 32]. Note that the slacks-based super-efficiency models of Tone [30] and Du *et al.* [32] are identical with respect to their constraints in that one can be obtained from the other using a simple transformation of variables. However, that applications where positive input and output data is a requirement, Du *et al.* [32] provided a variant of the model solved in stage 3 to accommodate this situation. In the next section, we shall use the above-described methodology to rank-order competing crude oil price volatility forecasting models.

25.3.4 Empirical Results from Slacks-Based Super-Efficiency DEA

In this subsection, we focus on the volatility of WTI crude oil daily spot prices. Our data covers the period ranging from 2 January 1986 to 28 May 2010, resulting in a total of 6157 observations. Note that we have chosen to consider several measures for each criterion to find out about the robustness of multicriteria rankings with respect to different measures.

Figure 25.4 provides the monocriterion rankings of 14 forecasting models of crude oil returns volatility based on nine measures of three criteria, namely, goodness-of-fit, biasedness and correct sign – this is a typical output presented by most existing forecasting studies [4–7]. These monocriterion rankings were devised by ranking the models from best to worst using the relevant measure of each of the criteria under consideration. Notice that different criteria led to different monocriterion rankings, which provides additional evidence of the problem resulting from the use of a monocriterion approach in a multicriteria setting as discussed in Section 25.1. For example, CGARCH(1, 1) outperforms SMA20 on measures of goodness-of-fit based on squared errors, whereas SMA20 performs better with respect to the biasedness criterion, as measured by both ME and MVolScE, and with respect to the correct-sign criterion, as measured by PCDCP. In order to remedy to these mixed performance results, one needs a single ranking that takes account of multiple criteria, which we provide using the proposed DEA framework.

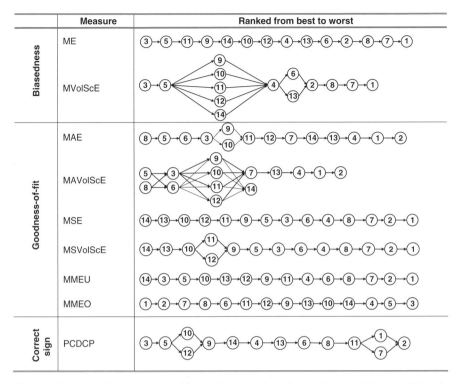

Figure 25.4 Unidimensional rankings of competing forecasting models – ranking in descending order of performance. 1, RW; 2, HM; 3, SMA20; 4, SMA60; 5, SES; 6, ARMA (1, 1); 7, AR(1); 8, AR(5); 9, GARCH(1, 1); 10, GARCH-M(1, 1); 11, EGARCH(1, 1); 12, TGARCH(1, 1); 13, PARCH(1, 1); 14, CGARCH(1, 1).

Figure 25.5 summarizes the multidimensional rankings of the 14 competing volatility-forecasting models for several combinations of performance measures, where the models are ranked from best to worst based on the corresponding super-efficiency scores obtained using both input-oriented and output-oriented radial super-efficiency DEA models – see Xu and Ouenniche [8]. Notice that, under VRS conditions, the rankings from input-oriented analysis and output-oriented analysis are different, on one hand, and the rankings from output-oriented analysis show more infeasibilities and ties, on the other hand. Figure 25.6 summarizes the multidimensional rankings of the volatility-forecasting models for several combinations of performance measures, where the models are ranked in descending order of the corresponding super-efficiency scores obtained using an orientation-free non-radial super-efficiency DEA model.

Figures 25.5 and 25.6 reveal that the rankings of forecasting models obtained by input-oriented super-efficiency DEA analysis, output-oriented super-efficiency DEA analysis and orientation-free super-efficiency DEA analysis are different.

These differences are mainly due to the fact that input-oriented analysis minimizes inputs for fixed amounts of output and output-oriented analysis maximizes outputs for fixed amounts of input, whereas orientation-free analysis optimizes both inputs and outputs simultaneously. In addition, input-oriented super-efficiency analysis and output-oriented super-efficiency analysis only take account of technical efficiency, whereas orientation-free super-efficiency analysis takes account of an additional performance component, namely, slacks. Notice that the efficient model SMA20 maintains its best position in the rankings regardless of whether the DEA analysis is input-oriented, output-oriented or orientation-free, because it is always on the efficient frontier and has zero slacks regardless of the performance measures used.

Input-oriented super-efficiency DEA scores-based rankings		
Inputs	**Output**	**Rankings from best to worst**
ME and MAE	PCDCP	(3)→(5)→(8)→(6)→(9)→(10)→(11)→(12)→(14)→(7)→(13)→(4)→(1)→(2)
ME and MAVolScE	PCDCP	(3)→(5)→(8)→(6)→{(9),(10),(11),(12)}→(14)→(7)→(13)→(4)→(1)→(2)
ME and MSE	PCDCP	(3)→(14)→(5)→(11)→(10)→(12)→(9)→(13)→(6)→(4)→(8)→(7)→(2)→(1)
ME and MSVolScE	PCDCP	(3)→(14)→(5)→(11)→(10)→(12)→(9)→(13)→(6)→(4)→(8)→(7)→(2)→(1)
ME and MMEU	PCDCP	(3)→(14)→(5)→(10)→(13)→(12)→(9)→(11)→(4)→(6)→(8)→(7)→(2)→(1)
ME and MMEO	PCDCP	(3)→(1)→(2)→(11)→(5)→(8)→(6)→(9)→(12)→(10)→(7)→(14)→(4)→(13)
Output-oriented super-efficiency DEA scores-based rankings		
ME and MAE	PCDCP	{(3),(5),(8)}→{(10),(12)}→(9)→(14)→(6)→(4)→(13)→(11)→(1)→(7)→(2)
ME and MAVolScE	PCDCP	{(3),(5)}→{(10),(12)}→(9)→(14)→(4)→(8)→(13)→(6)→(11)→(1)→(7)→(2)
ME and MSE	PCDCP	(3)→(14)→(5)→(10)→(12)→(9)→(13)→(4)→(6)→(11)→(8)→{(1),(7)}→(2)
ME and MSVolScE	PCDCP	(3)→(14)→(5)→(10)→(12)→(9)→(13)→(4)→(6)→(11)→(8)→{(1),(7)}→(2)
ME and MMEU	PCDCP	{(3),(14)}→(5)→{(10),(12)}→(9)→(4)→(13)→(6)→(8)→(11)→(1)→(7)→(2)
ME and MMEO	PCDCP	(3)→(2)→(1)→(11)→(5)→(8)→(6)→(9)→(12)→(10)→(7)→(14)→(4)→(13)

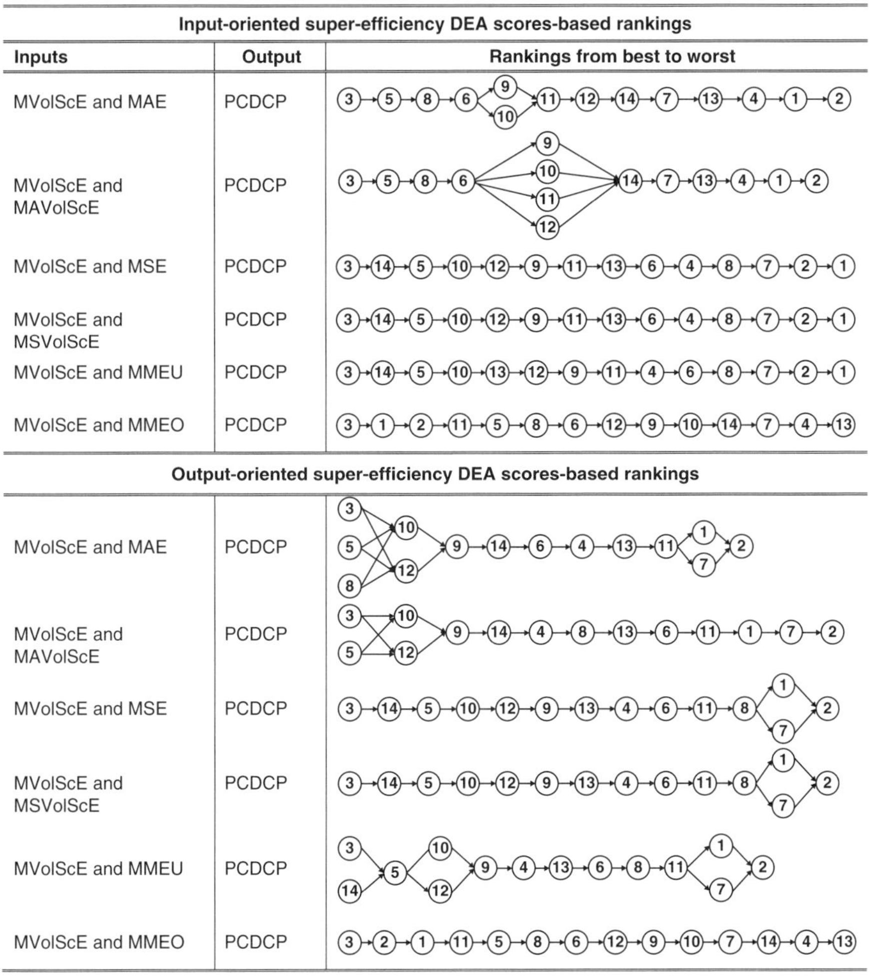

Input-oriented super-efficiency DEA scores-based rankings		
Inputs	**Output**	**Rankings from best to worst**
MVolScE and MAE	PCDCP	③→⑤→⑧→⑥→(⑨⑩)→⑪→⑫→⑭→⑦→⑬→④→①→②
MVolScE and MAVolScE	PCDCP	③→⑤→⑧→⑥→(⑨⑩⑪⑫)→⑭→⑦→⑬→④→①→②
MVolScE and MSE	PCDCP	③→⑭→⑤→⑩→⑫→⑨→⑪→⑬→⑥→④→⑧→⑦→②→①
MVolScE and MSVolScE	PCDCP	③→⑭→⑤→⑩→⑫→⑨→⑪→⑬→⑥→④→⑧→⑦→②→①
MVolScE and MMEU	PCDCP	③→⑭→⑤→⑩→⑬→⑫→⑨→⑪→④→⑥→⑧→⑦→②→①
MVolScE and MMEO	PCDCP	③→①→②→⑪→⑤→⑧→⑥→⑫→⑨→⑩→⑭→⑦→④→⑬

Output-oriented super-efficiency DEA scores-based rankings		
MVolScE and MAE	PCDCP	(③⑤⑧)→(⑩⑫)→⑨→⑭→⑥→④→⑬→⑪→(①⑦)→②
MVolScE and MAVolScE	PCDCP	(③⑤)→(⑩⑫)→⑨→⑭→④→⑧→⑬→⑥→⑪→①→⑦→②
MVolScE and MSE	PCDCP	③→⑭→⑤→⑩→⑫→⑨→⑬→④→⑥→⑪→⑧→(①⑦)→②
MVolScE and MSVolScE	PCDCP	③→⑭→⑤→⑩→⑫→⑨→⑬→④→⑥→⑪→⑧→(①⑦)→②
MVolScE and MMEU	PCDCP	(③⑭)→⑤→(⑩⑫)→⑨→④→⑬→⑥→⑧→⑪→(①⑦)→②
MVolScE and MMEO	PCDCP	③→②→①→⑪→⑤→⑧→⑥→⑫→⑨→⑩→⑦→⑭→④→⑬

Figure 25.5 Super-efficiency DEA scores-based multidimensional rankings of volatility forecasting models. 1, RW; 2, HM; 3, SMA20; 4, SMA60; 5, SES; 6, ARMA(1, 1); 7, AR(1); 8, AR(5); 9, GARCH(1, 1); 10, GARCH-M(1, 1); 11, EGARCH(1, 1); 12, TGARCH(1, 1); 13, PARCH(1, 1); 14, CGARCH(1, 1).

With respect to the orientation-free super-efficiency analysis, a close look at Figure 25.6 reveals that whether one measures biasedness by ME or MVolScE and one measures goodness-of-fit by MAE or MAVolScE, the ranks of the best models (e.g. SMA20, SES and AR(5)) and the worst models (e.g. RW, HM and AR(1)) remain the same; that is, they are robust to changes in measures. On the other hand, whether one measures biasedness by ME or MVolScE and one measures goodness-of-fit by MSE or MSVolScE, the ranks of the best models (e.g. SMA20, SES and

Inputs	Output	Models ranked from best to worst
ME and MAE	PCDCP	③→⑤→⑧→⑥→⑨→⑩→⑫→⑭→⑪→④→⑬→⑦→①→②
ME and MAVolScE	PCDCP	③→⑤→⑧→⑨→⑩→⑫→⑭→⑪→④→⑥→⑬→⑦→①→②
ME and MSE	PCDCP	③→⑭→⑤→⑪→⑨→⑩→⑫→⑬→④→⑥→⑧→⑦→①→②
ME and MSVolScE	PCDCP	③→⑭→⑤→⑪→⑨→⑩→⑫→⑬→④→⑥→⑧→⑦→①→②
ME and MMEU	PCDCP	③→⑭→⑤→⑨→⑩→⑫→⑪→⑬→④→⑥→⑧→⑦→②→①
ME and MMEO	PCDCP	③→②→①→⑪→⑤→⑧→⑥→⑦→⑨→⑫→⑩→⑭→④→⑬
MVolScE and MAE	PCDCP	③→⑤→⑧→⑥→⑩→⑨→⑫→⑭→⑪→④→⑬→⑦→①→②
MVolScE and MAVolScE	PCDCP	③→⑤→⑧→{⑩,⑫}→⑨→⑭→⑪→④→⑥→⑬→⑦→①→②
MVolScE and MSE	PCDCP	③→⑭→⑤→⑩→⑫→⑨→⑪→⑬→④→⑥→⑧→⑦→①→②
MVolScE and MSVolScE	PCDCP	③→⑭→⑤→⑩→⑫→⑨→⑪→⑬→④→⑥→⑧→⑦→①→②
MVolScE and MMEU	PCDCP	③→⑭→⑤→⑩→⑫→⑨→⑬→⑪→④→⑥→⑧→⑦→②→①
MVolScE and MMEO	PCDCP	③→②→①→⑪→⑤→⑧→⑥→⑫→⑨→⑦→⑩→⑭→④→⑬

Figure 25.6 Slacks-based super-efficiency DEA scores-based multidimensional rankings of volatility forecasting models. 1, RW; 2, HM; 3, SMA20; 4, SMA60; 5, SES; 6, ARMA(1, 1); 7, AR(1); 8, AR(5); 9, GARCH(1, 1); 10, GARCH-M(1, 1); 11, EGARCH(1, 1); 12, TGARCH(1, 1); 13, PARCH(1, 1); 14, CGARCH(1, 1).

CGARCH(1, 1)) and the worst models (e.g. RW, HM and AR(1)) remain the same. These rankings suggest that, for our dataset, AR(5) tends to produce large errors and CGARCH(1, 1) tends to produce small errors, as their ranks are sensitive to whether or not one penalizes large errors more than small ones. Finally, whether one measures biasedness by ME or MVolScE and one measures goodness-of-fit by MMEU (or by MMEO), the ranks of the best models such as SMA20 and CGARCH(1,1) (or RW, HM and SMA20, respectively) and the worst models such as RW, HM and AR(1) (or SMA60 and PARCH(1, 1), respectively) remain the same. Notice that the rankings under MMEU and MMEO differ significantly, which suggests for example that the performance of models such as RW, HM and CGARCH(1, 1) is very sensitive to whether one penalizes negative errors more than positive ones (that is, the decision-maker prefers models that underestimate the forecasts) or vice versa. In general, however, when underestimated forecasts are penalized, most GARCH types of models tend to perform well – suggesting that they often produce forecasts that are overestimated. On the other hand, when overestimated forecasts are penalized, averaging models such as RW, HM and SES tend to perform very well – suggesting that these models often produce forecasts that are underestimated.

Last but not least, given our dataset and the measures under consideration, the numerical results suggest that, with the exception of CGARCH, the family of GARCH models have an average performance compared with smoothing models such as SMA20 and SES – this suggests that the data generation process has a relatively long memory, which obviously gives an advantage to models such as SMA20 and SES as compared with GARCH(1, 1), GARCH-M(1, 1), EGARCH(1, 1), TGARCH(1, 1) and PARCH(1, 1), which are short-memory models. Similar findings on the GARCH type of models were reported by Kang *et al.* [21].

25.4 CONCLUSION

Nowadays, forecasts play a crucial role in driving our decisions and shaping our future plans in many application areas, such as economics, finance and investment, marketing, and the design and operational management of supply chains, among others. Obviously, prediction problems differ with respect to many dimensions; however, regardless of how one defines the prediction problem, a common issue faced by both academics and professionals is related to the performance evaluation of competing prediction models. Although most studies tend to use several performance criteria and, for each criterion, one or several metrics to measure each criterion, the assessment of the relative performance of competing forecasting models is generally restricted to a ranking of them by measure, which usually leads to different monocriteria rankings. The lack of a multicriteria framework for performance evaluation of competing prediction models has motivated the present line of research, in which we have proposed several frameworks based on both DEA analysis and MCDA analysis. In order to discuss the operationalization of the DEA-based relative performance evaluation frameworks in the area of forecasting, we have surveyed and classified the literature on performance criteria and their measures, including some statistical tests, commonly used in evaluating and selecting forecasting models. To illustrate the use of the proposed frameworks, we have used forecasting of crude oil prices and their volatility as an application area. We assessed the relative performance of competing prediction models of crude oil prices and volatility based on three criteria which are commonly used in the forecasting community, namely, the goodness-of-fit, biasedness and correct-sign criteria. We considered several measures for each criterion to find out about the robustness of multicriteria rankings with respect to different measures.

The main conclusions of our predictions of crude oil price volatility may be summarized as follows. First, models that are on the efficient frontier and have zero slacks regardless of the performance measures used (e.g. SMA20) maintain their rank regardless of whether the DEA analysis is input-oriented, output-oriented or orientation-free. Second, the multicriteria rankings of the best and the worst models seem to be robust to changes in most performance measures; however, SMA20 seems to be the best across the board. Third, when underestimated forecasts are penalized, most GARCH types of models tend to perform well – suggesting that they often produce forecasts that are overestimated. In contrast, when overestimated forecasts are penalized, averaging models such as RW, HM and SES tend to perform very well – suggesting that these models often produce forecasts that are underestimated. Finally, our empirical results seem to suggest that, with the exception of CGARCH, the family of GARCH models have an average performance compared with smoothing models such as SMA20 and SES, which suggests that the data generation process has a relatively long memory.

REFERENCES

[1] Xu, B. (2009) Multidimensional approaches to performance evaluation of competing forecasting models. PhD thesis. University of Edinburgh.

[2] Xu, B. and Ouenniche, J. (2011) A multidimensional framework for performance evaluation of forecasting models: Context-dependent DEA. *Applied Financial Economics*, **21**, 1873–1890.

[3] Day, T.E. and Lewis, C.M. (1993) Forecasting futures market volatility. *Journal of Derivatives*, **1**, 33–50.

[4] Sadorsky, P. (2005). Stochastic volatility forecasting and risk management. *Applied Financial Economics*, **15**, 121–135.

[5] Sadorsky, P. (2006). Modelling and forecasting petroleum futures volatility. *Energy Economics*, **28**, 467–488.

[6] Agnolucci, P. (2009) Volatility in crude oil futures: A comparison of the predictive ability of GARCH and implied volatility models. *Energy Economics*, **31**, 316–321.

[7] Marzo, M. and Zagaglia, P. (2010) Volatility forecasting for crude oil futures. *Applied Economics Letters*, **17**, 1587–1599.

[8] Xu, B. and Ouenniche, J. (2012) A data envelopment analysis-based framework for the relative performance evaluation of competing crude oil prices' volatility forecasting models. *Energy Economics*, **34**, 576–583.

[9] Xu, B. and Ouenniche, J. (2012) Performance evaluation of competing forecasting models: A multidimensional framework based on MCDA. *Expert Systems with Applications*, **39**, 8312–8324.

[10] Ouenniche, J., Xu, B. and Tone, K. (2014) Forecasting models evaluation using a slacks-based context-dependent DEA framework. *Journal of Applied Business Research*, **30**(5), 1477–1484.

[11] Ouenniche, J., Xu, B. and Tone, K. (2014) Relative performance evaluation of competing crude oil prices' volatility forecasting models: A slacks-based super-efficiency DEA model. *American Journal of Operations Research*, **4**(4), 235–245.

[12] Abhyankar, A., Xu, B. and Wang, J. (2013) Oil price shocks and the stock market: Evidence from Japan. *Energy Journal*, **34**(2), 199–222.

[13] Kilian, L. (2008) The economic effects of energy price shocks. *Journal of Economic Literature*, **46**, 871–909.

[14] Xu, B. (2015) Oil prices and UK industry-level stock returns. *Applied Economics*, **47**(25), 2608–2627.

[15] Pindyck, R.S. (1991) Irreversibility, uncertainty, and investment. *Journal of Economics Literature*, **29**, 1110–1148.

[16] Hamilton, J.D. (1988) A neoclassical model of unemployment and the business cycle. *Journal of Political Economics*, **96**, 593–617.

[17] Hamilton, J.D. (2009) Understanding crude oil prices. *Energy Journal*, **30**, 179–206.

[18] Bentzen, J. (2007). Does OPEC influence crude oil prices? Testing for co-movements and causality between regional crude oil prices. *Applied Economics*, **39**, 1375–1385.

[19] Kilian, L. (2009) Not all oil price shocks are alike: Disentangling demand and supply shocks in the crude oil market. *American Economics Review*, **99**, 1053–1069.

[20] BBC News (2005) Katrina set to raise oil prices, 28 August 2005, http://news.bbc.co.uk/1/hi/business/4192528.stm.

[21] Kang, S.H., Kang, S.M. and Yoon, S.M. (2009) Forecasting volatility of crude oil markets. *Energy Economics*, **31**, 119–125.

[22] Andersen, T.G. and Bollerslev, T. (1998) Answering the skeptics: Yes, standard volatility models do provide accurate forecasts. *International Economic Review*, **39**, 885–905.

[23] Wang, Y.D. and Wu, C.F. (2012) Forecasting energy market volatility using GRCH models: Can multivariate models beat univariate models? *Energy Economics*, **34**, 2167–2181.

[24] Fong, W.M. and See, K.H. (2002) A Markov switching model of the conditional volatility of crude oil futures prices. *Energy Economics*, **24**, 71–95.

[25] Nomikos, N.K. and Pouliasis, P.K. (2011) Forecasting petroleum futures markets volatility: The role of regimes and market conditions. *Energy Economics*, **33**, 321–337.

[26] Ghysels, E., Harvey, A. and Renault, E. (1996) Stochastic volat, in *Statistical Methods in Finance* (eds Maddala, G.S. and Rao, C.R.), Handbook of Statistics, Vol. **14**, Elsevier Science, Amsterdam, p. 119.

[27] Poon, S.H. and Granger, C.W.J. (2003) Forecasting financial market volatility: A review. *Journal of Economic Literature*, **41**, 478–539.

[28] Banker, R.D., Cooper, W.W., Seiford, L.M., Thrall, R.M. and Zhu, J. (2004) Returns to scale in different DEA models. *European Journal of Operational Research*, **154**, 345–362.

[29] Tone, K. (2001) A slacks-based measure of efficiency in data envelopment analysis. European Journal of Operational Research, **130**, 498–509.

[30] Tone, K. (2002) A slacks-based measure of super-efficiency in data envelopment analysis. *European Journal of Operational Research*, **143**, 32–41.

[31] Andersen, P. and Petersen, N.C. (1993) A procedure for ranking efficient units in data envelopment analysis. *Management Science*, **39**, 1261–1294.

[32] Du, J., Liang, L. and Zhu, J. (2010) A slacks-based measure of super-efficiency in data envelopment analysis: Comment. *European Journal of Operational Research*, **204**, 694–697.

26

PREDICTIVE EFFICIENCY ANALYSIS: A STUDY OF US HOSPITALS

ANDREW L. JOHNSON

Department of Industrial and Systems Engineering, Texas A&M University, College Station, TX, USA

CHIA-YEN LEE

Institute of Manufacturing Information and Systems, National Cheng Kung University, Tainan City, Taiwan

26.1   INTRODUCTION

In 2012, the United States' expenditure on health accounted for 16.9% of GDP, which is 7.5 percentage points above the OECD average for the same year [1]. Thirty-one percent of US healthcare expenditure is spent solely on hospital care, or approximately 5% of GDP [2]. Estimates of the excess cost in the system consistently exceed $750 billion and range as high as half of all healthcare expenditure [3]. These estimates motivate use to quantify the efficiency of hospitals. Because hospitals make up such a large portion of healthcare expenditure, hospitals are a potential large source of cost savings.

Cost-control and cost-efficiency analyses are familiar to the hospital industry, where concerns over rising costs have been present since the 1950s and 1960s [4–6]. It has been more than 25 years since accountability and assessment were hailed as the next revolution in medical care [7]. Valdez *et al.* [8] emphasized the role that potential operational improvements and improved efficiency can play in cost savings. Yet the best models for efficiency measurement in hospitals suffer from serious limitations and are rarely applied in practice.

Existing methods for analyzing the efficiency of hospitals (for a review, see Rosko and Mutter [9]) rely primarily on standard applications of data envelopment analysis

Advances in DEA Theory and Applications: With Extensions to Forecasting Models,
First Edition. Edited by Kaoru Tone.

(DEA) or stochastic frontier analysis (SFA). A particular limitation of these methodologies is that they assume hospitals will be able to perfectly predict customer demands for hospital services or that hospitals can adjust input resources without any time delays. Based on this assumption, these methods do not attempt to separate the quality of the forecasts for hospital services from the operational performance of the hospital [10]. Therefore, when a hospital is found to be inefficient, the analysis does not provide insight into whether that inefficiency comes from a poor forecast or it is the result of poor operational performance.

In this chapter, we build on the insights of Lee and Johnson [11, 12], who defined an effectiveness measure which complements the efficiency measure. Here, the effective input is defined as the input resource used in the production system that generates the forecasted output level with efficient operations. Furthermore, to measure effectiveness, we use the input-truncated production function, defined as the minimum resources needed in a hospital to generate the expected outputs. A hospital is achieving effective production if its input levels are equal to the effective input levels identified by the input-truncated production function.

A low effectiveness measure implies that the hospital used more inputs in a particular year than can be justified by efficient operations and forecasted growth for the industry. Persistent low effectiveness would indicate that the hospital is expanding resources faster than the forecasted demand is expanding, consistent with a medical arms race.

26.2 MODELING OF PREDICTIVE EFFICIENCY

In a typical productivity study, we estimate efficiency via a production function which defines the maximum outputs that a production system can produce with given input resources. Let x be a vector of input variables quantifying the input resources, y be a single output variable generated from the production system, and $y^{PF} = f(x)$ represent the maximum output level for the given inputs. Consider a multiple-input and multiple-output production process. Let $x \in \mathbb{R}_+^{|I|}$ denote the vector of input variables and $y \in \mathbb{R}_+^{|J|}$ denote the vector of output variables for the production system. The production possibility set (PPS) T is defined as $T = \{(x, y) : x \text{ can produce } y\}$. Let $i \in I$ be the input index, $j \in J$ be the output index, and $k \in K$ be the firm index. X_{ik} is the data for the ith input resource, Y_{jk} is the amount of the jth production output, and λ_k is the multiplier for the kth firm. Thus, the PPS can be estimated by a convex function enveloping all observations as shown in the model (26.1):

$$\tilde{T} = \left\{ (x, y) \mid \sum_k \lambda_k Y_{jk} \geq y_j, \forall j; \sum_k \lambda_k X_{ik} \leq x_i, \forall i; \sum_k \lambda_k = 1; \lambda_k \geq 0, \forall k \right\} \quad (26.1)$$

Then, the efficiency, θ, can be measured using the variable-returns-to-scale (VRS) DEA estimator. The input-oriented technical efficiency is defined as the distance function $D_I(x, y) = \inf\{\theta | (\theta x, y) \in \tilde{T}\}$. If $\theta = 1$, then the firm is efficient; otherwise, it is inefficient when $\theta < 1$.

To separate the effects of forecasting from operational performance, we need to make some assumptions about timing. Specifically, we will assume that a hospital manager knows the production function for period t and the forecast for growth in services required when they determine the input levels for period $t + 1$. Thus, our timing assumptions eliminate the concerns about endogeneity that are common in the econometrics literature [13–15]. Related to this issue, we have assumed that all inputs are adjustable once a year, but that after the level of inputs has been selected at the beginning of the year, the input levels are held fixed.

An input-truncated production function is defined based on the input demand function, which transforms the expected output to the input level in the current period. To maintain generality, the expected outputs are hospital-specific, each firm can have a different forecasted demand, and the input-truncated production function is defined as the production function truncated by the optimal inputs used by a specific hospital. Let d^{t+1} be the expected output in period $t + 1$. The effective input, $x^{E(t+1)}$, is the inverse of the production function in period t. The function $x^{E(t+1)}$ is formulated as in (26.2), where $f_t^{-1}(\cdot)$ is the inverse production function with respect to period t:

$$x^{E(t+1)} = f_t^{-1}\left(d^{t+1}\right) = D_I\left(x, d^{t+1}\right)x \tag{26.2}$$

Figure 26.1 illustrates the effective input for a single-input, single-output case. For an observation (production unit A in the figure), the effective input $X_A^{E(t+1)}$ is calculated from the production function $f_t(\cdot)$ and its expected output level d_A^{t+1} in period $t + 1$.

To measure the effectiveness, let $x^E \in \mathbb{R}_+^J$ denote an effective input vector estimated *from the previous period's production function*. The input-truncated production possibility set (PPS^E) is

$$T^E = \left\{ \left(\max\left(x^E, x\right), y\right) : \max\left(x^E, x\right) \text{ can produce } y \text{ in current period}\right\}$$

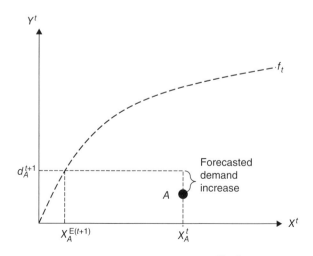

Figure 26.1   Effective input $x^{E(t+1)}$.

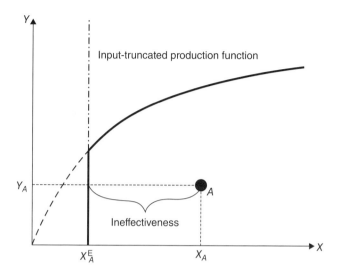

Figure 26.2   Effectiveness measure.

This can be estimated by a piecewise linear concave function truncated by the effective input level as shown in (26.3):

$$\tilde{T}^{\mathrm{E}} = \left\{ (\boldsymbol{x}, \boldsymbol{y}) \mid \sum_{k} \lambda_k Y_{jk} \geq y_j, \forall j; \; \sum_{k} \lambda_k X_{ik} \leq x_i, \forall i; \; X_i^{\mathrm{E}} \leq x_i, \forall i; \; \sum_{k} \lambda_k = 1; \; \lambda_k \geq 0, \forall k \right\}$$
(26.3)

Then, the effectiveness, θ^{E}, can be measured by the distance function $D_1(\boldsymbol{x}, \boldsymbol{y}) = \inf \left\{ \theta^{\mathrm{E}} \mid (\theta^{\mathrm{E}} \boldsymbol{x}, \boldsymbol{y}) \in \tilde{T}^{\mathrm{E}} \right\}$. If $\theta^{\mathrm{E}} \geq 1$, then the firm is effective in using the input resource; otherwise, it is ineffective when $\theta^{\mathrm{E}} < 1$, as illustrated in Figure 26.2. Note that if $\theta^{\mathrm{E}} > 1$, then the production unit is achieving the forecasted output with fewer resources than were believed to be needed based on the previous year's production function.

Let the index $r \in K$ be an alias of the index k. The effectiveness θ_r^{E} of one production unit r can be estimated by solving the following optimization problem:

$$D_1(\boldsymbol{x}, \boldsymbol{y}) = \min_{\lambda_k} \theta_r^{\mathrm{E}}$$

$$\text{s.t.} \sum_{k} \lambda_k Y_{jk} \geq Y_{jr}, \forall j$$

$$\sum_{k} \lambda_k X_{ik} \leq \theta_r^{\mathrm{E}} X_{ir}, \forall i$$
(26.4)

$$X_{ir}^{\mathrm{E}} \leq \theta_r^{\mathrm{E}} X_{ir}, \forall i$$

$$\sum_{k} \lambda_k = 1$$

$$\lambda_k \geq 0, \forall k$$

In order to solve for the effectiveness of all hospitals in one shot, we propose a combined optimization problem as follows. Let λ_{rk} be the multiplier describing production unit k's contribution to the benchmark for the rth production unit to calculate the effectiveness for production unit r. The effectiveness of all production units is estimated by solving the following formulation:

$$\min_{\lambda_k} \sum_{r} \theta_r^{E}$$

$$\text{s.t.} \sum_{k} \lambda_{rk} Y_{jk} \geq Y_{jr}, \forall j, r$$

$$\sum_{k} \lambda_{rk} X_{ik} \leq \theta_r^{E} X_{ir}, \forall i, r \qquad (26.5)$$

$$X_{ir}^{E} \leq \theta_r^{E} X_{ir}, \forall i, r$$

$$\sum_{k} \lambda_{rk} = 1, \forall r$$

$$\lambda_{rk} \geq 0, \forall k, r$$

26.3   STUDY OF US HOSPITALS

In order to examine the effectiveness of US hospitals, we used the 2009–2011 Nationwide Inpatient Sample from the Agency for Healthcare Research and Quality (AHRQ) Healthcare Cost and Utilization Project (HCUP); this is a dataset which contains all discharges from an approximately 20% sample (~1000 hospitals) of US community hospitals as defined by the American Hospital Association. The number of discharges (x_1) was a single input. We followed [16,17] and modeled the outputs using a four-dimensional vector including minor diagnostic procedures (y_1), major diagnostic procedures (y_2), minor therapeutic procedures (y_3), and major therapeutic procedures (y_4), categorized by the International Classification of Diseases, Clinical Modification codes. The distinguishing characteristic between the minor and major procedures of each type is the use of an operating room. For example, irrigation of a ventricular shunt is a minor therapeutic procedure, whereas an aorta–renal bypass is a major therapeutic procedure; a CT scan is a minor diagnostic procedure, whereas a brain biopsy is a major diagnostic procedure. In addition, we collected Centers for Medicare and Medicaid Services (CMS) reports which gave future projections regarding national health expenditure. For example, in 2009, these reports predicted the future hospital industry costs for 2010–2020, in 2010 they predicted costs for 2011–2021, and so forth. We used the expenditure projections to generate the expected output. That is, we took the distribution of outputs for 2009 and multiplied by the expenditure growth projection for 2010 and we had the forecasted 2010 output.

To measure the effectiveness, we estimated the optimal input level (x^E) given the expected 2010 output with respect to the 2009 frontier; this defined an input truncation level. Then we considered the observed outputs and inputs for 2010. We used all the data from 2010 to construct a frontier, and the hospital-specific truncation, x^E, was estimated for the observations observed in both 2009 and 2010. We then had the input-truncated production function and could calculate the effectiveness. We performed the same analysis using the observed data for 2010 and the expected 2011 output to define the input truncation level for 2011. Thus, when the observed input level was larger than the input truncation level (i.e., $x > x^E$), we had overusage of input and the effectiveness was less than 1; otherwise, when the observed number of discharges was less than (or equal to) the forecasted efficient input level, we had effective production and the effectiveness was greater than (or equal to) 1.

We did this analysis for two pairs of adjacent years, 2009–2010 and 2010–2011. Note that we did not observe the same hospitals each year, owing to the 10% sampling of the hospitals; thus, we assumed that the sample collected was representative and thereby that the distribution of effectiveness characterized the general population of hospitals. The summary statistics for the full sample and the hospitals that were observed in adjacent years are reported in Tables 26.1 and 26.2, respectively.

The results for effectiveness and efficiency regarding 2009–2010 are shown in Figures 26.3 and 26.4. Because the dataset was an unbalanced panel, there were

TABLE 26.1 Summary statistics for the full sample: number of hospitals and mean of input and output data for years 2009, 2010, and 2011.

Year	Number of hospitals	Sample mean				
		X_1	Y_1	Y_2	Y_3	Y_4
2009	1050	7439	2538	5777	120	3311
2010	1051	7422	2566	6016	117	3328
2011	1049	7649	2567	6223	118	3417

TABLE 26.2 Summary statistics for hospitals observed in consecutive periods: number of hospitals and input and output data for years 2010 and 2011.

Year	Number of hospitals	Sample mean				
		X_1	Y_1	Y_2	Y_3	Y_4
2010	279	7459	2524	5924	116	3539
2011	256	7524	2361	6040	114	3243

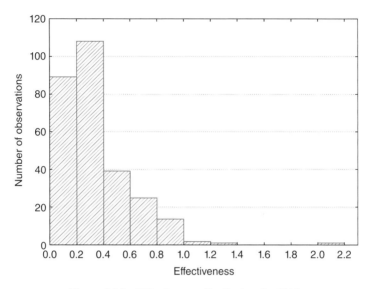

Figure 26.3 Effectiveness distribution for 2010.

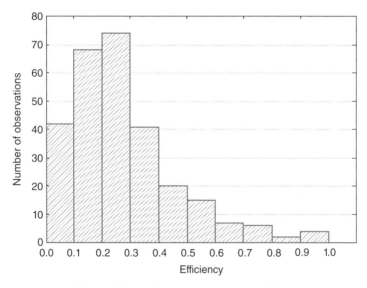

Figure 26.4 Efficiency distribution for 2010.

279 observations for the adjacent years 2009–2010 and therefore we could only calculate the effectiveness for those 279 observations. The expected growth rate of the output is 4.6% for 2010, the average of the effectiveness is 52% weighted by the observed inputs for 2010, and the average of the efficiency is 40%.

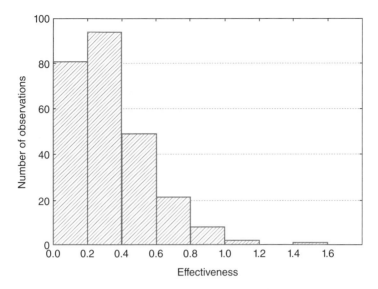

Figure 26.5   Effectiveness distribution for 2011.

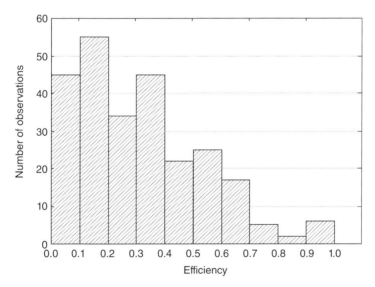

Figure 26.6   Efficiency distribution for 2011.

The results for effectiveness and efficiency for 2010–2011 are shown in Figures 26.5 and 26.6, respectively. There were 256 observations present in the adjacent years 2010–2011. The expected growth rate of the output is 4.3% for 2011, the average of the effectiveness is 50% weighted by the observed inputs for 2010, and the average of the efficiency is 49%.

26.4   FORECASTING, BENCHMARKING, AND FRONTIER SHIFTING

26.4.1   Effect of Forecast on Effectiveness

The distributions of efficiency and effectiveness are similar for both 2010 and 2011. The differences between the two metrics are due to the truncation of the production possibility set at the efficient input level associated with the forecasted output level. Thus, the difference between efficiency and effectiveness is driven by the forecast. Therefore, we investigated how effectiveness changes as the forecast changes.

Each hospital is likely to grow at its own rate. However, if we calculate the product-specific growth rates for each hospital between 2009 and 2010 and between 2010 and 2011 and use the actual growth rates as the forecasted growth rates, then each hospital will be effective. Therefore, we performed a sensitivity analysis on the forecasted growth rate in the CMS reports. The growth rates predicted in the CMS reports were 4.6% and 4.3% for 2010 and 2011, respectively. In Table 26.3, we start with a growth rate of 10% and consider 10% increments up to 110% growth. In 2010, the effectiveness is approximately 1 when the output growth rate is 110%. However, in 2011, even with a forecasted growth rate of 110%, the average effectiveness is still just 86%.

26.4.2   Benchmarks

While estimating the effectiveness or efficiency, the linear programming calculation also constructs benchmarks to measure effectiveness or efficiency. In particular, λ_{rk} is the multiplier for the kth hospital to investigate the effectiveness of one specific hospital r, that is, λ_{rk} implies the weight in effective benchmarks for the kth hospital. Note that we use the term "effective benchmark" for the effectiveness measure and "efficient benchmark" for the efficiency measure.

Given the expected growth rates of the output of 4.6% in 2009–2010 and 4.3% in 2010–2011, the results for the benchmarks are shown in Tables 26.4 and 26.5. These

TABLE 26.3   Effectiveness against output growth rate for 2010 and 2011.

Output growth rate	Effectiveness in 2010	Effectiveness in 2011
0.1	0.55	0.51
0.2	0.61	0.53
0.3	0.66	0.54
0.4	0.72	0.57
0.5	0.78	0.61
0.6	0.84	0.64
0.7	0.90	0.68
0.8	0.94	0.73
0.9	0.97	0.79
1	0.98	0.84
1.1	1.00	0.86

TABLE 26.4 Summary of effective/efficient benchmarks for 279 hospitals in 2009–2010.

Effectiveness in 2009–2010

Reference hospital #	1	3	5	6	11	107	139	277	450	471	582	606	625	638	739	848	937	943
Sum of λ_{rk}	12.4	0.00	4.11	0.68	1.51	4.01	0.27	0.51	126	0.46	6.49	2.90	28.8	27.7	15.8	1.42	33.9	11.7
Count ($\lambda_{rk} > 0$)	49	1	19	8	5	14	4	1	175	8	30	48	253	41	124	22	50	56

Efficiency in 2010

Reference hospital #	11	48	107	139	277	283	450	471	582	606	625	638	739	848	937	943
Sum of λ_{rk}	5.51	0.18	11.5	0.00	0.51	0.00	42.1	1.63	18.7	4.38	22.1	31.7	20.0	0.95	82.9	36.8
Count ($\lambda_{rk} > 0$)	19	1	38	1	1	0	65	17	80	43	255	45	119	27	118	121

TABLE 26.5 Summary of effective/efficient benchmarks regarding 256 hospitals in 2010–2011.

Effectiveness in 2010–2011

Reference hospital #	1	4	5	15	64	79	107	266	279	304	369	427	602	629	646	763	797	855	902	926	929	1045	1050
Sum of λ_{rk}	0.03	0.01	0.09	2.41	0.86	4.28	26.7	8.55	1.00	0.81	12.1	1.35	5.94	1.35	0.03	1.88	0.08	28.2	8.41	28.3	25.8	15.5	82.5
Count ($\lambda_{rk} > 0$)	2	2	2	7	18	21	106	108	1	8	40	8	36	6	7	3	7	186	33	99	87	68	116

Efficiency in 2011

Reference hospital #	15	64	79	107	266	279	304	369	427	602	629	763	797	855	902	926	929	983	1045
Sum of λ_{rk}	2.66	0.86	4.28	26.5	8.68	1.00	0.89	11.9	1.35	9.73	1.33	1.96	0.02	27.5	10.4	29.4	29.6	73.0	15.0
Count ($\lambda_{rk} > 0$)	6	16	21	91	96	1	8	33	8	57	4	4	1	186	51	109	110	108	57

tables show how frequently certain hospitals are identified as effective benchmarks and as efficient benchmarks by calculating the sum of λ_{rk} and the count of $\lambda_{rk} > 0$. For 2009–2010, there are 279 hospitals for which we can calculate effectiveness, and the results show that Hospital # 625 is part of the benchmark for 253 hospitals, or approximately 90% of the hospitals; when measuring efficiency, Hospital # 625 is part of the benchmark for 255 hospitals. The five hospitals (i.e., Hospitals # 450, 625, 739, 937, and 943) that are most often included in the effectiveness benchmarks are among the six hospitals (i.e., Hospitals # 450, 582, 625, 739, 937, and 943) that are most often included in the efficiency benchmarks. For 2010–2011, there are 256 hospitals for which we can calculate effectiveness, and the results show that Hospital # 855 is part of the benchmark for 186 hospitals, or approximately 72% of the hospitals; when measuring efficiency, Hospital # 855 is still the hospital that is most commonly part of the benchmarks. Again, for 2010–2011, the five hospitals (i.e., Hospitals # 107, 266, 855, 926, and 929) that are most often included in the effectiveness benchmarks are among the six hospitals (i.e., Hospitals # 107, 266, 855, 926, 929, and 983) that are most often included in the efficiency benchmarks. The hospitals that are part of the benchmarks are different between 2010 and 2011 because there was significant technical progress.

26.4.3 Technical Progress

The Malmquist productivity index (MPI) measures productivity change and its components. Färe *et al.* [18] proposed nonparametric methods to estimate the MPI. Estimating the MPI between period t and period $t+1$ requires an additional distance function to measure the cross-period distance function of an observation in period $t+1$ relative to the reference technology in period t as in the following equation:

$$D_1^t\left(x^{t+1}, y^{t+1}\right) = \inf\left\{\theta \,\big|\, \left(\theta x^{t+1}, y^{t+1}\right) \in \tilde{T}^t\right\} \qquad (26.6)$$

where $\tilde{T}^t$ is the PPS estimated for period t. Thus, the MPI can be estimated by following equation:

$$
\begin{aligned}
\text{MPI} &= \left[\frac{D_1^t\left(x^{t+1}, y^{t+1}\right)}{D_1^t\left(x^t, y^t\right)} \times \frac{D_1^{t+1}\left(x^{t+1}, y^{t+1}\right)}{D_1^{t+1}\left(x^t, y^t\right)}\right]^{1/2} \\
&= \frac{D_1^{t+1}\left(x^{t+1}, y^{t+1}\right)}{D_1^t\left(x^t, y^t\right)} \left[\frac{D_1^t\left(x^{t+1}, y^{t+1}\right)}{D_1^{t+1}\left(x^{t+1}, y^{t+1}\right)} \times \frac{D_1^t\left(x^t, y^t\right)}{D_1^{t+1}\left(x^t, y^t\right)}\right]^{1/2} \qquad (26.7) \\
&= \text{Efficiency change (EC)} \times \text{Frontier shift (FS)}
\end{aligned}
$$

A typical MPI can decomposed into an efficiency change and a frontier shift. The EC describes the change in technical efficiency, while FS characterizes the change in technology, that is, the shift of the production frontier. The MPI, EC, and FS are each interpreted as achieving progress, no change, and regress when the values of their

TABLE 26.6 Malmquist productivity index and its decomposition from 2009 to 2011.

	MPI	EC	FS
2009–2010	1.01	0.84	1.23
2010–2011	0.97	1.39	0.74

estimates are greater than 1, equal to 1, and less than 1, respectively. Note, however, that the frontier shift is also influenced by changes in demand levels. In periods in which forecasted growth is low and therefore relatively few resources are acquired, if the demands incurred are high, this will cause an increase in productivity levels, making the production function shift up. However, in periods in which forecasted growth is high relative to observed growth, then relatively many resources are acquired and the demand incurred is lower than expected, causing a decrease in productivity levels, making the production function shift down. Therefore hospitals, because of the relatively high uncertainty between forecasted demand and actual demand, illustrate how the frontier shift component of the Malmquist productivity index is also affected by demand. Often the frontier shift is interpreted as technical change, implying that the industry has developed new methods of production; however, this interpretation focuses on the supply side without consideration of the fact that products for which there is no demand do not get produced.

The results for the MPI and its decomposition are shown in Table 26.6. The MPI is close to 1 for both 2009–2010 and 2010–2011. For 2009–2010, the standard decomposition shows that the frontier shifted out by 23%, and the average technical efficiency dropped by 16%. However, for 2010–2011, the analysis indicates that the frontier contracted by 26%, and the average technical efficiency increased by 39%.

These large fluctuations in both the frontier shift and the efficiency change are hard to justify based solely on operational changes during this three-year period. We believe that demand fluctuations and random noise play an important role in these results. Specifically, the DEA estimator has no model of noise included. Therefore, a random shock or measurement error that causes a single observation or set of observations to be significantly above the true production function will cause a DEA-type estimator to overestimate the production frontier for that year. If those random shocks or measurement errors do not occur in prior or later periods, a sudden expansion followed by a sudden contraction of the frontier may be estimated and observed, when the true frontier has not changed at all. However, if this does occur, but the majority of the data still remain clustered around the production function, then we would expect to see results similar to those in Table 26.6.

Another alternative explanation is that demand in 2010 for healthcare services provided in hospitals could have increased significantly. These changes would typically be associated with population dynamics or policy changes. However, while the number of senior citizens and Medicare recipients in the US is rising with the retirement of the baby boomer generation following World War II, leading to an increase in demand in 2010, this would not explain the decrease then observed in 2011.

26.5 CONCLUSIONS

The efficient operation of hospitals is critical to controlling the costs associated with healthcare in the US. An extensive literature exists on measuring efficiency from the inputs consumed and outputs produced by a hospital. For the purposes of evaluating operational performance, this sort of efficiency measure combines the effects of forecasting and operational performance. To measure the performance of production units relative to forecasted demand, Lee and Johnson [11] introduced the concept of effectiveness and the truncated production function. We have applied these concepts to investigate the performance the US hospital industry.

We find that hospitals measured in terms of efficiency or effectiveness have distributions that are skewed towards having mostly inefficient and ineffective hospitals, with a small tail performing relatively well. Having low efficiency and effectiveness scores indicates that it is not primarily differences between the forecasted and observed demand that are driving the high inefficiency-level results; instead, it appears that operational inefficiency is more systematic. This is in part due to the random nature of the demand for hospital services that requires resources to be available at all times for emergency situations. However, the classical assumption in the stochastic frontier literature is that the inefficiency distribution has a mode of zero and that the probability decreases monotonically at higher inefficiency levels. This sort of assumption is typically motivated by the efficient market hypothesis [19, 20]. However, it is unlikely that the markets in which hospitals compete are efficient, because of the mixture of public, private, and not-for-profit hospitals and the government programs that subsidize various hospitals or services. Thus, these results add to the growing evidence that efficiency analyses which allow for the possibility that the inefficiency distribution may have a mode other than zero are important lines of research. Currently, the most popular efficiency models in this group are models which assume a two-parameter distribution for inefficiency such as the gamma distribution [21, 22].

Using an envelopment estimator such as DEA, we find that the average efficiency and effectiveness levels are quite low. This may be in part because inefficiency in our model captures noise, inefficiency, and any other unmodeled variables. The challenges of using a deterministic estimator are particularly apparent in the Malmquist productivity analysis. Such wide variations in the frontier shifts and technical efficiency change would not seem possible to justify by technical progress or changes in the operational efficiency of hospitals. Rather, we believe that the deterministic model and fluctuations in demand are the primary contributors to these large variations. Therefore, we propose to use a generalization of DEA to the stochastic setting that models noise separately from inefficiency. Specifically, the stochastic nonparametric envelopment of data (StoNED) estimator [23–25] is one common method to incorporate noise into the estimation of production frontiers.

The HCUP data provide extremely detailed information on the procedures and services that hospitals provide to customers. However, information about the resources used by hospitals is lacking. In this research, we have used the number of discharges as a proxy for the inputs consumed by a hospital. Finding other data sources with better information regarding the resources used by hospitals would allow more detailed modeling of resource consumption and could potentially lead to better estimates of efficiency and effectiveness.

In future research, we plan to investigate alternative methods for forecasting. In the work presented in this chapter, we used the CMS National Health Expenditure Projections reports; however, the hospitals in our sample may be expected to grow at different rates and therefore we should use alternative forecasts to CMS. These rates would be driven by local population growth and aging.

REFERENCES

[1] OECD (2014) OECD Health Statistics 2014: How does the United States compare? Available at: http://www.oecd.org/unitedstates/Briefing-Note-UNITED-STATES-2014.pdf [accessed 5 May 2015].

[2] Henry J. Kaiser Family Foundation (2012) Health care cost: A primer. Available at: http://kff.org/report-section/health-care-costs-a-primer-2012-report/[accessed 16 September 2015].

[3] PricewaterhouseCoopers (2009) The price of excess: Identifying waste in healthcare spending. Health Research Institute. Available at: http://www.oss.net/dynamaster/file_archive/080509/59f26a38c114f2295757bb6be522128a/The Price of Excess - Identifying Waste in Healthcare Spending - PWC.pdf [accessed 16 September 2015].

[4] Sheps, M.C. (1955) Approaches to the quality of hospital care. *Public Health Reports*, **70**(9), 877–886.

[5] Dowling, W.L. (1976) *Hospital Production: A Linear Programming Model*, Lexington Books, Boston, MA.

[6] Griffin, J.R., Hancock, W.M., and Munson, F.C. (1976) *Cost Control in Hospitals*, Health Administration Press, Ann Arbor, MI.

[7] Relman, A.S. (1988) Assessment and accountability: The 3rd revolution in medical care. *New England Journal of Medicine*, **319**(18), 1220–1222.

[8] Valdez, R.S., Ramly, E., and Brennan, P.F. (2010) Industrial and systems engineering and health care: Critical areas of research. Final Report, AHRQ Publication No. 10-0079. Agency for Healthcare Research and Quality, Rockville, MD.

[9] Rosko, M.D. and Mutter, R.L. (2011) What have we learned from the application of stochastic frontier analysis to U.S. Hospitals?, *Medical Care Research and Review*, **68**(1), 75S–100S.

[10] Lee, C.-Y. and Johnson, A.L. (2012) Two-dimensional efficiency decomposition to measure the demand effect in productivity analysis. *European Journal of Operational Research*, **216**(3), 584–593.

[11] Lee, C.-Y. and Johnson, A.L. (2014) Proactive data envelopment analysis: Effective production and capacity expansion in stochastic environments. *European Journal of Operational Research*, **232**(3), 537–548.

[12] Lee, C.-Y. and Johnson, A.L. (2015) Effective production: Measuring of the sales effect using data envelopment analysis. *Annals of Operations Research*, **235**(1), 453–486.

[13] Zellner, A., Kmenta, J., and Dreze, J. (1966) Specification and estimation of Cobb–Douglas production function models. *Econometrica*, **34**, 784–795.

[14] Marschak, J. and Andrews, W. (1944) Random simultaneous equations and the theory of production. *Econometrica*, **12**, 143–205.

[15] Olley, S. and Pakes, A. (1996) The dynamics of productivity in the telecommunications equipment industry. *Econometrica*, **64**, 1263–1298.

[16] Pope, B. and Johnson, A.L. (2013) Returns to scope: A metric for production synergies demonstrated for hospital production. *Journal of Productivity Analysis*, **40**(2), 239–250.

[17] Sarmento, M., Johnson, A.L., Preciado Arreola, J.L., and Ferrier, G.D. (2016) Cost efficiency of U.S. hospitals: A semi-parametric Bayesian analysis. Working paper.

[18] Färe, R., Grosskopf, S., Lindgren, B., and Roos, P. (1992) Productivity changes in Swedish pharmacies 1980–1989: A non-parametric Malmquist approach. *Journal of Productivity Analysis*, **3**(1–2), 85–101.

[19] Fama, E.F. (1965) Random walks in stock market prices. *Financial Analysts Journal*, **21**(5), 55–59 .

[20] Fama, E.F. and French, K.R. (1993) Common risk factors in the returns on stocks and bonds. *Journal of Financial Economics*, **33**(1), 3–56.

[21] Tsionas, E.G. (2000) Full likelihood inference in normal-gamma stochastic frontier models. *Journal of Productivity Analysis*, **13**, 179–201.

[22] Greene, W.H. (1990) A gamma distributed stochastic frontier model. *Journal of Econometrics*, **46**, 141–163.

[23] Kuosmanen, T. and Kortelainen, M. (2012) Stochastic non-smooth envelopment of data: Semi-parametric frontier estimation subject to shape constraints. *Journal of Productivity Analysis*, **38**(1), 11–28.

[24] Kuosmanen, T., and Johnson, A.L. (2010) Data envelopment analysis as nonparametric least squares regression. *Operations Research*, **58**(1), 149–160.

[25] Kuosmanen, T., Johnson, A.L., and Saastamoinen, A. (2015) Stochastic nonparametric approach to efficiency analysis: A unified framework, in *Handbook on Data Envelopment Analysis* (ed. J. Zhu), Vol. **2**, Springer, pp. 191–244.

27

EFFICIENCY PREDICTION USING FUZZY PIECEWISE AUTOREGRESSION[1]

Ming-Miin Yu

Department of Transportation Science, National Taiwan Ocean University, Keelung, Taiwan

Bo Hsiao

Department of Information Management, Chang Jung Christian University, Taiwan

27.1 INTRODUCTION

Forecasting methodologies for efficiency are still rarely applied in predicting productivity and efficiency in real-world applications, even though an analytic framework has already been proposed for several business functions, such as production, marketing, research and development, and finance. Existing forecasting methodologies have mostly focused on predicting the output from the input. However, when the forecasting methodology is applied to relative efficiency, the selection of data becomes more difficult because previous approaches have used absolute historical data or efficiency scores. Therefore, the conventional forecasting approaches cannot be used for relative concepts, such as time series of efficiency.

---

[1] This chapter is extended from the paper "Efficiency predictions by fuzzy piecewise auto-regression," by Bo Hsiao, Ching-Chin Chern, Ming-Miin Yu, and Gwo-Hsiung Tzeng, which was published in *Journal of International of Management Science* (Vol. 17, 2010, pp. 197–220). This chapter is authorized by the same journal.

Advances in DEA Theory and Applications: With Extensions to Forecasting Models,
First Edition. Edited by Kaoru Tone.
© 2017 John Wiley & Sons Ltd. Published 2017 by John Wiley & Sons Ltd.

Efficiency analysis consists of two competing paradigms. The first paradigm uses mathematical programming techniques, such as data envelopment analysis (DEA), which is popular in the operations research field. The second paradigm employs a regression approach, such as stochastic frontier analysis (SFA), which is widely accepted in the econometrics field. These methodologies have specific characteristics and limitations. On the one hand, DEA does not require explicit assumptions regarding the functional structure of the stochastic frontier. On the other hand, SFA imposes an explicit and overly restrictive frontier function upon models. In other words, DEA is based on a nonparametric approach, whereas SFA is based on a parametric one. Therefore, DEA cannot provide mechanisms for prediction, whereas parametric and frontier functions cannot easily be defined in SFA.

A new hybrid approach that comprises a catching-up efficiency index (CIE) and fuzzy piecewise autoregression analysis will be presented in this chapter to illustrate the prediction of efficiency and to show how it reinforces the prediction ability of DEA. The CIE is a measure of technical efficiency change during the period analyzed (the catching-up effect, or movement toward the frontier). This measure ignores the input-versus-output relationship and combines the inputs and outputs into an index. Developed by Yu *et al.* [1,2], fuzzy piecewise regression analysis provides information that can be used for grasping the dynamics of variable data and forecasting the efficiency when two regression estimation models are used simultaneously. A two-stage process is then used to predict efficiency. The CIE is calculated by efficiency evaluation in the first stage, whereas validation and/or prediction is performed in the second stage. DEA techniques are used in the first stage to evaluate efficiency scores for a number of periods and to transfer those efficiency scores to CIE indices. In the second stage, fuzzy piecewise autopiecewise regression is performed to calculate the CIE index data and forecast the relevant values, which fall into two ranges. The first range is provided by the possibility estimation model, which suggests that the predicted values must be included in the regression range. The second range is provided by the necessity estimation model, which suggests that the predicted values must be excluded from the regression range.

Additional details of the implementation of these concepts are explained in the following sections. Section 27.2 reviews the related literature on efficiency prediction. Section 27.3 describes the problem and presents the theoretical framework. Section 27.4 presents a case based on data from 17 Taiwanese train firms. Finally, Section 27.5 offers our conclusions and suggestions for readers.

27.2 EFFICIENCY PREDICTION

The DEA approach is suitable for analyzing institutional data, such as those from governments [3,4], schools [5,6], hospitals [7,8], and banks [9,10]. Although suitable for evaluating efficiency, this approach is inapplicable for prediction and forecasting. Traditional DEA studies focus on "one-shot state" efficiency analyses. A few approaches (e.g., SFA) predict efficiency either by modeling the production relationship or by using soft computing techniques. However, modeling the production/frontier function or framing the structure of an analyzed environment also has many limitations and is difficult to achieve.

Generally, econometricians tend to favor regression-based or other sophisticated approaches, whereas management scientists favor DEA approaches for evaluating performance issues [11–13]. Thanassoulis [13] found that DEA was suitable for regression analysis. By contrast, Schmidt [12] proposed that DEA lacked a statistical basis. The relevance and credibility of these conflicting results cannot easily be established, and the fundamental difference between regression analysis and DEA is unclear. To understand the characteristics of DEA, its two major advantages must be understood. First, DEA is based on ratio concepts instead of absolute input-versus-output relationships, and second, the efficiency score is relative to the frontier instead of the scores [14].

Ratios provide scale invariance characteristics such it is possible to ignore the influence of scale on the performance results. Therefore, these characteristics can be extrapolated in an evaluation. Despite some limitations, many techniques, such as the use of key business performance measurements, apply ratio analysis to simultaneously evaluate income and balance sheet financial statements. Each projected metric in the analysis has its respective goal value that is tied to the strategic vision of the business. For example, financial ratio analysis is used for performance evaluation [15,16] and can only measure one input and output simultaneously. However, this analysis faces several challenges, such as a lack of accredited financial ratio models and weight selection. Therefore, ratio analysis rules must be constructed using complicated computations with higher-order equations to achieve a more flexible analysis. DEA can work with simple rules (i.e., input/output) and allows the evaluation of multiple outputs and inputs [14,15]. Ratio analysis requires complex data for evaluation, whereas DEA does not require a large sample size [14,17].

Ratio analysis refocuses resources toward "the goal" (i.e., the efficiency should be 1) and does not reflect actual scenarios (i.e., compared with other decision-making units, or DMUs). DEA is a relative concept in that the efficiency of a specific DMU is dependent on best practices or frontiers and not on itself. Therefore, conventional evaluation techniques cannot fully fit requirements that originate from the inherent characteristics of DEA. Under such conditions, if conventional approaches must be implemented, they must combine many relationship constraints to satisfy these requirements.

DEA can deal simultaneously with ratio and ordinal-scale data, but regression analysis is difficult to implement. Moreover, DEA approaches lack any requirements for assumptions about any pre-specified functional form of the production function and tend to avoid the problem of parameter measures [14].

However, these advantages of DEA also lead to disadvantages, such as a lack of frontier functions. The absence of requirements for assumptions about any prespecified functional form of the production/frontier function implies that DEA is incapable of prediction. This limitation is apparent in other mathematical models, such as in regression analysis and prediction approaches. Models must be able to estimate efficiency predictions over time. The efficiency predictions of DEA do not have such a capability because that technique cannot simultaneously handle both negative values (e.g., data representing decay) in the dataset and a frontier shift over time [17].

Some studies have enhanced the prediction efficiency of DEA by combining this method with other prediction techniques. Productivity change, which is explained in terms of technical change, has recently become widely accepted in the field of predicting efficiency change. Such change can be simplified into an uncomplicated forecasting functionality (frontier shift) to some degree. The Malmquist index, which was introduced by Caves *et al.* [15] to predict productivity change, has an important role in supporting such discussion. Färe *et al.* [18] decomposed productivity change into efficiency and technical changes, as well as constructing a nonparametric mathematical programming model to provide a solution. Caves *et al.* [15] and Färe *et al.* [18] showed that under certain conditions, the Malmquist index approximates the Törnqvist [19] and Fisher [15,20] indices, which are generally accurate and easily computed but yield biased estimates in the presence of inefficiency [21,22]. The Malmquist index may be incapable of providing a full picture, however, as this measure only considers the productivity change between two periods. However, this index can be extended to multiple periods by multiplication. The Malmquist index, by its nature, is based on only two adjacent periods and may ignore performance over more than two previous periods.

To restructure a strategy of the Japanese Petroleum Company, Sueyoshi [23] proposed a stochastic DEA that was formulated by use of chance constraint programming and estimated via a program evaluation and review technique or the critical path method. Sueyoshi used stochastic efficiency (aspiration level) and conventional efficiency (risk criterion) to decide future efficiency. However, he proposed several assumptions about the stochastic variables of the output for computational convenience and assumed a normal distribution for a stochastic variable when conducting a statistical test. Stochastic DEA also predicts efficiency according to data from only one period. Therefore, the prediction capacity of this technique depends heavily on its required assumptions (i.e., the standard deviation of the error terms is equal to zero).

By contrast, Kao and Liu [24] introduced fuzzy concepts to forecast efficiency based on uncertain data that were represented by a range instead of a single value. The prediction results were presented as a range. Kao and Liu [24] adapted fuzzy concepts for use in DEA, relaxed the assumptions of Sueyoshi [23] about the error term variances of output variables (equal to zero), and assumed that the output probability had a beta distribution. Similarly to that of Sueyoshi [23], the model of Kao and Liu [24] considers only a single state and does not base its predictions on the past performance of DMU.

Yeh *et al.* [25] proposed a novel model for integrating rough set theory (RST) and support vector machines (SVM) techniques to enhance the accuracy of prediction of business failure. In their model, DEA was employed to evaluate input/output efficiency, remove the redundant attributes in an RST approach, and reduce the number of independent variables without losing important information. They later used such information as a preprocessor to improve the prediction accuracy through SVM.

Wu [26] integrated DEA and neural networks (NN) to examine and forecast the relative efficiency of Canadian bank branches. Tsai *et al.* [27] constructed a consumer loan default prediction model by conducting DEA–discriminate analysis (DA) and using NN. However, Wu and Tsai *et al.*'s models require a longer computing time and a larger amount of computing resources. These models classify data into two patterns, namely, good examples (positive data) and failed examples (negative data), during the training process. The regression results in these methods can be determined either by structure error minimization [25] or empirical error minimization [26,27]. However, if a specific DMU greatly outperforms its previous performance, that DMU may be viewed as an outlier and its performance will be ignored in these two models.

Edirisinghe and Zhang [22] proposed a complicated multistep heuristic algorithm with random sampling and local search that automatically selected a combination of inputs and outputs, in which the emerging DEA measure of financial strength is maximally correlated with stock performance. The algorithm generated a relative financial strength indicator that was demonstrated to be predictive of stock returns. As its major contribution, this method demonstrates flexibility and automation in the selection of input and output parameters to maximize the predictive ability of the emerging DEA estimation of stock performance. Although this approach uncovers the "black box" of the forecasting mechanism (i.e., NN), this approach cannot easily determine a "suitable" solution beforehand.

Efficiency evaluation through DEA has been widely applied in numerous empirical cases. However, this technique does not determine the extent to which asymmetric information is still relevant in efficiency prediction, which has been rarely questioned empirically. Previous approaches to efficiency prediction have not considered the appropriate forecasting method and prediction variables and have consequently suffered from the influences of variable variance [23], computing resources/efficiency (e.g., [22,25]), and data challenges (e.g., extension of the Malmquist index to the forecasting problem). The present chapter demonstrates a model that assures prediction efficiency using fuzzy piecewise autoregression and the catching-up index of Yu *et al.* [2,28].

27.3 MODELING AND FORMULATION

27.3.1 Notation

Before describing the notation, we describe the concepts of the methodology as follows. First, any measurement technique based on DEA can evaluate the efficiency performance of each DMU in each period. Therefore, the efficiency score of each DMU in each period is computed. To calculate the improvement or decay of the efficiency score, we determine the CIE for two adjacent periods. If the CIE of a specific DMU is larger than one, this CIE represents an improvement in the performance efficiency of that DMU from the base period to the calculation period. Otherwise, this

CIE represents decay in performance efficiency. Afterwards, the CIE of each DMU for these periods is forecasted according to the previous CIE efficiency performance. These CIE datasets are used as the input to a fuzzy piecewise autoregression. This autoregression identifies two ranges for future forecasts using two specific regression models for each DMU. The possibility estimation model suggests that the predicted values must be included in the regression range. The necessity estimation model suggests that the predicted values must be excluded from the regression range. After calculating the two ranges from these two regression models, we can obtain four CIE coefficients within the two ranges for each DMU. Using these coefficients allows us to forecast the future efficiency performance for each DMU.

Following the above methods, the approach illustrated here can be implemented in four phases. In the first phase, DEA is used to evaluate the efficiency score (e.g., the amount of deposits (input) that need to be invested to produce a given output) of each DMU in each period. In the second phase, the efficiency score of each DMU is applied to calculate the CIE. In the third phase, two regression models are built using fuzzy piecewise autoregression. In the fourth phase, the calculations and forecasting are validated and subsequently applied in the regression model obtained in the third phase. Table 27.1 shows the notation for the proposed model.

This chapter proposes a forecasting method that uses fuzzy piecewise autoregression and the CIE to predict efficiency and aid in strategic decision making. This section introduces the modeling concepts that are employed for efficiency prediction/forecasting, including those that focus on fuzzy piecewise regression and the CIE.

27.3.2 Phase I: Efficiency Evaluation

Based on the notation in Section 27.3.1, we assume T periods and N DMUs, with each DMU (DMU$_j$, where $j \in R_+^N$) having an input $X_{aj}^{(t)}$ ($a \in R_+^{n_a}$) in period t ($t \in R_+^T$) and an output $Y_{bj}^{(t)}$ ($b \in R_+^{n_b}$) in period t. The technology can then be described generally by the output sets as shown in the model (27.1):

$$P^{(t)}(x) = \left\{ \left(X^{(t)}, Y^{(t)} \right) : \sum_{j=1}^{N} \lambda_j^{(t)} Y_{bj}^{(t)} \geq y_{br}^{(t)}, b = 1, \ldots, n_b; r = 1, \ldots, N \right.$$

$$\sum_{j=1}^{N} \lambda_j^{(t)} X_{aj}^{(t)} \leq x_{ar}^{(t)}, a = 1, \ldots, n_a; r = 1, \ldots, N$$

$$\left. \lambda_j^{(t)} \geq 0, j = 1, \ldots, N; t = 1, \ldots, T \right\} \tag{27.1}$$

where $(X_{ar}^{(t)}, Y_{br}^{(t)})$ represents DMU$_r$, the ath input, and the bth output in the tth period. $\lambda_j^{(t)}$ is an intensity variable that shrinks or expands the individually observed activities of DMU$_k$ and constructs the convex combinations of the observed input and output in the tth period. The CCR measure is depicted in the model (27.2):

TABLE 27.1 Description of notation.

Variable/notation	Definition/item
N	Number of DMUs
T	Number of periods
P	Number of change points
n_a	Number of input variables
n_b	Number of output variables.
$X_{aj}^{(t)}$	Vector of ath specific input variables of the jth DMU for the tth period
$Y_{bj}^{(t)}$	bth specific output variable of the jth DMU for the tth period
$j\ (j=1,\ldots,N)$	Indices for DMUs
$k\ (k=1,\ldots,N)$	Indices for DMUs
$t\ (t=1,\ldots,T)$	Indices for periods
$a\ (a=1,\ldots,n_a)$	Indices for input variables
$b\ (b=1,\ldots,n_b)$	Indices for output variables
$p\ (p=1,\ldots,P)$	Indices for change points
$p_j^{(t)}$	Efficiency score of the jth DMU for the tth period
$\lambda_j^{(t)}$	Vector for projecting DMU$_j$ for the tth period
$\delta_j^{t,t-1}$	Catching-up index of the jth DMU for the tth and $(t-1)$th periods
ρ_j^{U}	Upper bound of the possibility regression prediction of the CIE values of the jth DMU
ρ_j^{L}	Lower bound of the possibility regression prediction of the CIE values of the jth DMU
π_j^{U}	Upper bound of the necessity regression prediction of the CIE values of the jth DMU
π_j^{L}	Lower bound of the necessity regression prediction of the CIE values of the jth DMU
$\xi_{j,t}^{(U)}$	Upper bound of the possibility regression prediction of the efficiency value of the jth DMU for the tth period
$\xi_{j,t}^{(L)}$	Lower bound of the possibility regression prediction of the efficiency value of the jth DMU for the tth period
$\psi_{j,t}^{(U)}$	Upper bound of the necessity regression prediction of the efficiency value of the jth DMU for the tth period
$\psi_{j,t}^{(L)}$	Lower bound of the necessity regression prediction of the efficiency value of the jth DMU for the tth period

$$\text{Min}\ \rho_k^{(t)}$$

$$\sum_{j=1}^{N}\lambda_j^{(t)}Y_{bj}^{(t)} \ge y_{bk}^{(t)}, b=1,\ldots,n_b;\ k=1,\ldots,N$$

$$\sum_{j=1}^{N}\lambda_j^{(t)}X_{aj}^{(t)} \le \rho_k^{(t)}x_{ak}^{(t)}, a=1,\ldots,n_a;\ k=1,\ldots,N \tag{27.2}$$

$$\lambda_j^{(t)} \ge 0, j=1,\ldots,N;\ t=1,\ldots,T$$

27.3.3   Phase II: CIE

After Phase I, we can calculate the periods from 1 to $t+1$. We use two periods here to demonstrate the procedure. Two periods, t and $t+1$, are defined after Phase I to measure the productivity change of DMU_k. Based on the model (27.2), the efficiency scores of DMU_k for the two periods, ρ_k^t and ρ_k^{t+1}, can be obtained. We then calculate the CIE between periods t and $t+1$ as shown in (27.3). The CIE is the ratio of the efficiency scores for periods t and $t+1$, which measures the change in technical efficiency for the periods analyzed (the catching-up effect or movement toward the frontier). CIE > 1 represents efficiency improvement, whereas CIE < 1 represents efficiency regression:

$$\delta_k^{t,t+1} = \frac{\rho_k^{(t+1)}}{\rho_k^{(t)}} \tag{27.3}$$

After Phase II, we obtain a number T of CIE data items. A number $T-1$ of these data items form the independent variables of the fuzzy piecewise regression, and the Tth data item forms the dependent variable.

27.3.4   Phase III: Fuzzy Piecewise Regression

Fuzzy regression analysis can be interpreted as an interval estimation of dependent variables [1,29,30]. Generally, an interval that covers all training data is calculated, and a membership function is constructed based on this interval. The effect of a quadratic function is the same as that of a linear one, and we adopt the linear form instead of quadratic programming (QP) in Phase III for the purposes of illustration [31]. After Phase II, let $\delta_k^{t-1,t}$ represent the dependent variables and let $\delta_k^{t-2,t-1}, \delta_k^{t-3,t-2}, \ldots, \delta_k^{1,2}$ represent the independent variables of a forecasting function for DMU_k. From the dataset for period T, we obtain one dependent variable $\delta_k^{t-1,t}$ and $T-2$ independent variables. The interval linear regression model for DMU_k with an output (dependent variables) that is calculated from all data (independent variables) is represented as follows:

$$\delta_k^{t-1,t} = A_0 + A_1 \delta_k^{t-2,t-1} + \cdots + A_{T-3} \delta_k^{1,2} \tag{27.4}$$

where $\delta_k^{t-1,t}$ is DMU_k, which is the predicted interval corresponding to the input vector $(\delta_k^{t-2,t-1}, \delta_k^{t-3,t-2}, \ldots, \delta_k^{1,2})$, and t is the index for time $(t=1,\ldots,T)$.

In short, $(\delta_k^{t-2,t-1}, \delta_k^{t-3,t-2}, \ldots, \delta_k^{1,2})$ is a one-dimensional input vector for DMU_k, which represents the CIE for two adjacent periods. An interval that is defined by an ordered pair in brackets is represented as follows:

$$A = [a_L, a_R] = [a : a_L < a < a_R] \tag{27.5}$$

where a_L denotes the left limit and a_R denotes the right limit of A.

Interval A is likewise denoted by its center and width (or radius) as shown below:

$$A = (a_c, a_w) = \{a : a_c - a_w \le a \le a_c + a_w\} \qquad (27.6)$$

where a_c denotes the center and a_w denotes the width (where, for example, it may have a radius of $a_w \ge 0$, similar to half the width of A).

From (27.5) and (27.6), the center and width of interval A can be calculated as in (27.7) and (27.8) [31]:

$$a_c = \frac{(a_R + a_L)}{2} \qquad (27.7)$$

$$a_w = \frac{(a_R - a_L)}{2} \qquad (27.8)$$

The linear model of (27.4) is represented in (27.9)–(27.11):

$$
\begin{aligned}
\delta_k^{t-1,t} &= A_o + A_1 \delta_k^{t-2,t-1} + \cdots + A_{T-2} \delta_k^{1,2} \\
&= (a_{0c,k}, a_{0w,k}) + (a_{1c,k}, a_{1w,k}) \delta_k^{t-2,t-1} + \cdots + (a_{T-2c.k}, a_{T-2w,k}) \delta_k^{1,2} \qquad (27.9) \\
&= (Y_{kc}, Y_{kw})
\end{aligned}
$$

$$Y_{kc} = a_{0c,k} + a_{1c,k} \delta_k^{t-2,t-1} + \cdots + a_{T-2c,k} \delta_k^{1,2} \qquad (27.10)$$

$$Y_{kw} = a_{0w,k} + a_{1w,k} \left| \delta_k^{t-2,t-1} \right| + \cdots + a_{T-2w,k} \left| \delta_k^{1,2} \right| \qquad (27.11)$$

where Y_{kc} represents the center and Y_{kw} represents the width of the predicted interval $\delta_k^{t-1,t}$ of DMU$_k$.

The two estimation models (i.e., the possibility and necessity estimation models) are considered for the input–output data $\left(\delta_k^{t-2,t-1}, \delta_k^{t-3,t-2}, \ldots, \delta_k^{1,2}; \delta_k^{t-1,t} \right)$. First, the possibility estimation model can be represented as in (27.12):

$$
\begin{aligned}
\left(\delta_k^{t-1,t} \right)^* &= A_0^* + A_1^* \left(\delta_k^{t-2,t-1} \right)^* + \cdots + A_{T-2}^* \left(\delta_k^{1,2} \right)^* \\
&= \left(a_{0c,k}^*, a_{0w,k}^* \right) + \left(a_{1c,k}^*, a_{1w,k}^* \right) \delta_k^{t-2,t-1} + \cdots + \left(a_{T-2c,k}^*, a_{T-2w,k}^* \right) \delta_k^{1,2} \\
&= \left(Y_{kc}^*, Y_{kw}^* \right)
\end{aligned}
$$

$$(27.12)$$

which satisfies the conditions of the following model:

$$\delta_k^{t-1,t} \subseteq \left(\delta_k^{t-1,t} \right)^*, \quad t = 1, \ldots, T \qquad (27.13)$$

where A_t^* is the interval of the possibility estimation model and $\delta_k^{t-1,t}$ is the observed interval for DMU_k.

The interval $\left(\delta_k^{t-1,t}\right)^*$ estimated by the possibility model always includes the observed interval $\delta_k^{t-1,t}$. The width of the predicted interval Y_{kw}^* in the possibility regression analysis is minimized and includes all observed data.

Second, the necessity estimation model can be represented as in (27.14):

$$
\begin{aligned}
\left(\delta_k^{t-1,t}\right)_* &= A_{0*} + A_{1*}\left(\delta_k^{t-2,t-1}\right)_* + \cdots + A_{T-2*}\left(\delta_k^{1,2}\right)_* \\
&= (a_{0c*,k}, a_{0w*,k}) + (a_{1c*,k}, a_{1w*,k})\delta_k^{t-2,t-1} + \cdots + (a_{T-2c*,k}, a_{T-2w*,k})\delta_k^{1,2} \\
&= (Y_{kc*}, Y_{kw*})
\end{aligned}
$$

(27.14)

which satisfies the conditions of the following model:

$$
\left(\delta_k^{t-1,t}\right)_* \subseteq \delta_k^{t-1,t}, \quad t = 1, \ldots, T
$$

(27.15)

A_{t*} is the interval of the necessity model and $\delta_k^{t-1,t}$ is the observed interval for DMU_k. The interval $\left(\delta_k^{t-1,t}\right)_*$ estimated by the necessity model must include the observed interval $\delta_k^{t-1,t}$. The width of the predicted interval Y_{kw*} is maximized in the necessity regression analysis and includes all observed data. The relation for DMU_k can be expressed as in the model (27.16):

$$
\left(\delta_k^{t-1,t}\right)_* \subseteq \delta_k^{t-1,t} \subseteq \left(\delta_k^{t-1,t}\right)^*
$$

(27.16)

The two regressions are also depicted in Figure 27.1.

Based on the discussion, the possibility estimation model can be formulated as the model (27.17):

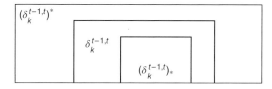

Figure 27.1   Relationships between possibility and necessity models.

$$\min \sum_{k=1}^{N} \left(a_{0\mathrm{w},k}^* + a_{1\mathrm{w},k}^* \left| \delta_k^{t-2,t-1} \right| + \cdots + a_{T-2\mathrm{w},k}^* \left| \delta_k^{1,2} \right| \right)$$

s.t.

$$\left(a_{0\mathrm{c},k}^* + \sum_{i=0}^{t-2} a_{i\mathrm{c},k}^* \delta_k^{t-i-2,t-i-1} \right) - \left(a_{0\mathrm{w},k}^* + \sum_{i=0}^{t-2} a_{i\mathrm{w},k}^* \delta_k^{t-i-2,t-i-1} \right) \le \delta_k^{t-1,t} - \varepsilon, \quad t=1,2,\ldots,T;$$

$$\left(a_{0\mathrm{c},k}^* + \sum_{i=0}^{t-2} a_{i\mathrm{c},k}^* \delta_k^{t-i-2,t-i-1} \right) + \left(a_{0\mathrm{w},k}^* + \sum_{i=0}^{t-2} a_{i\mathrm{c},k}^* \delta_k^{t-i-2,t-i-1} \right) \ge \delta_k^{t-1,t} + \varepsilon, \quad t=1,2,\ldots,T;$$

$$\forall a_{i\mathrm{w},k}^* \ge 0; \quad i=0,1,\ldots,t-2; \quad k=1,\ldots,N \tag{27.17}$$

where ε is represented by a small non-Archimedean quantity.

The necessity regression analysis introduced in (27.14) can be rewritten as the model (27.18):

$$\min \sum_{k=1}^{N} \left(a_{0\mathrm{w}*,k} + a_{1\mathrm{w}*,k} \left| \delta_k^{t-2,t-1} \right| + \cdots + a_{T-2\mathrm{w}*,k} \left| \delta_k^{1,2} \right| \right)$$

s.t.

$$\left(a_{0\mathrm{c}*,k} + \sum_{i=0}^{t-2} a_{i\mathrm{c}*,k} \delta_k^{t-i-2,t-i-1} \right) + \left(a_{0\mathrm{w},k*} + \sum_{i=0}^{t-2} a_{i\mathrm{w}*,k} \delta_k^{t-i-2,t-i-1} \right) \le \delta_k^{t-1,t} - \varepsilon, \quad t=1,2,\ldots,T;$$

$$\left(a_{0\mathrm{c}*,k} + \sum_{i=0}^{t-2} a_{i\mathrm{c}*,k} \delta_k^{t-i-2,t-i-1} \right) - \left(a_{0\mathrm{w}*,k} + \sum_{i=0}^{t-2} a_{i\mathrm{c}*,k} \delta_k^{t-i-2,t-i-1} \right) \ge \delta_k^{t-1,t} + \varepsilon, \quad t=1,2,\ldots,T;$$

$$\forall a_{i\mathrm{w}*,k} \ge 0; \quad i=0,1,\ldots,t-2; \quad k=1,\ldots,N \tag{27.18}$$

The linear programming formulation of the necessity analysis may be infeasible because of large fluctuations in the data. The fuzzy regression is then extended to fuzzy piecewise regression and adapted to our framework. Fuzzy piecewise regression analysis was developed and validated by Yu et al. [1,28]. We follow the description and notation in Section 27.3.1. A linear programming formulation is used to determine the necessity area. The piecewise linear interval model is presented in this subsection for linear piecewise regression, which is commonly observed in forecasting, as shown in (27.19). The possibility of the piecewise linear model is presented in (27.20):

$$\left(\delta_k^{t-1,t-2} \right)_* = h \left(\delta_k^{t-2,t-1} \right) + \sum_{p=1}^{P-1} \left\{ \frac{B_p^*}{2} \left(\left| \delta_k^{t-2-p,t-1-p} - P_p \right| + \delta_k^{t-2-p,t-1-p} - P_p \right) \right\} \tag{27.19}$$

$$\left(\delta_k^{t-1,t-2} \right)^* = h \left(\delta_k^{t-2,t-1} \right) + \sum_{p=1}^{P-1} \left\{ \frac{B_{p*}}{2} \left(\left| \delta_k^{t-2-p,t-1-p} - P_p \right| + \delta_k^{t-2-p,t-1-p} - P_p \right) \right\} \tag{27.20}$$

where $h\left(\delta_k^{t-2,t-1}\right) = a_{0*} + a_{1*}\delta_k^{t-2,t-1}$, and B_p^* is the interval of the necessity model of B_p. $B_p = \left(B_{pc}, B_{pw}\right)$ represents the center and radius of B_p.

If P_p is a change point, then

$$\frac{\left(|\delta_k^{t-2-p,t-1-p} - P_p| + \delta_k^{t-2-p,t-1-p} - P_p\right)}{2} = \begin{cases} \delta_k^{t-2-p,t-1-p} - P_p, & \text{if } \delta_k^{t-2-p,t-1-p} \geq P_p \\ 0 & , \text{if } \delta_k^{t-2-p,t-1-p} < P_p \end{cases}$$

(27.21)

where $P_p = \{P_1, \ldots, P_p, \ldots, P_{N-2}\}$ $(p \in R_+^{N-2})$ are the values of variables $\delta_k^{t-2-p,t-1-p}$ and are subject to an ordering constraint $P_1 < P_2 < \ldots < P_p$ $(p \leq N-2)$.

The necessity and possibility of the piecewise expression for the data are represented in (27.22) and (27.23):

$$\sum_{p=1}^{P-1}\left\{\frac{B_{p*}}{2}\left(\left|\delta_k^{t-2-p,t-1-p} - P_p\right| + \delta_k^{t-2-p,t-1-p} - P_p\right)\right\}$$

$$= \sum_{p=1}^{P-1}\left\{\frac{B_{pc*}}{2}\left(\left|\delta_k^{t-2-p,t-1-p} - P_p\right| + \delta_k^{t-2-p,t-1-p} - P_p\right)\right\}$$

(27.22)

$$+ \sum_{p=1}^{P-1}\left\{\frac{B_{pw*}}{2}\left(\left|\delta_k^{t-2-p,t-1-p} - P_p\right| + \delta_k^{t-2-p,t-1-p} - P_p\right)\right\}$$

$$\sum_{p=1}^{P-1}\left\{\frac{B_p^*}{2}\left(\left|\delta_k^{t-2-p,t-1-p} - P_p\right| + \delta_k^{t-2-p,t-1-p} - P_p\right)\right\}$$

$$= \sum_{p=1}^{P-1}\left\{\frac{B_{pc}^*}{2}\left(\left|\delta_k^{t-2-p,t-1-p} - P_p\right| + \delta_k^{t-2-p,t-1-p} - P_p\right)\right\}$$

(27.23)

$$+ \sum_{p=1}^{P-1}\left\{\frac{B_{pw}^*}{2}\left(\left|\delta_k^{t-2-p,t-1-p} - P_p\right| + \delta_k^{t-2-p,t-1-p} - P_p\right)\right\}$$

Following the previous discussion, the fuzzy piecewise QP formula for analysis is represented by the model (27.24):

$$\min \sum_{K=1}^{N}\left\{a_{0w*} + a_{1w*}\delta_k^{t-2,t-1} + \sum_{p=1}^{P-1}\left\{B_{pw*}\frac{\left(|\delta_k^{t-2-p,t-1-p} - P_p| + \delta_k^{t-2-p,t-1-p} - P_p\right)}{2}\right\}\right\}^2$$

subject to
(Possibility constraints)

$$a_{0c}^* + a_{1c}^* \delta_k^{t-2,t-1} + \sum_{p=1}^{P-1} \left\{ \frac{B_{pc}^*}{2} \left(\left| \delta_k^{t-2-p,t-1-p} - P_p \right| + \delta_k^{t-2-p,t-1-p} - P_p \right) \right\}$$

$$- \left\{ a_{0w}^* + a_{1w}^* \delta_k^{t-2,t-1} + \sum_{p=1}^{P-1} \left\{ \frac{B_{pw}^*}{2} \left(\left| \delta_k^{t-2-p,t-1-p} - P_p \right| + \delta_k^{t-2-p,t-1-p} - P_p \right) \right\} \right\} \leq \delta_k^{t-1,t-2} + \varepsilon,$$

$$a_{0c}^* + a_{1c}^* \delta_k^{t-2,t-1} + \sum_{p=1}^{P-1} \left\{ \frac{B_{pc}^*}{2} \left(\left| \delta_k^{t-2-p,t-1-p} - P_p \right| + \delta_k^{t-2-p,t-1-p} - P_p \right) \right\}$$

$$+ \left\{ a_{0w}^* + a_{1w}^* \delta_k^{t-2,t-1} + \sum_{p=1}^{P-1} \left\{ \frac{B_{pw}^*}{2} \left(\left| \delta_k^{t-2-p,t-1-p} - P_p \right| + \delta_k^{t-2-p,t-1-p} - P_p \right) \right\} \right\} \geq \delta_k^{t-1,t-2} - \varepsilon,$$

$$P \leq N-2, \forall k = 1, \ldots, N$$

(Necessity constraints)

$$a_{0c*} + a_{1c*} \delta_k^{t-2,t-1} + \sum_{p=1}^{P-1} \left\{ \frac{B_{pc*}}{2} \left(\left| \delta_k^{t-2-p,t-1-p} - P_p \right| + \delta_k^{t-2-p,t-1-p} - P_p \right) \right\}$$

$$- \left\{ a_{0w*} + a_{1w*} \delta_k^{t-2,t-1} + \sum_{p=1}^{P-1} \left\{ \frac{B_{pw*}}{2} \left(\left| \delta_k^{t-2-p,t-1-p} - P_p \right| + \delta_k^{t-2-p,t-1-p} - P_p \right) \right\} \right\} \geq \delta_k^{t-1,t} + \varepsilon,$$

$$a_{0c*} + a_{1c*} \delta_k^{t-2,t-1} + \sum_{p=1}^{P-1} \left\{ \frac{B_{pc*}}{2} \left(\left| \delta_k^{t-2-p,t-1-p} - P_p \right| + \delta_k^{t-2-p,t-1-p} - P_p \right) \right\}$$

$$+ \left\{ a_{0w*} + a_{1w*} \delta_k^{t-2,t-1} + \sum_{p=1}^{P-1} \left\{ \frac{B_{pw*}}{2} \left(\left| \delta_k^{t-2-p,t-1-p} - P_p \right| + \delta_k^{t-2-p,t-1-p} - P_p \right) \right\} \right\} \leq \delta_k^{t-1,t} - \varepsilon$$

$$P \leq N-2, \forall k = 1, \ldots, N \qquad (27.24)$$

27.3.5 Phase IV: Validating and Forecasting

We then calculate a_{0c}^*, a_{1c}^*, B_{pc}^*, and B_{pw}^* ($p \in R_+^P$, $P \leq N-1$). By substitution using (27.25) and (27.26), we determine two values for DMU_k, namely, the upper bound ρ_k^U (PRY) and the lower bound ρ_k^L (PLY). Any $\delta_k^{t-1,t-2}$ will depend on $[\rho_k^U, \rho_k^L]$:

$$\rho_k^{U} = \left(a_{0c}^* + a_{1c}^* \delta_k^{t-2,t-1} + \sum_{p=1}^{P-1} B_{pc}^* \delta_k^{t-2-p,t-1-p} \right) + \left\{ a_{0w}^* + a_{1w}^* \delta_k^{t-2,t-1} + \sum_{p=1}^{P-1} B_{pw}^* \delta_k^{t-2-p,t-1-p} \right\}$$

(27.25)

$$\rho_k^{L} = \left(a_{0c}^* + a_{1c}^* \delta_k^{t-2,t-1} + \sum_{p=1}^{P-1} B_{pc}^* \delta_k^{t-2-p,t-1-p} \right) - \left\{ a_{0w}^* + a_{1w}^* \delta_k^{t-2,t-1} + \sum_{p=1}^{P-1} B_{pw}^* \delta_k^{t-2-p,t-1-p} \right\}$$

(27.26)

Similarly, we calculate a_{0c*}, a_{1c*}, B_{pc*}, and B_{pw*} ($p \in R_+^P$, $P \le N - 1$). By substitution using (27.27) and (27.28), we determine two values for DMU_k, namely, the upper bound π_k^{U} (NRY) and the other lower bound π_k^{L} (NLY). For any $\delta_k^{t-1,t-2}$, $\delta_k^{t-1,t-2} \notin [\pi_k^{U}, \pi_k^{L}]$:

$$\pi_k^{U} = a_{0c*} + a_{1c*} \delta_k^{t-2,t-1} + \sum_{p=1}^{P-1} B_{pc*} \delta_k^{t-2-p,t-1-p} - \left\{ a_{0w*} + a_{1w*} \delta_k^{t-2,t-1} + \sum_{p=1}^{P-1} B_{pw*} \delta_k^{t-2-p,t-1-p} \right\}$$

(27.27)

$$\pi_k^{L} = a_{0c*} + a_{1c*} \delta_k^{t-2,t-1} + \sum_{p=1}^{P-1} B_{pc*} \delta_k^{t-2-p,t-1-p} + \left\{ a_{0w*} + a_{1w*} \delta_k^{t-2,t-1} + \sum_{p=1}^{P-1} B_{pw*} \delta_k^{t-2-p,t-1-p} \right\}$$

(27.28)

For any DMU_k, we check if the four values satisfy $\rho_k^{U} \ge \pi_k^{U} \ge \pi_k^{L} \ge \rho_k^{L}$. If satisfied, the efficiency values for period $p - 1$, ($p_k^{(t-1)}$) to ρ_k^{U}, π_k^{U}, π_k^{L}, and ρ_k^{L}, are multiplied to obtain the four efficiency values ($\xi_{k,t}^{(U)}$, $\psi_{k,t}^{(U)}$, $\psi_{k,t}^{(L)}$, and $\xi_{k,t}^{(L)}$) for the tth period. These values are represented as in (27.29)–(27.32):

$$\xi_{k,t}^{(U)} = p_k^{(t-1)} \times \rho_k^{U}$$

(27.29)

$$\xi_{k,t}^{(L)} = p_k^{(t-1)} \times \rho_k^{L}$$

(27.30)

$$\psi_{k,t}^{(U)} = p_k^{(t-1)} \times \pi_k^{U}$$

(27.31)

$$\psi_{k,t}^{(L)} = p_k^{(t-1)} \times \pi_k^{L}$$

(27.32)

After obtaining the four values, we check whether $p_k^{(t)} \in \left[\xi_{k,t}^{(U)}, \psi_{k,t}^{(U)} \right]$ (marked with "+" in the results to represent efficiency improvement) or $p_k^{(t)} \in \left[\psi_{k,t}^{(L)}, \xi_{k,t}^{(L)} \right]$ (marked with "−" to represent efficiency decay), and whether $p_n \notin \left(\psi_{k,t}^{(U)}, \psi_{k,t}^{(L)} \right)$. Other cases are marked "F" to represent failure of the analysis model. After validation, the time horizon is shifted from t to $t+1$ as shown in (27.33)–(27.36) to forecast the efficiency of each DMU:

$$\xi_{k,t+1}^{(U)} = p_k^{(t)} \times \left\{ \left(a_{0c}^* + a_{1c}^* \delta_k^{t-1,t} + \sum_{p=1}^{P-1} B_{pc}^* \delta_k^{t-1-p,t-p} \right) + \left\{ a_{0w}^* + a_{1w}^* \delta_k^{t-1,t} + \sum_{p=1}^{P-1} B_{pw}^* \delta_k^{t-1-p,t-p} \right\} \right\}$$

(27.33)

$$\xi_{k,t+1}^{(L)} = p_k^{(t)} \times \left\{ \left(a_{0c}^* + a_{1c}^* \delta_k^{t-1,t} + \sum_{p=1}^{P-1} B_{pc}^* \delta_k^{t-1-p,t-p} \right) + \left\{ a_{0w}^* + a_{1w}^* \delta_k^{t-1,t} + \sum_{p=1}^{P-1} B_{pw}^* \delta_k^{t-1-p,t-p} \right\} \right\}$$

(27.34)

$$\psi_{k,t+1}^{(U)} = p_k^{(t)} \times \left\{ a_{0c*} + a_{1c*} \delta_k^{t-1,t} + \sum_{p=1}^{P-1} B_{pc*} \delta_k^{t-1-p,t-p} - \left\{ a_{0w*} + a_{1w*} \delta_k^{t-1,t} + \sum_{p=1}^{P-1} B_{pw*} \delta_k^{t-1-p,t-p} \right\} \right\}$$

(27.35)

$$\psi_{k,t+1}^{(L)} = p_k^{(t)} \times \left\{ a_{0c*} + a_{1c*} \delta_k^{t-1,t} + \sum_{p=1}^{P-1} B_{pc*} \delta_k^{t-1-p,t-p} - \left\{ a_{0w*} + a_{1w*} \delta_k^{t-1,t} + \sum_{p=1}^{P-1} B_{pw*} \delta_k^{t-1-p,t-p} \right\} \right\}$$

(27.36)

27.4 ILLUSTRATING THE APPLICATION

A panel dataset was obtained from the 1997 to 2002 annual statistical reports of the National Federation of Bus Passenger Transportation of the Republic of China, which included 17 bus transit firms. All observations were referred to the largest possibility for Taiwanese bus transit firms. The two output variables were the passenger-kilometers (y_1, in 1000 passenger-km) and the vehicle-kilometers (y_2, in 1000 vehicle-km). Three inputs were used, namely, the number of drivers (x_1, in persons), the number of vehicles (x_2, in vehicles), and the liters of fuel (x_3, in 1000 liters). The descriptive statistics of all of the variables are presented in Table 27.2.

27.4.1 Efficiency Evaluations

The efficiency scores based on the evaluation of the model (27.2) in Section 27.3.1 are summarized in Table 27.3. Only DMUs 1 and 11 are efficient from 1997 to 2002.

Table 27.4 reports the CIE and the results of application of (27.3) from 1992 to 2002. If the cell values are larger than 1, they represent an adjacent period of efficiency improvement. Otherwise, they show a decay.

TABLE 27.2 Summary statistics of inputs and outputs by year.

Period	Variable	Mean	St. dev.	Max	Min	Period	Variable	Mean	St. dev.	Max	Min
1997	y_1	9 523	14 131	49 785	126	1998	y_1	10 534	16 979	63 997	107
	y_2	255 310	244 516	1 039 163	17 393		y_2	248 970	255 577	1 124 677	19 287
	x_1	75	79	271	2		x_1	93	122	500	4
	x_2	66	63	195	2		x_2	77	89	351	4
	x_3	1 616	2 097	7 451	4		x_3	1 625	2 102	7 463	44
1999	y_1	12 019	22 242	87 148	106	2000	y_1	18 443	30 479	101 132	83
	y_2	236 691	296 873	1 294 168	18 641		y_2	165 138	143 453	520 935	17 353
	x_1	92	126	494	3		x_1	140	199	716	3
	x_2	78	95	351	3		x_2	115	141	514	3
	x_3	2 139	3 595	14 271	23		x_3	3 201	4 998	15 805	43
2001	y_1	20 187	32 451	105 092	77	2002	y_1	21 214	32 602	103 387	69
	y_2	160 460	138 732	495 947	10 821		y_2	160 260	132 251	455 990	6 397
	x_1	153	214	776	3		x_1	168	244	875	2
	x_2	122	154	557	3		x_2	132	172	628	2
	x_3	3 530	5 202	16 363	44		x_3	3 863	5 693	17 445	49

TABLE 27.3 Efficiency evaluations.

DMU (k)	$\rho_k^{(1997)}$	$\rho_k^{(1998)}$	$\rho_k^{(1999)}$	$\rho_k^{(2000)}$	$\rho_k^{(2001)}$	$\rho_k^{(2002)}$
1	1.0000	1.0000	1.0000	1.0000	1.0000	1.0000
2	0.9232	1.0000	0.9825	0.9925	1.0000	0.8152
3	0.8836	1.0000	1.0000	1.0000	1.0000	0.8388
4	0.6414	1.0000	0.1515	0.5953	0.3435	0.4687
5	0.8296	0.6974	0.8428	1.0000	1.0000	0.8500
6	1.0000	1.0000	1.0000	1.0000	0.9267	0.9918
7	0.6111	0.5632	0.6031	0.9099	0.8245	0.9624
8	0.8054	0.7429	0.8265	0.8488	0.9722	1.0000
9	1.0000	0.9342	0.9978	1.0000	1.0000	0.8429
10	1.0000	1.0000	0.6540	1.0000	1.0000	0.9797
11	1.0000	1.0000	1.0000	1.0000	1.0000	1.0000
12	0.4744	0.4987	0.5709	0.5620	0.5360	0.2999
13	0.4761	0.3573	0.5128	0.3164	0.5435	0.3793
14	0.6281	0.9346	1.0000	1.0000	0.7798	1.0000
15	0.2962	0.2391	0.3550	0.3566	0.3350	0.2567
16	0.5097	0.6085	0.5572	0.6670	0.8382	0.6319
17	0.4394	0.6034	0.4688	0.4989	0.6340	0.6704

TABLE 27.4 Catching-up index.

DMU (k)	$\delta_k^{1997,1998}$	$\delta_k^{1998,1999}$	$\delta_k^{1999,2000}$	$\delta_k^{2000,2001}$	$\delta_k^{2001,2002}$
1	1.0000	1.0000	1.0000	1.0000	1.0000
2	1.0832	0.9825	1.0102	1.0075	0.8152
3	1.1318	1.0000	1.0000	1.0000	0.8388
4	1.5590	0.1515	3.9286	0.5770	1.3645
5	0.8407	1.2083	1.1866	1.0000	0.8500
6	1.0000	1.0000	1.0000	0.9267	1.0702
7	0.9216	1.0708	1.5088	0.9062	1.1672
8	0.9224	1.1125	1.0270	1.1454	1.0286
9	0.9342	1.0680	1.0022	1.0000	0.8429
10	1.0000	0.6540	1.5291	1.0000	0.9797
11	1.0000	1.0000	1.0000	1.0000	1.0000
12	1.0511	1.1448	0.9845	0.9537	0.5595
13	0.7505	1.4352	0.6171	1.7175	0.6979
14	1.4878	1.0700	1.0000	0.7798	1.2824
15	0.8071	1.4847	1.0045	0.9395	0.7663
16	1.1938	0.9157	1.1971	1.2567	0.7539
17	1.3733	0.7770	1.0642	1.2709	1.0573

27.4.2 Validation

The CIE ranges used in the validation are summarized in Table 27.5. The ranges for $\delta_k^{1997,1998}$, $\delta_k^{1998,1999}$, and $\delta_k^{1999,2000}$ ($k = 1, \ldots, 17$) were taken as the dependent variables, and $\delta_k^{2001,2000}$ ($k = 1, \ldots, 17$) was taken as an independent variable. The possibility estimation and necessity estimation models were obtained based on the model (27.25) as follows:

$$
\begin{aligned}
\delta_k^{20012000} &= [0.623587, 0, 0.130406] + [0,0,0.016027] * \delta_k^{1997,1998} \\
&\quad + [0.220767, 0, 0.012891] \delta_k^{1998,1999} \\
&\quad + [0,0,0.000676] * \delta_k^{1999,2000} + [\sigma_k, \nu_k, \upsilon_k]
\end{aligned}
\tag{27.37}
$$

where [0.623587, 0, 0.130406] represents a center that is located at 0.623587. The radius in the necessity estimation model is equal to zero, whereas the radius of the possibility estimation model is equal to 0.13406. In (27.37), the necessity model provides crisp values instead of ranges, which indicates that the upper- and lower-bound necessities are similar. Following the previous discussion, the results for $[\sigma_k, \nu_k, \upsilon_k]$ can be represented as follows:

TABLE 27.5   Validation of the CIE (PRY, NRY, NRL, and PLY).

DMU (k)	ρ_k^{U}	π_k^{U}	π_k^{L}	ρ_k^{L}	$\delta_k^{2001,2002}$	Trend
1	1.1667	1.0067	1.0067	0.8467	1.0000	+
2	1.2052	1.0441	1.0441	0.8829	1.0075	−
3	1.2009	1.0388	1.0388	0.8767	1.0000	−
4	0.8170	0.6570	0.6570	0.4970	0.5770	−
5	1.2403	1.0800	1.0800	0.9197	1.0000	−
6	1.1667	1.0067	1.0067	0.8467	0.9267	−
7	1.1462	0.9862	0.9862	0.8262	0.9062	−
8	1.2256	1.0654	1.0654	0.9052	1.1454	+
9	1.1857	1.0259	1.0259	0.8661	1.0000	−
10	1.1199	0.9640	0.9640	0.8081	1.0000	+
11	1.1667	1.0067	1.0067	0.8467	1.0000	−
12	1.1964	1.0337	1.0337	0.8710	0.9537	−
13	1.7988	1.6375	1.6375	1.4761	1.7175	+
14	1.0285	0.8598	0.8598	0.6911	0.7798	−
15	1.1826	1.0195	1.0195	0.8563	0.9395	−
16	1.3389	1.1767	1.1767	1.0145	1.2567	+
17	1.3541	1.1909	1.1909	1.0278	1.2709	+

$$[\sigma, \nu, \upsilon] = \left\{ \begin{array}{l} \sum_{p=1}^{P-1} \dfrac{B_{pc}}{2} \left(\left| \delta_k^{t-1-p} {}^{t-2-p} - P_p \right| + \delta_k^{t-1-p} {}^{t-2-p} - P_p \right), \sum_{p=1}^{P-1} \dfrac{B_{pw}^*}{2} \left(\left| \delta_k^{t-1-p} {}^{t-2-p} - P_p \right| + \delta_k^{t-1-p} {}^{t-2-p} - P_p \right), \\[2em] \sum_{p=1}^{P-1} \dfrac{B_{pw*}}{2} \left(\left| \delta_k^{t-1-p} {}^{t-2-p} - P_p \right| + \delta_k^{t-1-p} {}^{t-2-p} - P_p \right) \end{array} \right\}$$

$$= \begin{bmatrix} 0.1624 & 0.0000 & 0.0000 \\ 0.2036 & 0.0000 & 0.0000 \\ 0.1944 & 0.0000 & 0.0000 \\ 0.0000 & 0.0000 & 0.0000 \\ 0.1896 & 0.0000 & 0.0000 \\ 0.1624 & 0.0000 & 0.0000 \\ 0.1262 & 0.0000 & 0.0000 \\ 0.1962 & 0.0000 & 0.0000 \\ 0.1665 & 0.0000 & 0.0000 \\ 0.1960 & 0.0000 & 0.0000 \\ 0.1624 & 0.0000 & 0.0000 \\ 0.1574 & 0.0000 & 0.0000 \\ 0.6970 & 0.0000 & 0.0000 \\ 0.0000 & 0.0000 & 0.0000 \\ 0.0681 & 0.0000 & 0.0000 \\ 0.3510 & 0.0000 & 0.0000 \\ 0.3598 & 0.0000 & 0.0000 \end{bmatrix}$$

$$(27.38)$$

The first four columns of Table 27.5 report the regression variables except for the first column, the fifth column shows the independent variable, and the sixth column shows the actual data from Table 27.4. In the last column of Table 27.5, if $\delta_k^{2001,2002}$ lies between ρ_k^{U} and π_k^{U}, this variable is represented by "+." In contrast, if $\delta_k^{2001,2002}$ lies between π_k^{L} and ρ_k^{L}, it is represented by "–." Figure 27.2 reports the range of the possibility and regression values $\delta_k^{2001,2002}$.

27.4.3 Forecasting

The 2002 efficiency scores needed to be forecasted after validation. Based on (27.38), the 2001–2002 period was moved such that its forecast was represented as in (27.39):

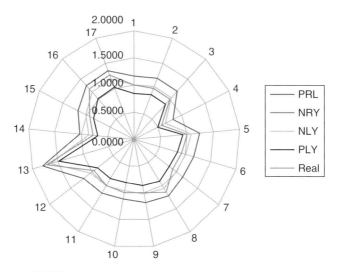

Figure 27.2 $\delta_k^{2001,2002}$, PRY, NRY, NLY, and PLY obtained by fuzzy piecewise autoregression.

$$p_k^{(2002)} = \left\{ \begin{array}{l} [0.623587,0,0.130406] + [0,0,0.016027]*\delta_k^{1998,1999} + [0.220767,0,0.012891]\delta_k^{1999,2000} \\ + [0,0,0.000676]*\delta_k^{2001,2002} + [\sigma_k,\nu_k,\upsilon_k] \end{array} \right\} *p_k^{(2001)}$$

(27.39)

As previously discussed, "+", "−", and "F" were used in the forecast results (Table 27.6) to represent the real efficiency, which lies between $\xi_{2002}^{(U)}$ and $\psi_{2002}^{(U)}$ or between $\psi_{2002}^{(L)}$ and $\xi_{k,2007}^{(L)}$, and in other cases. Table 27.6 shows that the accuracy rate was approximately 87%. Figure 27.3 shows the forecasting results. The values for validation and forecasting must have the same shape, except for the DMUs that are marked with "F" in Table 27.6.

27.5 DISCUSSION

A forecasting method using a hybrid of the CIE and fuzzy piecewise autoregression to resolve issues in the selection of variables and methodologies has been presented in this chapter. The CIE in the case of variable selection is used as a dependent or independent variable to forecast actual scenarios in place of absolute variables. For 2002, $\xi_{2002}^{(U)}$ denotes the optimal efficiency score of DMU_k, $\xi_{2002}^{(L)}$ denotes the pessimistic efficiency score of DMU_k, and $\psi_{2002}^{(U)}$ (or $\psi_{2002}^{(L)}$) denotes the highest possibility efficiency score of DMU_k. Table 27.6 shows that in our approach, 10 DMUs have reached the frontier in the optimal view, but only six DMUs have reached the frontier in the actual

TABLE 27.6   Forecasting efficiency for 2002: PRY, NRY, NRY, and NLY.

DMU (k)	$\xi_{2002}^{(U)}$	$\psi_{2002}^{(U)}$	$\psi_{2002}^{(L)}$	$\xi_{2002}^{(L)}$	$p_k^{(2002)}$	Trend
1	1.0000	1.0000	1.0000	0.8467	1.0000	–
2	1.0000	1.0000	1.0000	0.8903	1.0000	–
3	1.0000	1.0000	1.0000	0.8788	1.0000	–
4	0.5753	0.5121	0.5121	0.4490	0.3435	F
5	1.0000	1.0000	1.0000	0.9094	1.0000	–
6	1.0000	0.9330	0.9330	0.7847	0.9267	–
7	1.0000	0.8928	0.8928	0.7546	0.8245	–
8	1.0000	1.0000	1.0000	0.8597	0.9722	–
9	1.0000	1.0000	1.0000	0.8503	1.0000	–
10	1.0000	1.0000	1.0000	0.9959	1.0000	–
11	1.0000	1.0000	1.0000	0.8467	1.0000	–
12	0.6220	0.5351	0.5351	0.4482	0.5360	+
13	0.8801	0.7918	0.7918	0.7034	0.5435	F
14	0.7840	0.6584	0.6584	0.5329	0.7798	F
15	0.3622	0.3060	0.3060	0.2498	0.3350	+
16	1.0000	1.0000	1.0000	0.9032	0.8382	F
17	0.8951	0.7953	0.7953	0.6955	0.6340	F

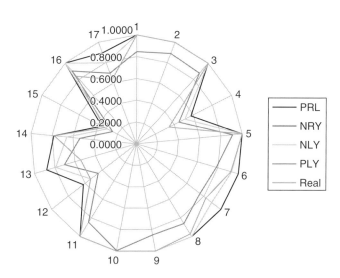

Figure 27.3 $p_k^{(2002)}$, PRY, NRY, NLY, and PLY obtained by fuzzy piecewise autoregression.

scenario (DMUs 8 and 16 are different in our analysis). The efficiency scores of these DMUs are close to our $\psi_{2002}^{(U)}$ (or $\psi_{2002}^{(L)}$) values.

Five DMUs (DMUs 4, 13, 14, 16, and 17) have failed the forecast. The efficiency scores of DMU 17 between 1997 and 2001 ranged from 0.4394 to 0.6340, but increased to 0.6704 in 2002. Fuzzy piecewise autoregression refers to past efficiency to regress the highest possible efficiency scores. Based on the historical data, the efficiency scores of DMU 17 cannot exceed 0.6704; the possibility regression predicts an optimal value of 0.8951 and the necessity regression predicts a pessimistic value of 0.6955. The two regressions cannot cover the actual value for DMU 17, which can be attributed to the poor performance of this DMU compared with the previous periods. The outputs of DMU 17 between 2001 and 2002 demonstrate that y_1 has increased by 13%, y_2 has increased by 0.4%, and x_3 has increased by 50%. The forecast for DMU 1 is always equal to one if the conventional regression approach is applied. The historical data for DMU 1 reflect the efficiency of DMU 5, which is always equal to one. However, the fuzzy piecewise autoregression provides a range from 0.8467 to 1, with values closer to 1 because of the catching-up effect. Although the CIE does not show input and output relationships, prior concepts about efficiency scores are compared with frontiers or best practices. In other words, the efficiency of DMU 1 may or may not be at the frontier. Therefore, these two ranges elaborate on the idea that the efficiency scores are compared with the frontier. The fuzzy ranges (i.e., between $\xi_{k,t}^{(U)}$ and $\xi_{k,t}^{(L)}$) provide us with relative concepts and the highest possible efficiency ($\xi_{k,t}^{(U)}$, $\xi_{k,t}^{(L)}$). We can also define the following features to highlight the advantages for future studies:

1. Efficiency range of possibility: $F_1 = \rho_k^U - \rho_k^L$.
2. Efficiency range of upper trend: $F_2 = \rho_k^U - \pi_k^U$.
3. Efficiency range of lower trend: $F_3 = \pi_k^L - \rho_k^L$.
4. Gap between real efficiency and lower-bound possibility efficiency: $F_4 = \lambda_k^{(t)} - \rho_k^L$.
5. Gap between upper-bound possibility efficiency and real efficiency: $F_5 = \rho_k^U - \lambda_k^{(t)}$.
6. Gap between real efficiency and necessity efficiency: $F_6 = \max\left(\lambda_k^{(t)} - \pi_k^U, \pi_k^L - \lambda_k^{(t)}\right)$.
7. Efficiency range of necessity: $F_7 = \pi_k^U - \pi_k^L$.

27.6   CONCLUSION

This chapter has discussed two hybrid methodological developments to show how efficiency in DEA can be used in forecasting. The proposed method has two advantages. First, the CIE shows the relative efficiency of two adjacent periods and avoids the direct usage of input and output variables. Therefore, the CIE not only provides a priori relative concepts about frontiers and best practices but also shows the possible efficiency. Second, historical data are used to regress the possibility and necessity estimation models in place of a random-error-type regression model. The four ranges obtained provide decision-makers with suggestions for specific DMUs (i.e., if a specific DMU does not reach the frontier in the current period, the DMU can exert more effort to reach the frontier in the future).

However, our analyses have several limitations. First, efficiency prediction can be divided two parts, namely, efficiency shift and efficiency movement. Efficiency shift is chiefly caused by changes in technique, whereas efficiency movement is caused by changes in the ratio of inputs and outputs. Although our analysis addresses efficiency shift, we have failed to address efficiency movement. This issue will be evaluated in future work. Second, we have not explained the external effect of certain variables, such as government power, on the evaluation results. The DEA method can be applied for purposes ranging from evaluation to planning techniques. Further research can examine other concepts regardless of the method.

REFERENCES

[1] Yu, J.R., Tzeng, G.H., and Li, H.L. (1999) A general piecewise necessity regression analysis based on linear programming. *Fuzzy Sets and Systems*, **105**, 429–436.

[2] Yu, J.R., Tzeng, G.H., and Li, H.L. (2001) General fuzzy piecewise regression analysis with auto-change point detection. *Fuzzy Sets and Systems*, **119**, 247–257.

[3] Hsu, F.M. and Hsueh, C.C. (2008) Measuring relative efficiency of government sponsored R&D projects: A three-stage approach. *Evaluation and Program Planning*, **32**, 178–186.

[4] Yunos, J.M. and Hawdon, D. (1997) The efficiency of the national electricity board in Malaysia: An inter country comparison using DEA. *Energy Economics*, **19**, 255–269.

[5] Soteriou, A.C., Karahanna, E., Papanastasiou, C., and Diakourakis, M.S. (1998) Using DEA to evaluate the efficiency of secondary school: The case of Cyprus. *International Journal of Educational Management*, **12**, 65–73.

[6] Tyagi, P., Yadav, S.P., and Singh, S.P. (2009) Relative performance of academic departments using DEA with sensitivity analysis. *Evaluation and Program Planning*, **32**, 168–177.

[7] Chilingerian, J.A. and Sherman, H.D. (1990) Managing physician efficiency and effectives in providing hospital services. *Health Service Management Research*, **3**, 3–15.

[8] Ozcan, Y.A. (1995) Efficiency of hospital service production in local markets: The balance sheet of U.S. medical armament. *Social Economic Planning Sciences*, **29**, 139–150.

[9] Barr, R.S., Killgo, K.A., Siems, T.F., and Zimmel, S. (2002) Evaluating the productive efficiency and performance of U.S. commercial banks. *Managerial Finance*, **28**(8), 3–25.

[10] Chen, T.Y. and Yeh, T.L. (2000) A measurement of bank efficiency, ownership and productivity changes in Taiwan. *Service Industries Journal*, **20**, 95–109.

[11] Bowlin, W.F., Charnes, A., Cooper, W.W., and Sherman, H.D. (1985) Data envelopment and regression approaches to efficiency estimation and evaluation. *Annals of Operational Research*, **2**, 113–138.

[12] Schmidt, P. (1986) Frontier production functions. *Econometric Reviews*, **2**, 289–328.

[13] Thanassoulis, E. (1993) A comparison of regression analysis and data envelopment analysis as alternative methods for performance assessments. *Journal of the Operational Research Society*, **44**, 1129–1144.

[14] Ganley, J.A. and Cubbin, J.S. (eds) (1992) *Public Sector Efficiency Measurement: Applications of Data Envelopment Analysis*, Elsevier, Amsterdam.

[15] Caves, D.W., Christensen, L.R., and Diewert, W.E. (1982) The economic theory of index numbers and the measurement of input, output and productivity. *Econometrica*, **50**, 1393–1414.

[16] Megginson, W., Nash, R., and van Randenborgh, M. (1994) The financial and operating performance of newly privatized firms: An international empirical analysis. *Journal of Finance*, **49**, 403–452.

[17] Cook, W.D. and Seiford, L.M. (2009) Data envelopment analysis (DEA) – thirty years on. *European Journal of Operational Research*, **19**, 1–17.

[18] Färe, R., Grosskopf, S., and Lovell, C.A.K. (eds) (1994) *Production Frontiers*, Cambridge University Press, Cambridge.

[19] Törnqvist, L. (1936) The bank of Finland's consumption price index. *Bank of Finland Monthly Bulletin*, **10**, 1–8.

[20] Fisher, I. (1922) *The Making of Index Numbers*, Houghton-Mifflin, Boston, MA.

[21] Coelli, T., Prasada Rao, D.S., and Battese, G.E. (eds) (1998) *An Introduction to Efficiency and Productivity Analysis*, Kluwer Academic, Boston, MA.

[22] Edirisinghe, N.C.P., and Zhang, X. (2007) Generalized DEA model of fundamental analysis and its application to portfolio optimization. *Journal of Banking & Finance*, **31**, 3311–3335.

[23] Sueyoshi T. (2000) Stochastic DEA for restructure strategy: An application to a Japanese petroleum company. *Omega*, **28**, 385–398.

[24] Kao, C. and Liu, S.T. (2004) Predicting bank performance with financial forecasts: A case of Taiwan commercial banks. *Journal of Banking & Finance*, **28**(10), 2353–2368.

[25] Yeh, C.C., Chi, D.J., and Hsu, M.F. (2010) A hybrid approach of DEA, rough set and support vector machines for business failure prediction. *Expert Systems with Applications*, **37**, 1535–1541.

[26] Wu, D. (2006) A note on DEA efficiency assessment using ideal point: An improvement of Wang and Luo's model. *Applied Mathematics and Computation*, **183**(2), 819–830.

[27] Tasi, M.C, Lin, S.P, Cheng, C.C., and Lin, Y.P. (2009) The consumer loan default prediction model: An application of DEA-DA and neural network. *Expert Systems with Applications*, **36**, 11682–11690.

[28] Yu, J.R., Tzeng, G.H., and Li, H.L. (2005) Interval piecewise regression model with automatic change-point detection by quadratic programming. *International Journal of Uncertainty, Fuzziness and Knowledge-Based Systems*, **13**, 347–361.

[29] Tanaka, H. and Ishibuchi, H. (1992) Possibilistic regression analysis based on linear programming, in *Studies in Fuzziness, Fuzzy Regression Analysis*, vol. **1** (eds J. Kacprzyk and M. Fedrizzi), Omnitech Press, Warsaw, pp. 47–60.

[30] Tanaka, H. and Watada, J. (1988) Possibilistic linear systems and their application to the linear regression model. *Fuzzy Sets and Systems*, **27**, 275–289.

[31] Huang, C.Y. and Tzeng, G.H. (2008) Multiple generation product life cycle predictions using a novel two-stage fuzzy piecewise regression analysis method. *Technological Forecasting and Social Change*, **75**, 12–31.

28

TIME SERIES BENCHMARKING ANALYSIS FOR NEW PRODUCT SCHEDULING: WHO ARE THE COMPETITORS AND HOW FAST ARE THEY MOVING FORWARD?

DONG-JOON LIM

Portland State University, Portland, OR, USA

TIMOTHY R. ANDERSON

Department of Engineering and Technology Management, Portland State University, Portland, OR, USA

28.1 INTRODUCTION

Consider the following questions that arise in the early stages of new product development. What should be the target market for proposed design concepts? Who will be the competitors, and how fast are they moving forward in terms of performance improvements? Ultimately, are the current design concept and targeted launch date feasible and competitive?

Product target setting is one of the most essential practices in the early stage of new product development to ensure that the firm pursues the right markets and product from a strategic viewpoint [1]. This involves decisions about the target market,

Advances in DEA Theory and Applications: With Extensions to Forecasting Models,
First Edition. Edited by Kaoru Tone.
© 2017 John Wiley & Sons Ltd. Published 2017 by John Wiley & Sons Ltd.

product mix, project prioritization, resource allocation, and technology selection [2]. However, in spite of the maturity of new product development disciplines, the risk analysis for new product scheduling has not received extensive attention, as opposed to project selection or resource allocation problems, in the literature [3–5]. In particular, target-setting practice in research and development has relied heavily on market research methods or heuristic ideation techniques [6, 7]. These classic approaches include brainstorming and Delphi [8], morphology (or morphological analysis) [9], conjoint analysis [10], and lead users analysis [11]. In addition, recent techniques such as the voice of the customer [12], probe and learn [13], empathic design [14], the fuzzy cognitive map [15], and crowdsourcing [16] have been used in an attempt to derive promising product concepts from consumers' perceptions and underlying behavior.

In contrast, attention to product categorization as an engineering approach has been enhanced mostly by benchmarking studies under the assumption that market segments can be identified by distinct combinations of product attributes into which customer value propositions may have been incorporated. Initial work related to this approach may be found in Doyle and Green's study [17], which used a widely known benchmarking technique, data envelopment analysis (DEA), to identify homogeneous product groups, that is, competitors, as well as market niches. Specifically, they applied DEA to classify printers by ordering them from broad to niche based on the number of times each printer appeared in others' reference sets. In a similar vein, Seiford and Zhu developed measures for the attractiveness and progress of products by separating context-dependent frontiers [18]. Furthermore, Po *et al.* showed how product-feature-based clustering can be used by decision-makers to allow them to know the changes required in product design so that the product can be classified into a desired cluster [19]. Amirteimoori and Kordrostami later extended this approach to take the size of products into account, thereby comparing products with groups of similar scale [20]. In addition, Amin *et al.* clarified the role of alternative optimal solutions in the clustering of multidimensional observations by a DEA approach [21]. Recently, Dai and Kuosmanen proposed a new approach that can take cluster-specific efficiency rankings as well as stochastic noise into account [22].

Although the above-mentioned approaches can shed light on target-setting practices for new products, there remains a need to integrate product positioning with the assessment of performance improvement so that analysts can have a measure of risk for their product launch planning. Consequently, this study presents how time series benchmarking analysis can be used to assist in scheduling new product releases by taking the rate of performance improvement expected in a target segment into consideration.

The rest of this chapter unfolds as follows. In the next section, Section 28.2, the notion of homogeneous product groups and the rate of performance improvement are introduced, with algebraic formulations. In Section 28.3, the proposed approach is illustrated by applying it to the development of commercial airplanes to demonstrate its possible usage. Finally, Section 28.4 summarizes the results and suggests possible future research directions.

28.2 METHODOLOGY

28.2.1 Preliminaries

What is generally expected from benchmarking is an identification of "best practices," from which current processes can learn and thereby ultimately improve their performance. The formation of the "best practice" frontier based on observed units is therefore the main focus of benchmarking studies. As in the traditional statistical literature, benchmarking models can be conveniently divided into two groups: parametric and nonparametric approaches. The former approach creates the frontier by fitting it to a predefined functional form, and therefore it tends to be robust to noise by filtering it with a predefined general pattern. The latter approach, by contrast, purely adapts the frontier to data without being shaped a priori, and hence it maximizes the flexibility to capture various production possibilities [23].

DEA, which is classified as a nonparametric frontier model, was originally proposed by Charnes *et al.* [24]. As the name "decision-making units" (DMUs) implies, the efficiency measure in DEA is defined as the ratio of the weighted sum of outputs to the weighted sum of inputs using a freely chosen weighting scheme for each DMU, and, as such, the efficiency measure will show those DMUs in the best possible light. The ratio (multiplier) form of the output-oriented variable-returns-to-scale DEA model can be presented as below:

$$\min \left\{ g_0 = \sum_i v_i x_{i0} + w \; \middle| \; \begin{array}{l} \sum_i v_i x_{ij} - \sum_r u_r y_{rj} + w \geq 0, \;\; \forall j \\ \sum_r u_r y_{rj} = 1, \;\; u_r, v_i \geq \varepsilon, \;\; w \text{ is free} \end{array} \right\} \qquad (28.1)$$

where g_0 denotes the output-oriented efficiency of the DMU being assessed, u_r the weight assigned to output r, v_i the weight assigned to input i, x_{ij} the ith input variable of DMU j, y_{rj} the rth output variable of DMU j, and w the returns-to-scale (RTS) parameter.

The above output-oriented multiplier model can be readily transformed into the envelopment model, which is shown below as a single-stage theoretical formulation:

$$\max \left\{ \varphi_o + \varepsilon \left(\sum_r s_r^+ + \sum_i s_i^- \right) \; \middle| \; \begin{array}{l} \sum_j \lambda_j y_{rj} - s_r^+ = \varphi_o y_{ro}, \;\; \forall r \\ \sum_j \lambda_j x_{ij} + s_i^- = x_{io}, \;\; \forall i \\ \sum_j \lambda_j = 1, \;\; s_r^+, s_i^-, \lambda_j \geq 0 \end{array} \right\} \qquad (28.2)$$

where ϕ_o denotes the output-oriented efficiency, λ_j denotes the intensity vector attached to DMU j, and s_r^+ and s_i^- denote the slacks, equal to the reduced cost of u_r and v_i, respectively. Note that if the optimal value of ϕ_o, that is, ϕ_o^*, is greater than 1, then

DMU_o is inefficient in that the model (28.2) will have identified another production possibility that secures at least the augmented output vector $\phi_o^* y_o$ using no more than the input vector x_o. Thus, ϕ_o^* is a measure of the radial output efficiency of DMU_o in that it reflects the largest radial proportion by which all of its outputs can be augmented pro rata given its output levels.

28.2.2   Conceptual Framework

As previously noted, it may be of interest to product development teams to know not only who the competitors are but also how fast they are moving forward in terms of performance improvement. This necessarily requires a time series application of benchmarking practices. In an early attempt, Inman developed a measure to quantify the rate of frontier expansion [25]. In his study, the rate of change (RoC) was defined as an annualized rate of efficiency change at which future technologies were expected to advance. Specifically, RoCs were obtained from technologies that were efficient, that is, located on the state-of-the-art (SOA) frontier, at the time of release but were later outpaced by new technologies. Thus, the aggregated RoC could be used either to estimate technical capabilities at a certain point in time or to forecast the time by which desired levels of technologies would be operational [26]. Lim *et al.* extended the model to identify segmented RoCs from each frontier facet so that technological progress in different product segments could be taken into account [27].

Figure 28.1 depicts how the local RoC and individualized RoC can be obtained. Product A was located on the SOA frontier in the past but later became obsolete with respect to the current SOA frontier formed by new competitive products B, C, and D. The fact that product A is compared with its virtual target, A', constituted by its peers B, C, and D, indicates that product A may have a similar mix of input–output levels to

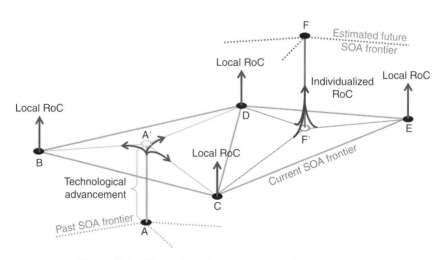

Figure 28.1   Illustration of segmented rate of change (RoC).

those peers although the absolute levels of attributes may vary; this means that one can classify them as homogeneous products [28]. Hence, the performance improvement, namely the performance gap between A and A′ in a given time period, can be represented by the peers as a form of local RoC with reference information about how close A′ is to B, C, and D. Thus, each local RoC indicates a growth potential for adjacent frontier facets based on the performance improvement observed from the related past products.

Once the local RoCs of current SOA products have been obtained, it becomes possible to compute individualized RoCs for new product concepts. Suppose some product developers have come up with a product concept F. Note that, by definition, a "better" product must be located beyond the current SOA frontier, as superseded products are enveloped by the current SOA frontier. It can be seen that the virtual target of F, that is, F′, is subject to the frontier facet constituted by the current SOA products C, D, and E. Therefore, the individualized RoC of F can be obtained by combining local RoCs with reference information about how close F′ is to C, D, and E. Notice that the technological advancement observed from product A may have affected the individualized RoC of F as the SOA products C and D are involved on both sides of the facets by having intermediate technological characteristics.

Benchmarking information in conjunction with the rate of performance improvement can give insight into product developers not only about who the major players in a target market are but also about how competitive the proposed design concept would be. In other words, it can provide a diagnostic of whether the proposed design concept is aggressive or conservative in terms of scheduled delivery to the market, considering the current rate of performance improvement expected in a target segment. One can also utilize this information to estimate the arrival of a competitor's design target as a product launch strategy.

28.2.3   Formulation

The algebraic formulation of the approach described above can be represented by the following processes. Suppose there are n DMUs, and let $x = (x_1, \ldots, x_m) \in \mathfrak{R}_+^M$ denote an input vector and $y = (y_1, \ldots, y_s) \in \mathfrak{R}_+^S$ an output vector. Following the minimum extrapolation principle [29], a production possibility set (PPS) can be formulated as in (28.3):

$$
\text{PPS} = \left\{ (x, y) \,\middle|\,
\begin{array}{l}
\sum_j \lambda_j x_j \leq x, \quad i = 1, \ldots, m \\
\sum_j \lambda_j y_r \geq y, \quad r = 1, \ldots, s \\
\lambda_j \geq 0, \quad j = 1, \ldots, n
\end{array}
\right\}
\tag{28.3}
$$

Note that the PPS constructed could change if the basic premises of the minimum extrapolation principle (e.g. convexity, free disposability, or constant returns to scale) were to change; see the alternate forms of the PPS in [28].

Having specified the PPS, the conventional DEA can be presented as in (28.4) for the input orientation or (28.5) for the output orientation:

$$1/z_k^* = \min\{ \theta_k : (\theta_k x_k, y_k) \in \text{PPS} \} \tag{28.4}$$

$$z_k^* = \max\{ \phi_k : (x_k, \phi_k y_k) \in \text{PPS} \} \tag{28.5}$$

The efficiency score z_k^* obtained is a measure of radial efficiency: DMU k is (at least weakly) efficient if $z_k^* = 1$, or inefficient if $z_k^* > 1$.

As previously discussed, the evolution of the technology frontier is captured by the efficiency changes of the DMUs. Following the notation of [30], let z_k^{t*} be an efficiency score obtained for DMU k from PPS^t including DMUs up to time t, let t_k be the release date of DMU k, and let T be the vantage point from which the RoC is captured. Then $z_k^{t_k*} = 1$ and $z_k^{T*} > 1$ indicates that DMU k was located on the technology frontier at the time of release but was later outperformed by the newly created technology frontier at T. By combining this information with the effective time, that is, the time gap between technology frontiers, as denoted in (28.6) [31], the RoC observed for each DMU can be obtained as formulated in (28.7).

Next, the local RoC is computed for a DMU or DMUs located on the technology frontier at T. Each local RoC therefore represents a growth pattern of adjacent frontier facets based on the efficiency changes observed from related past technologies [32]. Consequently, this enables an identification of how much frontier expansion has been caused by each benchmark technology among the others. This is represented in (28.8):

$$E_k = \frac{\sum_j \lambda_{jk}^{T*} \cdot (t_j - t_k)}{\sum_j \lambda_{jk}^{T*}}, \ \forall k \,|\, z_k^{t_k*} = 1, \, z_k^{T*} > 1 \tag{28.6}$$

$$\gamma_k^T = \left(z_k^{T*} \right)^{1/E_k}, \ \forall k \,|\, z_k^{t_k*} = 1, \, z_k^{T*} > 1 \tag{28.7}$$

$$\delta_j^T = \frac{\sum_k \lambda_{jk}^{T*} \cdot \gamma_k^T}{\sum_{k, \, \gamma_k^T > 0} \lambda_{jk}^{T*}}, \ \forall j \,|\, z_j^{T*} = 1 \tag{28.8}$$

Lastly, the "auspicious" arrival time of proposed design concepts can be estimated by consideration of how superior they are from the vantage point of the current frontier (at T), as well as how much performance improvement is expected in corresponding segments. The super-efficiency[1] and individualized RoC, respectively, contain these two types of information and the latter is computed by combining the local RoCs of

---

[1] Super-efficiency of the proposed design concept might be infeasible, and in such cases alternate measures of efficiency can be employed as discussed in [33].

the SOA product j that constitutes the frontier facet onto which product concept k is being projected, as in (28.9):

$$t_k^{\text{forecast}} = \frac{\ln\left(\frac{1}{z_k^{T^*}}\right)}{\ln\left(\frac{\sum_j \lambda_{jk}^{T^*} \cdot \delta_j^T}{\sum_j \lambda_{jk}^{T^*}}\right)} + \frac{\sum_j \lambda_{jk}^{T^*} \cdot t_j}{\sum_j \lambda_{jk}^{T^*}}, \ \forall k \mid t_k > T \qquad (28.9)$$

28.3 APPLICATION: COMMERCIAL AIRPLANE DEVELOPMENT

28.3.1 Research Framework

To illustrate the use of the presented method, this section assumes a scenario in which commercial airplane developers are examining four design concepts in 2007 (T). They have collected data including 24 aircraft introduced to the market in the last 40 years, and are attempting to identify which market segment the proposed design concepts are appropriate to and when an auspicious time for delivery as competitive products would be, considering the rate of technological advancement observed until 2007.

Note that we have adopted the performance characteristics used in the earlier study by Lamb *et al.* [6]. In the original study, those authors attempted to develop technology assessment models based on a multiple-regression analysis. However, the resulting model was confined to only two predictors owing to insufficient statistical significance, which resulted in a high unexplained variability [34]. This study revisits and updates the dataset not only to incorporate the latest information but also to investigate the industry dynamics, with consideration of different SOA trends as suggested in the previous study (see Table 28.1).

28.3.2 Analysis of the Current (2007) State of the Art

The commercial aircraft industry has important niches, with segmented levels of competition from regional jets to jumbo jets. Following the scenario, Table 28.2 records the local RoCs of six SOA airplanes from the vantage point of 2007. The third column lists dominated airplanes, that is, past airplanes for which the airplane in the first column has been appointed as a benchmark. As previously discussed, one can notice that airplanes with similarities in their specifications, which characterize distinct segments observed in 2007, are grouped together. While the frontier is five-dimensional in this application, the airplanes in the first column are equivalent to products B, C, D, and E in Figure 28.1, and the airplanes in the third column are obsolete airplanes such as A.

The Boeing 747 series, as its nickname "jumbo jet" suggests, has been recognized as the most successful series of wide-body commercial aircraft [35]. In particular, despite their large bodies, the advanced aerodynamic design still allows the 747-300 and 747-400 to reach a cruising speed of up to 902 km/h [36]. These

TABLE 28.1 Dataset.[a]

Airplane	EIS[b] (year)	Travel range (1000 km)	Passenger capacity (3rd class)	PFE[c] (passengers km/L)	Cruising speed (km/h)	Maximum speed (km/h)
DC8-55	1965	9.205	132	13.721	870	933
DC8-62	1966	9.620	159	16.646	870	965
747-100	1969	9.800	366	19.559	893	945
747-200	1971	12.700	366	23.339	893	945
DC10-30	1972	10.010	250	18.199	870	934
DC10-40	1973	9.265	250	16.844	870	934
L1011-500	1979	10.200	234	19.834	892	955
747-300	1983	12.400	412	25.652	902	945
767-200ER	1984	12.200	181	24.327	849	913
767-300ER	1988	11.065	218	26.575	849	913
747-400	1989	13.450	416	25.803	902	977
MD-11	1990	12.270	293	24.595	870	934
A330-300	1993	10.500	295	31.877	870	913
A340-200	1993	15.000	261	25.252	870	913
A340-300	1993	13.700	295	27.335	870	913
MD-11ER	1996	13.408	293	24.939	870	934
777-200ER	1997	14.305	301	25.155	892	945
777-300	1998	11.120	365	23.713	892	945
A330-200	1998	12.500	253	22.735	870	913
A340-600	2002	14.600	380	28.323	881	913
A340-500	2003	16.700	313	24.334	881	913
777-300ER	2004	14.685	365	29.568	892	945
777-200LR	2006	17.370	301	28.841	892	945
A380-800	2007	15.200	525	24.664	902	945

[a] For reproducible results, the dataset and proposed model are included in the R package DJL (version 1.7 or higher). The complete source code can be found at https://github.com/tgno3/TONE.2016.ARTS using the following commands:

```
> library(DJL)
> d <- dataset.airplane.2017
```

[b] EIS: entry into service.
[c] PFE: passenger fuel efficiency.

TABLE 28.2 Local rate of change (RoC) of SOA airplanes in the frontier year of 2007.[a]

SOA airplane (j)	Local RoC (δ_j^T)	Dominated airplanes (k)
747-300	1.000949	DC8-55, 747-100/200, L1011-500
747-400	1.001404	DC8-55/62, 747-100/200, L1011-500, A340-200
A330-300	1.002188	767-300ER, A340-300
777-300ER	1.002561	767-300ER, A340-200/300/600
777-200LR	1.004606	A340-200/500
A380-800	1.003989	A340-500/600

[a] Once the package and dataset are loaded, local RoCs can be obtained using the following commands in R:

```
> t <- subset(d, select = 2)
> x <- data.frame(Frew = rep(1, 28))
> y <- subset(d, select = 3:7)
> roc.dea(x, y, t, 2007, "vrs", "o", "min")$roc_local
```

characteristics can be identified from the dominated airplanes, which include not only the predecessor 747 s (747-100 and 747-200) but also the Douglas DC8 series and Lockheed L-1011, which were also known as fast-cruising airplanes. However, gradual technology advancement is observed from the relatively slow local RoC of the 747 aircraft, which is consistent with the fact that they had been a dominant design for a long time until Airbus created a strong market rival [35].

The Airbus series (A3X0) can be best characterized as long-range airplanes. In fact, the company has primarily targeted the growing demand for high capacity and transcontinental flights. In addition, they have focused their efforts on enhancing the structural design using advanced winglets and working on aerodynamic improvements for higher fuel efficiency [37]. For example, two recent long-range airplanes, the twinjet A330 and the four-engine A340, became popular for their efficient wing design [38]. Meanwhile, the Airbus A340-500 has an operating range of 16 700 km, which is the second longest range of any commercial jet after the Boeing 777-200LR (a range of 17 370 km). Therefore, it is not surprising that the A330-300 has been selected as a benchmark of not only the A340-300, from the same family of airplanes, but also the Boeing 767-300ER, which is also a relatively long-range (11 065 km) airplane with high passenger fuel efficiency (26.575 passenger km/L). Additionally, the Airbus A380-800 became the world's largest passenger airplane, with a seating capacity of 525 [39]. One can also relate this feature to the reference set which consists of its predecessors: A340-500 and A340-600, with relatively high passenger capacities as well. This type of long-range, wide-body airplane has emerged as a fast-growing segment as airlines have emphasized transcontinental aircraft capable of directly connecting any two cities in the world. This has indeed initiated a series of introductions of the A340 family by Airbus to compete with Boeing [40], which is consistent with the fast local RoCs, indicating a very competitive segment of the market with rapid improvement.

The Boeing 777 series ranks as one of Boeing's best-selling aircraft family because of their high fuel efficiency, which enables long-range routes [41]. In particular, the 777-300ER is an extended-range version of the 777-300, which has a maximum range of 14 685 km, made possible by a superior passenger fuel efficiency of 29.568 passenger km/L. These exceptional characteristics allowed not only the preceding 767-300ER but also the Airbus series that pursued higher fuel efficiency (A340-200/300/600) to have the 777-300ER appointed as a benchmark for their performance evaluation. Likewise, the 777-200LR was selected as a benchmark for long-range airplanes that have relatively smaller passenger capacities: the A340-200 and A340-500. Because of demanding energy-saving regulations, airlines have asked for a fuel-efficient alternative and have increasingly deployed these aircraft on long-haul transoceanic routes [42]. This has driven engineering efforts more toward energy-efficient aircraft, which is reflected in the fast local RoCs of the Boeing 777 series.

28.3.3 Risk Analysis

We now turn to the strategic planning for the proposed airplane concepts (see Table 28.3). In particular, the planning team would like to identify the relevant engineering targets for each design concept as well as the corresponding rate of

TABLE 28.3   Four airplane concepts in 2007.

Design concept	Travel range (1000 km)	Passenger capacity (3rd class)	PFE (passengers km/L)	Cruising speed (km/h)	Maximum speed (km/h)	Delivery target (year)
1	14.816	467	28.950	917	988	2010
2	15.750	280	34.851	913	954	2010
3	15.000	315	34.779	903	945	2013
4	14.800	369	35.008	903	945	2015

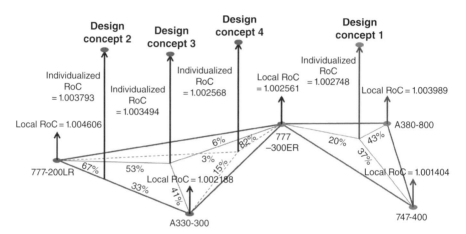

Figure 28.2   2007 state-of-the-art frontier with regard to four design concepts.[2,3]

technological advancement, that is, the individualized RoC, so that they can examine the feasibility of proposed design concepts in terms of their delivery target.

As SOA airplanes at the frontier of 2007 represent different types of past airplanes, one can classify future airplanes, namely design concepts, by the characteristics of their reference airplanes identified on the 2007 frontier. This allows the model to compute an individualized RoC under which each future airplane is expected to be released. Figure 28.2 summarizes the results.

The first design concept aims for a large commercial aircraft carrying 467 passengers while having a fast cruising speed of 917 km/h. As noted earlier, these characteristics are also reflected in its reference airplanes: the 747-400, 777-300ER, and A380-800. That is, this design concept would compete with these three airplanes in the current (2007) market with the given specifications. The individualized RoC of this design concept can therefore be obtained by interpolating local RoCs in conjunction with reference information. Here, the individualized RoC obtained was 1.002748, which suggests a more rapid technology development in its category

---

[2] This figure depicts conceptualized frontier facets relevant to the four design concepts under discussion.
[3] Individualized RoCs can be obtained using the following command in *R*:

```
> target.arrival.dea(x, y, t, 2007, "vrs", "o", "min")$roc_ind
```

compared with the average RoC of 1.002149. This is about 28% faster and resulted in an estimated EIS for the current design concept of 2011.49. Therefore, one may consider the delivery target of 2010 to be an aggressive goal that might encounter technical challenges in outpacing the rate of technological advancement of the past.

In a similar manner, the characteristics of the second design concept's long range of 15 750 km with an outstanding passenger fuel efficiency of 34.851 passenger km/L are consistent with the nature of its identified reference airplanes, the A330-300 and 777-200LR. As implied by the local RoCs of the 777-200LR (1.004606) with its reference information (67%), this concept is associated with one of the fastest-advancing technology clusters seeking a high fuel efficiency. Consequently, it was expected that with the very fast individualized RoC of 1.003793, this level of specification could be achieved by 2013.45. Similarly to the case for the first design concept, this indicates that the delivery target of 2010 may involve a significant technical risk, since it requires exceeding the past rate of technological advancement.

The third design concept is similar to the second one in that it is also aimed at a long-range, fuel-efficient aircraft; however, it also aims to achieve a large passenger capacity of 315. This feature is reflected in the reference set, which additionally includes the 777-300ER, which has a passenger capacity of 365. The relatively slow local RoC of the 777-300ER and the A330-300 may imply a difficulty in technological advancement with respect to travel range and passenger capacity. As a result, the individualized RoC for this design concept was found to be 1.003494, giving a forecasted EIS of 2012.45. Given the delivery target of 2013, the current design concept might be regarded as a feasible goal; however, on the other hand, this possibly entails a modest market risk of lagging behind in the performance competition.

The last design concept is a variation of the third design concept, aiming for a much larger airplane but with a shortened travel range. Not surprisingly, this different blend of the same peers results in a virtual target of this design concept that is positioned closer to the 777-300ER than to the long-range features represented by the 777-200LR and A330-300, which results in more conservative prospects for the current design concept. Consequently, the individualized RoC was found to be 1.002568, giving a forecasted EIS of 2020.16. This indicates that the delivery target of 2015 may be an overly optimistic goal, and there could be a postponement due to the technical risks involved.

28.3.4 Proof of Concept

We now come back to the present and validate the performance of the method presented here (see Table 28.4).

The first design concept was the Boeing 747-8, which began deliveries in 2012 [43]. In fact, this airplane faced two years of delay since its original planned EIS of 2010, owing to assembly and design problems followed by contractual issues [44].

The second design concept was another Boeing airplane, the 787-9, which made its maiden flight in 2013, and delivery began in July 2014 [45]. In line with the results, the originally targeted EIS in 2010 could not be met, because of multiple delays due to technical problems in addition to a machinists' strike [46].

TABLE 28.4   Results summary.[a]

Design concept	Reference airplanes (competitors)	Planned EIS	Estimated EIS	Delayed EIS
1 (747-8)	747-400, 777-300ER, A380-800	2010	2011.49	2012
2 (787-9)	A330-300, 777-200LR	2010	2013.45	2014
3 (A350-900[b])	A330-300, 777-300ER, 777-200LR	2013	2012.45	2014
4 (A350-1000)	A330-300, 777-300ER, 777-200LR	2015	2020.16	2017

[a] Forecasted arrivals can be obtained using the following command in R:
```
> target.arrival.dea(x, y, t, 2007, "vrs", "o", "min")$arrival_seg
```
[b] Initial design.

The third design concept was the initial design target of the Airbus A350-900, which has been changed and rescheduled to enter service at the end of the year 2014 [47]. The delay was caused mainly by a strategic redesign of the A350, the so-called XWB (extra-wide-body) program, that allows for a maximum seating capacity of 440 with a 10-abreast high-density seating configuration as well as a reinforced fuselage design [48]. It is interesting to note that Airbus has made a strategic decision to delay the A350-900's delivery while improving the specifications to compete with the Boeing 777 series in the jumbo jet segment; the need for this was recognized in the analysis results seven years ago.

Similarly, the last design concept was the Airbus A350-1000, which has also had its EIS rescheduled, to 2017 [49]. This airplane is the largest variation of the A350 family and is designed to compete with the Boeing 777-300ER, as can also be seen from the reference information. Nevertheless, the postponed delivery target of 2017 may still be an aggressive goal considering the technological advancement observed in this segment.

28.4   CONCLUSION AND MATTERS FOR FUTURE WORK

The motivation for this chapter stems from a practical question, "When might be the ideal time to release a new product?" To answer this question, one needs to know not only what type the new product is but also how competitive the corresponding segments are. This subject area can be translated into the research topic of integrating product positioning with the assessment of performance improvement over time, which has rarely been addressed in the literature on either new product development or management science. The use of time series benchmarking analysis as presented here makes it possible to estimate an "auspicious" time by which a proposed design concept will be available as a competitive product by taking into account the rate of performance improvement expected in the target segment.

An empirical illustration of commercial airplane development has shown that the method presented provides valuable information such as dominating designs, distinct segments, and the potential rate of performance improvement, which can be utilized in the early stages of new product development. In particular, six dominant airplanes classifying the rest of the 18 past airplanes considered were identified together with their local RoCs, and, inter alia, the technological advancement toward long-range and wide-body airplanes represented very competitive segments of the market with rapid changes. The resulting individualized RoCs could be used to estimate the arrival times of four different design concepts, and the results are consistent with what has happened since 2007 in the commercial airplane industry. In this chapter, we employed a scenario set in 2007 to demonstrate the possible use of the method presented, considering the general development lifecycle of the commercial airplane industry. Obviously, the predictive power could be improved by updating the rate of performance improvement with up-to-date data.

As a direction of future research, one could consider developing the use of a risk index as a measure of innovativeness. When there is a need to quantify the innovativeness of a product independent of market factors, such a method could suggest how much a certain product has contributed to accelerating the rate of performance improvement or has advanced the product release date compared with the expected date. An alternate approach could also investigate possible modifications to product designs to reduce the risk arising from a predetermined release date.

Another future research topic might be to consider incorporating stochastic characteristics into the model. DEA is, by definition, a deterministic model, which renders the method presented here confined to capturing the rate of performance improvement from the evolution of the SOA frontier. This might provide an aggressive estimation unless the best-performing products on the market are being sought. Stochastic measurements might be able to complement this aspect such that the rate of performance improvement could be obtained from diverse levels of products, thereby yielding a risk distribution for each design concept instead of a point estimation.

REFERENCES

[1] Krishnan, V. and Ulrich, K.T. (2001) Product development decisions: A review of the literature. *Management Science*, **47**(1), 1–21.

[2] Mansfield, M.V. and Wagner, K. (1975) Organizational and strategic factors associated with probabilities of success in industrial R&D. *Journal of Business*, **48**, 179–198.

[3] Schulze, A. and Hoegl, M. (2008) Organizational knowledge creation and the generation of new product ideas: A behavioral approach. *Research Policy*, **37**(10), 1742–1750.

[4] Kim, J. and Wilemon, D. (2002) Focusing the fuzzy front-end in new product development. *R&D Management*, **32**(4), 269–279.

[5] Engwall, M. and Jerbrant, A. (2003) The resource allocation syndrome: The prime challenge of multi-project management? *International Journal of Project Management*, **21**(6), 403–409.

[6] Lamb, A.-M., Anderson, T.R., and Daim, T. (2012) Research and development target-setting difficulties addressed through the emergent method: Technology forecasting using data envelopment analysis. *R&D Management*, **42**(4), 327–341.

[7] Schirr, G.R. (2012) Flawed tools: The efficacy of group research methods to generate customer ideas. *Journal of Product Innovation Management*, **29**(3), 473–488.

[8] Dalkey, N. and Helmer, O. (1963) An experimental application of the Delphi method to the use of experts. *Management Science*, **9**(3), 458–467.

[9] Zwicky, F. (1962) *Morphology of Propulsive Power*, Society for Morphological Research.

[10] Green, P.E. and Srinivasan, V. (1978) Conjoint analysis in consumer research: Issues and outlook. *Journal of Consumer Research*, **5**(2), 103.

[11] Von Hippel, E. (1986) Lead users: A source of novel product concepts. *Management Science*, **32**(7), 791–805.

[12] Griffin, A. and Hauser, J. (1993) The voice of the customer. *Marketing Science*, **12**(1), 1–27.

[13] Lynn, G.S., Morone, J.G., and Paulson, A.S. (1996) Marketing and discontinuous innovation: The probe and learn process. *California Management Review*, **38**(3), 8–37.

[14] Leonard, D. and Rayport, J.F. (1997) Spark innovation through empathic design. *Harvard Business Review*, **75**(6), 102–113.

[15] Jetter, A.J. (2003) Educating the guess: Strategies, concepts and tools for the fuzzy front end of product development. Proceedings of PICMET, July 20–24, 2003, Portland, OR, IEEE.

[16] Sethi, R., Pant, S., and Sethi, A. (2003) Web-based product development systems integration and new product outcomes: A conceptual framework. *Journal of Product Innovation Management*, **20**(1), 37–56.

[17] Doyle, J.R. and Green, R.H. (1991) Comparing products using data envelopment analysis. *Omega*, **19**(6), 631–8.

[18] Seiford, L.M. and Zhu, J. (2003) Context-dependent data envelopment analysis: Measuring attractiveness and progress. *Omega*, **31**(5), 397–408.

[19] Po, R.-W., Guh, Y.-Y., and Yang, M.-S. (2009) A new clustering approach using data envelopment analysis. *European Journal of Operational Research*, **199**(1), 276–284.

[20] Amirteimoori, A. and Kordrostami, S. (2011) An alternative clustering approach: A DEA-based procedure. *Optimization*, 1–14.

[21] Amin, G.R., Emrouznejad, A., and Rezaei, S. (2011) Some clarifications on the DEA clustering approach. *European Journal of Operational Research*, **215**(2), 498–501.

[22] Dai, X. and Kuosmanen, T. (2014) Best-practice benchmarking using clustering methods: Application to energy regulation. *Omega*, **42**(1), 179–188.

[23] Bogetoft, P. and Otto, L. (2010) *Benchmarking with DEA, SFA, and R*, Springer Science & Business Media.

[24] Charnes, A., Cooper, W.W., and Rhodes, E. (1978) Measuring the efficiency of decision making units. *European Journal of Operational Research*, **2**(6), 429–44.

[25] Inman, O.L. (2004) Technology forecasting using data envelopment analysis. Dissertation. Portland State University.

[26] Lim, D.-J., Runde, N., and Anderson, T.R. (2013) Applying technology forecasting to new product development target setting of LCD panels, in *Advances in Business and*

Management Forecasting, vol. **9** (ed. K.D. Lawrence), Emerald Group, Bingley, pp. 137–152.

[27] Lim, D.-J., Jahromi, S.R., Anderson, T.R., and Tudorie, A.-A. (2015) Comparing technological advancement of hybrid electric vehicles (HEV) in different market segments. *Technological Forecasting and Social Change*, **97**, 140–153.

[28] Fried, H.O., Lovell, C.A.K., and Schmidt, S.S. (2008) *The Measurement of Productive Efficiency and Productivity Growth*, Oxford University Press, New York.

[29] Banker, R.D., Charnes, A., and Cooper, W.W. (1984) Some models for estimating technical and scale inefficiencies in data envelopment analysis. *Management Science*, **30**(9), 1078–1092.

[30] Lim, D.-J. (2016) Inverse DEA with frontier changes for new product target setting. *European Journal of Operational Research*, **254**(2), 510–516.

[31] Anderson, T., Färe, R., Grosskopf, S., *et al.* (2002) Further examination of Moore's law with data envelopment analysis. *Technological Forecasting and Social Change*, **69**(5), 465–477.

[32] Lim, D.-J., Anderson, T.R., and Shott, T. (2015) Technological forecasting of supercomputer development: The march to exascale computing. *Omega*, **51**, 128–135.

[33] Lim, D.-J. (2015) Technology forecasting using DEA in the presence of infeasibility. *International Transactions in Operational Research*, in press.

[34] Lamb, A.-M., Daim, T.U., and Anderson, T.R. (2010) Forecasting airplane technologies. *Foresight*, **12**(6), 38–54.

[35] Irwin, D.A. and Pavcnik, N. (2004) Airbus versus Boeing revisited: International competition in the aircraft market. *Journal of International Economics*, **64**(2), 223–245.

[36] Boeing (2013) Technical characteristics – Boeing 747-400. Boeing 747 family, http://www.boeing.com/resources/boeingdotcom/company/about_bca/startup/pdf/freighters/747-400f.pdf (accessed 1 June 2015).

[37] Greene, D.L. (1992) Energy-efficiency improvement potential of commercial aircraft. *Annual Review of Energy and the Environment*, **17**(1), 537–573.

[38] Briere, D., and Traverse, P. (1993) Airbus A320/A330/A340 electrical flight controls – A family of fault-tolerant systems. Proceedings of the 23rd International Symposium on Fault Tolerant Computing, September 23, 1993, Toulouse, France. IEEE.

[39] Michaels, J. (2007) The Airbus A380: The giant on the runway. *The Economist* (Oct. 11), 81–83.

[40] Berrittella, M., La Franca, L., Mandina, V., and Zito, P. (2007). Modelling strategic alliances in the wide-body long-range aircraft market. *Journal of Air Transport Management*, **13**(3), 139–148.

[41] Williams, J.C. and Starke, E.A. (2003) Progress in structural materials for aerospace systems. *Acta Materialia*, **51**(19), 5775–5799.

[42] Senzig, D.A., Fleming, G.G., and Iovinelli, R.J. (2009) Modeling of terminal-area airplane fuel consumption. *Journal of Aircraft*, **46**(4), 1089–1093.

[43] Boeing (2013) Technical characteristics – Boeing 747-8, http://www.boeing.com/farnborough2014/pdf/BCA/bck-747-8%20Family.pdf (accessed 1 June 2015).

[44] Steve, W. (2011) Reports hint at what may be behind Boeing 747-8 delay. *Puget Sound Business Journal* (Sep. 16).

[45] Harriet, M. (2014) Boeing delays decision on 787-10 production site. *Reuters* (Mar. 10).

[46] Dominic, G. (2008) Simmering Boeing strike scorching both sides. *The Seattle Times* (Sep. 29).

[47] Scott, H. (2014) Smaller seats, fee rises and new planes? 2014: The year ahead in air travel. CNN (Jun. 1).

[48] Max, K. (2008) 10-abreast A350 XWB "would offer unprecedented operating cost advantage." [Picture.] *Flight* (May 19).

[49] Airbus (2014) A350XWB family A350-1000, http://www.airbus.com/aircraftfamilies/passengeraircraft/a350xwbfamily/a350-1000/(accessed 1 June 2015).

29

DEA SCORE CONFIDENCE INTERVALS WITH PAST–PRESENT AND PAST–PRESENT–FUTURE-BASED RESAMPLING[1]

KAORU TONE

National Graduate Institute for Policy Studies, Tokyo, Japan

JAMAL OUENNICHE

Business School, University of Edinburgh, Edinburgh, UK

29.1 INTRODUCTION

Data envelopment analysis (DEA) is a non-parametric methodology for performance evaluation and benchmarking. Since the publication of the seminal paper by Charnes, Cooper and Rhodes [1], DEA has witnessed numerous developments, some of which have been motivated by theoretical considerations and others by practical considerations. The focus of this chapter is on practical considerations related to data variations. The first practical issue is the lack of a statistical foundation for DEA. This problem was first discussed by Banker [2], who proved that DEA models could be viewed as maximum likelihood estimation models under specific conditions, and then by Banker and Natarajan [3], who proved that DEA provides a consistent estimator of arbitrary

---

[1] Part of the material in this chapter is adapted from the *American Journal of Operations*, 2016, **6**, 121–135, with permission from Scientific Research.

---

monotone and concave production functions when the (one-sided) deviations from such a production function are degraded by stochastic variations in technical ineffi- ciency. Subsequently, the treatment of data variations in DEA has taken a variety of forms. Several authors have investigated the sensitivity of DEA scores to variations of the data in the inputs and/or outputs using sensitivity analysis and super-efficiency analysis. For example, Charnes and Neralić [4] and Neralić [5] used conventional lin- ear-programming-based sensitivity analysis with additive and multiplicative changes in inputs and/or outputs to investigate the conditions under which the efficiency status of an efficient decision-making unit (DMU) is preserved (i.e. the basis remains unchanged), whereas Zhu [6] performed sensitivity analysis using various super- efficiency DEA models in which a test DMU is not included in the reference set. This sensitivity analysis approach simultaneously considers input and output data pertur- bations in all DMUs, namely, changes both in the test DMU and in the remaining DMUs. On the other hand, several authors have investigated the sensitivity of DEA scores to the estimated efficiency frontier. For example, Simar and Wilson [7,8] used a bootstrapping method to approximate the sampling distributions of DEA scores and to compute confidence intervals (CIs) for such scores. Barnum et al. [9] provided an alternative methodology based on panel data analysis for com- puting CIs of DEA scores; in sum, they complemented Simar and Wilson's bootstrap- ping by using panel data along with generalized least squares models to correct CIs for any violations of the standard statistical assumptions (i.e. that the DEA scores are independent and identically distributed, and normally distributed), such as the presence of contemporaneous correlation, serial correlation and heteroscedasticity. Note, however, that Simar and Wilson [7,8] did not take account of data variations in the inputs and outputs. Also, although Barnum et al. [9] took account of data variations in the inputs and outputs by considering panel data and computing DEA scores separately for each cross-section of the data, the reliability of their approach depends on the amount of data available for estimating the generalized least squares models.

In this chapter, we follow the principles set out by Cook et al. [10], and we believe that DEA performance measures are relative, not absolute, and are frontier-dependent. DEA scores undergo changes depending on the choice of inputs, outputs, DMUs and the DEA models by which the DMUs are evaluated. In the study presented here, we compute efficiency scores or, equivalently, solve the frontier problem using the non- oriented slacks-based super-efficiency model. Our approach deals with variations in both the estimated efficiency frontier and the input and output data directly by resam- pling from historical data over two different time frames (i.e. past–present and past– present–future); thus, the production possibility set for the entire DMUs differs for every sample.[2] In addition, our approach works for both small and large sets of data

---

[2] Throughout this chapter, we assume that the dataset is free from outliers and is homogeneous in the kind of DMUs (e.g. hospitals, banks or universities in the same category). For outlier detection, see Yang et al. [11] and references therein.

and does not make any parametric assumptions. Hence, our approach presents another alternative for computing confidence intervals of DEA scores.

This chapter unfolds as follows. Section 29.2 presents a generic methodological framework to estimate the confidence intervals of DEA scores in a past–present time frame and extends it to the past–present–future time frame. Section 29.3 presents a healthcare application to illustrate the proposed resampling framework. Finally, Section 29.4 concludes the chapter.

29.2   PROPOSED METHODOLOGY

In this section, we propose a generic methodological framework to estimate the confidence intervals of DEA scores in a past–present time frame. This framework is generic in that its implementation requires a number of decisions to be made, as will be discussed below. Then, we extend the use of this framework to the past–present–future time frame.

29.2.1   Past–Present-Based Framework

The first framework is designed for when past–present information on say m inputs and s outputs of a set of n DMUs is available: that is, $(X^t, Y^t) = \left\{ \left(x_{i,j}^t, y_{r,j}^t \right); i = 1, \ldots, m, r = 1, \ldots, s, j = 1, \ldots n, \right\}, t = 1, \ldots, T$, where period T denotes the present and periods 1 to $T-1$ represent the past. The proposed framework can be summarized as follows.

Initialization step

Choose an appropriate DEA model for computing the efficiency scores of DMUs.

Use the chosen DEA model to estimate the DEA scores of DMUs based on the present information, that is, (X^T, Y^T). Let $\delta_j^T, j = 1, \ldots, n$ denote these scores – in the iterative step, we gauge the confidence interval of $\delta_j^T, j = 1, \ldots, n$ using replicas of historical data $(X^t, Y^t), t = 1, \ldots, T$.

Choose an appropriate scheme, say w, to weight the available information about the past and the present.

Choose a confidence level $1 - \alpha$.

Choose the number of replicas or samples to draw from the past, say B, along with any properties that they should satisfy before being considered appropriate to use for generating the sampling distributions of $\delta_j^T, j = 1, \ldots, n$ and computing their confidence intervals.

Set an indicator variable, say *property_status*, that reflects whether or not the B replicas satisfy the required properties to false;

Iterative step

```
WHILE (property _status = false) DO
{
    Draw randomly and with replacement B replicas or samples
    from the past-present, check whether they satisfy the
    required properties and update property_status
    accordingly.
    IF property _status = true THEN
    {
        Use the weighted version of the chosen DEA model to
        estimate the DEA scores of the DMUs in each of the
        B samples.
        FOR j = 1 TO n DO
        {
          Given the sampling distribution of δⱼᵀ estimated above,
          compute the confidence interval of δⱼᵀ at the pre-
          specified confidence level 1 - α.
        }
    }
}
```

The generic nature of this framework requires a number of decisions to be made about its implementation for any particular application. In what follows, we shall discuss how one might make such decisions.

29.2.1.1 Choice of a DEA Model In principle, one might choose from a relatively wide range of DEA models; however, given the nature of this exercise, we recommend the use of the non-oriented super-slacks-based measure model [12,13] under the relevant returns-to-scale (RTS) set-up (e.g. constant, variable, increasing or decreasing) as suggested by an RTS analysis of the dataset one is dealing with. This model is an extension of the slacks-based measure (SBM) model of Tone [14] – see also [15]. Although one could use other models (e.g. radial or oriented), our recommendation is based on the following reasons. First, as a non-radial model, the SBM model is appropriate for taking account of input and output slacks which affect efficiency scores directly, whereas radial models are mainly concerned with proportional changes in inputs or outputs. Thus, SBM scores are more sensitive to data variations than the scores from radial models. Second, the non-oriented SBM model can deal with input surpluses and output shortfalls within the same scheme. Finally, as most DEA scores are bounded by unity (≤1 or ≥1), difficulties may be encountered in comparing efficient DMUs; therefore, we recommend using the super-efficiency version of the non-oriented SBM as it removes such unity bounds.

29.2.1.2 *Choice of a Weighting Scheme for Past–Present Information*
Many different weighting schemes can be used to weight information about the past and the present, that is, $\left(x_{i,j}^t\right), i = 1,\ldots,m, j = 1\ldots,n, t = 1,\ldots,T$ and $\left(y_{r,j}^t\right), r = 1,\ldots,s,$ $j = 1\ldots,n, t = 1,\ldots,T$. The choice of the weighting scheme should reflect the decision-makers' perspective and knowledge of the application area with respect to how the past should influence the present. In this chapter, we set the weights w_t of the periods t so that the weights increase with t; in sum, we assume that more recent periods carry information that is more relevant to estimating efficiency scores in the present time. Thus, the following Lucas number series $(l_1, \ldots, l_T)$, a variant of the Fibonacci series, is a candidate, where $l_{t+2} = l_t + l_{t+1}, t = 1,\ldots,T-2, l_1 = 1, l_2 = 2$. Let L denote the sum of the series, that is, $L = \sum_{t=1}^{T} l_t$. We define the weights w_t as l_t/L for $t = 1,\ldots,T$. For example, when $T = 5$, we have $w_1 = 0.0526$, $w_2 = 0.1053$, $w_3 = 0.1579$, $w_4 = 0.2631$ and $w_5 = 0.4211$. Thus, the influence of past periods fades away gradually as we approach the present.

29.2.1.3 *Choice of the Replication Process and the Number of Replicas*
In this chapter, we regard historical data $(X^t, Y^t) = \left\{\left(x_{i,j}^t, y_{r,j}^t\right); i = 1,\ldots,m, r = 1,\ldots,s,\right.$ $\left. j = 1,\ldots n,\right\}, t = 1,\ldots,T$ as discrete events with probability w_t and cumulative probability $W_t = \sum_{k=1}^{t} w_k, t = 1,\ldots,T$. We propose a replication process based on bootstrapping. First proposed by Efron [16], bootstrapping nowadays refers to a collection of methods that resample randomly with replacement from the original sample. Thus, in bootstrapping, the population is to the sample what the sample is to the bootstrapped sample. Bootstrapping can be either parametric or non-parametric. Parametric bootstrapping is concerned with fitting a parametric model, which in our case would be a theoretical distribution, to the data and sampling from such a fitted distribution. This is a viable approach for large datasets where the distribution of each input and each output can be approximated reasonably by a specific theoretical distribution. However, when no theoretical distribution could serve as a good approximation to the empirical one or when the dataset is small, non-parametric bootstrapping is the way to proceed. Non-parametric bootstrapping does not make any assumptions except that the sample distribution is a good approximation to the population distribution or, equivalently, that the sample is representative of the population. Consequently, datasets with different features require different resampling methods that take account of such features and thus generate representative replicas.

For a non-correlated and homoscedastic dataset, one could for example use smooth bootstrapping or Bayesian bootstrapping, where smooth bootstrapping generates replicas by adding small amounts of zero-centred random noise (usually normally distributed) to the resampled observations, whereas Bayesian bootstrapping generates replicas by reweighting the initial dataset according to a randomly generated weighting scheme. In this chapter, we recommend the use of a variant of Bayesian bootstrapping whereby the weighting scheme consists of weights w_t based on the Lucas number series presented above, because it is more appropriate when one is resampling over a past–present time frame and more recent information is considered more valuable. For a non-correlated and homoscedastic dataset, our data generation process may be summarized as follows. First, a random number ρ is drawn from the uniform distribution over the interval $[0, 1]$, and then whichever cross-section data (X^t, Y^t) is such that $W_{t-1} < \rho \leq W_t$ is resampled, where $W_0 = 0$. This process is repeated as many times as necessary to produce the required number of valid replicas or samples.

On the other hand, for a correlated and/or heteroscedastic dataset, one could use a block bootstrapping method, where replicas are generated by splitting the dataset into non-overlapping blocks (simple block bootstrap) or into overlapping blocks of the same or different lengths (moving block bootstrap), sampling such blocks with replacement and then aligning them in the order in which they were drawn. The main idea of all block bootstrap procedures consists of dividing the data into blocks of consecutive observations of length ℓ, say $[(X^t, Y^t)$, $(X^{t+1}, Y^{t+1}), \ldots, (X^{t+\ell-1}, Y^{t+\ell-1})]$, and sampling the blocks randomly with replacement from all possible blocks – for an overview of bootstrapping methods, the reader is referred to [17]. The block bootstrap procedure with blocks of non-random length can be summarized as follows:

Input: A block length $\ell \in \mathbb{N}$ such that $\ell \ll T$.

Step 1: Draw block labels, say $b_1, b_2, \ldots, b_{R+1}$, randomly and independently from the set of labels, say L, where $R = [T/\ell]$, $L = \{1, \ell+1, 2\ell+1, \ldots, (R-1)\ell+1\}$ if non-overlapping blocks are considered, and $L = \{1, 2, \ldots, T-\ell+1\}$ if overlapping blocks are considered.

Step 2: Lay the blocks $[(X^{b_k}, Y^{b_k}), (X^{b_k+1}, Y^{b_k+1}), \ldots, (X^{b_k+\ell-1},$ $Y^{b_k+\ell-1})]$, $k = 1, \ldots, R+1$, end to end in the order sampled together and discard the last $\ell - T + R\ell$ observations to form a bootstrap series $(\hat{X}^1, \hat{Y}^1), (\hat{X}^2, \hat{Y}^2), \ldots, (\hat{X}^T, \hat{Y}^T)$.

Output: A bootstrap sample $(\hat{X}^1, \hat{Y}^1), (\hat{X}^2, \hat{Y}^2), \ldots, (\hat{X}^T, \hat{Y}^T)$.

As to the choice of the number of replicas B, there is no universal rule except that the larger the value of B, the more stable the results. However, one should take computational requirements into consideration; therefore, in practice, one should keep increasing the value of B until the simulation converges, that is, the results from a run do not change when more iterations are added.

29.2.1.4 *Choice of the Properties the Replicas Should Satisfy*   As replicas are required to be representative of the dataset under consideration, one has to perform a preliminary analysis of the data to find out about their features, namely, whether or not the data are correlated and whether or not they are heteroscedastic, using

statistical tests such as the ones used in [9]. For a correlated and/or heteroscedastic dataset, the same relevant statistical tests have to be used to find whether or not the replicas are representative. When the replicas are not representative, one has to reject them and resample again. However, for a non-correlated and homoscedastic dataset, one can use hypothesis tests or confidence intervals based on Fisher's z transformation to compare correlation patterns in past and present data. For example, for data for the present time period, one can compute the correlation coefficient between all pairs of inputs, outputs and input–output combinations over all DMUs. Then, one computes their $\zeta\%$ confidence intervals, for example 95%, using Fisher's z transformation [18]. If the corresponding correlation of the resampled data is out of the range of this interval, we discard this resampled data. Thus, inappropriate samples with unbalanced inputs and outputs relative to the inputs and outputs for the last period are excluded from resampling. The 95% confidence interval mentioned above is not essential. The narrower the interval, the closer the resample will be to the data for the last period.

29.2.2 Past–Present–Future Time-Based Framework

In the previous subsection, we utilized historical data (X^t, Y^t), $t = 1, \ldots, T$ to gauge the confidence interval of the last period's scores. In this section, we forecast the 'future', namely (X^{T+1}, Y^{T+1}), by using 'past–present' data (X^t, Y^t), $t = 1, \ldots, T$, and forecast the efficiency scores of the future DMUs along with their confidence intervals. In order to avoid repetition, we shall discuss here how the past–present time-based framework can be extended to the past–present–future context. First, we have to forecast the future; to be more specific, given the observed historical data $\left(x_{i,j}^t, y_{r,j}^t\right)$, $t = 1, \ldots, T$ for a certain input i $(i = 1, \ldots, m)$ and output r $(r = 1, \ldots, s)$ of a DMU j $(j = 1, \ldots n)$, we wish to forecast $\left(x_{i,j}^{T+1}, y_{r,j}^{T+1}\right)$. There are several forecasting engines available for this purpose. Once these forecasts are obtained, we then estimate the super-efficiency score of the 'future' DMU (X^{T+1}, Y^{T+1}) using the non-oriented super-slacks-based measure model. Finally, given the past–present–future intertemporal dataset (X^t, Y^t), $t = 1, \ldots, T+1$, we apply the resampling scheme proposed in the previous section and obtain confidence intervals.

29.3 AN APPLICATION TO HEALTHCARE

In this study, we utilized a dataset concerning 19 Japanese municipal hospitals from 2007 to 2009 to illustrate how the proposed framework works. There are approximately 1000 municipal hospitals in Japan and there is a large amount of heterogeneity amongst them. We selected 19 municipal hospitals with more than 400 beds. Therefore, this sample may represent larger acute-care hospitals with homogeneous functions. The data were collected from the *Annual Databook of Local Public Enterprises* published by the Ministry of Internal Affairs and Communications. For illustration purposes, we chose two inputs for this study, namely, Doctor ((I) Doc) and Nurse ((I) Nur), and two outputs, namely, Inpatient ((O) In) and Outpatient ((O) Out). Table 29.1 shows the data, and Table 29.2 shows the main statistics. The data are the yearly averages of the fiscal-year data, as we had no daily or monthly data; the Japanese government's fiscal year begins on 1 April and ends on 31 March. As can be seen, the data for the inputs and outputs fluctuate between years, which suggests the need for an analysis of data variation.

We solved the non-oriented super-slacks-based measure model year by year and obtained the super-efficiency scores shown in Table 29.3, along with their graphical representation shown in Figure 29.1. As can be seen, the scores fluctuate between years. Once again, this suggests the need for an analysis of data variation. If we had daily data, this could be done. However, we only had fiscal-year data and hence we needed to resample the data in order to gauge the confidence interval of the efficiency scores. So, we merged the datasets for all years and evaluated the efficiency scores relative to 57 (=19 × 3) DMUs, as shown in Table 29.4 and Figure 29.2. Comparing the averages for these three years, we found that the average of 0.820 for the year 2007 was better than those for 2008 (0.763) and 2009 (0.732). We also performed a non-parametric Wilcoxon rank-sum test and the results indicated that the null hypothesis, that is, that 2007 and 2008 had the same distribution of efficiency scores, was rejected at a significance level of 1%; therefore, 2007 outperforms 2008. Similarly, 2007 outperforms 2009. However, we cannot see a significant difference between 2008 and 2009.

29.3.1   Illustration of the Past–Present Framework

We applied the proposed procedure to the historical data listed in Table 29.1 for the 19 hospitals for the two years 2008–2009. We excluded the data for the year 2007, because they belong to a different population from the data for 2009 (see Table 29.4 and Figure 29.2). Note that historical data may be affected by accidental or exceptional events, for example oil shocks, earthquakes, financial crises, environmental changes and so forth. We must exclude these effects from the data. Also, if some data are subject to age depreciation, we must adjust them properly. In this study, we used Lucas weights for the past and present data. However, we could have used other weighting schemes (e.g. exponential) instead.

Table 29.5 shows the correlation matrix of the observed year 2009 data shown in Table 29.1, and the Fisher 95% confidence intervals are shown in Table 29.6. For example, the correlation coefficient between Doc and Outpatient is 0.5178, and its 95% lower and upper bounds are 0.0832 and 0.7869, respectively. In addition, we report the Fisher 20% confidence lower and upper bounds in Table 29.7. These intervals are considerably narrowed down compared with the Fisher 95% case.

Table 29.8 presents results obtained with 500 replicas, where the column 'DEA' shows the efficiency scores for the last period (2009) and 'Average' indicates the average score over the 500 replicas. The column 'Rank' shows the ranking of the

TABLE 29.1 The data.

DMU	2007 (I) Doc	(I) Nur	(O) In	(O) Out	2008 (I) Doc	(I) Nur	(O) In	(O) Out	2009 (I) Doc	(I) Nur	(O) In	(O) Out
H1	108	433	606	1239	114	453	617	1244	116	545	603	1295
H2	125	448	642	1363	133	499	638	1310	136	482	618	1300
H3	118	567	585	1072	121	600	569	1051	125	616	561	1071
H4	138	541	699	1210	138	531	704	1194	140	554	679	1182
H5	138	613	653	1195	142	616	644	1147	137	633	622	1147
H6	99	569	716	1533	106	592	701	1478	109	613	651	1457
H7	94	498	540	1065	103	494	551	1067	101	491	540	1067
H8	106	461	496	1051	118	490	504	1033	133	479	505	1081
H9	109	450	483	851	119	483	487	877	121	501	486	904
H10	102	540	581	1268	106	558	565	1278	148	611	586	1321
H11	92	495	490	1217	101	497	501	1146	102	501	479	1113
H12	148	721	771	1637	147	710	723	1657	158	737	743	1714
H13	103	593	679	2011	106	673	642	1883	120	697	634	1872
H14	101	500	613	1868	110	519	617	1894	116	517	623	2009
H15	159	793	964	2224	160	801	906	2148	166	817	877	2155
H16	77	354	410	1047	68	359	391	916	81	378	406	897
H17	111	663	717	1674	112	645	702	1774	112	663	709	1733
H18	62	388	480	913	64	385	467	907	63	381	463	872
H19	98	323	508	1192	95	314	483	1018	95	320	490	1034

TABLE 29.2 Main statistics.

	2007						2008						2009					
	(I) Doc	(I) Nur	(O) In	(O) Out			(I) Doc	(I) Nur	(O) In	(O) Out			(I) Doc	(I) Nur	(O) In	(O) Out
Min	62	323	410	851			64	314	391	877			63	320	406	872
Max	159	793	964	2224			160	801	906	2148			166	817	877	2155
Avg	110	524	612	1349			114	538	601	1317			120	555	593	1328
StdDev	23.75	120.41	130.51	378.24			24.15	121.43	119.57	380.07			25.58	126.78	113.05	389.49

TABLE 29.3   Super-SBM scores by cross-section (year).

	2007	2008	2009
H1	0.883	0.905	0.754
H2	0.875	0.801	0.779
H3	0.623	0.615	0.592
H4	0.700	0.765	0.680
H5	0.619	0.620	0.604
H6	1.004	0.942	0.848
H7	0.719	0.732	0.725
H8	0.676	0.651	0.631
H9	0.588	0.583	0.568
H10	0.758	0.764	0.631
H11	0.757	0.740	0.698
H12	0.711	0.741	0.714
H13	1.034	1.025	0.831
H14	1.039	1.107	1.145
H15	0.858	0.857	0.811
H16	0.831	0.847	0.742
H17	0.847	0.948	0.937
H18	1.034	1.050	1.074
H19	1.071	1.072	1.100
Avg	0.822	0.830	0.782

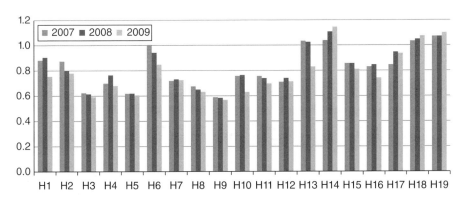

Figure 29.1   Super-SBM scores by cross-section (year).

average scores. We applied a Fisher 95% threshold and found no out-of-range samples. Figure 29.3 shows the 95% confidence intervals of the DEA scores for the last period (2009) along with the average scores. The average of the 95% confidence intervals for all hospitals is 0.10.

In the Fisher 95% (ζ95) case, we found no discarded samples, whereas in the Fisher 20% (ζ20) case, 1945 samples were discarded before 500 replicas were obtained.

TABLE 29.4   Super-SBM scores for panel data (all years).

	2007	2008	2009
H1	0.883	0.833	0.727
H2	0.875	0.750	0.745
H3	0.623	0.584	0.571
H4	0.700	0.712	0.654
H5	0.619	0.590	0.584
H6	1.004	0.860	0.783
H7	0.719	0.696	0.699
H8	0.676	0.620	0.613
H9	0.588	0.556	0.551
H10	0.758	0.726	0.610
H11	0.757	0.703	0.672
H12	0.711	0.704	0.688
H13	1.034	0.871	0.794
H14	1.024	0.950	1.020
H15	0.858	0.812	0.779
H16	0.831	0.798	0.715
H17	0.847	0.872	0.855
H18	1.028	0.929	0.922
H19	1.042	0.920	0.924
Avg.	0.820	0.763	0.732

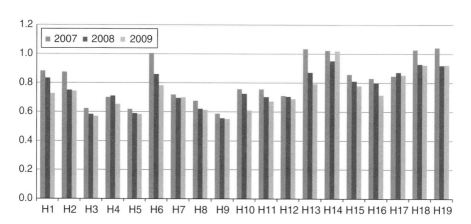

Figure 29.2   Super-SBM scores for panel data (all years).

Table 29.9 shows a comparisons of the scores calculated with both thresholds, where we cannot see significant differences.

Note that one resample produces one efficiency score for each DMU. We compared results for 500 and 5000 replicas and obtained the 95% confidence intervals

TABLE 29.5   Correlation matrix.

	Doc	Nurse	Inpatient	Outpatient
Doc	1	0.7453	0.7372	0.5178
Nurse	0.7453	1	0.8610	0.7387
Inpatient	0.7372	0.8610	1	0.8264
Outpatient	0.5178	0.7387	0.8264	1

TABLE 29.6   Fisher 95% confidence lower/upper bounds for correlation matrix.

		Lower bounds			
		Doc	Nurse	Inpatient	Outpatient
Upper bounds	Doc		0.4400	0.4255	0.0832
	Nurse	0.8961		0.6681	0.4281
	Inpatient	0.8926	0.9455		0.5959
	Outpatient	0.7869	0.8932	0.9311	

TABLE 29.7   Fisher 20% confidence lower/upper bounds for correlation matrix.

		Lower bounds			
		Doc	Nurse	Inpatient	Outpatient
Upper bounds	Doc		0.71578	0.70695	0.46998
	Nurse	0.77214		0.8437	0.70854
	Inpatient	0.76482	0.87652		0.80525
	Outpatient	0.56266	0.76614	0.84547	

as shown in Table 29.10. As can be seen, the difference is negligibly small. So, 500 replicas may be acceptable in this case. However, the number of replicas depends on the numbers of inputs, outputs and DMUs. Hence, we need to check the variations in the scores by increasing the number of replicas.

As to the comparison of individual hospitals, on looking at Hospitals 1 and 2 in Table 29.8 and Figure 29.3, it is difficult to judge which hospital exhibits better performance. In fact, the 2009 score and the average score are reversed (H1 (2009) = 0.754, H1 (average) = 0.8047, H2 (2009) = 0.7789, H2 (average) = 0.7865), and the confidence intervals overlap. We applied the Wilcoxon rank-sum test and found that Hospital 1 outperforms Hospital 2 at a significance level of 1%. In this way, we can compare individual hospitals by using efficiency measurements.

Finally, we would like to draw the reader's attention to the fact that, in some applications, one might put weights on the inputs and outputs. If the costs for inputs and incomes from outputs are available, we can evaluate the comparative cost

TABLE 29.8   DEA scores and confidence intervals obtained with 500 replicas.

	97.50%	DEA (2009)	Average	2.50%	Rank (avg)
H1	0.9228	0.7540	0.8047	0.7240	8
H2	0.8279	0.7787	0.7865	0.7415	9
H3	0.6285	0.5918	0.5999	0.5730	18
H4	0.7574	0.6802	0.7090	0.6694	14
H5	0.6375	0.6042	0.6088	0.5792	17
H6	0.9384	0.8475	0.8758	0.8159	6
H7	0.7620	0.7250	0.7284	0.6998	11
H8	0.6902	0.6311	0.6365	0.6002	16
H9	0.6030	0.5681	0.5732	0.5452	19
H10	0.7963	0.6308	0.6818	0.6032	15
H11	0.7433	0.6985	0.7116	0.6808	13
H12	0.7684	0.7140	0.7237	0.6849	12
H13	1.0465	0.8310	0.8978	0.8081	5
H14	1.1564	1.1448	1.1329	1.1037	1
H15	0.8692	0.8107	0.8277	0.7886	7
H16	0.8792	0.7418	0.7782	0.7140	10
H17	1.0142	0.9368	0.9542	0.9076	4
H18	1.0837	1.0745	1.0708	1.0497	3
H19	1.1194	1.0996	1.0897	1.0618	2

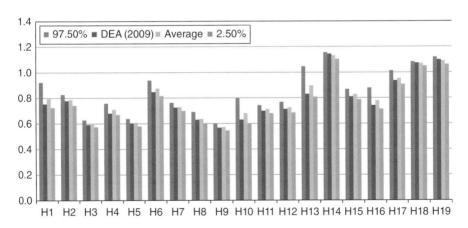

Figure 29.3   95% confidence intervals.

performance of DMUs. In the absence of such information, we can instead put weights on the inputs and outputs. For example, the weights of Doc and Nurse may be assumed to be in the range 5 to 1 (on average), and those of Outpatient and Inpatient in the range 1 to 10 (on average). We can solve this problem via the weighted-SBM model, which should enhance the reliability and applicability of our approach.

TABLE 29.9 Comparison of Fisher 20% (ζ20) and 95% (ζ95) thresholds.

ζ20	97.50%	DEA	2.50%	ζ95	97.50%	DEA	2.50%	ζ20 – ζ95, 97.50%	ζ20 – ζ95, 2.50%
H1	0.9061	0.7540	0.724	H1	0.9228	0.754	0.724	-0.017	0.000
H2	0.8247	0.7787	0.7419	H2	0.8279	0.7787	0.7415	-0.003	0.000
H3	0.6279	0.5918	0.5757	H3	0.6285	0.5918	0.573	-0.001	0.003
H4	0.7476	0.6802	0.6684	H4	0.7574	0.6802	0.6694	-0.010	-0.001
H5	0.6375	0.6042	0.5832	H5	0.6375	0.6042	0.5792	0.000	0.004
H6	0.9382	0.8475	0.8168	H6	0.9384	0.8475	0.8159	0.000	0.001
H7	0.7611	0.7250	0.6989	H7	0.7620	0.7250	0.6998	-0.001	-0.001
H8	0.6905	0.6311	0.6011	H8	0.6902	0.6311	0.6002	0.000	0.001
H9	0.6023	0.5681	0.5467	H9	0.6030	0.5681	0.5452	-0.001	0.001
H10	0.7903	0.6308	0.6044	H10	0.7963	0.6308	0.6032	-0.006	0.001
H11	0.7469	0.6985	0.6808	H11	0.7433	0.6985	0.6808	0.004	0.000
H12	0.7670	0.7140	0.6828	H12	0.7684	0.7140	0.6849	-0.001	-0.002
H13	1.0445	0.831	0.8081	H13	1.0465	0.831	0.8081	-0.002	0.000
H14	1.1568	1.1448	1.1041	H14	1.1564	1.1448	1.1037	0.000	0.000
H15	0.8670,	0.8107	0.7886	H15	0.8692	0.8107	0.7886	-0.002	0.000
H16	0.8747	0.7418	0.7222	H16	0.8792	0.7418	0.7140	-0.004	0.008
H17	1.0121	0.9368	0.9058	H17	1.0142	0.9368	0.9076	-0.002	-0.002
H18	1.0837	1.0745	1.0491	H18	1.0837	1.0745	1.0497	0.000	-0.001
H19	1.1195	1.0996	1.063	H19	1.1194	1.0996	1.0618	0.000	0.001

TABLE 29.10 Comparison of 5000 and 500 replicas (Fisher 95%).

	500 replicas			5000 replicas				Difference	
500	97.50%	DEA	2.50%	5000	97.50%	DEA	2.50%	97.50%	2.50%
H1	0.9228	0.7540	0.724	H1	0.9184	0.7540	0.7227	0.0044	0.0013
H2	0.8279	0.7787	0.7415	H2	0.8266	0.7787	0.7412	0.0013	0.0003
H3	0.6285	0.5918	0.573	H3	0.6291	0.5918	0.5719	-0.0006	0.0011
H4	0.7574	0.6802	0.6694	H4	0.7581	0.6802	0.6679	-0.0007	0.0015
H5	0.6375	0.6042	0.5792	H5	0.6379	0.6042	0.5801	-0.0004	-0.0009
H6	0.9384	0.8475	0.8159	H6	0.9423	0.8475	0.8164	-0.0039	-0.0005
H7	0.7620	0.7250	0.6998	H7	0.7615	0.7250	0.6985	0.0005	0.0013
H8	0.6902	0.6311	0.6002	H8	0.6907	0.6311	0.5998	-0.0005	0.0004
H9	0.6030	0.5681	0.5452	H9	0.6030	0.5681	0.5456	0	-0.0004
H10	0.7963	0.6308	0.6032	H10	0.7942	0.6308	0.6055	0.0021	-0.0023
H11	0.7433	0.6985	0.6808	H11	0.7447	0.6985	0.6808	-0.0014	0
H12	0.7684	0.7140	0.6849	H12	0.7684	0.7140	0.6828	0	0.0021
H13	1.0465	0.8310	0.8081	H13	1.0460	0.8310	0.8081	0.0005	0
H14	1.1564	1.1448	1.1037	H14	1.1565	1.1448	1.1026	$-1E-04$	0.0011
H15	0.8692	0.8107	0.7886	H15	0.8726	0.8107	0.7886	-0.0034	0
H16	0.8792	0.7418	0.714	H16	0.8785	0.7418	0.7198	0.0007	-0.0058
H17	1.0142	0.9368	0.9076	H17	1.0141	0.9368	0.9051	$1E-04$	0.0025
H18	1.0837	1.0745	1.0497	H18	1.0837	1.0745	1.0459	0	0.0038
H19	1.1194	1.0996	1.0618	H19	1.1193	1.0996	1.0618	$1E-04$	0
							Max	0.0044	0.0038
							Min	-0.0039	-0.0058

TABLE 29.11   2009 forecasts: linear trend model.

DMU	(I) Doc	(I) Nurse	(O) Inpatient	(O) Outpatient
H1	120	473	628	1249
H2	141	550	634	1257
H3	124	633	553	1030
H4	138	521	709	1178
H5	146	619	635	1099
H6	113	615	686	1423
H7	112	490	562	1069
H8	130	519	512	1015
H9	129	516	491	903
H10	110	576	549	1288
H11	110	499	512	1075
H12	146	699	675	1677
H13	109	753	605	1755
H14	119	538	621	1920
H15	161	809	848	2072
H16	59	364	372	785
H17	113	627	687	1874
H18	66	382	454	901
H19	92	305	458	844

29.3.2   Illustration of the Past–Present–Future Framework

Here, we present numerical results for the past–present–future framework. In this case we regard 2007–2008 as the past–present and 2009 as the future. In our application, we used three simple prediction models to forecast the future, namely, a linear-trend analysis model, a weighted average model with Lucas weights and a hybrid model that consists of averaging the predictions of the latter two models.

Table 29.11 reports the forecasts for 2009 obtained from the linear-trend analysis model. Table 29.12 shows the forecast DEA scores and confidence intervals along with the actual super-SBM scores for 2009. Figure 29.4 shows the 97.5% and 2.5% confidence intervals, the forecast scores, and the actual scores. It can be observed that, out of the 19 hospitals, the actual 2009 scores of 16 are included in the 95% confidence interval. The average of forecast minus actual over the 19 hospitals was 0.063 (6.3%).

Table 29.13 reports 2009 forecasts obtained from the weighted average model with Lucas weights, and Table 29.14 shows the actuals and the forecasts of the 2009 scores along with the confidence intervals. In this case, only four hospitals are included in the 95% confidence interval. The average of forecast minus actual over the 19 hospitals is 0.056 (5.6%). Although we have not reported the results from the average of the trend and Lucas cases, the results are similar to those for the Lucas case. We compared the number of fails for the three forecast models where the actual score was outside the 97.5% and 2.5% intervals. The results are shown in Table 29.15. 'Trend' gives the best performance among the three in this example.

TABLE 29.12   **Forecast DEA scores, actual (2009) scores and confidence intervals: forecasts by linear trend model.**

DMU	97.50%	Forecast (2009)	Actual (2009)	2.50%
H1	1.0237	0.9338	0.754	0.8245
H2	1.0027	0.787	0.7787	0.722
H3	0.6649	0.6148	0.5918	0.5641
H4	0.8816	0.8581	0.6802	0.7319
H5	0.6814	0.6421	0.6042	0.5771
H6	1.0213	0.8768	0.8475	0.8062
H7	0.8292	0.7586	0.7250	0.6945
H8	0.7641	0.6725	0.6311	0.6066
H9	0.6983	0.6213	0.5681	0.539
H10	0.8422	0.7781	0.6308	0.7111
H11	0.8425	0.7206	0.6985	0.6679
H12	0.8136	0.7716	0.714	0.7068
H13	1.0814	1	0.831	0.8276
H14	1.1575	1.0909	1.1448	1.0281
H15	0.9467	0.8541	0.8107	0.7902
H16	1.0376	0.9444	0.7418	0.7258
H17	1.0387	1.0348	0.9368	0.8982
H18	1.0899	1.0537	1.0745	0.9692
H19	1.1354	1.0594	1.0996	1.0113

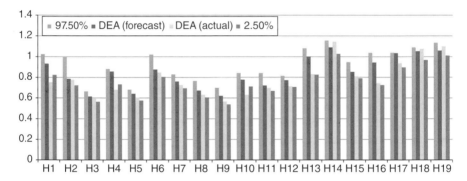

Figure 29.4   Confidence intervals, forecast scores and actual 2009 scores: forecasts by linear trend model.

29.4   CONCLUSION

DEA, originated by Charnes *et al.* [1], is a non-parametric mathematical programming methodology that deals directly with input/output data. Using the data, DEA can evaluate the relative efficiency of DMUs and propose a plan to improve the inputs/outputs of inefficient DMUs. This function is difficult to achieve with similar models using

TABLE 29.13   2009 forecasts: Lucas-weighted average model.

DMU	(I) Doc	(I) Nurse	(O) Inpatient	(O) Outpatient
H1	112	446	613	1242
H2	130	482	639	1328
H3	120	589	574	1058
H4	138	534	702	1199
H5	141	615	647	1163
H6	104	584	706	1496
H7	100	495	547	1066
H8	114	480	501	1039
H9	116	472	486	868
H10	105	552	570	1275
H11	98	496	497	1170
H12	147	714	739	1650
H13	105	646	654	1926
H14	107	513	616	1885
H15	160	798	925	2173
H16	71	357	397	960
H17	112	651	707	1741
H18	63	386	471	909
H19	96	317	491	1076

TABLE 29.14   DEA score and confidence interval forecasts: Lucas-weighted average model.

	97.50%	Forecast (2009)	Actual (2009)	2.50%
H1	1.0001	0.8974	0.7540	0.8469
H2	0.9329	0.8527	0.7787	0.797
H3	0.6448	0.6218	0.5918	0.5987
H4	0.7855	0.7618	0.6802	0.7303
H5	0.6584	0.6400	0.6042	0.6200
H6	1.0101	0.9604	0.8475	0.9123
H7	0.7813	0.7347	0.7250	0.7006
H8	0.7201	0.6867	0.6311	0.6596
H9	0.6578	0.6177	0.5681	0.5894
H10	0.8109	0.7829	0.6308	0.7441
H11	0.8101	0.7573	0.6985	0.7171
H12	0.7623	0.7336	0.7140	0.712
H13	1.0590	1.0286	0.8310	1
H14	1.1306	1.0868	1.1448	1.0409
H15	0.9120	0.8665	0.8107	0.8263
H16	0.9296	0.8488	0.7418	0.7869
H17	0.9731	0.9427	0.9368	0.8984
H18	1.0686	1.0443	1.0745	1.0115
H19	1.1075	1.0769	1.0996	1.0417

TABLE 29.15   Number of fails.

	Trend	Lucas	Average of trend and Lucas
No. of fails	3	15	15

statistics, for example stochastic frontier analysis. DEA scores are not absolute but relative. They depend on the choice of inputs, outputs and DMUs as well as on the choice of the model for assessing DMUs. DEA scores are subject to change, and thus data variations should be taken into account in DEA. This subject should be discussed from the perspective of itemized input/output variations. From this point of view, we have proposed two models. The first model utilizes historical data for the data generation process, and hence this model resamples data from a discrete distribution. It is expected that, if the historical data are widely volatile, the confidence intervals will prove to be very wide, even when the Lucas weights decrease in the past–present periods. In such cases, application of the moving-average method is recommended. Rolling simulations will be useful for deciding on the choice of the length of the historical span. However, too large an amount of past-year data is not recommended, because environments such as healthcare service systems change rapidly. The second model aims to forecast the future efficiency and its confidence intervals. For forecasting, we used three models, namely, a linear trend model, a weighted average and the average of the results of the latter two models. On this subject, the work of Xu and Ouenniche [19,20] may be useful for the selection of forecasting models, and that of Chang et al. [21] may provide useful information about the estimation of pessimistic and optimistic probabilities in forecasts of future input/output values.[3]

REFERENCES

[1] Charnes, A., Cooper W.W. and Rhodes, E. (1978) Measuring the efficiency of decision making units. *European Journal of Operational Research*, **2**, 429–444.

[2] Banker, R. (1993) Maximum likelihood, consistency and data envelopment analysis: A statistical foundation. *Management Science*, **39**, 1265–1273.

[3] Banker, R. and Natarajan, R. (2004) Statistical test based on DEA efficiency scores, in *Handbook on Data Envelopment Analysis* (eds W.W. Cooper, L.M. Seiford and J. Zhu), Springer, Chapter 11.

[4] Charnes, A. and Neralić, L. (1990) Sensitivity analysis of the additive model in data envelopment analysis. *European Journal of Operational Research*, **48**, 332–341.

[5] Neralić, L. (1998) Sensitivity analysis in models of data envelopment analysis. *Mathematical Communications*, **3**, 41–59.

[3] Software for resampling models is included in DEA-Solver Pro V13 (http://www.saitech-inc.com). See also Appendix A.

[6] Zhu, J. (2001) Super-efficiency and DEA sensitivity analysis. *European Journal of Operational Research*, **129**, 443–455.

[7] Simar, L. and Wilson, P. (1998) Sensitivity of efficiency scores: How to bootstrap in nonparametric frontier models. *Management Science*, **44**(1), 49–61.

[8] Simar, L. and Wilson, P.W. (2000) A general methodology for bootstrapping in non-parametric frontier models. *Journal of Applied Statistics*, **27**, 779–802.

[9] Barnum, D.T., Gleason, J.M., Karlaftis, M.G., Schumock, G.T., Shields, K.L., Tandon, S. and Walton, S.M. (2011) Estimating DEA confidence intervals with statistical panel data analysis. *Journal of Applied Statistics*, **39**(4), 815–828.

[10] Cook, W.D., Tone, K. and Zhu, J. (2014) Data envelopment analysis: Prior to choosing a model. *Omega*, **44**, 1–4.

[11] Yang, M., Wan, G. and Zheng, E. (2014) A predictive DEA model for outlier detection. *Journal of Management Analytics*, **1**(1), 20–41.

[12] Tone, K. (2002) A slacks-based measure of super-efficiency in data envelopment analysis. *European Journal of Operational Research*, **143**, 32–41.

[13] Ouenniche, J., Xu, B. and Tone, K. (2014) Relative performance evaluation of competing crude oil prices' volatility forecasting models: A slacks-based super-efficiency DEA model. *American Journal of Operations Research*, **4**, 235–245.

[14] Tone, K. (2001) A slacks-based measure of efficiency in data envelopment analysis. *European Journal of Operational Research*, **130**, 498–509.

[15] Cooper, W.W., Seiford, L.M. and Tone, K. (2007) *Data Envelopment Analysis: A Comprehensive Text with Models, Applications, References and DEA-Solver Software*, 2nd edn, Springer.

[16] Efron, B. (1979) Bootstrap methods: Another look at the jackknife. *Annals of Statistics*, **7**, 1–26.

[17] Efron, B. and Tibshirani, R. (1993) *An Introduction to the Bootstrap*. Chapman & Hall/CRC Press, New York.

[18] Fisher, R.A. (1915) Frequency distribution of the values of the correlation coefficient in samples from an indefinitely large population. *Biometrika*, **10**(4), 507–521.

[19] Xu, B. and Ouenniche, J. (2011) A multidimensional framework for performance evaluation of forecasting models: Context-dependent DEA. *Applied Financial Economics*, **21**(24), 1873–1890.

[20] Xu, B. and Ouenniche, J. (2012) A data envelopment analysis-based framework for the relative performance evaluation of competing crude oil prices' volatility forecasting model. *Energy Economics*, **34**, 576–583.

[21] Chang, T.S., Tone, K. and Wu, C.H. (2014) Past–present–future intertemporal DEA models. *Journal of the Operational Research Society*, **214**, 73–98.

30

DEA MODELS INCORPORATING UNCERTAIN FUTURE PERFORMANCE[1]

TSUNG-SHENG CHANG

Department of Transportation and Logistics Management, National Chiao Tung University, Hsinchu, Taiwan

KAORU TONE

National Graduate Institute for Policy Studies, Tokyo, Japan

CHEN-HUI WU

Department of Accounting and Information Technology, National Chung Cheng University, Chia-yi County, Taiwan

30.1 INTRODUCTION

Companies in most, if not all, industries are operating in a volatile world. The pace of change in various business environments has been nothing short of remarkable. Therefore, companies can expect to experience a sustained level of volatility over the next few years. For example, crude oil prices and currency exchange rates have been exhibiting high volatility recently, due to both natural and human causes, and will continue to do so. It is evident that all companies, regardless of the industry they operate in, are

---

[1] Part of the material in this chapter is adapted from the *European Journal of Operational Research* (2016), doi.10.1016/j.ejor.2016.04.005 [1], with permission from Elsevier Science.

inevitably affected to varying degrees by crude oil prices and/or currency exchange rates. Of particular interest in this chapter are the industries that are highly sensitive to macroeconomic indices such as crude oil prices and currency exchange rates. That is, the entirety of company performance in those industries depends tightly on the future volatility of macroeconomic indices. It follows that to thoroughly evaluate such companies' performance, an evaluator must assess not only their past and present records but also their future potential. Obviously, it is very challenging to evaluate a company's performance when it involves a past–present–future time span. Hence, the research presented in this chapter aims to tackle the problem of how to fully evaluate company performance in highly volatile future environments.

Data envelopment analysis (DEA) has been well recognized as a powerful evaluation tool, and has been applied to a wide variety of practical evaluation applications. It is a non-parametric linear programming technique that measures the relative efficiency of decision-making units (DMUs) by capturing the interaction among a common set of multiple inputs and outputs. It should be noted that conventional DEA models are designed for measuring the productive efficiency of DMUs based merely on historical data. However, such past results are not sufficient for evaluating a DMU's performance in highly volatile operating environments such as those which include highly volatile crude oil prices and currency exchange rates. It is evident that, in such environments, if a DMU's future performance is ignored in the evaluation process when it is sensitive to crude oil price volatility and/or currency fluctuations, then its whole performance may be seriously distorted. Hence, performance evaluation techniques that explicitly take future volatility into account are unavoidable and indispensable in practice.

However, despite its importance, to the best of our knowledge, there have been no DEA models proposed in the literature that take future performance volatility into account. We believe that the work of Chang *et al.* [2] is the only research study so far that simultaneously takes past, present and future performance indicators into account. Their proposed DEA models are, however, most suitable for conducting performance evaluations of DMUs in which future potential, for example R&D expenses, plays a vital role in their competitive success. That is, those DEA models are not designed for evaluating a DMU's performance that is sensitive to macroeconomic indices such as crude oil prices and currency exchange rates, as mentioned above. Therefore, the present research study seeks to develop a new system of DEA models that incorporate the DMUs' uncertain future performance, and thus can be applied to fully measure the efficiency of DMUs in volatile environments. To empirically demonstrate the advantages of the proposed new DEA models over conventional DEA models that ignore future efficiency, our proposed DEA models have been applied to evaluate the performance of high-tech IC design companies in Taiwan.

The remainder of this chapter is organized as follows. Section 30.2 depicts the generalized dynamic evaluation structures used. Section 30.3 introduces the future performance forecasts. In Section 30.4, generalized dynamic DEA models are constructed, and different types of efficiency are defined. Section 30.5 describes the empirical study which was conducted. Section 30.6 presents conclusions based on this research.

30.2   GENERALIZED DYNAMIC EVALUATION STRUCTURES

Consider a past–present–future intertemporal evaluation structure that consists of $(T+k)$ terms $(1,2,\ldots,T+k)$, where the terms $(1,\ldots,T-1)$, T and $(T+1,\ldots,T+k)$, represent the past, present and future time structures, respectively. Figure 30.1 demonstrates such an evaluation structure. As shown in the figure, the past and present terms $(1,2,\ldots,T)$ exhibit a typical dynamic structure; however, the future terms $(T+1,\ldots,T+k)$ show a non-typical dynamic structure. Therefore, this past–present–future intertemporal evaluation structure is referred to as a *generalized dynamic* structure in this chapter. In addition, it should be noted that this evaluation structure is an integration of three different single-term structures that correspond to term t $(t=1,\ldots,T)$, term $T+1$ and term l $(l=T+2,\ldots,T+k)$, respectively. Therefore, in what follows, we first introduce the three single-term evaluation structures. Then, based on these single-term structures, we construct the complete evaluation structure. However, to begin with, we need to define the carry-over activities between two consecutive terms that play a critical role in constructing DEA models that can measure an efficiency that changes over time. Here, we classify the carry-overs into two types to explicitly reflect their actual characteristics: discretionary (free) and non-discretionary (fixed) carry-overs. DMUs can freely handle free carry-overs such as current assets. By contrast, DMUs cannot control fixed carry-overs such as non-current assets. Note that in the generalized dynamic structure, there are carry-overs between pairs of terms $(t,t+1)$, $t=1,\ldots,T$; however, there are no intermediate carry-overs between pairs of future terms $(t,t+1)$, $t=1,\ldots,T,\,T+(k-1)$, owing to the difficulty of forecasting the related values.

First, the evaluation structure with respect to term $t(t=1,\ldots,T)$ is associated with an input set t, output set t, incoming carry-over t and outgoing carry-over t; however, it should be noted that the incoming carry-over 1 from the initial term 0 is usually unknown and is thus omitted (see [3]). Second, the non-typical dynamic evaluation structure with respect to the future term $T+1$ comprises h subterms, denoted by $T+1(l)$, $l=1,\ldots,h$. That is, it is assumed that there are h possible states associated with future term $T+1$; for example, there could be h possible crude oil prices or US dollar currency exchange rates in term $T+1$. Each subterm $T+1(l)(l=1,\ldots,h)$ is associated with a transition probability (weight) from the present term T to the subterm $T+1(l)$ denoted by p_l^{T+1}, such that $\sum_{l=1}^{k}p_l^{T+1}=1$. How to determine p_l^{T+1}, $l=1,\ldots,h$ is detailed in the next section. In addition, each subterm $T+1(l)(l=1,\ldots,h)$ is associated with an input set $T+1(l)$, output set $T+1(l)$ and incoming carry-over $T+1(l)$ with weight p_l^{T+1}. Third, the structure associated with the future terms $T+2,T+3,\ldots,T+k$ is slightly different from that which is associated with the future term $T+1$. More precisely, the only difference between the two structures is that there are no incoming carry-over activities with respect to the future terms $T+2,T+3,\ldots,T+k$ because of the difficulty of forecasting their values. However, two consecutive terms in the future terms $T+2,T+3,\ldots,T+k$ are still connected

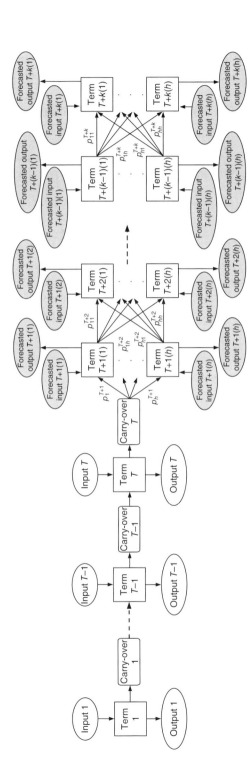

Figure 30.1 Generalized dynamic evaluation structure.

by the conditional probability of occurrence. That is, there is a transition probability (weight) from subterm $T + g(z)(z = 1,\ldots,h)$ of the future term $T + g(g = 1,\ldots,k-1)$ to subterm $T + (g + 1)(l)(l = 1,\ldots,h)$ of the future term $T + (g + 1)$, denoted by $p_{zl}^{T + (g + 1)}$. How to determine these transition probabilities is also detailed in the next section. Furthermore, each subterm $T + g(l)(l = 1,\ldots,h)$ of the future term $T + g(g = 2,\ldots,k)$ is associated with an input set $T + g(l)$ and output set $T + g(l)$ with weight $\sum_{z=1}^{h} p_{zl}^{T + g}$. Note that the assumption here that there are also h possible states associated with the future terms $T + 2, T + 3,\ldots,T + k$ is just for convenience of presentation, and is not a requirement.

Lastly, Figure 30.1 demonstrates the complete generalized dynamic evaluation structure, displaying a time spanning past, present and future periods that are constructed based on the three single-term evaluation structures described above. It is evident that DEA models building on the generalized dynamic structure shown in Figure 30.1 can more accurately evaluate a DMU's performance by explicitly taking its forecasted future performance into account. However, to our knowledge, most of the DEA models, if not all, that have been proposed in the literature do not deal with such a concern.

30.3   FUTURE PERFORMANCE FORECASTS

Notice that the forecasted inputs (e.g. cost of sales) and outputs (e.g. net revenue) depicted in Figure 30.1 are actually functions of variables (e.g. crude oil prices and currency exchange rates) that are sensitive to highly volatile operating environments. It is quite possible, and common, that different DMUs will have different degrees of sensitivity to the variables. Therefore, in such circumstances, to completely evaluate the DMUs, the evaluator must take future performance volatility into account, which is exactly the major point of this research. In addition, each of these variables, for example currency exchange rates, may be measured in several different currencies. For example, a DMU may procure resources (input costs) from and sell products (output revenues) to different countries so that it faces different currencies and thus varying currency exchange rates. Theoretically, a variable that involves n different currencies should be treated as n different variables. However, if this is done, the numbers of inputs and outputs, and thus the size of the generalized dynamic evaluation structure shown in Figure 30.1, will increase exponentially and dramatically. It follows that the differentiation power of the corresponding generalized dynamic DEA models will decrease significantly. Hence, in this instance, we use a single currency to measure the variables by converting other currencies into that currency. For example, we can consider crude oil prices or currency exchange rates based on US dollars by converting other foreign currencies into US dollars.

There exist a variety of forecasting methods to predict the values of the above variables [4]. However, none of them can be considered to be superior to the others in every respect (see e.g. [5, 6]). Nonetheless, there are some well-accepted principles, such as that short-term forecasts are generally more accurate than medium- and long-term ones; that aggregate forecasts are generally more precise than single ones; and that simple methods are preferable to complex methods because they are easier to understand and explain. It should be noted that the development and the choice of forecasting techniques are themselves big research areas, which are not the focus of this research. This study utilized the moving average method (see e.g. [4]) to estimate the future performance forecasts because the moving average method is one of the most well-known and established forecasting methods in practice [5,7]. Furthermore, this research directly applied data from public-domain resources, which generally do not provide detailed information. Under such circumstances, the concept of entropy in information theory offers a feasible way to measure the uncertainty in the probability distributions of random variables (the future inputs and outputs in this research) (see e.g. [8]). Kapur [8, p. 11] stated that 'We should take all given information into account and we should scrupulously avoid taking into account any information that is not given to us.' This leads to the renowned maximum entropy principle, which 'aims to give us as uniform or as broad a distribution as possible, subject to the constraints being satisfied' [8, p. 11]. Moreover, based on data availability, future inputs and outputs were treated in the present study as discrete random variables that take a finite number of values.

The above analysis suggests that we should utilize the maximum entropy approach to determine p_l^{T+1}, the transition probability from the present term T to the subterm $T+1(l)$ of the future term $T+1$, and $p_{zl}^{T+(g+1)}$, the transition probability from the subterm $T+g(z)$ of the future term $T+g$ to the subterm $T+(g+1)(l)$ of the future term $T+(g+1)$, which were described in the preceding section. First, the determination of p_l^{T+1}, $l=1,\ldots,h$ according to the principle of maximum entropy can be formulated as the following mathematical problem (P1):

$$\max \quad -\sum_{l=1}^{h} p_l^{T+1} \ln p_l^{T+1}$$

s.t.

(P1)
$$\sum_{l=1}^{h} p_l^{T+1} q_l^{T+1} = E^{T+1}$$

$$\sum_{l=1}^{h} p_l^{T+1} = 1$$

$$p_l^{T+1} \geq 0 \quad l=1,\ldots,h$$

where q_l^{T+1} denotes the forecast of the target variable q (e.g. crude oil prices or currency exchange rates) associated with subterm $T+1(l)$ of the future term $T+1$, and E^{T+1} denotes the expected value of q in the future term $T+1$. E^{T+1} is used here to confine the values of p_l^{T+1}, $l=1,\ldots,h$. The reason for adding the expected-value constraint in problem (P1) is that, as indicated by Hyndman and Athanasopoulos [9], 'When we talk about the "forecast", we usually mean the average value of the forecast distribution'. That is, after the forecasting process described above, the expected value

E^{T+1} is obtained and fixed. Hence, the expected-value constraint must be added into problem (P1). Note also that, in the absence of the expected-value constraint, the solution to the mathematical problem (P1) is the uniform probability distribution. Therefore, the maximum entropy principle can be considered as an extension of Laplace's principle of insufficient reasoning (see e.g. [10]). Next, the problem of determining $p_{zl}^t, z, l = 1, \ldots, h (t = T+2, \ldots, T+k)$ according to the principle of maximum entropy has the following mathematical form (P2):

$$\max \quad -\sum_{l=1}^{h}\sum_{z=1}^{h} p_{zl}^t \ln p_{zl}^t$$

s.t.

$$\sum_{l=1}^{h} p_{zl}^t = 1 \quad z = 1, \ldots, h$$

(P2)
$$\left(\sum_{z=1}^{h} p_{zl}^t\right)/h = p_l^t \quad l = 1, \ldots, h$$

$$\sum_{l=1}^{h} p_l^t q_l^t = E^t$$

$$\left(\left|q_l^t - q_z^{t-1}\right| - \left|q_l^t - q_m^{t-1}\right|\right)\left(p_{zl}^t - p_{ml}^t\right) \le 0 \quad l, z, m = 1, \ldots, h$$

$$p_{zl}^t \ge 0 \quad z, l = 1, \ldots, h$$

$$p_l^t \ge 0 \quad l = 1, \ldots, h$$

where q_l^t denotes the forecast of the target variable q (e.g. the crude oil price or a foreign exchange rate) associated with subterm $t(l)$ of the future term $t(t = T+2, \ldots, T+k)$, and E^t denotes the expected value of q in the future term t that is known from public domain resources. In addition, the fourth set of constraints ensure that if q_l^t is closer to q_z^{t-1} than to q_m^{t-1}, then the transition probability p_{zl}^t is no less than the transition probability p_{ml}^t. That is, the constraints ensure that the decisions made in term t should take the information in term $t-1$ into account. Note that the mathematical problems (P1) and (P2) are concave programming problems with linear constraints, which can be solved directly using commercial optimization solvers such as LINGO. We refer the reader to, for example, Kapur [8] and Fang et al. [10] for methods for solving entropy optimization problems.

In addition, let $\bar{p}_l^{T+1}, l = 1, \ldots, h$ and $\bar{p}_{zl}^t, z, l = 1, \ldots, h, t = T+2, \ldots, T+k$ define the solutions after solving problems (P1) and (P2), respectively. If, however, the decision-makers have their own subjective weight perception of $\bar{p}_l^{T+1}, l = 1, \ldots, h$ and $\bar{p}_{zl}^t, z, l = 1, \ldots, h (t = T+2, \ldots, T+k)$ denoted, respectively, by $\vartheta_l^{T+1}, l = 1, \ldots, h$ and $\phi_{zl}^t, z, l = 1, \ldots, h (t = T+2, \ldots, T+k)$, then they can modify the solution values by using the following formulas, which are similar to the one proposed by Yoon and Hwang [11].

First, with respect to $\bar{p}_l^{T+1}, l = 1, \ldots, h$,

$$w_l^{T+1} = \frac{\bar{p}_l^{T+1} \, \vartheta_l^{T+1}}{\displaystyle\sum_{i=1}^{h} \bar{p}_i^{T+1} \, \vartheta_i^{T+1}}, \quad l = 1, \ldots, h$$

Then, with respect to $\bar{p}_{zl}^t, z, l = 1, \ldots, h (t = T+2, \ldots, T+k)$,

$$w_{zl}^t = \frac{\bar{p}_{zl}^t \, \phi_{zl}^t}{\displaystyle\sum_{j=1}^{h} \bar{p}_{zj}^t \, \phi_{zj}^t}, \quad z, l = 1, \ldots, h$$

$$w_l^t = \left(\sum_{z=1}^{h} w_{zl}^t \right) / h, \quad l = 1, \ldots, h$$

It follows that w_l^t can be used to denote the weight associated with subterm $t(l)$ of every future term $t(t = T+1, \ldots, T+k)$. We note, however, that decision-makers may have different subjective weight perceptions of inputs and outputs. Therefore, we separate w_l^t further into w_l^{-t} and w_l^{+t}, which correspond to inputs and outputs, respectively.

30.4 GENERALIZED DYNAMIC DEA MODELS

This research proposes a new system of DEA models which embed the generalized dynamic structure described in Section 30.2. However, dynamic DEA models with a typical dynamic structure such as those proposed by Tone and Tsutsui [3] can be used as building blocks to develop generalized dynamic DEA models that incorporate the uncertain future performance of DMUs.

To construct these generalized dynamic DEA models, it is assumed that there are n DMUs $(j = 1, \ldots, n)$ over $(T+k)$ terms $(t = 1, \ldots, T+k)$. In each term $t(t = 1, \ldots, T)$, the DMUs have m common inputs $(i = 1, \ldots, m)$ and s common outputs $(i = 1, \ldots, s)$. On the other hand, in each term $t(t = T+1, \ldots, T+k)$, the DMUs have r common inputs $(i = 1, \ldots, r)$ and/or d common outputs $(i = 1, \ldots, d)$. That is, it is important to note that, depending on the problem considered, the future terms $T+1, \ldots, T+k$ may not simultaneously associate both inputs and outputs. Furthermore, let $x_{ijt}(i = 1, \ldots, m)$ and $y_{ijt}(i = 1, \ldots, s)$ represent the input and output, respectively, of DMU j in term $t(t = 1, \ldots, T)$, and let $u_{ijtl}(i = 1, \ldots, r)$ and $v_{ijtl}(i = 1, \ldots, d)$ represent the input and output, respectively, of DMU j in subterm $t(l)$ of the future term $t(t = T+1, \ldots, T+k)$. Recall that both the input u_{ijtl} and the output v_{ijtl} are functions of variables, such as crude oil prices and currency exchange rates, that are measured in a common currency, for example the US dollar.

In addition, recall that it is assumed that each future term $t(t = T + 1, \ldots, T + k)$ comprises h subterms (possible states) $t(l)$, $l = 1, \ldots, h$. Moreover, we denote the free and fixed carry-overs (links) by z_{ijt}^{free} $(i = 1, \ldots, nfree; j = 1, \ldots n; t = 1, \ldots T)$ and z_{ijt}^{fix} $(i = 1, \ldots, nfix; j = 1, \ldots n; t = 1, \ldots T)$, respectively, where $nfree$ and $nfix$ are the numbers of free and fixed links, respectively. Recall that there are no carry-over activities with respect to future terms owing to the high degree of difficulty of forecasting.

30.4.1 Production Possibility Sets

Based on the notation defined above, the production possibility set $\left\{ \left(x_{it}, y_{it}, u_{itl}, v_{itl}, z_{it}^{\text{free}}, z_{it}^{\text{fix}} \right) \right\}$ with respect to the generalized dynamic DEA models is defined as follows:

$$x_{it} \geq \sum_{j=1}^{n} x_{ijt}\lambda_j^t \quad (i = 1, \ldots, m; t = 1, \ldots, T)$$

$$y_{it} \leq \sum_{j=1}^{n} y_{ijt}\lambda_j^t \quad (i = 1, \ldots, s; t = 1, \ldots, T)$$

$$u_{itl} \geq \sum_{j=1}^{n} u_{ijtl}\delta_{jl}^t \quad (i = 1, \ldots, r; t = T + 1, \ldots, T + k; l = 1, \ldots, l)$$

$$v_{itl} \geq \sum_{j=1}^{n} v_{ijtl}\delta_{jl}^t \quad (i = 1, \ldots, d; t = T + 1, \ldots, T + k; l = 1, \ldots, l)$$

$$z_{it}^{\text{free}} \quad \text{unrestricted} \quad (i = 1, \ldots, nfree; t = 1, \ldots, T)$$

$$z_{it}^{\text{fix}} = \sum_{j=1}^{n} z_{ijt}^{\text{fix}}\lambda_j^t \quad (i = 1, \ldots, nfix; t = 1, \ldots, T)$$

$$\sum_{j=1}^{n} \lambda_j^t = 1 \quad (t = 1, \ldots, T)$$

$$\sum_{j=1}^{n} \delta_{jl}^t = 1 \quad (t = T + 1, \ldots, T + k; l = 1, \ldots, h)$$

$$\lambda_j^t \geq 0 \quad (j = 1, \ldots, n; t = 1, \ldots, T)$$

$$\delta_{jl}^t \geq 0 \quad (j = 1, \ldots, n; l = 1, \ldots, h; t = T + 1, \ldots, T + k)$$

In the above production possibility set, $\lambda^t \in R^n (t = 1, \ldots, T)$ and $\delta_l^t \in R^n$ $(l = 1, \ldots, h; t = T + 1, \ldots, T + k)$ are the intensity vectors, and the third and fourth to last constraints correspond to the variable-returns-to-scale assumption (if the constraints are omitted, then the production possibility set is associated with the assumption of constant returns to scale). Furthermore, it should be noted that x_{ijt} and y_{ijt} on the right-hand side of the above constraints are observed positive data, u_{ijtl} and v_{ijtl} are forecasted positive data, and x_{it}, y_{it}, u_{itl} and v_{itl} on the left-hand side are variables. Moreover, notice that the constraints in the production possibility set are defined separately for each term. Hence, to ensure the continuity of link flows (carry-overs) between two consecutive terms of the past terms $(1, \ldots, T-1)$, the present term (T) and the first future term $(T+1)$, we need to include the following conditions:

$$\sum_{j=1}^{n} z_{ijt}^{\text{free}} \lambda_j^t = \sum_{j=1}^{n} z_{ijt}^{\text{free}} \lambda_j^{t+1} \quad (\forall i = 1, \ldots, nfree; t = 1, \ldots, T-1)$$

$$\sum_{j=1}^{n} z_{ijt}^{\text{fix}} \lambda_j^t = \sum_{j=1}^{n} z_{ijt}^{\text{fix}} \lambda_j^{t+1} \quad (\forall i = 1, \ldots, nfix; t = 1, \ldots, T-1)$$

$$\sum_{j=1}^{n} z_{ijT}^{\text{free}} \lambda_j^T = \sum_{l=1}^{h} p_l^{T+1} \left(\sum_{j=1}^{n} z_{ijT}^{\text{free}} \delta_{jl}^{T+1} \right) \quad (\forall i = 1, \ldots, nfree)$$

$$\sum_{j=1}^{n} z_{ijT}^{\text{fix}} \lambda_j^T = \sum_{l=1}^{h} p_l^{T+1} \left(\sum_{j=1}^{n} z_{ijT}^{\text{fix}} \delta_{jl}^{T+1} \right) \quad (\forall i = 1, \ldots, nfix)$$

30.4.2 DEA Models Incorporating Uncertain Future Performance

Based on the production possibility set constructed in the preceding subsection, we now develop DEA models that incorporate uncertain future performance. It is emphasized that all the proposed models are non-radial slacks-based measure (SBM) models [12]. That is, these models consider the excesses associated with inputs and/or the shortfalls associated with outputs as the main targets of the evaluation. In addition, because of this, and depending on the problem considered, the future terms $T+1, \ldots, T+k$ may not simultaneously associate both inputs and outputs. This research considers input-oriented, output-oriented and non-oriented models, which are introduced in that sequence in the following subsections.

However, for convenience of modelling, we first denote DMU$_o$ $(o = 1, \ldots, n)$ as follows:

$$x_{iot} = \sum_{j=1}^{n} x_{ijt} \lambda_j^t + s_{iot}^- \quad (i = 1, \ldots, m; t = 1, \ldots, T) \tag{30.1}$$

$$y_{iot} = \sum_{j=1}^{n} y_{ijt} \lambda_j^t - s_{iot}^+ \quad (i = 1, \ldots, s; t = 1, \ldots, T) \tag{30.2}$$

$$z_{iot}^{\text{free}} = \sum_{j=1}^{n} z_{ijt}^{\text{free}} \lambda_j^t + s_{iot}^{\text{free}} \quad (i = 1, \ldots, nfree; t = 1, \ldots, T) \tag{30.3}$$

$$z_{iot}^{\text{fix}} = \sum_{j=1}^{n} z_{ijt}^{\text{fix}} \lambda_j^t \quad (i = 1, \ldots, nfix; t = 1, \ldots, T) \tag{30.4}$$

$$\sum_{j=1}^{n} z_{ijt}^{\text{free}} \lambda_j^t = \sum_{j=1}^{n} z_{ijt}^{\text{free}} \lambda_j^{t+1} \quad (\forall i = 1, \ldots, nfree; t = 1, \ldots, T-1) \tag{30.5}$$

$$\sum_{j=1}^{n} z_{ijt}^{\text{fix}} \lambda_j^t = \sum_{j=1}^{n} z_{ijt}^{\text{fix}} \lambda_j^{t+1} \quad (\forall i = 1, \ldots, nfix; t = 1, \ldots, T-1) \tag{30.6}$$

$$\sum_{j=1}^{n} \lambda_j^t = 1 \quad (t = 1, \ldots, T) \tag{30.7}$$

$$\lambda_j^t \geq 0 \quad (\forall j, t) \tag{30.8}$$

$$s_{iot}^- \geq 0 \quad (\forall i, t) \tag{30.9}$$

$$s_{iot}^+ \geq 0 \quad (\forall i, t) \tag{30.10}$$

$$s_{iot}^{\text{free}} : \text{unrestricted in sign} \quad (\forall i, t) \tag{30.11}$$

$$u_{iotl} = \sum_{j=1}^{n} u_{ijtl} \delta_{jl}^t + e_{iotl}^- \quad (i = 1, \ldots, r; t = T+1, \ldots, T+k; l = 1, \ldots, h) \tag{30.12}$$

$$v_{iotl} = \sum_{j=1}^{n} v_{ijtl} \delta_{jl}^t - e_{iotl}^+ \quad (i = 1, \ldots, d; t = T+1, \ldots, T+k; l = 1, \ldots, h) \tag{30.13}$$

$$\sum_{j=1}^{n} z_{ijT}^{\text{free}} \lambda_j^T = \sum_{l=1}^{h} p_l^{T+1} \left(\sum_{j=1}^{n} z_{ijT}^{\text{free}} \delta_{jl}^{T+1} \right) \quad (\forall i = 1, \ldots, nfree) \tag{30.14}$$

$$\sum_{j=1}^{n} z_{ijT}^{\text{fix}} \lambda_j^T = \sum_{l=1}^{h} p_l^{T+1} \left(\sum_{j=1}^{n} z_{ijT}^{\text{fix}} \delta_{jl}^{T+1} \right) \quad (\forall i = 1, \ldots, nfix) \tag{30.15}$$

$$\sum_{j=1}^{n} \delta_{jl}^t = 1 \quad (t = T+1, \ldots, T+k; l = 1, \ldots, h) \tag{30.16}$$

$$\delta_{jl}^t \geq 0 \quad (\forall j, l, t) \tag{30.17}$$

$$e_{iotl}^- \geq 0 \quad (\forall i, l, t) \tag{30.18}$$

$$e_{iotl}^+ \geq 0 \quad (\forall i, l, t) \tag{30.19}$$

30.4.2.1 *Input-Oriented Efficiency* θ_o^*   The input-oriented generalized dynamic DEA model corresponding to DMU$_o$ $(o = 1, \ldots, n)$ can be expressed as follows:

$$\theta_o^* = \min \frac{1}{\sum_{t=1}^{T+k} \alpha^t} \left[\sum_{t=1}^{T} \alpha^t \left[1 - \frac{1}{m} \left(\sum_{i=1}^{m} \frac{\rho_i^- s_{iot}^-}{x_{iot}} \right) \right] + \sum_{t=T+1}^{T+k} \alpha^t \sum_{l=1}^{h} w_l^{-t} \left[1 - \frac{1}{r} \left(\sum_{i=1}^{r} \frac{\mu_i^- e_{iotl}^-}{u_{iotl}} \right) \right] \right]$$

$$\tag{30.20}$$

subject to (30.1)–(30.19), where α^t is the *term weight* corresponding to term $t(t=1,\ldots,T+k)$ that is specified by the evaluator; w_l^{-t}, defined in Section 30.3, is the evaluator-specified *future subterm input weight* corresponding to subterm $t(l)(l=1,\ldots,h)$ of the future term $t(t=T+1,\ldots,T+k)$; and ρ_i^-, μ_i^- are the evaluator-specified *past–present input weight* and *future input weight* that correspond, respectively, to the past–present input $i(i=1,\ldots,m)$ and the future input $i(i=1,\ldots,r)$. In addition, the weights are set to satisfy the following conditions:

$$\sum_{l=1}^{h} w_l^{-t} = 1 \ (t=T+1,\ldots,T+k), \ \sum_{i=1}^{m}\rho_i^- = m \text{ and } \sum_{i=1}^{r}\mu_i^- = r.$$

It is evident that the objective function involves $T+hk$ efficiency-related scores measured by the relative slacks of the inputs, where T scores are related to the T past–present terms, and hk scores are related to the k future terms, with each consisting of h subterms. That is, the objective function is defined as the weighted average of $T+hk$ efficiency-related scores measured by the relative slacks of inputs. Note that each score is units-invariant and has a value less than or equal to 1 (the latter is realized when all the corresponding slacks are zero). It follows that the objective function value is less than or equal to 1. Recall that the future subterm input weight $w_l^{-t}(l=1,\ldots,h; t=T+1,\ldots,T+k)$ in the objective function (30.20) is derived from $\bar{p}_l^{T+1}(l=1,\ldots,h)$ and $\bar{p}_{zl}^t(z,l=1,\ldots,h; t=T+2,\ldots,T+k)$, which are the solutions after solving problems (P1) and (P2), respectively, in Section 30.3.

Let the optimal solution to the above model be $\lambda_j^{t*}, \delta_{jl}^{t*}, s_{iot}^{-*}, s_{iot}^{+*}$, $s_{iot}^{\text{free}*}, e_{iotl}^{-*}, e_{iotl}^{+*}$ ($\forall i,j,t,l$). It is important to note that, since s_{iot}^{free} is unrestricted in sign (i.e. if $s_{iot}^{\text{free}} > 0$, then the current value z_{iot}^{free} is excessive and if $s_{iot}^{\text{free}} < 0$, then z_{iot}^{free} is deficient), slacks in the free links are not considered in the objective function of the input-oriented past–present DEA model. However, as shown by Tone and Tsutsui [3], the slacks can be taken into account in either of the following two ways: (i) the *ex post* way and (ii) the binary mixed integer fractional programming approach. We refer the reader to Tone and Tsutsui [3] for the latter approach, and consider only the former method. That is, let $s_{iot}^{\text{free}*-} = \max\{0, s_{iot}^{\text{free}*}\}$ and $s_{iot}^{\text{free}*+} = -\min\{0, s_{iot}^{\text{free}*}\}$.

Then, we can define the input-oriented overall efficiency θ_o^* as

$$\theta_o^* = \frac{1}{\displaystyle\sum_{t=1}^{T+k}\alpha^t}\left[\sum_{t=1}^{T}\alpha^t\left[1-\frac{1}{m+nfree}\left(\sum_{i=1}^{m}\frac{\rho_i^- \, s_{iot}^{-*}}{x_{iot}}+\sum_{i=1}^{nfree}\frac{s_{iot}^{\text{free}*-}}{z_{iot}^{\text{free}}}\right)\right]\right.$$
$$\left. +\sum_{t=T+1}^{T+k}\alpha^t\sum_{l=1}^{h}w_l^{-t}\left[1-\frac{1}{r}\left(\sum_{i=1}^{r}\frac{\mu_i^- \, e_{iotl}^{-*}}{u_{iotl}}\right)\right]\right]$$

Besides, in such a generalized dynamic evaluation structure, θ_o^* is actually a weighted average of $T+hk$ efficiency scores that are represented by $\theta_{ot}^*, t=1,\ldots,T$ and $\theta_{otl}^*, t=T+1,\ldots,T+k, l=1,\ldots,h$. That is,

$$\theta_{ot}^* = 1 - \frac{1}{m + nfree}\left(\sum_{i=1}^{m} \frac{\rho_i^- \, s_{iot}^{-*}}{x_{iot}} + \sum_{i=1}^{nfree} \frac{s_{iot}^{free*-}}{z_{iot}^{free}} \right), \ (t=1,\dots,T)$$

$$\theta_{otl}^* = 1 - \frac{1}{r}\left(\sum_{i=1}^{r} \frac{\mu_i^- \, e_{iotl}^{-*}}{u_{iotl}} \right), \ (t=T+1,\dots,T+k, l=1,\dots,h)$$

Therefore, the input-oriented overall efficiency, that is, θ_o^*, can be defined as follows:

$$\theta_o^* = \frac{1}{\displaystyle\sum_{t=1}^{T+k} \alpha^t} \left(\sum_{t=1}^{T} \alpha^t \theta_{ot}^* + \sum_{t=T+1}^{T+k} \alpha^t \sum_{l=1}^{h} w_l^{-t} \theta_{otl}^* \right)$$

Definition 30.1 (Input-oriented term efficient)

If $\theta_{ot}^*(t=1,\dots,T)=1$ or $\theta_{otl}^*(t=T+1,\dots,T+k, l=1,\dots,h)=1$, then DMU$_o$ is referred to as *input-oriented term efficient* with respect to the past–present term $t(t=1,\dots,T)$ or to the subterm $t(l)(l=1,\dots,h)$ of the future term $t(t=T+1,\dots,T+k)$, respectively.

Definition 30.2 (Input-oriented overall efficient)

If $\theta_o^*=1$, then DMU$_o$ is referred to as *input-oriented overall efficient*.

Theorem 30.1   DMU$_o$ is *input-oriented overall efficient* if and only if all $T+hk$ terms are *input-oriented term efficient*, that is, $\theta_{ot}^*=1, t=1,\dots,T$ and $\theta_{otl}^*=1, t=T+1,\dots,T+k, l=1,\dots,h$.

Proof. The proof of Theorem 30.1 is straightforward (see e.g. [3]), and is thus omitted here.

30.4.2.2 *Output-Oriented Efficiency* τ_o^* The output-oriented generalized dynamic DEA model corresponding to DMU$_o$ $(o=1,\dots,n)$ can be expressed as follows:

$$\frac{1}{\tau_o^*} = \max \frac{1}{\displaystyle\sum_{t=1}^{T+k} \alpha^t} \left[\sum_{t=1}^{T} \alpha^t \left[1 + \frac{1}{s}\left(\sum_{i=1}^{s} \frac{\rho_i^+ \, s_{iot}^+}{y_{iot}} \right) \right] + \sum_{t=T+1}^{T+k} \alpha^t \sum_{l=1}^{h} w_l^{+t} \left[1 + \frac{1}{d}\left(\sum_{i=1}^{d} \frac{\mu_i^+ \, e_{iotl}^+}{v_{iotl}} \right) \right] \right]$$

$$(30.21)$$

subject to (30.1)–(30.19), where α^t has the same definition as in Section 30.4.2.1, w_l^{+t} is the evaluator-specified *future subterm output weight*, and ρ_i^+, μ_i^+ are the evaluator-specified *past–present output weight* and *future output weight* that correspond, respectively, to the past–present output $i(i=1,\dots,s)$ and the future output $i(i=1,\dots,d)$. In addition, the weights, w_l^{+t}, ρ_i^+ and μ_i^+ are set to satisfy the following conditions:

$$\sum_{l=1}^{h} w_l^{+t} = 1 \ (t = T+1, ..., T+k), \ \sum_{i=1}^{s} \rho_i^+ = s \text{ and } \sum_{i=1}^{d} \mu_i^+ = d$$

The objective function is clearly defined as a weighted average of $T + hk$ efficiency-related scores measured by the relative slacks of the outputs. Each score is units-invariant and has a value greater than or equal to 1 (the latter is realized when all the corresponding slacks are zero). It follows that the objective function value is greater than or equal to 1. Recall that the future subterm output weight $w_l^{+t}(l = 1, ..., h; t = T+1, ..., T+k)$ in the objective function (30.21) is derived from $\bar{p}_l^{T+1}(l = 1, ..., h)$ and $\bar{p}_{zl}^{t}(z, l = 1, ..., h; t = T+2, ..., T+k)$, which are the solutions after solving problems (P1) and (P2), respectively, in Section 30.3.

Let the optimal solution to the above model be $\lambda_j^{t*}, \delta_{jl}^{t*}, s_{iot}^{-*}, s_{iot}^{+*}, s_{iot}^{\text{free}*},$ $e_{iotl}^{-*}, e_{iotl}^{+*} \ (\forall i, j, t, l)$, where $s_{iot}^{\text{free}*}$ is dealt with in the same way as in Section 30.4.2.1. Then, we can define the output-oriented overall efficiency τ_o^* as

$$\tau_o^* = \cfrac{1}{\cfrac{1}{\sum\limits_{t=1}^{T+k} \alpha^t} \left[\sum\limits_{t=1}^{T} \alpha^t \left[1 + \cfrac{1}{s + nfree} \left(\sum\limits_{i=1}^{s} \cfrac{\rho_i^+ s_{iot}^{+*}}{y_{iot}} + \sum\limits_{i=1}^{nfree} \cfrac{s_{iot}^{\text{free}*+}}{z_{iot}^{\text{free}}} \right) \right] + \sum\limits_{t=T+1}^{T+k} \alpha^t \sum\limits_{l=1}^{h} w_l^{+t} \left[1 + \cfrac{1}{d} \left(\sum\limits_{i=1}^{d} \cfrac{\mu_i^+ e_{iotl}^{+*}}{v_{iotl}} \right) \right] \right]}$$

Similarly to the analysis in Section 30.4.2.1, τ_o^* here is a weighted average of $T + hk$ efficiency scores that are represented by $1/\tau_{ot}^*$, $t = 1, ..., T$ and $1/\tau_{otl}^*$, $t = T+1, ..., T+k$, $l = 1, ..., h$. That is,

$$\frac{1}{\tau_{ot}^*} = 1 + \frac{1}{s + nfree} \left(\sum_{i=1}^{s} \frac{\rho_i^+ s_{iot}^{+*}}{y_{iot}} + \sum_{i=1}^{nfree} \frac{s_{iot}^{\text{free}*+}}{z_{iot}^{\text{free}}} \right), \ (t = 1, ..., T)$$

$$\frac{1}{\tau_{otl}^*} = 1 + \frac{1}{d} \left(\sum_{i=1}^{d} \frac{\mu_i^+ e_{iotl}^{+*}}{v_{iotl}} \right), \ (t = T+1, ..., T+k, l = 1, ..., h)$$

Therefore, the output-oriented overall efficiency, that is, τ_o^*, can be defined as follows:

$$\tau_o^* = \cfrac{1}{\cfrac{1}{\sum\limits_{t=1}^{T+k} \alpha^t} \left(\sum\limits_{t=1}^{T} \alpha^t \cfrac{1}{\tau_{ot}^*} + \sum\limits_{t=T+1}^{T+k} \alpha^t \sum\limits_{l=1}^{h} w_l^{+t} \cfrac{1}{\tau_{otl}^*} \right)}$$

Definition 30.3 (Output-oriented term efficient)
If $\tau_{ot}^*(t = 1, ..., T) = 1$ or $\tau_{otl}^*(t = T+1, ..., T+k, l = 1, ..., h) = 1$, then DMU$_o$ is referred to as *output-oriented term efficient* with respect to the past–present term $t(t = 1, ..., T)$ or to subterm $t(l)$ $(l = 1, ..., h)$ of the future term $t(t = T+1, ..., T+k)$, respectively.

Definition 30.4 (Output-oriented overall efficient)

If $\tau_o^* = 1$, then DMU$_o$ is referred to as *output-oriented overall efficient*.

Theorem 30.2   DMU$_o$ is *output-oriented overall efficient* if and only if all $T + hk$ terms are *output-oriented term efficient*, that is, $\tau_{ot}^* = 1$, $t = 1, \ldots, T$ and $\tau_{otl}^* = 1$, $t = T + 1, \ldots, T + k$, $l = 1, \ldots, h$.

Proof. See the proof of Theorem 30.1.

30.4.2.3 *Non-oriented Efficiency* ϖ_o^*

The non-oriented generalized dynamic DEA model corresponding to DMU$_o$ $(o = 1, \ldots, n)$ can be expressed as follows:

$$
\varpi_o^* = \min \frac{\dfrac{1}{\displaystyle\sum_{t=1}^{T+k} \alpha^t} \left[\displaystyle\sum_{t=1}^{T} \alpha^t \left[1 - \dfrac{1}{m} \left(\displaystyle\sum_{i=1}^{m} \dfrac{\rho_i^- \, s_{iot}^-}{x_{iot}} \right) \right] + \displaystyle\sum_{t=T+1}^{T+k} \alpha^t \displaystyle\sum_{l=1}^{h} w_l^{-t} \left[1 - \dfrac{1}{r} \left(\displaystyle\sum_{i=1}^{r} \dfrac{\mu_i^- \, e_{iotl}^-}{u_{iotl}} \right) \right] \right]}{\dfrac{1}{\displaystyle\sum_{t=1}^{T+k} \alpha^t} \left[\displaystyle\sum_{t=1}^{T} \alpha^t \left[1 + \dfrac{1}{s} \left(\displaystyle\sum_{i=1}^{s} \dfrac{\rho_i^+ \, s_{iot}^+}{y_{iot}} \right) \right] + \displaystyle\sum_{t=T+1}^{T+k} \alpha^t \displaystyle\sum_{l=1}^{h} w_l^{+t} \left[1 + \dfrac{1}{d} \left(\displaystyle\sum_{i=1}^{d} \dfrac{\mu_i^+ \, e_{iotl}^+}{v_{iotl}} \right) \right] \right]}
$$

$$(30.22)$$

subject to (30.1)–(30.19), where α^t, w_l^{-t}, ρ_i^-, μ_i^-, w_l^{+t}, ρ_i^+ and μ_i^+ have the same definitions as in Sections 30.4.2.1 and 30.4.2.2.

Note that the non-oriented generalized dynamic DEA model (i.e. (30.1)–(30.19), (30.22)) is a fractional (non-linear) program. Therefore, we transform the model into a linear program by using the Charnes–Cooper transformation (see e.g. [13]). Let the optimal solution to the model (30.1)–(30.19), (30.22) be λ_j^{t*}, δ_{jl}^{t*}, s_{iot}^{-*}, s_{iot}^{+*}, $s_{iot}^{\text{free}*}$, e_{iotl}^{-*}, e_{iotl}^{+*} $(\forall i, j, t, l)$, where $s_{iot}^{\text{free}*}$ is dealt with in the same way as in Section 30.4.2.1. Then, we can define the non-oriented predicted overall efficiency ϖ_o^* as

$$
\varpi_o^* = \frac{\displaystyle\sum_{t=1}^{T} \alpha^t \left[1 - \dfrac{1}{m + nfree} \left(\displaystyle\sum_{i=1}^{m} \dfrac{\rho_i^- \, s_{iot}^{-*}}{x_{iot}} + \displaystyle\sum_{i=1}^{nfree} \dfrac{s_{iot}^{\text{free}*-}}{z_{iot}^{\text{free}}} \right) \right] + \displaystyle\sum_{t=T+1}^{T+k} \alpha^t \displaystyle\sum_{l=1}^{h} w_l^{-t} \left[1 - \dfrac{1}{r} \left(\displaystyle\sum_{i=1}^{r} \dfrac{\mu_i^- \, e_{iotl}^{-*}}{u_{iotl}} \right) \right]}{\displaystyle\sum_{t=1}^{T} \alpha^t \left[1 + \dfrac{1}{s + nfree} \left(\displaystyle\sum_{i=1}^{s} \dfrac{\rho_i^+ \, s_{iot}^{+*}}{y_{iot}} + \displaystyle\sum_{i=1}^{nfree} \dfrac{s_{iot}^{\text{free}*+}}{z_{iot}^{\text{free}}} \right) \right] + \displaystyle\sum_{t=T+1}^{T+k} \alpha^t \displaystyle\sum_{l=1}^{h} w_l^{+t} \left[1 + \dfrac{1}{d} \left(\displaystyle\sum_{i=1}^{d} \dfrac{\mu_i^+ \, e_{iotl}^{+*}}{v_{iotl}} \right) \right]}
$$

In addition, we define the non-oriented term efficiency as

$$
\varpi_{ot}^* = \frac{1 - \dfrac{1}{m + nfree} \left(\displaystyle\sum_{i=1}^{m} \dfrac{\rho_i^- \, s_{iot}^{-*}}{x_{iot}} + \displaystyle\sum_{i=1}^{nfree} \dfrac{s_{iot}^{\text{free}*-}}{z_{iot}^{\text{free}}} \right)}{1 + \dfrac{1}{s + nfree} \left(\displaystyle\sum_{i=1}^{s} \dfrac{\rho_i^+ \, s_{iot}^{+*}}{y_{iot}} + \displaystyle\sum_{i=1}^{nfree} \dfrac{s_{iot}^{\text{free}*+}}{z_{iot}^{\text{free}}} \right)}
$$

with respect to the past–present term $t(t=1,\ldots,T)$ and

$$\varpi_{otl}^* = \frac{1-\dfrac{1}{r}\left(\displaystyle\sum_{i=1}^{r}\dfrac{\mu_i^-\ e_{iotl}^{-*}}{u_{iotl}}\right)}{1+\dfrac{1}{d}\left(\displaystyle\sum_{i=1}^{d}\dfrac{\mu_i^+\ e_{iotl}^{+*}}{v_{iotl}}\right)}$$

with respect to subterm $t(l)(l=1,\ldots,h)$ of the future term $t(t=T+1,\ldots,T+k)$.

Definition 30.5 (Non-oriented term efficient)
If $\varpi_{ot}^* = 1(t=1,\ldots,T)$ or $\varpi_{otl}^* = 1$ $(t=T+1,\ldots,T+k,l=1,\ldots,h)$, then DMU$_o$ is referred to as *non-oriented term efficient* with respect to the past–present term $t(t=1,\ldots,T)$ or to subterm $t(l)$ $(l=1,\ldots,h)$ of the future term $t(t=T+1,\ldots,T+k)$, respectively.

Definition 30.6 (Non-oriented overall efficient)
If $\varpi_o^* = 1$, then DMU$_o$ is referred to as *non-oriented overall efficient*.

Theorem 30.3   DMU$_o$ is *non-oriented overall efficient* if and only if all $T+hk$ terms are *non-oriented term efficient*, that is, $\varpi_{ot}^* = 1$, $t=1,\ldots,T$ and $\varpi_{otl}^* = 1$, $t=T+1,\ldots,T+k$, $l=1,\ldots,h$.
  Proof. See the proof of Theorem 30.1.

30.5 EMPIRICAL STUDY

The proposed generalized dynamic DEA models are new to the DEA literature. Therefore, we conducted an empirical study to analyse and evaluate this new system of DEA models. To begin with, we use the flow chart shown in Figure 30.2 to summarize the whole procedure introduced in the previous sections for conducting the empirical study; that is, (i) choosing inputs (e.g. cost of sales), outputs (e.g. net revenue) and carry-overs (e.g. current assets and non-current assets); (ii) identifying a macroeconomic index (e.g. a currency exchange rate) that affects the situation; (iii) using a forecasting method (e.g. a moving average method) to estimate the values of the macroeconomic index; (iv) specifying the states of the nature of the macroeconomic index, and computing their corresponding transition probabilities by using, for example, the maximum entropy approach; (v) computing forecasts of inputs and outputs from the forecasts of the macroeconomic index (recalling that the forecasted inputs (e.g. cost of sales) and outputs (e.g. net revenue) are functions of the macroeconomic index (e.g. currency exchange rate)); (vi) specifying input, output and term weights; (vii) solving the generalized dynamic DEA models for each DMU; and (viii) analysing efficiency scores and providing managerial guidelines.
  The empirical study was conducted based on actual data concerning high-tech IC design companies in Taiwan. It is well known that the IC design industry is extremely

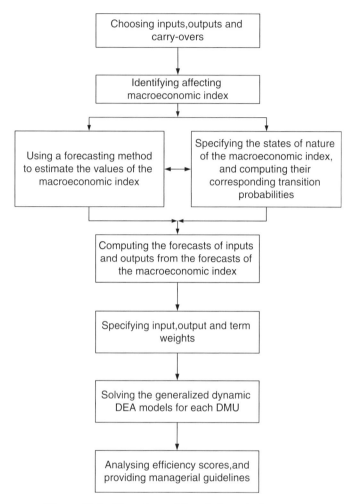

Figure 30.2 Flow chart of the whole analysis process.

competitive. An IC design company usually procures raw materials from a few different countries, seeking to lower its operational costs. And, at the same time, it seeks to sell its products to as many countries as possible to increase profits. Hence, the performance of an IC design company's operations is very sensitive to today's highly volatile international currency exchange rates. To conduct this empirical study, we extracted empirical data, comprising 40 IC design companies, from the *Taiwan Economic Journal* (TEJ) database, utilizing only the latest periods, year 2010 to year 2014 (i.e. $T = 5$), owing to concerns over the ineffectiveness of prior data. That is, short-term forecasts are generally more accurate than medium- and long-term ones in highly volatile operating environments.

To justify the efficacy of the proposed new DEA models, this study used a cross-validation technique to benchmark their performance against realized outcomes. That is, we separated the 2010–2014 data into training and testing sets, in which the 2010–2013 data were used for training and the 2014 data were used for testing. Then, we applied the moving average method to predict year 2014 forecasts (i.e. considering a single future term, and thus $k = 1$) based on the TEJ data from years 2010 to 2013. That is, here, 2010–2012 (term 1–term 3) represents past terms, 2013 (term 4) represents the present term and 2014 (term 5) represents future terms. In addition, to develop benchmarks, this research used the SBM DEA models proposed by Tone [12] and the dynamic SBM DEA models proposed by Tone and Tsutsui [3] to measure the productive efficiency of the 40 IC design companies based on known historical data. These benchmarks are detailed below.

In the following, we first explain the selected inputs, outputs, carry-overs and future performance indicators for the IC design firms illustrated, and then show and analyse the empirical results.

30.5.1   Data Analysis

This study considered *cost of sales* as the single input indicator associated with past, present and future terms; that is, $m = r = 1$. The cost of sales normally includes materials, labour and allocated overheads, and thus contributes directly to the revenue of a firm. More precisely, the cost of sales measures the cost of goods or services supplied in a particular period. On the other hand, this research considered *net revenue* as the single output indicator associated with past, present and future terms; that is, $s = d = 1$. The net revenue is defined by the International Accounting Standards (IAS 18) as the gross inflow of economic benefits during a period arising in the course of the ordinary activities of an entity when those inflows result in increases in equity, other than increases relating to contributions from equity participants. The net revenue arises primarily from the sale of goods or the performance of services. The realized input and output data with respect to the past (years 2010–2012), present (year 2013) and future (year 2014) terms are shown in Tables 30.1–30.5, and comprise 40 companies; the unit of currency used in these tables is the new Taiwan dollar (NT$). To maintain confidentiality when the data were used in this research, the names of the companies have been replaced by Arabic numerals. The realized 2014 'future' data in Table 30.5 were used as a test dataset; recall that the 2014 forecasts were obtained by using the moving average method, which is discussed in more detail below. (Owing to space limitations, please see the original paper [1] for detailed data on the 40 DMUs for which data are presented in Tables 30.1–30.6.)

Furthermore, this research treated a firm's total assets as carry-over activities that connect two consecutive terms. Total assets were classified further here into free carry-over and fixed carry-over because assets are economic resources owned by a firm that will probably be used to produce future economic benefits. More precisely, free carry-over (current assets) denotes the assets that a firm expects to convert to cash or use up within one year (or one operating cycle, whichever is longer). Current assets

TABLE 30.1   Term 1 (2010) data.

DMU	Cost of sales	Net revenue	Current assets	Non-current assets
Average	4 175 227	6 535 401	4 867 534	3 512 612
Max	32 726 157	71 988 430	59 573 161	74 902 743
Min	32 554	48 375	28 564	2 757
S.D.	7 729 137	13 247 030	10 246 537	11 847 992

TABLE 30.2   Term 2 (2011) data.

DMU	Cost of sales	Net revenue	Current assets	Non-current assets
Average	3 900 691	5 653 919	4 234 236	3 949 479
Max	31 773 236	53 842 366	42 508 698	95 386 304
Min	17 656	30 115	36 970	2 992
S.D.	7 560 933	11 047 726	8 036 473	15 030 950

here may be cash and cash equivalents, receivables, inventories, and short-term investments. In contrast, fixed carry-over (non-current assets) denotes the assets that are not easily converted into cash or used up within one year (or one operating cycle, whichever is longer). That is, non-current assets here may be property, plant and equipment, long-term investments, and intangible assets. Note that managers usually have greater discretion to handle current assets, but have less discretion to dispose of non-current assets, which are thus treated here as fixed carry-overs. Note also that each of the past and present terms is associated with a free and a fixed carry-over; that is, $nfree = nfix = 1$. The free and fixed carry-overs with respect to the past (years 2010–2012), present (year 2013) and future (year 2014) terms are also shown in Tables 30.1–30.5.

Moreover, it should be noted that the data in the TEJ database are actually aggregate data. That is, the data do not show the details of the IC design companies' sources and the corresponding quantities for costs, nor the details of the outlets and the corresponding quantities for sales. In addition, the cost of sales and net revenue are all transformed into the same currency, the new Taiwan dollar. We could therefore only use this aggregate data by applying the moving average method to predict the 2014 cost of sales and 2014 net revenue. Nonetheless, aggregate forecasts are generally more precise than single ones. Besides, we adapted the so-called three-sigma rule of thumb to take future uncertainty into account, because it is well known that nearly all values can be taken to lie within three standard deviations of the mean. That is, it was assumed here that each future term was associated with three inputs corresponding to the cost of sales, and three outputs corresponding to the net revenue (i.e. $h = 3$). The three inputs and outputs were defined as the expected value, minimum and maximum of the cost of sales and the net revenue, respectively, for the future term. The minimum and maximum were defined as $\max\{\mu - 3\sigma, \varsigma\}$ and $\min\{\mu + 3\sigma, \xi\}$, respectively, where μ, σ, ς and ξ denote the mean, standard deviation, minimum and

TABLE 30.3   Term 3 (2012) data.

DMU	Cost of sales	Net revenue	Current assets	Non-current assets
Average	4 166 437	6 099 471	4 622 551	5 319 996
Max	40 770 355	63 474 029	49 299 361	148 457 717
Min	5 259	9 554	10 339	6 584
S.D.	8 583 664	12 471 173	9 244 663	23 372 952

TABLE 30.4   Term 4 (2013) data.

DMU	Cost of sales	Net revenue	Current assets	Non-current assets
Average	4 546 435	7 034 350	5 176 708	5 930 613
Max	54 894 385	96 230 064	70 707 646	158 547 310
Min	12 959	33 882	31 303	13 071
S.D.	10 379 908	16 917 176	12 152 507	24 998 370

TABLE 30.5   Term 5 (2014) data.

DMU	Cost of sales	Net revenue	Current assets	Non-current assets
Average	5 160 648	8 417 632	7 484 314	6 395 470
Max	67 990 658	136 265 018	149 267 002	167 574 152
Min	17 190	29 930	45 596	9 090
S.D.	12 669 581	23 148 425	24 024 721	26 540 097

maximum, respectively, of the corresponding raw data (i.e. for the years 2010–2013). Note that μ also denotes E^{T+1} ($= E^5$ here) in problem (P1), which was used to derive the transition probability p^{T+1} ($= p^5$ here) according to the principle of maximum entropy. The resulting forecasted inputs and outputs with respect to the future (year 2014) term are shown in Table 30.6.

Finally, the term weights α^t, $t = 1, \ldots, 5$ were set as follows: $\alpha^1 = 0.7$, $\alpha^2 = 0.8$, $\alpha^3 = 0.9$, $\alpha^4 = 1$, $\alpha^5 = 0.9$; that is, it was assumed that the importance of the information decreases over time. Also, we set the past–present input weight $\rho_1^- = 1$, the future input weight $\mu_1^- = 1$, the past–present output weight $\rho_1^+ = 1$ and the future output weight $\mu_1^+ = 1$. Lastly, based on the data in Table 30.6, we estimated the future subterm input weight, the future subterm output weight and the transition probability as follows. First, we solved the mathematical problem (P1) 40 times with respect to the forecasted inputs and forecasted outputs to obtain 40 possible future subterm input weights and 40 possible future subterm output weights, with each corresponding to a DMU. Then, the future subterm input weights $w_i^{-5}, i = 1, 2, 3$ and future subterm output weights $w_i^{+5}, i = 1, 2, 3$ were set to the averages of the 40 possible future subterm input weights and 40 possible future subterm output weights, respectively. That is, $w_1^{-5} = 0.358734$,

TABLE 30.6   Term 5 (2014) forecasts.

	Cost of sales			Net revenue		
DMU	Minimum	Average (E^5)	Maximum	Minimum	Average (E^5)	Maximum
Average	3 689 589	4 197 198	4 971 864	5 412 220	6 330 785	7 642 849
Max	31 773 236	4 0041 033	54 894 385	53 842 366	71 383 722	96 230 064
Min	5 259	17 107	32 554	9 554	32 157	51 596
S.D.	7 493 897	8 500 767	10 422 141	11 078 441	13 351 848	16 829 011

$w_2^{-5} = 0.339835$, $w_3^{-5} = 0.301431$, and $w_1^{+5} = 0.37963$, $w_2^{+5} = 0.341791$, $w_3^{+5} = 0.278579$ correspond to the forecasted inputs and forecasted outputs, respectively. The transition probabilities $p_i^5 (i = 1, 2, 3)$ were then obtained by taking the average of $w_i^{-5}(i = 1, 2, 3)$ and $w_i^{+5}(i = 1, 2, 3)$; that is, $p_1^5 = 0.369182$, $p_2^5 = 0.340813$, $p_3^5 = 0.290005$.

30.5.2   Analysis of Empirical Results

This section analyses mainly two types of empirical results. First, we contrast past–present efficiency scores with past–present–future (overall) ones. Second, we benchmark the outcomes obtained from the proposed generalized dynamic DEA models with the realized (true) ones to justify the efficacy of these new proposed DEA models.

30.5.2.1   Comparison of Past–Present and Predicted Overall Efficiency Scores

In this section, we compare two sets of efficiency scores of the 40 IC design companies (i.e. overall and past–present efficiency scores) obtained by solving the generalized dynamic DEA models and by solving the generalized dynamic DEA models but ignoring the future terms, which is equivalent to solving the dynamic SBM DEA models proposed by Tone and Tsutsui [3]. The resulting term (past, present and future), past–present and overall efficiency scores corresponding to the input-, output- and non-oriented DEA models are shown in Tables 30.7, 30.8 and 30.9, respectively. These three tables clearly show that the same 12 of the 40 DMUs are input-, output- and non-oriented *overall efficient*. Note that, according to Theorems 30.1–30.3, a DMU is (input-, output- or non-oriented) *overall efficient* if and only if all of the DMU's terms are (input-, output- or non-oriented) *term efficient*. Figure 30.3 shows graphically the input-, output- and non-oriented overall efficiency scores for the 40 DMUs. It turns out that DMU 24 has the greatest difference of 0.409451 among the three scores, which are 0.723425 (input-oriented), 0.360383 (output-oriented) and 0.313974 (non-oriented).

TABLE 30.7 Term efficiency versus overall efficiency (input-oriented case).

DMU	Term efficiency							Past–present efficiency	Overall efficiency
	Term 1	Term 2	Term 3	Term 4	Term 5-1	Term 5-2	Term 5-3		
1	1	1	1	1	1	1	1	1	1
2	1	1	1	1	1	1	1	1	1
3	1	1	1	1	1	1	1	1	1
4	0.579968	1	1	1	0.787473	0.735852	0.937195	1	0.961292
5	1	0.711427	0.629671	0.633289	0.528037	0.526168	0.479886	0.639739	0.613189
6	1	1	1	1	1	1	1	1	1
7	0.864309	0.909265	1	1	0.853187	0.937637	0.986407	0.950714	0.944713
8	0.823282	0.806143	0.840762	0.840527	0.66218	0.665583	0.624407	0.828949	0.791902
9	0.82209	1	0.895437	0.95855	0.885026	0.853936	0.798325	0.923502	0.907767
10	1	1	1	1	1	1	1	1	1
11	0.939729	1	0.985488	1	1	0.978622	0.920359	0.98375	0.980606
12	1	1	1	1	1	1	1	1	1
13	1	1	1	0.914905	0.55406	1	1	0.974972	0.946728
14	0.685954	0.701095	0.57836	0.751402	0.242477	0.240696	0.314155	0.680285	0.593046
15	0.737062	0.923108	0.862162	0.891092	0.614525	0.764685	0.826998	0.859255	0.832118
16	0.65553	0.659839	1	1	1	1	1	0.849042	0.880638
17	1	1	1	1	0.869463	0.746489	1	1	0.972167
18	1	1	1	1	1	1	1	1	1
19	1	1	1	1	0.353378	0.471415	0.685876	1	0.894034
20	0.765205	0.75752	0.817763	0.814016	0.60439	0.627089	0.646916	0.791665	0.756766
21	0.773914	0.878073	0.708645	0.770624	0.327362	0.270755	0.276157	0.780177	0.678145
22	0.878732	0.909403	1	0.965336	0.914668	0.892526	0.859248	0.943521	0.93241
23	0.512471	0.625966	0.762957	0.671913	0.243191	0.234005	0.265056	0.652376	0.567458
24	0.79317	0.717623	0.739548	0.910054	0.320544	0.464478	0.590553	0.795578	0.723425

(continued overleaf)

TABLE 30.7 (*continued*)

DMU	Term efficiency							Past–present efficiency	Overall efficiency
	Term 1	Term 2	Term 3	Term 4	Term 5-1	Term 5-2	Term 5-3		
25	0.742608	0.733516	0.761965	0.711925	0.476522	0.497237	0.502913	0.736568	0.685279
26	0.812016	0.777273	0.875834	0.809555	0.608531	0.5964	0.57701	0.82001	0.772896
27	1	1	1	1	1	1	1	1	1
28	0.818823	0.866214	0.960667	0.94361	0.921145	0.925338	0.85635	0.904223	0.903975
29	0.741116	0.615064	0.726857	0.619101	0.136539	0.168654	0.498592	0.671795	0.584891
30	1	1	1	1	1	1	1	1	1
31	0.621424	0.656665	0.611587	0.543204	0.410257	0.293403	0.310712	0.604106	0.548941
32	1	1	1	1	1	1	1	1	1
33	1	1	0.725296	0.624574	1	0.550374	0.78362	0.816865	0.809563
34	0.660426	0.83274	1	1	1	1	1	0.890732	0.913602
35	1	1	1	1	1	1	1	1	1
36	0.781958	1	1	1	1	1	1	0.955109	0.964505
37	0.872062	0.86376	0.858984	0.888531	0.73241	0.740161	0.686242	0.871491	0.840019
38	0.778174	0.894862	0.964935	0.946716	0.748536	0.713548	0.724994	0.904638	0.867992
39	0.81392	0.849169	0.803093	0.828593	0.838056	0.834191	0.780895	0.823664	0.822795
40	1	1	1	1	1	1	1	1	1
Average	0.861849	0.892218	0.90275	0.900938	0.765799	0.768231	0.798322	0.891318	0.867272
Max	1	1	1	1	1	1	1	1	1
Min	0.512471	0.615064	0.57836	0.543204	0.136539	0.168654	0.265056	0.604106	0.548941
S.D.	0.142645	0.130266	0.130943	0.135486	0.273526	0.268211	0.240355	0.12171	0.144161

TABLE 30.8 Term efficiency versus overall efficiency (output-oriented case).

DMU	Term efficiency							Past–present efficiency	Overall efficiency
	Term 1	Term 2	Term 3	Term 4	Term 5-1	Term 5-2	Term 5-3		
1	1	1	1	1	1	1	1	1	1
2	1	1	1	1	1	1	1	1	1
3	1	1	1	1	1	1	1	1	1
4	1	1	1	1	0.8645	0.73716	0.935011	1	0.95968
5	0.79342	0.7125	0.709344	0.631492	0.642317	0.639993	0.628841	0.699964	0.68595
6	1	1	1	1	1	1	1	1	1
7	0.926314	0.914889	1	1	0.869357	0.942271	0.998738	0.963144	0.955425
8	0.709645	0.670342	0.674178	0.738584	0.702596	0.702307	0.676305	0.698334	0.697628
9	0.873133	1	0.973656	0.906037	0.905167	0.883252	0.853474	0.936698	0.924876
10	1	1	1	1	1	1	1	1	1
11	0.958926	1	0.991575	1	1	0.990972	0.923638	0.989054	0.985976
12	1	1	1	1	1	1	1	1	1
13	1	1	1	1	0.43972	0.793915	0.877296	1	0.88655
14	0.536579	0.53418	0.467487	0.668554	0.266921	0.817651	0.914543	0.546349	0.527379
15	0.915622	0.927081	0.781864	0.614505	0.699158	0.857695	0.822959	0.771701	0.773681
16	0.906725	0.815069	1	1	1	1	1	0.930609	0.944324
17	1	1	1	1	0.903485	1	1	1	0.991583
18	1	1	1	1	1	1	1	1	1
19	1	1	1	1	0.265664	0.42325	1	1	0.759235
20	0.797735	0.735872	0.70643	0.686582	0.63542	0.677412	0.715603	0.724154	0.712239
21	0.320485	0.21431	0.265039	0.3278	0.237665	0.389201	0.82439	0.275004	0.288677
22	0.982774	0.929613	1	0.969318	0.931155	0.914806	0.913481	0.970182	0.959361
23	0.714046	0.571878	0.616943	0.53685	0.333298	0.755131	0.899289	0.596412	0.580138
24	0.561392	0.373531	0.65416	0.208	0.198332	0.800585	0.992208	0.355198	0.360383
25	0.695693	0.631298	0.749941	0.715865	0.661392	0.666776	0.664996	0.698091	0.690721

(continued overleaf)

TABLE 30.8 (continued)

DMU	Term efficiency							Past–present efficiency	Overall efficiency
	Term 1	Term 2	Term 3	Term 4	Term 5-1	Term 5-2	Term 5-3		
26	0.67523	0.619556	0.570778	0.513661	0.711628	0.70313	0.668293	0.581044	0.601877
27	1	1	1	1	1	1	1	1	1
28	0.975773	0.937463	0.927158	0.913171	0.938902	0.945715	0.885761	0.934954	0.933004
29	0.516266	0.303093	0.409813	0.366318	0.344533	0.617527	0.798214	0.381109	0.400923
30	1	1	1	1	1	1	1	1	1
31	0.734463	0.631158	0.803585	0.640064	0.404034	0.46445	0.465227	0.693465	0.618724
32	1	1	1	1	1	1	1	1	1
33	1	1	0.904474	0.753471	0.616785	0.467349		0.88953	0.813605
34	0.980339	0.832905	1	1	1	1	1	0.951173	0.960994
35	1	1	1	1	1	1	1	1	1
36	0.896009	1	1	1	1	1	1	0.976663	0.981457
37	0.866441	0.743105	0.702283	0.733733	0.764974	0.771826	0.774194	0.750735	0.75466
38	0.876349	0.892481	0.670492	0.623782	0.804524	0.775864	0.917129	0.732669	0.749768
39	0.924528	0.90082	0.904343	0.793479	0.862769	0.854609	0.822647	0.87164	0.866687
40	1	1	1	1	1	1	1	1	1
Average	0.878447	0.847279	0.862089	0.833532	0.775107	0.839821	0.899306	0.847947	0.834138
Max	1	1	1	1	1	1	1	1	1
Min	0.320485	0.21431	0.265039	0.208	0.198332	0.389201	0.465227	0.275004	0.288677
S.D.	0.169468	0.215522	0.196698	0.222315	0.266968	0.184889	0.134252	0.203734	0.200232

TABLE 30.9 Term efficiency versus overall efficiency (non-oriented case).

DMU	Term efficiency							Past–present efficiency	Overall efficiency
	Term 1	Term 2	Term 3	Term 4	Term 5-1	Term 5-2	Term 5-3		
1	1	1	1	1	1	1	1	1	1
2	1	1	1	1	1	1	1	1	1
3	1	1	1	1	1	1	1	1	1
4	1	1	1	1	0.837113	0.73716	0.937195	1	0.959337
5	0.579968	0.711427	0.629671	0.633289	0.528037	0.526168	0.479886	0.639739	0.613189
6	1	1	1	1	1	1	1	1	1
7	0.864309	0.909265	1	1	0.853187	0.937637	0.986407	0.950714	0.944713
8	0.786675	0.719729	0.771352	0.806894	0.66218	0.665583	0.624407	0.771898	0.748179
9	0.821407	1	0.895437	0.90002	0.885026	0.853936	0.798325	0.906007	0.894117
10	1	1	1	1	1	1	1	1	1
11	0.939729	1	0.985488	1	1	0.978622	0.920359	0.98375	0.980606
12	1	1	1	1	1	1	1	1	1
13	1	1	1	0.913613	0.43972	0.793915	0.877296	0.974592	0.868739
14	0.491054	0.471775	0.517568	0.647316	0.242477	0.221785	0.261621	0.533486	0.481747
15	0.743704	0.755085	0.633233	0.55596	0.615485	0.764685	0.826998	0.642995	0.657255
16	0.65553	0.659839	1	1	1	1	1	0.849042	0.880638
17	1	1	1	1	0.869463	0.746489	1	1	0.972167
18	1	1	1	1	1	1	1	1	1
19	1	1	1	1	0.248265	0.423012	0.461516	1	0.74217
20	0.765205	0.75752	0.775963	0.710261	0.60439	0.627089	0.646916	0.748815	0.724007
21	0.333753	0.235387	0.269027	0.354477	0.203974	0.270755	0.192906	0.288476	0.276378
22	0.878732	0.909403	1	0.946911	0.914668	0.892526	0.859248	0.938152	0.92821
23	0.50908	0.594973	0.734988	0.55662	0.184214	0.275799	0.257516	0.601275	0.520049
24	0.531393	0.294102	0.571836	0.208	0.193904	0.448031	0.897165	0.310752	0.313974

(continued overleaf)

TABLE 30.9 (continued)

| DMU | Term efficiency | | | | | | | | Past-present efficiency | Overall efficiency |
	Term 1	Term 2	Term 3	Term 4	Term 5-1	Term 5-2	Term 5-3		
25	0.668944	0.619532	0.761965	0.711925	0.476522	0.497237	0.502913	0.690989	0.651309
26	0.743992	0.623685	0.706596	0.651557	0.608531	0.5964	0.57701	0.676862	0.662166
27	1	1	1	1	1	1	1	1	1
28	0.818823	0.866214	0.958475	0.94361	0.921145	0.925338	0.85635	0.903676	0.903543
29	0.583162	0.323263	0.536952	0.445697	0.136539	0.168654	0.488398	0.452845	0.422947
30	1	1	1	1	1	1	1	1	1
31	0.62857	0.556202	0.734229	0.531352	0.229832	0.285929	0.385448	0.60377	0.518516
32	1	1	1	1	1	1	1	1	1
33	1	1	0.853519	0.622659	0.884826	0.467349	0.617251	0.833064	0.774106
34	0.660426	0.83274	1	1	1	1	1	0.890732	0.913602
35	1	1	1	1	1	1	1	1	1
36	0.781958	1	1	1	1	1	1	0.955109	0.964505
37	0.823704	0.767399	0.730465	0.779295	0.73241	0.740161	0.686242	0.771618	0.762032
38	0.778174	0.887303	0.70402	0.656271	0.748536	0.713548	0.724994	0.735314	0.734294
39	0.81392	0.849169	0.803093	0.828593	0.838056	0.834191	0.780895	0.823664	0.822795
40	1	1	1	1	1	1	1	1	1
Average	0.830055	0.8336	0.864347	0.835108	0.746463	0.7598	0.791182	0.836933	0.815882
Max	1	1	1	1	1	1	1	1	1
Min	0.333753	0.235387	0.269027	0.208	0.136539	0.168654	0.192906	0.288476	0.276378
S.D.	0.185068	0.221181	0.183961	0.216239	0.299742	0.26883	0.24864	0.200395	0.208309

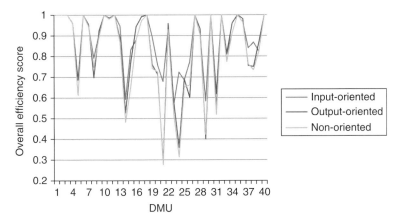

Figure 30.3   Input-, output- and non-oriented overall efficiency scores.

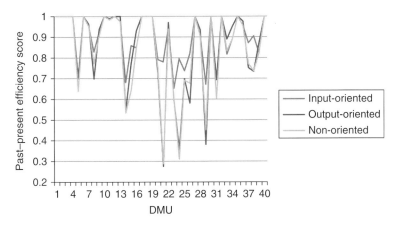

Figure 30.4   Input-, output- and non-oriented past–present efficiency scores.

Moreover, Tables 30.7–30.9 also show that 15, 16 and 15 out of the 40 DMUs are input-, output- and non-oriented *past–present efficient*, respectively. Figure 30.4 shows graphically the input-, output- and non-oriented past–present efficiency scores for the 40 DMUs. These results indicate that three (DMUs 4, 17 and 19), four (DMUs 4, 13, 17 and 19) and three (DMUs 4, 17 and 19) DMUs, corresponding to input-, output- and non-oriented DEA models, respectively, which are efficient based on past–present performance, are turned into inefficient DMUs once future performance is considered in the performance evaluation. By contrast, the results also show that DMUs 16, 34 and 36 are inefficient based on their past–present performance, but are all efficient according to their future performance. Such a fact evidently signals the importance of taking future performance into account when evaluating a DMU's

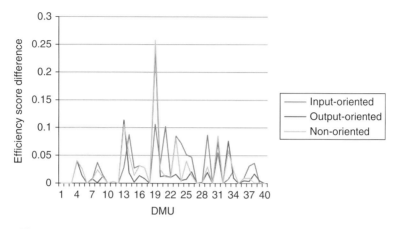

Figure 30.5   Difference between past–present and overall efficiency scores.

overall performance. Figure 30.5 shows graphically the difference between the *past–present* and *overall* efficiency scores with respect to the 40 DMUs. The greatest (average) differences between the two scores are 0.105966 (0.02724), 0.240765 (0.019351) and 0.25783 (0.025118), corresponding to the input-, output- and non-oriented DEA modes, respectively. DMU 19 turns out to be the one that possesses the greatest difference between the two scores in all of the three DEA models. Here, it is important to note that, as shown in Tables 30.7–30.9, there is no specific (positive or negative) relationship between past–present and overall performance. That is, the empirical results presented in Tables 30.7–30.9 show that if future performance indicators are omitted when conducting a performance evaluation, then the DMUs' performance may be either overestimated or underestimated.

30.5.2.2 Performance Benchmarking   This research used the SBM DEA models proposed by Tone [12] and the dynamic SBM DEA models proposed by Tone and Tsutsui [3] to develop benchmarks. To be more specific, the SBM DEA models were used to measure the present (i.e. year 2013) efficiency. On the other hand, the dynamic SBM DEA models were used to measure both the past–present (i.e. years 2010–2013) efficiency, and the realized overall (i.e. years 2010–2014) efficiency. In contrast, the proposed generalized dynamic DEA models were used to measure the (predicted) overall (i.e. years 2010–2014) efficiency. The resulting realized overall, present, past–present and (predicted) overall efficiency scores corresponding to the input-, output- and non-oriented DEA models are shown in Tables 30.10, 30.11 and 30.12, respectively. Tables 30.10–30.12 clearly show that both the past–present and the (predicted) overall performance significantly outperform the present performance. Therefore, in what follows, we focus more on analysing the empirical results obtained from the DEA models that provide the past–present and the (predicted) overall efficiency scores.

TABLE 30.10   Efficiency differences (input-oriented case).

DMU	Realized overall efficiency (a)	Present efficiency (b)	Past–present efficiency (c)	(Predicted) overall efficiency (d)	$\lvert(a)-(b)\rvert$	$\lvert(a)-(c)\rvert$	$\lvert(a)-(d)\rvert$
1	1	0.2008	1	1	0.7992	0	0
2	1	0.9746	1	1	0.0254	0	0
3	1	0.696	1	1	0.304	0	0
4	0.9142	0.7133	1	0.9613	0.2009	0.0858	0.0471
5	0.5388	0.5002	0.6397	0.6132	0.0386	0.1009	0.0744
6	1	1	1	1	0	0	0
7	0.9709	0.9852	0.9507	0.9447	0.0143	0.0202	0.0262
8	0.7177	0.6599	0.8289	0.7919	0.0578	0.1112	0.0742
9	0.9501	0.7978	0.9235	0.9078	0.1523	0.0266	0.0423
10	1	0.7829	1	1	0.2171	0	0
11	0.9891	0.9509	0.9838	0.9806	0.0382	0.0053	0.0085
12	1	0.8289	1	1	0.1711	0	0
13	0.9964	0.856	0.975	0.9467	0.1404	0.0214	0.0497
14	0.5832	0.3696	0.6803	0.593	0.2136	0.0971	0.0098
15	0.8918	0.2234	0.8593	0.8321	0.6684	0.0325	0.0597
16	0.9394	0.3006	0.849	0.8806	0.6388	0.0904	0.0588
17	1	0.1866	1	0.9722	0.8134	0	0.0278
18	1	0.3449	1	1	0.6551	0	0
19	0.8583	0.4267	1	0.894	0.4316	0.1417	0.0357
20	0.6377	0.606	0.7917	0.7568	0.0317	0.154	0.1191
21	0.4885	0.2449	0.7802	0.6781	0.2436	0.2917	0.1896
22	0.9483	0.8149	0.9435	0.9324	0.1334	0.0048	0.0159
23	0.3917	0.2782	0.6524	0.5675	0.1135	0.2607	0.1758
24	0.8444	0.523	0.7956	0.7234	0.3214	0.0488	0.121
25	0.6561	0.4877	0.7366	0.6853	0.1684	0.0805	0.0292
26	0.677	0.5751	0.82	0.7729	0.1019	0.143	0.0959
27	1	0.8765	1	1	0.1235	0	0
28	0.919	0.8783	0.9042	0.904	0.0407	0.0148	0.015
29	0.3109	0.2001	0.6718	0.5849	0.1108	0.3609	0.274
30	1	1	1	1	0	0	0
31	0.8041	0.3527	0.6041	0.5489	0.4514	0.2	0.2552
32	1	1	1	1	0	0	0
33	0.999	0.4593	0.8169	0.8096	0.5397	0.1821	0.1894
34	0.9279	0.2314	0.8907	0.9136	0.6965	0.0372	0.0143
35	1	0.8802	1	1	0.1198	0	0
36	0.9778	0.2245	0.9551	0.9645	0.7533	0.0227	0.0133
37	0.9138	0.6721	0.8715	0.84	0.2417	0.0423	0.0738
38	0.9143	0.6625	0.9046	0.868	0.2518	0.0097	0.0463
39	0.8131	0.7528	0.8237	0.8228	0.0603	0.0106	0.0097
40	1	0.6855	1	1	0.3145	0	0
Average	0.8643	0.6051	0.8913	0.8673	0.26	0.0649	0.0538
Max	1	1	1	1	0.8134	0.3609	0.274
Min	0.3109	0.1866	0.6041	0.5489	0	0	0
S.D.	0.187	0.273241	0.1217	0.1442	0.2492	0.0899	0.072

TABLE 30.11 Efficiency differences (output-oriented case).

DMU	Realized overall efficiency (a)	Present efficiency (b)	Past–present efficiency (c)	(Predicted) overall efficiency (d)	$|(a)-(b)|$	$|(a)-(c)|$	$|(a)-(d)|$
1	1	0.3534	1	1	0.6466	0	0
2	1	0.9752	1	1	0.0248	0	0
3	1	0.7484	1	1	0.2516	0	0
4	0.8782	0.7542	1	0.9597	0.124	0.1218	0.0815
5	0.6688	0.5964	0.7	0.686	0.0724	0.0312	0.0172
6	1	1	1	1	0	0	0
7	0.9728	0.9863	0.9631	0.9554	0.0135	0.0097	0.0174
8	0.7508	0.6801	0.6983	0.6976	0.0707	0.0525	0.0532
9	0.9623	0.8255	0.9367	0.9249	0.1368	0.0256	0.0374
10	1	0.7854	1	1	0.2146	0	0
11	0.9929	0.9551	0.9891	0.986	0.0378	0.0038	0.0069
12	1	0.8511	1	1	0.1489	0	0
13	0.996	0.8754	1	0.8866	0.1206	0.004	0.1095
14	0.5932	0.3229	0.5463	0.5274	0.2703	0.0469	0.0658
15	0.9232	0.3017	0.7717	0.7737	0.6215	0.1515	0.1495
16	0.9598	0.2671	0.9306	0.9443	0.6927	0.0292	0.0155
17	1	0.4169	1	0.9916	0.5831	0	0.0084
18	1	0.3076	1	1	0.6924	0	0
19	0.7523	0.3739	1	0.7592	0.3784	0.2477	0.0069
20	0.6952	0.6289	0.7242	0.7122	0.0663	0.029	0.017
21	0.4677	0.3076	0.275	0.2887	0.1601	0.1927	0.179
22	0.9638	0.8423	0.9702	0.9594	0.1215	0.0064	0.0044
23	0.5888	0.3492	0.5964	0.5801	0.2396	0.0076	0.0087
24	0.855	0.2973	0.3552	0.3604	0.5577	0.4998	0.4946
25	0.7346	0.6314	0.6981	0.6907	0.1032	0.0365	0.0439
26	0.7558	0.6646	0.581	0.6019	0.0912	0.1748	0.1539
27	1	0.7972	1	1	0.2028	0	0
28	0.9323	0.8962	0.935	0.933	0.0361	0.0027	0.0007
29	0.5356	0.346	0.3811	0.4009	0.1896	0.1545	0.1347
30	1	1	1	1	0	0	0
31	0.6993	0.4272	0.6935	0.6187	0.2721	0.0058	0.0806
32	1	1	1	1	0	0	0
33	0.9962	0.4528	0.8895	0.8136	0.5434	0.1067	0.1826
34	0.9578	0.4852	0.9512	0.961	0.4726	0.0066	0.0032
35	1	0.8862	1	1	0.1138	0	0
36	0.9884	0.3244	0.9767	0.9815	0.664	0.0117	0.0069
37	0.9451	0.6953	0.7507	0.7547	0.2498	0.1944	0.1904
38	0.9397	0.714	0.7327	0.7498	0.2257	0.207	0.1899
39	0.8526	0.7821	0.8716	0.8667	0.0705	0.019	0.0141
40	1	0.6895	1	1	0.3105	0	0
Average	0.884	0.63985	0.8479	0.8341	0.2448	0.0595	0.0568
Max	1	1	1	1	0.6927	0.4998	0.4946
Min	0.4677	0.2671	0.275	0.2887	0	0	0
S.D.	0.1539	0.247974	0.2037	0.2002	0.22	0.1015	0.0953

TABLE 30.12 Efficiency differences (non-oriented case).

DMU	Realized overall efficiency (a)	Present efficiency (b)	Past–present efficiency (c)	(Predicted) overall efficiency (d)	$|(a)-(b)|$	$|(a)-(c)|$	$|(a)-(d)|$
1	1	0.1933	1	1	0.8067	0	0
2	1	0.9746	1	1	0.0254	0	0
3	1	0.696	1	1	0.304	0	0
4	0.8782	0.7132	1	0.9593	0.165	0.1218	0.0811
5	0.5388	0.5002	0.6397	0.6132	0.0386	0.1009	0.0744
6	1	1	1	1	0	0	0
7	0.9708	0.9852	0.9507	0.9447	0.0144	0.0201	0.0261
8	0.7131	0.6599	0.7719	0.7482	0.0532	0.0588	0.0351
9	0.95	0.7978	0.906	0.8941	0.1522	0.044	0.0559
10	1	0.7829	1	1	0.2171	0	0
11	0.9891	0.9509	0.9838	0.9806	0.0382	0.0053	0.0085
12	1	0.8289	1	1	0.1711	0	0
13	0.996	0.856	0.9746	0.8687	0.14	0.0214	0.1273
14	0.4993	0.3258	0.5335	0.4817	0.1735	0.0342	0.0176
15	0.8862	0.2074	0.643	0.6573	0.6788	0.2432	0.2289
16	0.9308	0.2671	0.849	0.8806	0.6637	0.0818	0.0502
17	1	0.1853	1	0.9722	0.8147	0	0.0278
18	1	0.3076	1	1	0.6924	0	0
19	0.6989	0.3739	1	0.7422	0.325	0.3011	0.0433
20	0.6373	0.606	0.7488	0.724	0.0313	0.1115	0.0867
21	0.3616	0.2266	0.2885	0.2764	0.135	0.0731	0.0852
22	0.9482	0.8149	0.9382	0.9282	0.1333	0.01	0.02
23	0.3913	0.2608	0.6013	0.52	0.1305	0.21	0.1287
24	0.8182	0.2972	0.3108	0.314	0.521	0.5074	0.5042
25	0.6555	0.4877	0.691	0.6513	0.1678	0.0355	0.0042
26	0.677	0.5751	0.6769	0.6622	0.1019	0.0001	0.0148
27	1	0.9999	1	1	0.0001	0	0
28	0.9174	0.8783	0.9037	0.9035	0.0391	0.0137	0.0139
29	0.3093	0.192	0.4528	0.4229	0.1173	0.1435	0.1136
30	1	1	1	1	0	0	0
31	0.6422	0.3362	0.6038	0.5185	0.306	0.0384	0.1237
32	1	1	1	1	0	0	0
33	0.8464	0.436	0.8331	0.7741	0.4104	0.0133	0.0723
34	0.9279	0.2303	0.8907	0.9136	0.6976	0.0372	0.0143
35	1	0.8802	1	1	0.1198	0	0
36	0.9778	0.2111	0.9551	0.9645	0.7667	0.0227	0.0133
37	0.914	0.6721	0.7716	0.762	0.2419	0.1424	0.152
38	0.9143	0.6625	0.7353	0.7343	0.2518	0.179	0.18
39	0.813	0.7528	0.8237	0.8228	0.0602	0.0107	0.0098
40	1	0.6855	1	1	0.3145	0	0
Average	0.8451	0.5953	0.8369	0.8159	0.2505	0.0645	0.0578
Max	1	1	1	1	0.8147	0.5074	0.5042
Min	0.3093	0.1853	0.2885	0.2764	0	0	0
S.D.	0.2007	0.2884	0.2004	0.2083	0.2542	0.1038	0.0925

Table 30.10 shows that, in the input-oriented case, 13, 3, 15 and 12 out of the 40 DMUs are overall, present, past–present and (predicted) overall efficient, respectively. It turns out that DMUs 4 and 19 were wrongly considered as efficient based on their past–present performance. By contrast, DMU 17 was wrongly considered as an inefficient DMU according to its (predicted) overall efficiency score. The absolute average (maximum) deviations between the realized overall efficiency scores and the present, past–present and (predicted) overall efficiency scores are 0.26 (0.8134), 0.0649 (0.3609) and 0.0538 (0.274), respectively. DMU 17 possesses the maximum deviation with respect to the input-oriented case, and DMU 29 possesses the maximum deviation with respect to both the output- and the non-oriented cases.

Table 30.11 shows that, in the output-oriented case, 13, 3, 16 and 12 out of the 40 DMUs are overall, present, past–present and predicted overall efficient, respectively. It is found that DMUs 4, 13 and 19 were wrongly considered as efficient based on their past–present performance. By contrast, DMU 17 was wrongly considered as an inefficient DMU according to its (predicted) overall efficiency score. The absolute average (maximum) deviations between the realized overall efficiency scores and the present, past–present and (predicted) overall efficiency scores are 0.2448 (0.6927), 0.0595 (0.4998) and 0.0568 (0.4946), respectively. DMU 16 possesses the maximum deviation with respect to the input-oriented case. DMU 24 possesses the maximum deviation with respect to both the output- and the non-oriented cases.

Table 30.12 shows that, in the non-oriented case, 13, 3, 15 and 12 out of the 40 DMUs are overall, present, past–present and (predicted) overall efficient, respectively. Similarly to the input-oriented case, DMUs 4 and 19 were wrongly considered as efficient based on their past–present performance, and DMU 17 was wrongly considered as an inefficient DMU according to its (predicted) overall efficiency score. The absolute average (maximum) deviations between the realized overall efficiency scores and the present, past–present and (predicted) overall efficiency scores are 0.2505 (0.8147), 0.0645 (0.5074) and 0.0578 (0.5042), respectively. DMU 17 possesses the maximum deviation with respect to the input-oriented case, but DMU 24 possesses the maximum deviation with respect to both the output- and the non-oriented cases. Figure 30.6 shows graphically the absolute average and the absolute maximum deviations of the present, past–present and (predicted) overall efficiency scores from the corresponding realized overall values with respect to the input-, output- and non-oriented cases.

Lastly, we applied ordinary least squares (OLS) regression to measure how well the forecasts of the 2014 input (cost of sales) and output (net revenue), obtained by using the moving average method, explained the corresponding realized 2014 values. The resulting R-squared (coefficient of determination) values for the mean, minimum and maximum forecasted cost of sales and net revenue data ranged from 0.8876 (minimum net revenue) to 0.9862 (maximum cost of sales). That is, the R-squared values suggest a good fit of the OLS lines. In addition, the results of a t-test showed that the null hypothesis of the forecasted outcomes being the same as the realized ones cannot be rejected. That is, the moving average method performs quite well on the empirical example considered, concerning high-tech IC design companies in Taiwan.

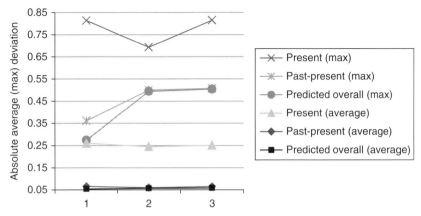

1: input-oriented; 2: output-oriented; 3: non-oriented

Figure 30.6   Absolute average and maximum deviations from realized overall efficiency scores.

In summary, taking the realized overall efficiency scores, obtained based on 2010–2014 realized data, as benchmarks, our proposed generalized dynamic DEA models (based on both 2010–2013 past–present data and predicted 2014 data) outperform the dynamic SBM DEA models (based on 2010–2013 past–present data) proposed by Tone and Tsutsui [3], and significantly outperform the SBM DEA models (based on only 2013 present data) proposed by Tone [12]. That is, in contrast to the realized 2014 outcomes, the empirical results clearly justify the efficacy of our proposed generalized dynamic DEA models.

30.6   CONCLUSIONS

This study proposes a new system of generalized dynamic DEA models that simultaneously and explicitly take DMUs' past, present and future actions into account to evaluate the DMUs' overall performance. The user can embed any forecasting technique into the models to predict future values of the inputs and outputs considered. We note that the accuracy of the forecasts has an impact on the performance of the generalized dynamic DEA models. However, we re-emphasize that this research addresses neither the development nor the choice of forecasting techniques, which are themselves big research areas. To date, there have been very limited DEA studies in the literature that consider a DMU's future performance. In fact, to the best of our knowledge, this study is the first to attempt to develop DEA models for evaluating a DMU's future performance in a highly volatile operating environment, including, for example, highly volatile crude oil prices and/or currency exchange rates. In addition, it is worth mentioning that this study applied the maximum entropy approach to deal with uncertain future circumstances. We believe that entropy theory can play an import role in developing past–present–future intertemporal DEA models.

In this research, the proposed generalized dynamic DEA models were applied to the IC design industry. The companies in this industry procure raw materials from and sell products to multiple countries, and are thus faced with highly dynamic currency exchange rates, which makes the industry a typical application area for the proposed new DEA models. Therefore, in this study we conducted an empirical study based on real data concerning high-tech IC design companies in Taiwan. The empirical results verify the importance of taking future performance into account when evaluating a DMU's overall performance. In addition, the empirical results also justify the efficacy of the generalized dynamic DEA models by using a cross-validation technique. That is, our proposed new DEA models outperform the well-known (dynamic) SBM DEA models in terms of both average and maximum deviations from the realized overall efficiency scores.

Finally, we would like to emphasize that, owing to data availability, we could not estimate the cost of sales (input) and net revenue (output) from forecasted currency exchange rates (the macroeconomic index). Recall that forecasted inputs and outputs should be functions of foreign exchange rates. Therefore, we had no choice but to apply the moving average method to directly forecast future inputs and outputs from historical data. That is, the forecasts cannot fully reflect the highly volatile operating environment. We believe that detailed data, if available, could further reveal the value of the proposed new past–present–future intertemporal DEA models.

REFERENCES

[1] Chang, T.-S., Tone, K. and Wu, C.-H. (2016) DEA models incorporating uncertain future performance. *European Journal of Operational Research*, **254**, 532–549.

[2] Chang, T.-S., Tone, K. and Wu, C.-H. (2015) Past–present–future intertemporal DEA models. *Journal of the Operational Research Society*, **66**, 16–32.

[3] Tone, K. and Tsutsui, M. (2010) Dynamic DEA: A slacks-based measure approach. *Omega*, **38**, 145–156.

[4] Montgomery, D.C., Johnson, L.A. and Gardiner, J.S. (1990) *Forecasting & Time Series Analysis*, 2nd edn, McGraw-Hill.

[5] Armstrong, J.S. (2001) Selecting forecasting methods, in *Principles of Forecasting: A Handbook for Researchers and Practitioners* (ed. J.S. Armstrong), Springer, pp. 365–386.

[6] Ouenniche, J., Xu, B. and Tone, K. (2014) Relative performance evaluation of competing crude oil prices' volatility forecasting models: A slacks-based super-efficiency DEA model. *American Journal of Operations Research*, **4**, 235–245.

[7] Sanders, N.R. and Manrodt, K.B. (1994) Forecasting practices in US corporations: Survey results. *Interfaces*, **24**(2), 92–100.

[8] Kapur, J.N. (1989) *Maximum-Entropy Models in Science and Engineering*. Wiley Eastern Ltd.

[9] Hyndman, R.J. and Athanasopoulos, G. (2013) The statistical forecasting perspective, in *Forecasting: Principles and Practice*, Section 1.7, http://otexts.org/fpp/1/7 (accessed 22 February 2015).

[10] Fang, S.-C., Rajasekera, J.R. and Tsao, H.-S.J. (1997) *Entropy Optimization and Mathematical Programming*. Kluwer Academic.

[11] Yoon, K. and Hwang, C.-L. (1980) *Multiple Attribute Decision Making – Methods and Applications: A State-of-the-Art Survey*. Springer.

[12] Tone, K. (2001) A slacks-based measure of efficiency in data envelopment analysis. *European Journal of Operational Research*, **130**, 498–509.

[13] Cooper, W.W., Seiford, L.M. and Tone, K. (2007) *Data Envelopment Analysis: A Comprehensive Text with Models, Applications, References and DEA-Solver Software*. Springer Science + Business Media, New York.

31

SITE SELECTION FOR THE NEXT-GENERATION SUPERCOMPUTING CENTER OF JAPAN

KAORU TONE

National Graduate Institute for Policy Studies, Tokyo, Japan

31.1 INTRODUCTION

The Next-Generation Supercomputer R&D Project was an endeavor to create a 10 petaflop/s system by 2012. It will cost about 115 billion yen (about US$ 1.3 billion at 2006 rate) and is considered to be one of the "key technologies of national importance to Japan." The Next-Generation Supercomputer Project's goals are (i) the development and installation of the most advanced high-performance supercomputer system; (ii) the development and wide use of application software to utilize the supercomputer to the maximum extent; (iii) the provision of a flexible computing environment by sharing the Next-Generation Supercomputer through connection with other supercomputers located at universities and research institutes; and (iv) the establishment of an Advanced Computational Science and Technology Center. The system will be tuned to 21 selected target applications. Nanotechnology and life sciences have been identified as "grand challenges." For Japan to maintain and improve its international competitiveness in science and technology, it is essential that it carries out top-quality R&D on supercomputing hardware and software. This is why the government's third Science and Technology Basic Plan called for the development and utilization of the Next-Generation Supercomputer, as a technological foundation of national importance that required major investment. A law supporting this

Advances in DEA Theory and Applications: With Extensions to Forecasting Models,
First Edition. Edited by Kaoru Tone.
© 2017 John Wiley & Sons Ltd. Published 2017 by John Wiley & Sons Ltd.

TABLE 31.1   **Candidates and their supporting bodies.**

Candidate	Supporting bodies
1. Sapporo	Hokkaido, Sapporo City, Hokkaido University, Hokkaido Federation of Economic Organizations
2. Hirosaki	Aomori Prefecture
3. Sendai	Sendai City, Tohoku University
4. Tsukuba	Ibaraki Prefecture, Tsukuba City, Tsukuba University
5. Wako	Saitama Prefecture, Wako City
6. Yokohama	Yokohama City
7. Suntogun	Shizuoka Prefecture
8. Nagano	Nagano City, Nagano Prefecture, Shinshuu University, Nagano Chamber of Commerce and Industry
9. Kusatsu	Shiga Prefecture, Kusatsu City
10. Kyoto	Kyoto Prefecture
11. Ikoma	Nara Prefecture
12. Osaka	Osaka City
13. Kobe	Kobe City, Hyogo Prefecture, Hyogo Chamber of Commerce and Industry
14. Sayo	Hyogo Prefecture
15. Fukuoka	Fukuoka City, Fukuoka Prefecture, Fukuoka University, Kyushu Chamber of Commerce and Industry

recommendation has now come into force. RIKEN (an independent administrative institute funded by the Japanese government) is responsible for the development and operation of the supercomputer under this law. One of RIKEN's important missions is to design and create the R&D facilities that are essential for the advancement of Japanese science and technology, and to take the lead in putting these facilities to use. They are putting their full efforts into the development of the Next-Generation Supercomputer. See Watanabe [1] for more details of this project.

In July 2006, RIKEN announced a public competition for the supercomputing site. The 15 cities listed in Table 31.1 and shown in Figure 31.1 applied in response to the announcement.

At the same time, RIKEN nominated a site selection committee for this project, consisting of 14 members. Six of them were supercomputer users, four were software engineers and two were system designers. The head of the committee was Dr. Kiyoshi Kurokawa, the President of the Science Council of Japan. The author of this chapter was engaged by the committee as a specialist in the field of decision methodology.[1] The mission of the committee was to reach a decision in such a way that the process of site selection would be rational, open to the public, and easily understandable, since this was a big national project using public money. As a

---

[1] The views expressed in this chapter are those of the author and are not necessarily indicative of those of RIKEN and the site selection committee.

Figure 31.1   The 15 candidate sites.

typical site selection problem, this problem had multiple criteria for comparing candidate sites; these included both quantitative and qualitative elements. Furthermore,
the criteria had a hierarchical structure, as described later. For such purposes,
Saaty's analytic hierarchy process (AHP) [2] is a practical and useful method for
group decision-making in multiple-criteria environments. However, one of the difficulties in making decisions when committee members have diverse preferences is
that decisions made using the group average are not always persuasive. We must
take into account the diversity of opinions. Hence, we applied data envelopment
analysis (DEA) [3,4] and, in particular, the assurance region model [5], in a two-
stage process. The combined use of AHP and DEA was discussed by Tone [6]
and Sinuany-Stern *et al.* [7]. These schemes were developed by Takamura and Tone
[5] for relocating Japanese government agencies out of Tokyo and are well recognized in the literature (see [8–12], among others). Using a combination of AHP and
DEA, the site selection committee for this project reached a conclusion and reported
it to RIKEN. On March 29, 2007, RIKEN announced to the public "The winner
is Kobe."

  This chapter is organized as follows. Section 31.2 describes AHP and its group
decision method. Section 31.3 presents the DEA assurance region model and its
use for finding relative positives and negatives of each candidate site. We present
the application of the above methods to this site selection problem in Section 31.4,
followed by the decision and a conclusion in Section 31.5.

31.2 HIERARCHICAL STRUCTURE AND GROUP DECISION BY AHP

In this section, we describe the hierarchical structure and group decision scheme for this project.

31.2.1 Hierarchical Structure

RIKEN specified two basic requirements for the candidate sites:

(A1) Maintenance
(A2) Utilization

These two factors were divided further into detailed criteria (C1–C24) hierarchically, including intermediate criteria (B1–B9):

(A1) Maintenance
    (B1) Natural disasters
        (C1) Earthquake
        (C2) Lightning
    (B2) Land and weather environment
        (C3) Weather
        (C4) Expandability of land
    (B3) Utilities
        (C5) Electric power
        (C6) Water
        (C7) Gas
        (C8) Communication network
    (B4) Neighborhood
        (C9) Accidents
        (C10) Radio interference
        (C11) Residential
(A2) Utilization
    (B5) Living environment
        (C12) Convenience
        (C13) Attractiveness
        (C14) Home environment
        (C15) Surroundings
        (C16) Internationality

(B6) Access

    (C17) Access from Japanese cities

    (C18) Access from abroad

(B7) Research environment

    (C19) Cooperation with universities and other research units

    (C20) Cooperation with private companies

    (C21) Infrastructure for cooperation

(B8) Collaboration with municipality

    (C22) Interpretive plan

    (C23) Utilization

(B9) Administration

    (C24) Administration

31.2.2 Evaluation of Candidate Sites with Respect to Criteria, and Importance of Criteria

We numbered the candidate sites from S1 to S15 and the criteria from C1 to C24. The score of site j $(j = 1, \ldots, 15)$ with respect to criterion i $(i = 1, \ldots, 24)$ was denoted by S_{ij}. The values of these scores were obtained from specialist teams in the related fields.

Concurrently, six evaluators from among the committee members estimated weights for each criterion using their own subjective judgment. For this purpose, AHP was useful in quantifying these subjective (or qualitative) judgments. The weight matrix obtained was denoted by (W_{ki}), where k is the index of the evaluator and i the index of the criterion.

31.2.3 Evaluation by Average Weights

Let the average weight of criterion i over the entire set of evaluators be $\bar{W}_i$. Applying this average weight to the score matrix $S = (S_{ij})$ leads to a comparison of the 15 sites; that is, we can obtain the score for site j as

$$\pi_j = \sum_{i=1}^{24} \bar{W}_i S_{ij} \ (j = 1, \ldots, 15) \tag{31.1}$$

However, using this average suggests that only one "virtual" evaluator was "representative" of all members' judgments. Thus, the variety of opinions across the six evaluators was not taken into account. The use of such an "average" or "median" of weights must be employed cautiously from the point of view of consensus-making. Another way to look at the above approach is that the weights are common to all sites.

We may call this a "fixed-weight" approach, as contrasted with the following "variable-weight" structure.

31.3   DEA ASSURANCE REGION APPROACH

In this section, we describe the method for evaluating the relative strengths and weaknesses of candidate sites using the assurance region model of DEA.

31.3.1   Use of Variable Weights

Given the score matrix $S = (S_{ij})$, we evaluated the total score of site $j = j_0$ using a weighted sum of S_{ij_0} as

$$\theta_{j_0} = \sum_i u_i S_{ij_0} \qquad (31.2)$$

with a nonnegative weight set (u_i). We assumed that the weights could vary from site to site in accordance with the principle that we chose for characterizing the sites. For this purpose, we employed the two extreme cases presented in the following sections.

31.3.2   Evaluation of the "Positives" of Each Site

In order to evaluate the positives of site j_0, we chose the weights (u_i) in (31.2) so that they maximized θ_{j_0} under the condition that the same weights were applied in evaluating all other sites, so that the site under consideration could be compared relative to them. This principle is in accordance with that of DEA and can be formulated as follows:

$$\max \theta_{j_0} = \sum_i u_i S_{ij_0}$$
$$\text{subject to} \sum_i u_i S_{ij} \le 1 \ (\forall j), \ u_i \ge 0 \ (\forall i) \qquad (31.3)$$

Furthermore, the weights had to reflect all evaluators' preferences regarding the criteria. This could be represented by a version of the assurance region model proposed by Thompson *et al.* [10]. For every pair (i_1, i_2) of criteria, the ratio u_{i_1}/u_{i_2} was bounded by $L_{i_1 i_2}$ and $U_{i_1 i_2}$ according to

$$L_{i_1 i_2} \le {}^{u_{i_1}}/_{u_{i_2}} \le U_{i_1 i_2} \qquad (31.4)$$

where the bounds were calculated by using the evaluators' weights (W_{ki}) as follows:

$$L_{i_1 i_2} = \min_k \frac{W_{ki_1}}{W_{ki_2}}, \ U_{i_1 i_2} = \max_k \frac{W_{ki_1}}{W_{ki_2}} \qquad (31.5)$$

Thus, we maximized θ_{j_o} in (31.3) subject to the constraints expressed by (31.3) and (31.4). We thus assigned the most preferable weight set to the target site within allowable ranges so that the "positives" of the site were evaluated. However, the same weights were utilized for the evaluation of all other sites when the target site was compared with them. If the optimal objective value $\theta_{j_o}^*$ satisfied $\theta_{j_o}^* = 1$, then the site j_o was be judged to be the best. If, on the other hand, $\theta_{j_o}^* < 1$, the site was inferior to the others with respect to some (or all) criteria.

31.3.3 Evaluation of the "Negatives" of Each Site

Turning to the opposite side, we wished also to evaluate each candidate site from the worst side. For this purpose, we sought the "worst" weights in the sense that the objective function in (31.3) was minimized. This principle can be formulated as follows:

$$\min \theta_{j_o} = \sum_i u_i S_{ij_o}$$

$$\text{subject to} \sum_i u_i S_{ij} \ge 1 \ (\forall j)$$

$$L_{i_1 i_2} \le {}^{u_{i_1}}/_{u_{i_2}} \le U_{i_1 i_2} (\forall (i_i, i_2))$$

$$u_i \ge 0 \ (\forall i)$$

(31.6)

By dint of the reversed inequality in (31.6), the optimal $\theta_{j_o}^*$ satisfied $\theta_{j_o}^* \ge 1$. If $\theta_{j_o}^* = 1$, then the site belonged to the group of worst performers; otherwise, if $\theta_{j_o}^* > 1$, it rated higher than that group. Each site was compared with these worst performers and was gauged by its efficiency "negatives," which were the ratios of distances from the "worst" frontiers in the same way as in ordinary DEA. (Yamada et al. [14] named this worst-side approach "inverted DEA.")

In order to make straightforward comparisons of the scores for the "negative" and "positive" cases, we inverted $\theta_{j_o}^*$ for the negative case as follows:

$$\tau_{j_o}^* = {}^1/\theta_{j_o}^*$$

(31.7)

We call this the "negatives" score.

31.4 APPLICATION TO THE SITE SELECTION PROBLEM

We applied the above method in two steps, that is, a preliminary and a final step, since detailed surveys of all 15 candidate sites would have been too demanding, especially with respect to the cost-related terms. As described in Section 31.2.2, a score matrix was obtained from a team of specialists in the subjects concerned. The scale was 1–5. Table 31.2 presents the score matrix for the finalist sites.[2]

---

[2] A similar score matrix was applied in the preliminary selection process.

TABLE 31.2 Evaluation of criteria – final.

A	B	C	Sendai	Wako	Yokohama	Osaka	Kobe
A1	B1	C1	4.52	4.10	3.26	3.42	4.11
		C2	4.20	3.92	3.93	3.35	3.68
	B2	C3	5.00	5.00	4.67	5.00	4.67
		C4	5.00	4.00	4.00	2.00	5.00
	B3	C5	4.75	4.75	5.00	5.00	5.00
		C6	5.00	5.00	5.00	5.00	5.00
		C7	5.00	5.00	5.00	5.00	5.00
		C8	4.94	4.89	4.89	5.00	4.89
	B4	C9	5.00	5.00	4.00	4.00	5.00
		C10	5.00	4.00	5.00	5.00	5.00
		C11	4.88	4.13	5.00	4.38	4.88
A2	B5	C12	2.75	3.25	2.50	5.00	2.75
		C13	4.00	3.50	3.25	4.25	3.75
		C14	4.00	3.75	3.25	4.25	4.25
		C15	4.25	4.00	3.50	4.75	4.50
		C16	3.75	3.75	4.00	4.75	4.50
	B6	C17	3.00	4.00	4.00	4.50	4.25
		C18	3.00	4.25	4.50	4.75	4.50
	B7	C19	4.50	4.50	3.25	4.25	3.75
		C20	3.50	4.00	4.25	4.50	4.25
		C21	4.75	3.75	3.75	4.75	4.00
	B8	C22	4.00	3.75	2.50	4.00	4.00
		C23	4.50	4.00	3.50	3.75	4.25
	B9	C24	3.67	4.50	3.33	3.17	4.50

31.4.1 Preliminary Selection

In the preliminary selection stage, we tried to draw a dividing line between superior and inferior groups of the 15 candidate sites. AHP results obtained using the average weights assigned by the evaluators are shown in Figure 31.2. We also applied the variable-weight approach and obtained the "positives" and "negatives" of the 15 candidate sites, as displayed in Figure 31.3.

The committee decided on five sites, A, B, C, D, and E, as finalists. These were Sendai, Wako, Yokohama, Osaka, and Kobe, in random order, as shown in Figure 31.4.

31.4.2 Final Selection

We examined the five finalists fully again regarding the criteria C1–C24 by on-the-spot visits and obtained the scores presented in Table 31.2. Using the averages of the six evaluators' weights on the criteria C1–C24, we calculated the score (31.1), as displayed in Figure 31.5. As can be seen, Kobe is at the top, followed by Wako, Sendai, Osaka, and Yokohama in that order.

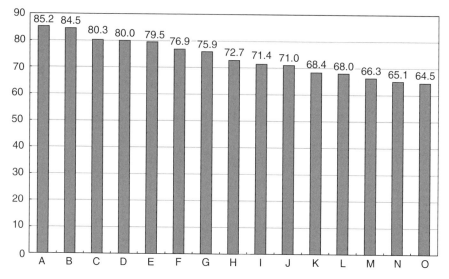

Figure 31.2   Preliminary AHP results.

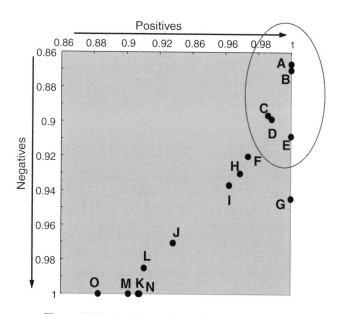

Figure 31.3   Positives and negatives of the 15 sites.

However, as we noted in Section 31.2, using the average suggests that only one "virtual" evaluator was "representative" of all members' judgments. Thus, the variety of opinions across evaluators was not taken into account. Table 31.3 reports the lower and upper bounds L and U of the ratio u_1/u_j $(j = 2, \ldots, 24)$ calculated from (31.5).

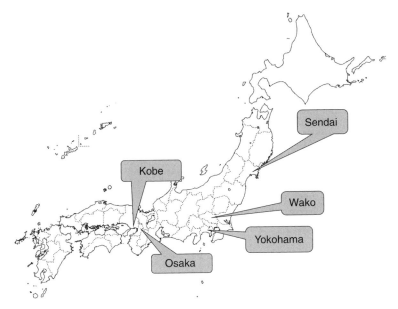

Figure 31.4   Finalists.

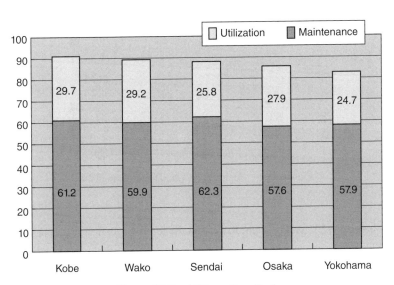

Figure 31.5   AHP results – final.

Although this is only a partial list, we can see big differences in the preferences of the evaluators.

We employed L and U for all pairs (i_1, i_2) $(i_1, i_2 = 1, \ldots, 24 : i_1 < i_2)$ (276 pairs in total) for evaluation of the "positives" and "negatives" of the finalists. Figure 31.6

TABLE 31.3   Lower and upper bounds (partial list).

L	$\leq u_1/u_j \leq$	U	L	$\leq u_1/u_j \leq$	U
1.000	$\leq u_1/u_2 \leq$	5.000	3.932	$\leq u_1/u_{13} \leq$	418.493
0.751	$\leq u_1/u_3 \leq$	34.396	2.919	$\leq u_1/u_{14} \leq$	184.396
0.361	$\leq u_1/u_4 \leq$	11.465	1.004	$\leq u_1/u_{15} \leq$	184.396
0.212	$\leq u_1/u_5 \leq$	2.918	2.423	$\leq u_1/u_{16} \leq$	184.396
0.461	$\leq u_1/u_6 \leq$	7.339	0.133	$\leq u_1/u_{17} \leq$	7.150
0.690	$\leq u_1/u_7 \leq$	29.708	0.663	$\leq u_1/u_{18} \leq$	35.751
0.212	$\leq u_1/u_8 \leq$	7.339	0.185	$\leq u_1/u_{19} \leq$	40.614
0.418	$\leq u_1/u_9 \leq$	7.052	1.670	$\leq u_1/u_{20} \leq$	84.481
3.017	$\leq u_1/u_{10} \leq$	42.891	0.430	$\leq u_1/u_{21} \leq$	19.525
0.870	$\leq u_1/u_{11} \leq$	22.500	0.650	$\leq u_1/u_{22} \leq$	68.992
0.669	$\leq u_1/u_{12} \leq$	74.152	3.252	$\leq u_1/u_{23} \leq$	68.992
			0.110	$\leq u_1/u_{24} \leq$	8.270

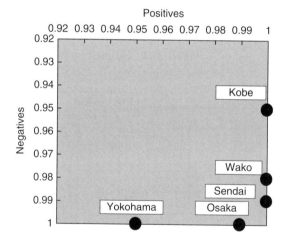

Figure 31.6   Positives and negatives of finalists.

shows the results. As can be seen, Kobe is at the top for "positives" and at the bottom for "negatives."

In addition, we compared the costs for the five finalists, which consisted of the initial and running costs. The initial cost included the costs of land acquisition, construction of the supercomputer and building, installation of utilities, and other costs. The running cost consisted of administrative and maintenance expenses, the costs of utilities (electric power, water, gas, and communications), and the operational expenses of the office. The total cost was calculated as the sum of the initial cost and the running cost over 10 years. However, we could not find significant differences in the total cost among the five finalists.

31.5   DECISION AND CONCLUSION

On March 29, 2007, RIKEN announced to the public "The winner is Kobe." The Supercomputer, called "K", is now in operation: see http://www.aics.riken.jp/en/ or RIKEN's homepage. Details of the site selection process can be found in [15]. As of 2013, "K" was the world's fourth-fastest computer. RIKEN is now improving "K" to make it rank top again, as it was in 2011.

In this chapter, we have presented an application of the combined use of AHP and DEA for selecting the site for the Next-Generation Supercomputing Center of Japan. Although the selection committee members were not familiar with OR/MS methodologies at the beginning of the process, they quickly acknowledged the strengths of AHP and DEA and reached a conclusion smoothly.

REFERENCES

[1] Watanabe, T. (2007) The Japan Next Generation Supercomputer Project. HPC Workshop, December 14, 2007.

[2] Saaty, T.L. (1980) *Analytic Hierarchy Process, McGraw-Hill.*

[3] Charnes, A., Cooper, W.W., and Rhodes, E. (1978) Measuring the efficiency of decision making units. *European Journal of Operational Research*, **2**, 429–444.

[4] Cooper, W.W., Seiford, L.M., and Tone, K. (2007) *Data Envelopment Analysis: A Comprehensive Text with Models, Applications, References and DEA-Solver Software*, 2nd edn, Springer.

[5] Takamura, Y. and Tone, K. (2003) A comparative site evaluation study for relocating Japanese government agencies out of Tokyo. *Socio-Economic Planning Sciences*, **37**, 85–102.

[6] Tone, K. (1989) A comparative study on AHP and DEA. *International Journal of Policy and Information*, **13**, 57–63.

[7] Sinuany-Stern, Z., Mehrez, A., and Hadad, Y. (2000) An AHP/DEA methodology for ranking decision making units. *International Transaction of Operations Research*, **7**, 109–124.

[8] Azadeh, A., Keramati. A., and Songhori, M.J. (2009) An integrated Delphi/VAHP/DEA framework for evaluation of information technology/information system (IT/IS) investments. *International Journal of Advanced Manufacturing Technology*, **45**(11–12), 1233–1251.

[9] Jyoti, D., Banwet, K., and Deshmukh, S.G. (2008) Evaluating performance of national R&D organizations using integrated DEA-AHP technique. *International Journal of Productivity and Performance Management*, **57**, 5, 370–388.

[10] Thompson, R.G., Singleton Jr., F.D., Thrall, R.M., and Smith, B.A. (1986) Comparative site evaluations for locating a high-energy physics lab in Texas. *Interface*, **16**, 35–44.

[11] Tikoria, J., Banwet, D.K., and Deshmukh, S.G. (2010) Ranking national R&D organisations: A data envelopment analysis. *International Journal of Society Systems Science*, **2**(2), 176–193.

[12] Wang, Y.-M., Liu, J., and Elhag, T.M.S. (2008) An integrated AHP–DEA methodology for bridge risk assessment. *Computers & Industrial Engineering*, **54**(3), 513–525.

[13] Sipahi, S. and Mehpare, T. (2010) The analytic hierarchy process and analytic network process: An overview of applications. *Management Decision*, **48**(5), 775–808.

[14] Yamada, Y., Matsui, T., and Sugiyama, M. (1994) An inefficiency measurement method for management systems. *Journal of the Operations Research Society of Japan*, **37**, 158–167.

[15] RIKEN (2007). Report on site selection for the Next-Generation Supercomputing Center. [In Japanese.]

Appendix A

DEA-SOLVER-PRO

KAORU TONE

National Graduate Institute for Policy Studies, Tokyo, Japan

A.1 INTRODUCTION

DEA-Solver was originally developed by the author of this chapter and was included in the books [1, 2] by Cooper, Seiford and Tone and [3] by Ozcan. In the original software package, the numbers of models and DMUs were limited. Subsequently, SAITECH, Inc. (www.saitech-inc.com) released DEA-Solver-Pro, which can deal with up to 52 clusters of models and 6000 DMUs.

A.2 PLATFORM

The platform for DEA-Solver-Pro is Microsoft Excel 2007 or later, running on a Windows PC or server.

A.3 NOTATION

DEA-Solver uses the following notation to describe DEA models:

Model Name-I or O -C, V or GRS

Advances in DEA Theory and Applications: With Extensions to Forecasting Models,
First Edition. Edited by Kaoru Tone.
© 2017 John Wiley & Sons Ltd. Published 2017 by John Wiley & Sons Ltd.

where 'I' or 'O' corresponds to the input or output orientation, and 'C' or 'V' to constant or variable returns to scale, respectively. For example, 'AR-I-C' means the input-oriented assurance region model under the constant-returns-to-scale assumption. In some cases, the 'I' or 'O' and/or 'C' or 'V' are omitted. For example, 'CCR-I' indicates the input-oriented CCR model, which is naturally under constant returns to scale. 'GRS' indicates the 'general' returns-to-scale model. Models with the GRS extension require the input of two parameters via the keyboard: one is the lower bound L of the sum of lambdas (λ) (see Chapter 1), and the other is its upper bound U. 'Bilateral', 'Congestion' and 'FDH' have no extensions.

A.4   DEA MODELS INCLUDED

Table A.1 lists the models included in DEA-Solver-Pro, where 'Chapter' indicates a chapter number in this book and 'CST' indicates a chapter in [2].

TABLE A.1   Models.

No.	Cluster	Model	Chapter in this book or CST
1	Assurance Region	AR-I-C, AR-I-V, AR-I-GRS, AR-O-C, AR-O-V, AR-O-GRS	Chapter 1, CST6
2	Assurance Region Global	ARG-I-C, ARG-I-V, ARG-I-GRS, ARG-O-C, ARG-O-V, ARG-O-GRS	CST6
3	BCC	BCC-I, BCC-O	Chapter 1
4	Bilateral	Bilateral-CCR-I, Bilateral-BCC-I, Bilateral-SBM-C, Bilateral-SBM-V	CST7
5	Bounded Variable	BND-I-C, BND-I-V, BND-I-GRS, BND-O-C, BND-O-V, BND-O-GRS	CST7
6	Categorical Variable	CAT-I-C, CAT-I-V, CAT-O-C, CAT-O-V	CST7
7	CCR	CCR-I, CCR-O	Chapter 1
8	Congestion	Congestion	CST12
9	Cost	Cost-C, Cost-V, Cost-GRS	CST8
10	Decreasing RTS	DRS-I, DRS-O	Chapter 5, CST5
11	DirectionalDistance-NonOriented	DD-C(V), SuperDD-C(V)	Chapter 3
12	DirectionalDistance-Oriented	DD-I(O)-C(V), SuperDD-I(O)-C(V)	Chapter 3
13	DynamicNetworkSBM-NonOriented	DNSBM-C(V)	Chapter 9

TABLE A.1   (*continued*)

No.	Cluster	Model	Chapter in this book or CST
14	DynamicNetworkSBM-Oriented	DNSBM-I(O)-C(V)	Chapter 9
15	DynamicSBM-NonOriented	DynamicSBM-C(V)	Chapter 8
16	DynamicSBM-Oriented	DynamicSBM-I(O)-C(V)	Chapter 8
17	EBM-NonOriented	EBM-C(V)	Tone and Tsutsui [4]
18	EBM-Oriented	EBM-I(O)-C(V)	Tone and Tsutsui [4]
19	FDH	FDH	CST4
20	Generalized RTS	GRS-I, GRS-O	CST5
21	Hybrid	Hybrid-C, Hybrid-V, Hybrid-I-C, Hybrid-I-V, Hybrid-O-C, Hybrid-O-V	CST4
22	Increasing RTS	IRS-I, IRS-O	Chapter 5
23	Malmquist	Malmquist-I-C, Malmquist-I-V, Malmquist-I-GRS, Malmquist-O-C, Malmquist-O-V, Malmquist-O-GRS, Malmquist-C, Malmquist-V, Malmquist-GRS	Chapter 6
24	Malmquist-Radial	Malmquist-Radial-I-C, Malmquist-Radial-I-V, Malmquist-Radial-I-GRS, Malmquist-Radial-O-C, Malmquist-Radial-O-V, Malmquist-Radial-O-GRS	Chapter 6
25	NetworkSBM-NonOriented	NetworkSBM-C(V)	Chapter 7
26	NetworkSBM-Oriented	NetworkSBM-I(O)-C(V)	Chapter 7
27	New-Cost	New-Cost-C, New-Cost-V, New-Cost-GRS	CST8
28	New-Profit	New-Profit-C, New-Profit-V, New-Profit-GRS	CST8
29	New-Revenue	New-Revenue-C, New-Revenue-V, New-Revenue-GRS	CST8
30	Non-Controllable	NCN-I-C, NCN-I-V, NCN-O-C, NCN-O-V	CST7
31	NonConvex-Radial	NonConvex-Radial-I(O)	Chapter 20
32	NonConvex-SBM	NonConvex-SBM-I(O, NonOriented)	Chapter 20

(*continued overleaf*)

TABLE A.1 (*continued*)

No.	Cluster	Model	Chapter in this book or CST
33	Non-Discretionary	NDSC-I-C, NDSC-I-V, NDSC-I-GRS, NDSC-O-C, NDSC-O-V, NDSC-O-GRS	CST7
34	Profit	Profit-C, Profit-V, Profit-GRS	CST8
35	ResamplePastPresent	Resampling-(Super)SBM, Resampling-(Super)Radial	Chapter 29
36	ResamplePastPresent Future	Resampling-(Super)SBM, Resampling-(Super)Radial	Chapter 29
37	ResampleTriangular	Resampling-(Super)SBM, Resampling-(Super)Radial	Tone [5]
38	ResampleTriangular Historical	Resampling-(Super)SBM, Resampling-(Super)Radial	Tone [5]
39	Revenue	Revenue-C, Revenue-V, Revenue-GRS	CST8
40	Revenue/Cost	Ratio-C, Ratio-V	CST8
41	SBM_Max	SBM_Max-I-C, SBM_Max-I-V, SBM_Max-O-C, SBM_Max-O-V, SBM_Max-C, SBM_Max-V	Chapter 22
42	SBM-NonOriented	SBM-C, SBM-V, SBM-GRS, SBM-AR-C, SBM-AR-V	Chapter 2
43	SBM-Oriented	SBM-I-C, SBM-I-V, SBM-I-GRS, SBM-O-C, SBM-O-V, SBM-O-GRS, SBM-AR-I-C, SBM-AR-I-V, SBM-AR-O-C, SBM-AR-O-V	Chapter 2
44	Scale Elasticity	Elasticity-I, Elasticity-O	Chapter 5
45	Super-Radial	Super-CCR-I, Super-CCR-O, Super-BCC-I, Super-BCC-O	Chapter 4
46	Super-SBM-NonOriented	Super-SBM-C, Super-SBM-V, Super-SBM-GRS	Chapter 4
47	Super-SBM-Oriented	Super-SBM-I-C, Super-SBM-I-V, Super-SBM-I-GRS, Super-SBM-O-C, Super-SBM-O-V, Super-SBM-O-GRS	Chapter 4
48	Super-SBM_Max	Super-SBM_Max-I-C, Super-SBM_Max-I-V, Super-SBM_Max-I-GRS, Super-SBM_Max-O-C, Super-SBM_Max-O-V, Super-SBM_Max-O-GRS, Super-SBM_Max-C, Super-SBM_Max-V	Tone [6]
49	Systems	SYS-I-C, SYS-I-V, SYS-O-C, SYS-O-V	CST7
50	Undesirable Outputs	BadOutput-C, BadOutput-V, BadOutput-GRS, NonSeparable-C, NonSeparable-V, NonSeparable-GRS	CST13

TABLE A.1   (*continued*)

No.	Cluster	Model	Chapter in this book or CST
51	WeightedSBM	WeightedSBM-C, WeightedSBM-V, WeightedSBM-I-C, WeightedSBM-I-V, WeightedSBM-O-C, WeightedSBM-O-V	Chapter 2
52	Window	Window-I-C, Window-I-V, Window-I-GRS, Window-O-C, Window-O-V, Window-O-GRS	CST9

TABLE A.2   Typical Excel data format.

	A	B	C	D	E	F
1	Hospital	(I)Doctor	(I)Nurse	(O)Outpatient	(O)Inpatient	
2	A	20	151	100	90	
3	B	19	131	150	50	
4	C	25	160	160	55	
5	D	27	168	180	72	
6	E	22	158	94	66	
7	F	55	255	230	90	
8	G	33	235	220	88	
9	H	31	206	152	80	
10	I	30	244	190	100	
11	J	50	268	250	100	
12	K	53	306	260	147	
13	L	38	284	250	120	
14						

A.5   TYPICAL DATA FORMAT

Table A.2 shows a typical example of data in Excel, where (I) and (O) indicate input and output items, respectively.

REFERENCES

[1] Cooper, W.W., Seiford, L.M. and Tone, K. (2006) *Introduction to Data Envelopment Analysis and Its Uses: With DEA-Solver Software and References,* Springer, New York.

[2] Cooper, W.W., Seiford, L.M. and Tone, K. (2007) *Data Envelopment Analysis: A Comprehensive Text with Models, Applications, References and DEA-Solver Software,* 2nd edn, Springer, New York.

[3] Ozcan Y.A. (2014) *Health Care Benchmarking and Performance Evaluation,* 2nd edn, Springer, New York.

[4] Tone, K. and Tsutsui, M. (2010) An epsilon-based measure of efficiency in DEA – A third pole of technical efficiency. *European Journal of Operational Research,* **207**, 1554–1563.

[5] Tone, K. (2013) Resampling in DEA. GRIPS Discussion Paper, http://id.nii.ac.jp/1295/00001133/.

[6] Tone, K. (2016) On the consistency of slacks-based measure-max model and super-slacks-based measure model. GRIPS Discussion Paper, http://id.nii.ac.jp/1295/00001533/.

INDEX

Advances in DEA Theory and Applications: With Extensions to Forecasting Models,
First Edition. Edited by Kaoru Tone.
© 2017 John Wiley & Sons Ltd. Published 2017 by John Wiley & Sons Ltd.